International Perspectives on the Teaching and Learning of Mathematical Modelling

This book series will publish various books from different theoretical perspectives around the world focusing on Teaching and Learning of Mathematical Modelling at Secondary and Tertiary level. The proceedings of the biennial conference called ICTMA, organised by the ICMI affiliated Study Group ICTMA (International Community of Teachers of Mathematical Modelling and Applications) will also be published in this series. These proceedings display the worldwide state-of-the-art in this field and will be of interest for a wider audience than the conference participants.

ICTMA is a worldwide unique group, in which not only mathematics educators aiming for education at school level are included but also applied mathematicians interested in teaching and learning modelling at tertiary level are represented. ICTMA discusses all aspects related to Teaching and Learning of Mathematical Modelling at Secondary and Tertiary Level, e.g. usage of technology in modelling, psychological aspects of modelling and its teaching, modelling competencies, modelling examples and courses, teacher education and teacher education courses.

More information about this series at http://www.springer.com/series/10093

Gloria Ann Stillman • Werner Blum
Maria Salett Biembengut
Editors

Mathematical Modelling in Education Research and Practice

Cultural, Social and Cognitive Influences

Springer

Editors
Gloria Ann Stillman
School of Education
Australian Catholic University
Ballarat, VIC, Australia

Werner Blum
University of Kassel
Kassel, Germany

Maria Salett Biembengut
Pontifical Catholic University
of Rio Grande do Sul - PUCRS
Porto Alegre, Brazil

ISSN 2211-4920 ISSN 2211-4939 (electronic)
International Perspectives on the Teaching and Learning of Mathematical Modelling
ISBN 978-3-319-18271-1 ISBN 978-3-319-18272-8 (eBook)
DOI 10.1007/978-3-319-18272-8

Library of Congress Control Number: 2015945819

Springer Cham Heidelberg New York Dordrecht London

Printed on acid-free paper

Springer International Publishing AG Switzerland is part of Springer Science+Business Media (www.springer.com)

Series Preface

Applications and modelling and their learning and teaching in schools and universities have become a prominent topic in the last decades in view of the growing worldwide relevance of the usage of mathematics in science, technology and everyday life. However, although there is consensus that modelling should play an important role in mathematics education, the situation in schools and universities is less than ideal in many educational jurisdictions. Given the worldwide impending shortage of students who are interested in mathematics and science, it is essential to discuss possible changes of mathematics education in school and tertiary education towards the inclusion of real-world examples and the competencies to use mathematics to solve real-world problems.

This innovative book series *International Perspectives on the Teaching and Learning of Mathematical Modelling* established by Springer aims at promoting academic discussion on the teaching and learning of mathematical modelling at various educational levels all over the world. The series will publish books from different theoretical perspectives from around the world dealing with Teaching and Learning of Mathematical Modelling in Schooling and at tertiary level. This series will also enable the *International Community of Teachers of Mathematical Modelling and Applications* (ICTMA), an International Commission on Mathematical Instruction-affiliated study group, to publish books arising from its biennial conference series. ICTMA is a unique worldwide group where not only mathematics educators dealing with education at school level are included but also applied mathematicians interested in teaching and learning modelling at tertiary level are represented. Three of these books published by Springer have already appeared.

The planned books will display the worldwide state of the art in this field, most recent educational research results and new theoretical developments and will be of interest for a wide audience. Themes dealt with in the books will be teaching and learning of mathematical modelling in schooling and at tertiary level including the usage of technology in modelling; psychological, social and cultural aspects of modelling and its teaching; modelling competencies; curricular aspects; modelling examples and courses; teacher education; and teacher education courses. The book series aims to support the discussion on mathematical modelling and its teaching internationally and will promote the teaching and learning of mathematical modelling and research of this field all over the world in schools and universities.

The series is supported by an editorial board of internationally well-known scholars, who bring their long experience in the field as well as their expertise to this series. The members of the editorial board are Maria Salett Biembengut (Brazil), Werner Blum (Germany), Helen Doerr (USA), Peter Galbraith (Australia), Toshikazu Ikeda (Japan), Mogens Niss (Denmark) and Jinxing Xie (China).

We hope this book series will inspire readers in the present and the future to promote the teaching and learning of mathematical modelling all over the world.

Ballarat, Australia — Gloria Ann Stillman
Hamburg, Germany — Gabriele Kaiser

Contents

Chapter 1
Cultural, Social, Cognitive and Research Influences on Mathematical Modelling Education

Gloria Ann Stillman, Werner Blum, and Maria Salett Biembengut

Abstract This contribution from the ICTMA community on the latest in research and teaching ideas in the area of mathematical modelling and applications education differs from previous volumes in that there is a much stronger emphasis on social and cultural influences on modelling education because of the location of the preceding conference in Brazil. However, another point of difference is the number of chapters that are influenced by cognitive perspectives as there are strong research teams taking this perspective internationally. This chapter situates the work in this volume within the field which has led to much research and evaluative studies in the last decade.

1.1 Introduction

According to Niss (2001), the majority of activity in the applications and modelling field in mathematics education was 'proto-research' up until the 1990s. Even in the decade from 1990 to 2000, he identified only 50 papers that he considered were genuine research. However, the next decade saw more consolidation of the field as the ICMI study was held and the post conference volume (Blum et al. 2007)

G.A. Stillman (✉)
Education VIC, Faculty of Education & Arts, Australian Catholic University,
1200 Mair Street, Ballarat, VIC 3350, Australia
e-mail: gloria.stillman@acu.edu.au

W. Blum
Institute of Mathematics, University of Kassel, Kassel 34109, Germany
e-mail: blum@mathematik.uni-kassel.de

M.S. Biembengut
Pontificia Universidade Católica do Rio Grande do Sul, Rua Soldado Mário Luiz Bertolini, 273, Blumenau, SC 89.053-180, Brazil
e-mail: mariasalettb@gmail.com

G.A. Stillman et al. (eds.), *Mathematical Modelling in Education Research and Practice*, International Perspectives on the Teaching and Learning of Mathematical Modelling, DOI 10.1007/978-3-319-18272-8_1

published. Since that time there has been an increasing number of research studies carried out and quite some diversification of the field. There has been, for example, an increase in the studies that take a cognitive perspective. The publication of this volume following ICTMA 16 in Brazil in 2013 is expected to not only add to this rich research output but also possibly stimulate new lines of inquiry. We hope that it will contribute to inspiring more research into social and cultural influences in the future.

Niss et al. (2007) noted as one imperative of the field of modelling and applications education the articulation of its relationship to the world we live in. This is in pursuit of the goal of "linking of the field of mathematics with some aspects of the world, with the purpose of enhancing knowledge, but also ensuring or advancing the sustainability of health, education and environmental well-being, and the reduction of poverty and disadvantage" (pp. 17–18). Others have attempted to look critically at the role of applications of mathematics and mathematical modelling itself in society. Such sentiments are in keeping with development during education of what is called a socio-critical competency that emphasises the social/political context of modelling (Stillman et al. 2013). It is a feature of this volume that several authors take this emphasis in their work (e.g., Luna, Souza and Lima and Villarreal, Esteley and Smith). Within this context this edited collection represents the current state of thinking and research in this diverse international community distilled and synthesised after the biennial conference. However, it is important that any published contributions stimulated by the conference be connected to the on-going body of research and scholarship in this field as well as previous outputs from the community (e.g., Kaiser et al. 2011; Lesh et al. 2010; Stillman et al. 2013). In this chapter we attempt a first pass on this task.

1.2 Innovative Practices in Modelling Education Research and Teaching

Many issues related to the successful solution of real world tasks and problems through mathematical modelling have been identified for some time now in the international literature (see Blum et al. 2007; Niss 2001, for an overview). It seems that in order to resolve these issues we need innovative methods and practices in research into modelling education as well as innovative pedagogical practices. Innovative research methods give us new tools to prise open old problems in order to look at these in a new light. At the same time, innovative classroom practices could bring more teachers into the fold who are able to realise the espoused potential benefits of modelling for students in allowing them to utilise mathematics as a means of "experiencing the world" (Chapman 2009).

A unique Latin American perspective to modelling is brought by the work of Ubi D'Ambrosio. In his chapter he discusses how knowledge is generated (i.e., cognition), how it is individually and socially organised (i.e., epistemology) and how it is confiscated by power structures, institutionalised and given back to the people who generated it through filters (i.e., politics). How knowledge is built up by individuals and within societal groups is examined. "The full cycle of knowledge includes its generation, individually and socially, its organisation, its expropriation, institutionalization, transmission and diffusion, through systems of education and different forms of filters (such as examinations, degrees, certifications)" (p. 42). Mathematical modelling is seen as a strategy for building up systems of knowledge in different cultural environments. To this end, mathematical modelling allows humans to understand, to explain and to cope with selected facts and phenomena of reality and ideally reality as a whole. Our natural limitations as human beings restrict our access to only selected facts and phenomena (cf Skovsmose 1994). Modelling, by its iterative nature, "allows a better understanding of the selected facts and phenomena of reality, which is the goal that justifies our practices as scientists" (p. 44).

The meaning assigned by students to the problem studied in mathematical modelling activities is considered by Almeida and Silva who adopt a Peircean semiotic (see Sebeok 1991) approach, where meaning is associated with interpretants produced by the students during activity development. The authors highlight the meaning attributed by students to elements of the problem when they attempt modelling activities, through analysing their actions constituted in signs used to suggest and represent the object being dealt with. Thus, Almeida and Silva contend a method to understand the meaning assigned to a problem is to analyse the interpretants of the students during the resolution of that problem as well as what is said about the problem after the signs are generated. A case study of one student's actions as his group attempts to model tree pruning so as to optimise street lighting is reported to explain this idea in detail. Analyses of the case reveal that familiarity with the real problem for the student and an intention to make meaning from a reference were critical. Almeida and Silva conclude that even though the problem is defined early in the development of a modelling activity, the assignment of meaning is consolidated throughout to the extent that the student defines, interprets and validates a solution. Thus modellers who are successful do not lose sight of the referents that make it meaningful for them in the real world (Nunes et al. 1993; Stillman 2000).

The Models and Modelling Perspective (MMP) (Lesh and Doerr 2003; Lesh and English 2005) for both teaching and researching modelling is now over 30 years old. Brady, Lesh and Sevis describe an innovative research effort to expand the reach of the MMP tradition, engaging questions about the interconnected models and modelling processes of students and teachers at larger, course-length scales. New tools (e.g., Learning Progress Maps and Concept Analysis Wheels) are discussed as well as new directions. Learning progressions using the metaphor of finding your way around a terrain or an evolutionary model are highlighted as possible new research ideas. In addition, further work on the idea of teaching problem solving and heuristics is mooted but reconceptualised within the MMP tradition.

Niss in his contribution makes a distinction between two kinds of mathematical modelling purposes and related modelling endeavours, *descriptive modelling* and *prescriptive modelling*. In descriptive modelling attempts are made to capture and come to grips with, and perhaps eventually act in, some extra-mathematical domain in the real world, in a field of practice or in an academic discipline. The processes of descriptive modelling are typically represented by a *modelling cycle*. When the purpose of modelling is not primarily to understand some existing part of the world, but "*to design, prescribe, organise or structure certain aspects of it*" (p. 69), Niss calls this *prescriptive modelling*. "In prescriptive modelling the ultimate aim is to pave the way for *taking action based on decisions* resulting from a certain kind of mathematical considerations, in other words 'to change the world' rather than only 'to understand the world'" (p. 69). Whilst descriptive modelling is usually the focus of practice, Niss notes that prescriptive modelling is hardly noticed, let alone investigated in mathematics education. Three substantive examples of descriptive modelling are presented and discussed with reference to the Niss (2010) modelling cycle. This includes the clear distinction of the discussion re modelling rather than models and the interweaving of the two types of modelling. In Blum and Niss (1991), the authors wrote about *descriptive models* rather than *descriptive modelling*, and also about *normative models* (not prescriptive modelling). The terms proposed in the present chapter represent a move from focusing on the *product* (the model) to focusing on the modelling *purposes*, whilst replacing 'normative' with 'prescriptive', following Davis (1991). Niss proceeds to argue successfully re the limitations of the usual modelling cycles with regard to adequately capturing all processes involved in prescriptive modelling as well as potential difficulties. Furthermore, he strongly advocates for the need for the ICTMA Community to engage themselves and their students in prescriptive modelling.

The development of a paper and pencil test instrument to assess mathematical modelling competency holistically in contrast to other instruments that measure sub-competencies (e.g., Crouch and Haines 2004) is the focus of Reit and Ludwig. The authors provide an innovative research method for determining the degree of difficulty of holistic mathematical tasks taking a cognitive approach which takes into account the order of thought operations and cognitive demand from the perspective of cognitive load theory (Sweller 2010). Basing their beginning ideas on the work of Briedenbach (1969), Reit and Ludwig use the number of simplex structures to be processed serially and whether a simultaneous performance of simplex structures is necessary to determine a measure of thought structure. Their theoretical ideas were applied to three items for Year 9 students and compared to solution rates. The empirical results were in agreement with thought structure complexity with a more complicated thought structure resulting in a lower solution rate and simpler thought structures having higher solution rates. Simultaneous processing (indicated by width of thought structure being higher) was seen to be associated with this lowering of solution rate as it was indicative of increased cognitive demand, a similar interpretation to Stillman (2001).

Resolution of long-standing issues (see e.g., Wijaya et al. 2014) related to problem formulation and specification and their successful mathematisation by novice modellers are taken up by Stillman, Brown and Geiger. Their point of departure is a theoretical construct – *implemented anticipation* – proposed and theoretically analysed by Niss (2010), and then they make use of this construct to investigate as well as to facilitate novice secondary school modellers' mathematical modelling activities. In so doing, the authors propose to extend the construct to also pertain to the affordances (Gibson 1977) of technology in Technology Rich Teaching and Learning Environments (TRTLEs), and they put forward a set of scaffolding questions that modellers might pose to themselves individually or in groups so as to "facilitate productive discussion that moves novices forward from a problem context focus to actualising model construction" (p. 102). The approach taken is illustrated by a "paradigmatic example" of a group of students working to determine if it is possible for a stunt car to cross a broken bridge and land on its wheels after having performed a rotation in the air. The main innovative aspect of this chapter is that it transforms and extends Niss's implemented anticipation within his cognitive model of ideal mathematisation into an operationalized scaffold so that it can be used by novice modellers. The framework relates to foreshadowing and feedback loops during (1) problem finding and specification, (2) mathematising and (3) problem solving. It facilitates group members asking each other scaffolding questions to develop a sense of direction for their modelling (Treilibs 1979). The framework both facilitates and exposes interthinking (Mercer 2000) in group modelling (i.e., social aspects) intended to enhance the effectiveness of discussion leading to model construction. Another strength of this chapter is that it succeeds in establishing connections with earlier and more general notions of anticipation (e.g., Dewey 1916, 1917) and its relatives. The chapter serves to consolidate and provide empirical evidence, adding to previous work (Stillman and Brown 2014), for a theoretical construct put forward by other researchers in the field.

Vos has contributed previously (Vos 2011) to the debate on authenticity (e.g., see Jablonka 2007) but this chapter brings new aspects, both theoretically and empirically. Firstly, she critiques word problems and classroom tasks which simulate or imitate real world situations before defining authenticity by considering it as a social construct following Durkheim (1982), with authenticity being established by a social process. For an aspect of mathematics education to be authentic according to Vos, it needs to have: (1) an out-of-school origin and (2) a certification of originality attested to. This aligns to some extent with a previous notion of authenticity put forward by Niss (1992); however, Vos sees her definition as broader with the origin being possibly mathematical and the "attesting experts" being others than those working in the field. Vos then validates her definition by analysing a particular experience, an excursion for secondary students. The interaction between definition and validation, which is offered in this chapter, represents by itself an interesting innovation. Vos offers ideas that can contribute to the analysis of mathematical modelling activities, contending that definitions of authenticity that list sets of features (e.g., Palm 2008) are using "implications of authentic aspects and not indicators or descriptors of authenticity" (p. 113). In a later chapter, Galbraith provides a different view (see Sect. 28.5).

Taking their experience in Chinese universities as a starting point, Wu, Wang and Duan aim at overviewing what they see as the teaching and the learning goals in a tertiary mathematical modelling course. From their viewpoint, a mathematical modelling course should provide a way of teaching its own systematic thinking pattern, not be only a collection of various methods. Thus in their chapter meta-knowledge about the thinking underpinning mathematical modelling is advocated as the main goal of teaching a mathematical modelling course, whilst the mathematical methods and modelling cases are considered the carriers of thought transmission and serve the goal of helping students understand mathematical thinking better. Examples are used as illustrations of why the authors believe intuitive insights into how the problem should be modelled, making connections, innovation through critique, choice based on common sense and cultivating inductive thinking are the essential ingredients in teaching such thinking. The chapter possesses interesting content for teachers, curriculum designers of university modelling courses, or learners of mathematical modelling.

1.3 Research into, or Evaluation of, Teaching Practice

As the ICTMA community is very much concerned with research into the practices of teaching mathematical modelling and applications, contemporary themes are usually reflected in the work that is highlighted at biennial conferences. Some of these are on-going involving several research groups internationally such as the theoretical and empirical work on modelling competencies whereas others make only brief appearances in ICTMA literature. Together, however, they form the corpus of research input that drives the modelling agenda forward. Linked to this are evaluations of teaching sequences that usually involve teaching experiments with theoretically inspired approaches or research based material design.

The chapter by Kaiser and Brand is an extension of previous work by the ICTMA research community in the area of fostering modelling competencies. The authors trace the beginnings of the appearance of the construct in the modelling discussion internationally through its many and varied interpretations and how it has been conceptualised and evaluated by various researchers. From this historical analysis, they identify four strands of modelling competencies research and development as shown and exemplified in Table 1.1 which is a synthesis of their text with some additional exemplification literature. The identified strands are the outputs of four geographically based research groups but with some cross-over (e.g., Kaiser using the work of Haines and Crouch). The work of the groups is not discrete. Kaiser and Brand divide the fours strands between two approaches (after Blomhøj and Højgaard Jensen 2003) to supporting modelling competences, namely holistic (requiring full scale modelling) and atomistic (concentrating on mathematisation and analysis of models) (see Table 1.1). The four strands are not comprehensive of research work in modelling competencies internationally (see section on competencies in Blum et al. 2007 for other work) but have been the basis of a good deal of further study in this area.

The identification of strands facilitates the raising of pertinent questions to be researched with respect to the nature and conceptualisation of modelling

Table 1.1 Perspectives on modelling competencies research

Approach	Strand characteristics	Measures	Exemplar literature
Holistic	I. Modelling competence as one of eight interdependent mathematical competencies	Quality of a competency is assessed via a three dimensional approach (degree of coverage, radius of action, technical level)	Niss (2003), Blomhøj and Højgaard Jensen (2007), Højgaard Jensen (2007), Niss and Højgaard (2011)
Analytic	II. Identification of indicators of competence – descriptors of students' modelling behaviours in terms of sub-competencies related to transitions in the modelling cycle	Multiple choice items mapped to each sub-competency	Haines et al. (1993), Haines and Izard (1995), Haines et al. (2001), Houston and Neil (2003), Izard et al. (2003)
	III. Development of a comprehensive concept of modelling competencies with sub-competencies connected to the modelling cycle including process-oriented and social sub-competencies	Levels of competency	Maaß (2004, 2006, 2007), Kaiser (2007), Henning and Keune (2007), Blum (2011)
	IV. Integration of meta-cognition into modelling competencies	Qualitative descriptions	Stillman and Galbraith (1998), Stillman (1998, 2002, 2004, 2011), Galbraith et al. (2007), Stillman et al. (2010)

competency, its measurement, development and classroom approaches to support its development. An exemplar study from the doctoral work of Brand is then presented in an effort to provide first answers to the question: Which description of modelling competency is more appropriate – holistic or analytic? The results of the study suggest an analytic description of the construct. With respect to which kind of promotion of modelling competency is more effective, the study showed that both approaches promoted overall modelling competency with strengths and weaknesses in both. The authors point to future directions for research on modelling competencies. Understandably these could be within or outside the four strands identified earlier which all take a cognitive view of modelling competencies.

Besser, Blum and Leiss report on a research study that aimed to investigate effective practice in providing formative feedback related to improving student achievement on mathematical modelling tasks. The study has attempted to address this aim by providing teacher training based on enhancing teacher Pedagogical Content Knowledge (PCK) and general pedagogical knowledge (PK) about formative assessment when dealing with modelling tasks. This is an important issue within mathematics education in general and mathematical modelling in particular. Preliminary results indicate that teachers trained in notions of formative assessment when dealing with modelling tasks specifically out-performed teachers merely trained in general didactical ideas of modelling and problem solving in competency-oriented mathematics on tests of PK and PCK at the end of their training program.

A meta-analysis of the use of theory in research literature on the teaching and learning of mathematical modelling is the focus of the chapter by Geiger and Frejd. The literature used as a basis for the analysis is pragmatically concentrated on the five volumes arising from the ICTMA conferences from 2001 to 2009 as well as the 14th ICMI Study volume from the conference held in 2004. Two different lenses were chosen for guiding analysis. English's *matrix of priorities* for mathematics education research (English 2002, p. 10) was used to determine the orientation of the field whereas the notions of *local theories* (Kaiser and Sriraman 2006) and *general theories* (Sriraman and English 2010) were used to characterise the diversity of the field. There were, however, research chapters where no theoretical underpinnings were evident in the work. Geiger and Frejd conclude that there was an increase over the selected time period in the use of theory to underpin chapters. Local theories (e.g., modelling competencies) were more frequently used than general theoretical approaches such as socio-cultural approaches. Both these results are attributed initially by the authors to the developing maturity of our research field. Another explanation raised is the increased number of chapters applying theories could be a result of a general "academisation" of the field as early career researchers (such as doctoral students) come into the field and apply theories to a well-defined problem as this format is taken to be equivalent with research quality in doctoral programs internationally. The field of research in the teaching and learning of mathematical modelling is an exemplary case of a field in mathematics education, which is developing "home grown theories"; therefore the focus is on "particular local theories" such as the modelling cycle and modelling competencies rather than general theories from outside the field. Geiger and Frejd point out that several general theoretical approaches which they had identified before beginning their analysis were not used in the selected sample, suggesting these could be mined in the future for greater theoretical diversity and richness. This clearly has already occurred in subsequent ICTMAs as is evident in Stillman et al. (2013) and this current volume.

Scaffolding methods for facilitating problem solving by novices during modelling tasks have been suggested and used by many both in research and in teaching. Those that are closely aligned to modelling cycles (e.g., the four step *Solution Plan* for the DISUM project, Blum 2011) are cyclic in nature whereas many that are aligned more closely with problem solving processes are linear (e.g., Greefrath and Leuders 2013). Similar linear schemes have been widely critiqued in the past when problem solving was receiving extensive research and classroom attention (see Törner et al. 2007 for an overview). Greefrath presents the results from a qualitative pilot study where three pairs of students in Year 4 and Year 6 respectively were introduced to the Greefrath and Leuders method using matching of given solution steps to the scheme for an identical task whether students were in Year 4 or 6 then the groups at each Year level were asked to solve a modelling task using the method. From the responses to interviews conducted after the modelling sessions, Greefrath concluded that the students' final solutions were able to be influenced by providing the scaffolding method particularly the written recording of the solution; however, actual solution routes differed markedly from each other (as has been found previously) and from the problem solving method they were asked to follow. Written records did not align very well with the actual problem solving processes students engaged in. A more sustained introduction to the method is suggested if further research into its use or another problem solving scaffold is to be contemplated for the future.

Quantitative reasoning with and about measurable attributes of objects and phenomena seems to be a central element of mathematical modelling of particular situations that afford use of measurement skills being "a central mechanism for iterative refinement of solution processes to real-world problems" (Larson 2010, p. 117). Little is known about the influence of what is called measurement sense (Shaw and Puckett-Cliatt 1989) on mathematical modelling. Hagena presents first quantitative insights from an intervention study into the interplay between measurement sense and mathematical modelling conducted with German pre-service teachers. Their mathematical modelling competencies were influenced positively by fostering their measurement sense in a short-term intervention. Hagena's results confirm that measurement sense seems to influence mathematical modelling that affords use of measurement skills.

Several different models (e.g., multidisciplinarity, Andresen and Petersen 2011, interdisciplinarity, Wineburg and Grossman 2000, and transdisciplinarity, Kaufman et al. 2003) have been proposed for connecting meaningfully across discipline areas in educational settings but there seems not to be sufficient research evidence (see English 2013) as one would expect for the broad support and claims made by education systems that advocate them. To address this gap partially at least, interdisciplinary project work in the Singapore context is the focus of a chapter by Ng and Stillman which reports findings from a research study in lower secondary schools (Years 7 and 8) which examined interconnectedness. The extent to which participation in a project work facilitates perception of interconnections between school disciplines, within mathematics and between school-based mathematics and real-world problem solving was measured using researcher developed scales. There was an overall increase in mean scores on the scales measuring perception of interconnectedness of mathematics and inter-subject learning and beliefs and efforts at making connections after project work; with a significant impact of the project work on the first two but not beliefs and efforts at making connections. One reason for the lack of significance for beliefs and efforts could be prior experience of project work in primary school already shaping positive beliefs about connections between mathematics and other subjects. Qualitative results revealed that these seemingly positive results disguised issues with students' ability to make the desired interconnections in a meaningful manner. Although students could state inter-subject connections from the project work they made little use of these during calculations and decision-making. In keeping with previous findings by Stillman (2000), some groups proposed their mathematical solutions without taking into account real-world constraints. Using mathematical concepts in real settings also challenged some groups where again the connection was lost between the calculations and the real context. These findings call into question the espoused benefits of interdisciplinary initiatives without an accompanying change in teacher practice.

The Dual Modelling Cycle Framework (DMCF) was first introduced by Saeki and Matsuzaki (2013) in an attempt to support students who display a wide diversity in their modelling progress on the same task. An *Oil Tank Task* and a *Toilet Roll Task* were used to illustrate the correspondence between single, double and dual modelling cycles. Matsuzaki and Saeki (2013) provided empirical evidence for the existence of their dual modelling cycle through evaluation of an activity using the two tasks with pre-service teachers. Kawakami, Saeki and Matsuzaki have a twofold purpose for their chapter in the current volume. Firstly, it serves to show how primary school students who could not solve an initial task independently used models shared by the teacher and other students and refined these models whilst being taught by a teacher using DMCF for teaching modelling. Secondly, they wanted to make suggestions from this experience for using DMCF in teaching with less successful students. Through examining students' worksheets and protocols of video/audio records of the lesson, the authors show that unsuccessful modellers could change/modify their own models and those of their classmates to progress their dual modelling cycle by taking advantage of different models that were shared with the class. Thus the authors conclude that a crucial point to facilitate this progression in the dual modelling cycle is sharing of various models used by other students related with both the initial task and the similar one, and to ensure there is a variety of ways to progress. The notion of double and dual modelling cycles, and the distinction between them, is potentially fruitful in supporting the development of modelling expertise, providing a structured approach within situations where previously known or more accessible models are relevant in addressing a new task.

Palharini and Almedia provide the reader with an example of how mathematical modelling can be analysed through Tall's Theory of Three Worlds of Mathematics (Tall 2004) namely, the conceptual-embodied world, the proceptual world and the formal-axiomatic world. The authors view mathematical modelling tasks from a cognitive perspective. They draw on part of a study by the first author to argue that mathematics students preparing to teach mathematics in middle and secondary schools engaged in mathematical modelling are likely to make the transition from elementary to advanced mathematical thinking, as well as engage in elementary mathematical thinking (presumably in areas of mathematics that are novel). Representation (visualisation, switching/translating, modelling) and abstraction (generalisation and synthesis) are said to be the most important of the cognitive processes of mathematical thinking. Modelling tasks allowed students in this study to complete a cycle of development, from one world to another, non-linearly, using cognitive processes related to mathematical thinking with advanced mathematical thinking occurring when multiple processes interacted. Based on their analysis, Palharini and Almedia assert that student engagement in

modelling tasks promoted this development of advanced mathematical thinking. They also consider the transitions between the mathematical worlds as sites for refinement of thought. Thus, it is possible for educators to perceive and study mathematical thinking processes with modelling tasks and, by being so informed, to create learning environments that explore the emergence of such processes.

Rivera, Londoño and Jaramillo present results from a case study conducted in a rural education institution in Colombia, aimed at analysing a mathematical modelling activity undertaken in the classroom but stimulated by the events in students' home lives. The study modelling activity involved a flood due to the overflowing of a local river in a high school as its context. In the chapter the authors wish to clarify an issue which has been raised theoretically and see its outworking in practice by analyzing the way students built mathematical models based on the measurement of area and volume resulting within this context. The issue is the timing of the introduction of new mathematical concepts necessary for modelling to proceed, a dilemma raised by others when researching teaching practice (e.g., Antonius et al. 2007; Oliveira and Barbosa 2013). Hein and Biembengut (2006) claim it is possible to teach curricular content necessary for the resolution of a modelling situation concurrently. In this instance evidence is provided of students being able to access mathematical content beyond that which they were taught through information technology and then the teacher taking up the opportunity to treat this new content in class. The use of software such as GeoGebra helped students visualize the effects of using unfamiliar mathematical functions needed to resolve their lack of known models. Geogebra also allowed them to visualize the effect of different parameters in the situation being modelled and the municipality. The authors conclude that the authenticity of the context enabled the deepening of mathematical concepts such as area and volume and their measurement in real situations thus not only allowing the students to associate mathematics with a phenomenon in their natural environment but also to propose solutions to reduce the social impact of floods on their community. The modelling activity thus contributed to both their mathematical and citizenship education.

Modelling conducted in a cultural and contextual setting is also the focus of the research study by Villa-Ochoa and Berrio. This chapter presents partial findings from a qualitative case study involving students from a rural educational institution in Colombia. The students were motivated by their teacher to make sense of mathematics in situations beyond the classroom in their everyday experiences in keeping with the goals and ideals imposed on Colombian Mathematics curricula by policy writers. In this context, the role of mathematical modelling goes beyond an emphasis on cognitive and conceptual processes of mathematics in order to link mathematics to critical societal issues and culture. This chapter focuses on the latter examining the project work of two pairs of Year 11 students modelling the mathematical relationships involved in coffee farming, which is not only part of the cultural context in which they live but also is of direct relevance to them as their own families are involved in coffee growing. During this in-class activity students initially did not use their knowledge of local agricultural techniques for where trees are planted until "given permission" by the teacher to do so. It would appear that the

didactic contract (Brousseau 1997) had to be changed so that they realized such knowledge would be valuable (and valued by the teacher) in the classroom. As Brown et al. (1989) point out, "many methods of didactic education assume a separation between knowing and doing, treating knowledge as an integral, self-sufficient substance, theoretically independent of the situation in which it is learned and used" (p. 32), so it should be no surprise that Year 11 students act in this manner. An alternative reading of this incident taken by the researcher is that the authority of the mathematics in the classroom activity overrode their prior knowledge from daily life and subsequently the need for sense making in the classroom activity. The teacher thus had to intervene and re-orient the task, and in this way students were able to recognise the associated meanings. The teacher's intervention in the students' mathematical modelling process was necessary in order to recognise the role of context as an establishing element of scientific knowledge at school with neither mathematics nor non-mathematical knowledge being subordinated one to the other. For teachers to be able to know *when to do* this, they must develop a *sense of reality* built through mathematical modelling as discussed in earlier work by Villa-Ochoa and López (2011).

The problem of the income and expenditure of an electric power company formed the basis for teaching material for mathematical modelling in a teaching experiment in Japan by Yoshimura as the third of three modelling activities used with lower secondary students (Years 7–9). This teaching experiment was fairly teacher led in the Japanese lesson fashion but provided students with the opportunity to make assumptions and validate results in mathematical modelling when posing their own solutions to the company's financial deficit problem. As a result, it was confirmed that students' modelling skills in handling uncertain numerical values and conditions were acquired through their discussion about the correct answers but they were not able to make assumptions appropriately from a complicated real-life problem. In addition, the teaching experiment demonstrated that the context was suitable to use in mathematical modelling teaching materials with Year 9 Japanese students. Short questionnaires were used to compare student responses before and after the activity and across the three tasks. Students were quite positive about the utility of mathematics in solving real problems and this increased after the task. The power company context was of interest to most students similar to the other contexts (pension taxes and bluefin tuna depletion). Participation in the modelling experience had a positive effect on eroding the notion of mathematics always involving the certainty of one answer.

1.4 Pedagogical Issues for Teaching and Learning of Modelling

Although the pedagogy of applications and modelling shares some practices that are part of general pedagogy, there are a range of practices that are different (Niss et al. 2007) and these are the subject of much research and experimentation in the

ICTMA community. Some of these studies are associated with the introduction of modelling into new contexts but more commonly now the focus is on how to address many long standing issues that these differences in practices bring to light for the teaching and learning of modelling.

As mathematical modelling is being introduced as a new product into the complex system of mathematics education in many countries (e.g., USA, see National Governors Association Center for Best Practices and Council of Chief State School Officers 2010), Pollak reminds us it is important to view this educational change within the *total system* as it has to fit with the existing parts and interfaces in this system. As one example, he considers the effect of mathematical modelling on the transition from secondary to tertiary education and whether the purpose is to teach modelling as a vehicle to teach mathematics or as content in its own right (Galbraith 2007b; Julie 2002). If mathematical modelling is of greater importance to the planners at either level of education than at the other, stresses may result. As a second example, he examines changes in teacher education necessitated by the introduction of mathematical modelling at the secondary level. Ideally, one might wish to prepare teachers to teach mathematical modelling by concentrating purely on modelling without the distraction of new mathematical ideas, but this cannot always be done as Rivera et al. have already shown. Finally, the effect of mathematical modelling on the relationship between mathematics education and mathematics itself is discussed. In response to Pollak's plenary address at ICTMA 16 in Brazil, Borromeo Ferri reflects at the meta-level on the basis of a "mathematical modelling map" which she created as a simple instrument opening a segue between applied and pure mathematics as well as between theoretical aspects and empirical results concerning teaching and learning of mathematical modelling. She discusses three key questions stemming from Pollak's lecture, such as "Why should modelling be taught and learnt?" and interprets the answers as moves within that map.

On the surface the issue that Araújo and Campos appear to approach in their chapter is that of the relationship between modelling and views of mathematics and how the diverse forms of this relationship impinge on the often, vexed relationship between mathematics education and mathematics. Pollak in his chapter has already suggested that the main purpose of teaching modelling might be as a vehicle for the teaching of mathematical content, as mathematical content in its own right or a mixture of the two. The authors appear to subscribe to the notion of modelling as content in its own right to progress real world problem solving that is of a socio-critical nature. Thus, the issue runs much deeper on closer examination to *fundamental ways in which actors in classrooms operate when a new approach disrupts the didactic contract* (Brousseau 1997). This was the first time the tertiary student group had modelled so coming to a shared understanding of the nature of mathematical modelling in a socio-political perspective (Kaiser and Sriraman 2006) with their teachers was understandably fraught with perturbations. A break down in meaning occurred; not because the students did not identify a real world societal problem with political implications, but rather because the data that they assembled right from the beginning needed no further mathematisation to describe the

situation or answer the question posed. Araújo and Campos use Barbosa's notion (2007) of *negotiation spaces* to describe the verbal interactions between student and teacher as the teacher uses the student's voicing of her dilemma with respect to mathematising the situation so as to produce a project proposal within the "rules of the game" (Bourdieu 1977) of what the teacher sees constitutes a modelling project in the context.

The issue of *moving from isolated modelling activities to a coherent program involving sequences of modelling tasks*, which is more likely to result in increased uptake in classrooms, is indeed an important one. In introducing the challenges involved in such a progression, Ärlebäck and Doerr reference two studies at university level and Galbraith and Clatworthy's 1990 study centred around the use of a sequence of modelling tasks in a 2 year course at senior secondary school level. The background discussion concerning models, modelling, and model development sequences provides an excellent synopsis of important ideas from the field. Following reference to the familiar model eliciting activities, the authors proceed to outline the nature and purpose of model exploration activities and model application activities from a *Variation Theory* perspective (Runesson 2005). There are strong links with Valsiner's *Zone of Promoted Action (ZPA),* and other aspects of his Zone Theory (Valsiner 1997) which Galbraith (2007b) has applied previously to "construction and analysis of teaching to achieve modelling goals" (p. 58), particularly with respect to the activities provided for student learning. Thus support for the approach Ärlebäck and Doerr are proposing is strengthened as theoretical underpinnings come from multiple sources. Tentative design principles for model exploration and model application activities within a model development sequence are enunciated, illustrated by learning excerpts and examples. This application of variation theory could be seen as providing a rather prescriptive taskmaster. The level of teacher design and control also appears heavy in the examples of contrast variation, variation of fusion, variation of separation and variation of generalisation. With respect to the authors' suggestion that "Multiple aspects of complex objects of learning need to be experienced simultaneously at the same time in order for the learner to fuse them together into a coherent whole" (p. 209), if the 'frames of reference' of different students are moving at different speeds regarding understanding, as has been found previously (e.g., Stillman 2010), 'simultaneous' may not have a viable meaning across a whole class. To sum up, Ärlebäck and Doerr have produced a stimulating piece of writing that suggests new directions and in doing so stimulates several questions to consider in furthering this initiative.

The issue of "*how to promote creative and flexible use of mathematical and scientific ideas within an interdisciplinary context where students solve substantive, authentic problems that address multiple core learning*" (English 2013, p. 494) has been brought to the fore previously in ICTMA literature as too has the solution of using modelling as the teaching approach to address this issue. Barboza, Bassani, Lewandoski and Abitante explore this issue in the Brazilian context by addressing the possibility of interdisciplinary integration through mathematical modelling of optical phenomena using a series of physical experiments they designed and presented to in-service and pre-service teachers in workshops. According to these

authors, their program of interdisciplinary activities contributed to the learning of both the mathematical and physical concepts involved through the manipulation of the well designed experiments targeting particular optical concepts related to mirror manipulation. The program also stimulated the interest of participants. These claims are in keeping with research findings of others such as Michelsen and Sriraman (2009). Michelsen and Sriraman (2009) found that upper secondary Danish students' interest in mathematics and science was able to be improved through interdisciplinary modelling activities.

Bossio, Londoño and Jaramillo present results from another Colombian case study analysing a mathematical modelling activity undertaken in the classroom but stimulated by the events in students' home lives. The authors' purpose was to address the issue of *how social and cultural factors engage students in using school mathematics.* The participants were a group of ninth-grade high school students whose life situations were used to generate linear models. The chapter illustrates how prior knowledge of real life situations of the learner becomes a resource for modelling activity in the mathematics classroom (see also Galbraith, Sect. 28.4). The authors select one student and show how this student's prior experiential knowledge is used as she models in the classroom developing a formal linear model first graphically then algebraically. They chose the situation in the context of pre-paid energy consumption as an exemplar from practice in the Colombian context to argue the case for mathematical modelling examples arising naturally in the classroom as students activate prior knowledge from their everyday experiences (cf. Stillman 2000). The authors conclude that in social and cultural contexts of rural communities, teachers do not need to invent modelling activities for the classroom as situations already exist where students can develop a process of mathematical modelling in the classroom, due to the richness of the meanings and relations constructed in their life experience, and their particular economic and social conditions. In addition, these classroom activities meet social objectives as well as the mathematical formation purposes of the curriculum as intended by the national statutory education authority.

Frejd raises the issue of *whether, and how, modelling should be taught in schools so as to be a link to workplace mathematics*, as advocated by many such as Li (2013) seeing that school mathematics and workplace mathematics are fundamentally different in purpose (Kaiser et al. 2013). This chapter examines how nine professional mathematical modellers learnt mathematical modelling and their opinions on mathematical modelling in upper secondary and general education. Frejd finds Salling Olesen's model (2008) for the analysis of workplace learning a useful means to analyse the responses of his interviewees. The participants mainly learnt mathematical modelling during their doctoral studies and through their occupation, by working with what they saw as 'real modelling'. In their opinion mathematical modelling should be a part of mathematics education in upper secondary school; in particular, modelling should be emphasized more as part of general education to develop students' critical views on how models are used in society. In addition, they suggested approaches to teach modelling and suitable modelling problems to work with from their own workplaces that were not idealised or reduced.

The chapter by Galbraith addresses issues of fundamental importance to the development of mathematical modelling, as a field, and for modelling as an important element within curriculum. The main issue is *the extent to which students leaving school are able to apply their knowledge proficiently to problems located in personal, work or civic contexts or to other knowledge areas whilst in school* falls well short of what was hoped for. An important aspect of this chapter is that it uses a distinction between "modelling as content" ("mathematics for the sake of modelling") and "modelling as a vehicle" ("modelling for the sake of mathematics") (see also Julie 2002) to point out what the author calls the epistemic fallacy between the nature of modelling for real world problem solving (ontology) and the curricular interpretations and approaches (epistemology) that are often encountered in which the ontology is distorted, if not replaced with something else. This can be seen as the source of much criticism of applications and modelling in mathematics education. Galbraith sees "privileging of perceived conservative classroom conceptions" (p. 344) as a source of some of these criticisms taking as an example the notion of "authenticity" and its ramifications in applications and modelling. He reminds the reader he has previously argued that the term has been used too broadly (Galbraith 2013) then discusses an example involving population prediction from the Australian media that has some doubtful in-built assumptions. Once the situation is modelled mathematically and this is discovered he then raises the question of what would be authentic actions that would flow given the intention of the espoused curriculum is to develop modelling as real world problem solving. For the modelling as content teacher, the modelling would continue with questions being posed from the implications of what has been discovered whereas the modelling as vehicle teacher would be pressed to simply move on by other imperatives such as time and need for curriculum coverage. This leads to Valsiner's Zone Theory (1997) being applied as a framework for theorising and structuring teaching and learning within the content/vehicle metaphor. What is provided within the Zone of Free Movement (ZFM) impacts the Zone of Promoted Action (ZPA) that is oriented to defining and promoting valued new skills. Within a content approach, teachers ensure the ZFM provides the requirements and support for the development of modelling as real world problem solving but within a modelling as vehicle approach the competing other imperatives of the curriculum are allowed to restrict the ZFM below the threshold that would allow the ZPA to support the development of modelling as curricular content in its own right. Galbraith suggests that for change in this situation to be effected, curriculum documents need explicit mandated objectives that require students to formulate and solve their own problems located in their world as was shown in more localised modelling curricula (Stillman and Galbraith 2011).

Ikeda and Stephens take up the issue of *how to think about models and modelling* considering the diversity of interpretations that have mushroomed over the years. They attempt to re-construct what is meant by a model and the process of its construction which is generally assumed to be built by translating a real world problem into a mathematical representation. Through this process they hope to elicit new perspectives on modelling. There are two ways learners can set up models according to Ikeda and Stephens: (1) as hypothetical working spaces, and

(2) as physical/mental entities for comparing and contrasting. These means of construction then morph into "roles in the context of the learner". By reflecting on the teaching of modelling from these two roles the authors draw attention to four different perspectives of modelling: (a) where modelling is interpreted as interactive translations among plural worlds (the real world, a representational world able to be manipulated, a geometric world subject to drawing and measuring, and a symbolic and algebraic world) not between two fixed worlds (presumably the real world and the mathematical world); (b) where models have the potential to incorporate scenarios beyond the initial problem situation; (c) where the mathematical world is used as a source of mental entities for comparing and contrasting; (d) where modelling competency means knowing how to balance between these different roles. Perspectives (a) and (d) are concerned with Role 1 and (b), (c) and (d) are concerned with Role 2.

Lamb and Visnovska address two pedagogical issues in their chapter reporting a project which aims at enhancing the teaching practices of teachers who teach at lower secondary level (Years 8–10) in Australia. The focus is set as the teaching of statistical literacy in the framework of "numeracy" in relation to real-life-problems. The issues in focus are *sense-making in real-life problems* and the *provision of quality professional development to foster quality in teaching practices.* The researchers note the themes of statistical literacy and modelling appeared to resonate with the teachers as being broadly connected since context was considered necessary to give meaning to statistics and to engage in decision making associated with data. The teachers worked in groups on a real-world problem with data provided and the sense of direction (Treilibs 1979) already posed in order to experience how students might develop key mathematical ideas in this phase of a lesson. The working-sessions of the teachers were orchestrated by the researchers who deliberately chose the order of presentation of solutions to the whole group to expose those that had taken a purely mathematically directed approach focusing on aspects of correctly displaying the mathematical object (a histogram) rather than a modelling approach involving sense making, to "press" these "problem solvers" to see the inadequacy in their approach and their lack of sense making. The researchers saw the order in which responses were presented as playing a pivotal role in the quality of the discussion that arose. These discussions highlighted the differences between doing mathematics for its own sake and using mathematical models to respond to, and interpret, contextualized situations that they see as an essential part of statistical literacy pedagogy.

Aspects of Cognitive Load Theory (CLT) (Sweller 2010), a psychological theory about the consequences of human memory structure for teaching, are discussed by Perrenet and Zwaneveld from the perspective of what it might contribute to mathematical modelling education particularly in secondary school. The authors characterise the teaching and doing of mathematical modelling in secondary education as a complex task involving a heavy cognitive load for both teachers and students. Perrenet and Zwaneveld suggest that some of the directives of CLT could prove useful in modelling education in reducing this *issue of extraneous cognitive load*, particularly the notion of redundancy if viewed in a

new light. As the authors state, CTL's criticism of constructivism is countered by an argument from Problem Based Learning, and applied to mathematical modelling education, this argument stresses the importance of structured support in using the mathematical modelling cycle (see Bruder and Prescott 2013 for a discussion of research evidence). The chapter contributes to the existing discussion on how theories from cognitive psychology can assist researchers in mathematics education, in better understanding students' work in solving modelling problems, and in better teaching modelling and applications.

Rosa and Orey highlight the importance of using the social-critical dimension of mathematical modelling to solve problem situations that afflict contemporary society. They raise two issues: (1) *the role of schools in promoting students' social-critical efficacy* and (2) *how current teaching practices in mathematics impact students' social-critical efficacy*. The idea of social-critical efficacy is novel encompassing students' critical analysis of the role of mathematics in the power structures of society informed by their critical reflection on their prior experiences related to social elements underpinning life in a globalised world. Research related to mathematical modelling with a social-critical dimension has redefined its objectives, and is developing a sense of its own form and the potential of research methods to legitimate its pedagogical action. The chapter deals with philosophical and epistemological issues supporting the inclusion of a social-critical component in the teaching of mathematical modelling. These span a re-articulation of critical mathematics education with its mathematics for action proposed by Skovsmose (1994), the inclusion of learner goals as proposed by Mellin-Olsen (1987) and Freire's recommendation of teaching based on a dialogical model rather than a banking model (Freire 2000). Three models of teaching for developing the critical consciousness of students are proposed. The authors extoll the importance of a learning environment that helps students develop their social-critical efficacy supported by the use of a socio-critical mathematical modelling cycle that allows them to construct solutions to problems that are of interest to them and their community but where the solution is of social significance to society more generally. An alternative approach to developing modelling integrated to critical aspects of reality, that is, by reflective mathematical modelling, is proposed by Scheller and colleagues (see Scheller et al., this volume).

Deciding to *use a modelling approach to teaching in an effort to engage students in learning mathematics because it is said to be more interesting and motivating for students* does not mean that the students will agree with your new approach. Ustra and Ustra present results of a qualitative evaluation conducted in Fundamentals of Calculus classes in four service courses from a federal university in Brazil. The objectives were to analyze the main difficulties the students had regarding mathematical modelling, especially with functions and the relationship of these difficulties with the different fields of study. The analysis of the results identified some teaching difficulties were related to contextualization in mathematical modelling, as well as the teaching of mathematical concepts for different courses. Students failed to appreciate the connection of modelling examples to their courses when they were first introduced and had even greater difficulty as examples with connection to future occupations were used. Ustra and Ustra's evaluation of their new

approach in these courses identified three categories of contextualisation of teaching examples that they saw as being able to both facilitate and impede the students' modelling. These were recognition of relevance to the area of training and the students' everyday world, translation of the situation through modelling (i.e., being able to handle the mathematics to do this) and personal factors associated with the student. It would seem imperative that these be considered if modelling is to "effectively promote views of mathematics that extend beyond transmissive techniques" as desired by many of these students and "to advance its role as a tool for structuring other areas of knowledge" (Niss et al. 2007, p. 20).

In an endeavour to address the issue of *being able to provide sufficient variation* (Marton and Tsui 2004) *in real world problem solving in engineering so that students develop problem solving skills and are capable of handling unknown problems in future occupations*, Wedelin and Adawi enunciate a set of seven design principles for small problems that capture the essential characteristics of larger and more complex real world problems. The potential strength of this chapter lies in the design of these "small realistic problems" and when, and how, such problems can be used to support the development of students' abilities to solve more complex, realistic problems. The authors emphasize that real problems require "learning and searching for existing theory" and caution against over burdening students with new theory to resolve problems. A set of examples that support the principles that are put forward are given. All problems position students in an exploratory mode where they are trained in metacognitive aspects. Students develop a case library of problems over the course as they develop in three dimensions of learning.

1.5 Influences of Technologies

The early embracing of computer technology within the ICTMA community (Berry et al. 1986; Moscardini et al. 1984; Prior and Moscardini 1984) and the professional modelling community generally (Aslaken and Santosa 2013) certainly led to the solution of many real world problems that had proved difficult to resolve with the conventional mathematical tools of the time. However within little more than a decade technology was being seen as more than just a means of carrying out repeated operations, calculations and graphing quickly, rather by drawing on cognitive development and sociocultural theory it was being suggested (e.g., Goos 1998) that computers were cognitive tools for socially supported learning where collaborative peer interaction would allow students to perform beyond the limits of their existing capabilities. This had obvious implications for pedagogical modelling in schooling, for example, as technology use would transform the tasks offered to students and mathematics lessons would need to be reoriented (Henn 1998) as what students were capable of doing in such environments would be transformed. Tikhomirov (1981, p. 271) had insightfully claimed earlier that technologies such as computers could lead to "a transformation of human activity" and this would lead to new forms of activity emerging. Both Brown and Soares take up this transformative role of technology in their chapters.

Many affordances (Gibson 1977) useful in facilitating the solution of real world tasks are available in Technology-Rich Teaching and Learning Environments (TRTLE's) where both teacher and students have access to a wide range of electronic technologies. Of particular use for modelling are those allowing visual image generation by technology (Pead et al. 2007). Whilst the TRTLE provides additional opportunities and approaches to engaging with the real world, thus potentially transforming human mathematical modelling activity (cf Tikhomirov 1981), but additional complexities also exist (Galbraith et al. 2007). One of the transformational powers of the technology is to produce technology-generated images to clarify and refine students' mental models of the situation; however, Brown asks: Is this power being realised? Using a grounded theory approach, Brown's study (see also Brown 2015) showed that students often did not take up the opportunities, such as the usefulness of the data plot on a technological device, to inform their choice of function model or to compare models with data or each other, even though they had the technological and mathematical knowledge to do so. Brown suggests that teachers and students must realise "the cognitive role of visualisation" (p. 440) in modelling, in particular how this supports their mathematisation (cf. Rivera et al.), which in turn should be supported by an anticipation of the role digital technology can play in a modelling context. After all, affordances of TRTLE's must first be perceived in order to be enacted. Unless this happens the transformational power of technologies in modelling will remain unlocked.

Rodríguez and Quiroz use a rich technological environment to support engineering students in the development of modelling competencies where the mathematical focus is DEs in a tertiary physics context. Electronic technologies in the form of calculators, sensors, physical materials, simulation and graphing software play an important role in the teaching and learning process presented. In this chapter, a complex cycle developed by the first author in her doctoral study (Rodríguez 2010) for the kind of physics-technical examples that the authors use forms the basis for analysis of competencies and choice of supporting technologies. These technologies allow the students working in teams to interact with the phenomena being studied and promote several competencies. Competencies are identified by what is required to successfully transition between stages in this modelling cycle. These all appear to be modelling competencies not technological nor social ones. As competencies are the focus of many studies (see Kaiser and Brand, Chap. 10) and are of particular interest to the modelling community this work supports other findings in this area.

Soares deals with the ways in which we can describe and understand students' work on interpreting, validating and varying already existing models in relation to mathematical modelling. This is an issue which, although relevant and significant, has not attracted as much attention in literature as have other aspects of applications and modelling. Soares uses the term *model analysis* to indicate what happens when students undertake investigation of models, especially in relation to coming to grips with old and new mathematical concepts with which they are not necessarily very familiar. She uses the software, Modellus, in the design of a teaching approach that involves model analysis. The chapter makes reference to, and use of, a classical two-population ODE model of the transmission of malaria, which was used as an exemplar for university students majoring in biology but taking a course in applied

mathematics. The chapter focuses on the role of the software in making the model solutions accessible to students. In attempting to place her analysis in terms of the aspects of the modelling cycle that may or may not be manifest with regard to model analysis, she came to the conclusion that her students did not complete the entire cycle but rather were developing a rudimentary cycle (Niss this volume, Chap. 5) focusing on the transition from a model situation to a mathematical model and its interpretation. She attributed this reduction of the modelling cycle to their use of Modellus to access the solutions of a mathematical model produced by others even though they did not know the necessary mathematics to produce the model and solve it analytically themselves. The software allowed them to experiment with effects of parameters on the phenomenon being modelled in a similar manner to how Rivera et al.'s high school students used GeoGebra with unfamiliar functions to model a flood situation. The offerings and constraints of the software influenced the students' reasoning and interactions with it in both these situations. These affordances of the technological environments are dependent on the particular technology used (Galbraith et al. 2007) and so lead to a reorganisation of the processes involved. In the case of Modellus, the biology students were able to focus completely on the behaviour of the solutions and not be distracted by doing calculations correctly or choosing correct procedures as they would have done with a by-hand analytical treatment. Soares sees this shift in focus to reasoning about the model rather than regulating mathematizing to produce the model as characteristic of a reorganisation of thinking with respect to technology use that was foreshadowed by Tikhomirov (1981), Goos (1998) and others. She argues that this reorganization of the analysis of the model results in a *hybrid approach* which is not purely modelling and not purely applications. Even though there is some convergence between model analysis and applications there is enough point of difference to support this claim as pre-existing models not within the scope of the models students already know are analysed using software to discuss new concepts. The modelling activity is thus changed by the presence of the software.

Souza Júnior et al. report the development of Learning Objects as mathematical models. Learning Objects have a clearly defined educational purpose, an element encouraging student reflection and the application must be transferable across contexts. The practice of developing Learning Objects is outlined where the role of mathematical modelling is implicit and the real world is thus able to be brought into the classroom through simulations. The designers of the Learning Objects form a multidisciplinary team including pre-service secondary mathematics teachers, computer programing students and academics from mathematics, teacher education and computer science. It seems, although not explicitly stated, that the authors are arguing that the benefits are twofold – for school students and for these pre-service teachers. The main argument of the chapter is that the process is a social production that empowers pre-service teachers to be supported to produce these materials and it is the mathematical model involved that links the disciplines: education, mathematics and computing. "The Learning Object becomes a means of expression, an act of mathematical modelling on the computer itself" (p. 469). This point is exemplified by the Learning Object associated with the predator–prey population dynamics models of fish populations in the Adriatic Sea during and after World War I.

1.6 Assessment in Schools and Universities

Although assessment of modelling in schools and universities continues to be a major issue (Galbraith 2007a), it does not seem to be a popular topic of research (Frejd 2013) within ICTMA literature in recent times. Using ideas and notions from ethnomethodology and the sociological study of work in science, the actual scripts of examinees were analysed by Julie to tease out the examinees' ways of working with a modelling and applications-like problem in a high-stakes school-leaving examination in South Africa. Such examinations are time-restricted and "definitely not suitable for assessing open modelling" (Antonius 2007) so research in such situations is rare in a research community focussing on quality practices in modelling education. In the South African context the examination items of interest take the form of guided applications to financial contexts supported by a formula sheet of common financial models. These are the applications found most difficult by candidates on these national examinations. To apply the models only the calculation parts of modelling are really in focus. Julie's analysis was anchored around the various agencies exerted by elements present in the context of high-stakes school-leaving examinations as represented in the lived work students engaged in as they produced answer responses to applications-like items. Three ways of working that characterised the candidates' ways of working are focused on: *Translation to familiar symbolism, Anticipation of adjustments to be made* and *U-turning*. The chapter demonstrates how the prevailing contexts of writing high-stakes examinations exercised agency for these ways of working.

Blum and Leiß (2007) have indicated that there are reciprocal processes between the real situation and the situation model in modelling and that these processes are important in understanding a modelling task. Matsuzaki and Kaneko surmise there might be the possibility that a modeller imagines situations from modelling tasks, and that these situations might be able to be changed. Firstly, a situation model could be formulated from the initial imagined situation, and formulation of a situation model occur. Secondly, a situation model could be reformulated from the initial situation model as modelling is proceeding. In previous studies Matsuzaki (2004) adopted drawing pictures as a method for gaining insight into these experiences in setting up the problem from the task situation in the real world. Matsuzaki and Kaneko have designed a pre-test and a post-test that can be used in the primary school setting with a modelling activity to see if the student (a) formulates an initial situation model and/or (b) reformulates the situation model during modelling.

1.7 Applicability at Different Levels of Schooling, Vocational Education, and in Tertiary Education

In all books in the ICTMA series there are several examples of demonstrations through documented implementations of the applicability at different levels of education of modelling and applications of mathematics. This book is no exception

but some of the examples bring points of difference from what has been demonstrated before. There are new contexts such as Youth and Adult Education and different emphases, for example, development of a socio-critical perspective in pre-service teacher education where the pre-service teachers take up social concerns through modelling activity.

In the tertiary education space, several examples are presented. According to Campos, Lombardo, Jacobini and Wodewotzki, blending statistics education, critical education (Giroux 1988), critical mathematics education (Skovmose 1994) and mathematical modelling allows the possibility of pedagogical projects that value interdisciplinarity and active participation of students in knowledge construction at the first year university level. They present what they call Critical Statistics Education which highlights the valuing of socio-political interfaces within undergraduate teaching. From this perspective it is essential that societal or environmental problems dealt with relate to "fundamental social conflicts" and students own these problems. Modelling is the tool that allows students to transform their reality whilst their lecturers promote the socio-political role played by statistics in society. In short, modelling is the link between espoused theory and its realisation in practice. In this chapter, a scenario is presented that occurred in the statistics discipline of an economics sciences course. By adopting critical education practices the class discusses and mathematises global warming, its causes and consequences for society.

Elaborating a modelling approach to the integration of physics and mathematics instruction in a first-year engineering course in a Mexican university is the focus of the chapter by Dominguez, Garza and Zavala. The authors' intention is to close a gap in science and mathematics; therefore they take certain situations that give rise to interesting activities in the mathematics/physics interface. The purpose is to improve students' abilities to make connections between these disciplines, to increase student motivation and to develop competencies including social competencies such as working collaboratively. In this context, models serve as conceptual resources to develop understanding of a variety of phenomena. The key to success is problem design that facilitates the application of the main pedagogical approach, Modelling Instruction (see Hestenes 2010 for an overview). The authors analyse a project undertaken by students in small groups to illustrate how modelling is applied in this context. Future developments mooted by the authors include use of more realistic situations for the problems and the gathering of evidence for claims that critical thinking, problem solving and argumentation competencies are facilitated by this application of modelling in a tertiary setting.

The chapter by Rodríguez presents an experience of an educational practice in the same Mexican university involving a different way to teach a Differential Equations course for future engineers based on a teaching proposal developed through mathematical modelling. Rodríguez acknowledges that the purpose of teaching mathematics is to prepare critical citizens but takes a realistic or applied perspective (Kaiser and Sriraman 2006) to the use of mathematical modelling in teaching. This proposal emphasizes that mathematics is a human activity (Jacobs 1994) that answers several problems of a different nature, and throughout this problem solving activity it is likely that the emergence of mathematical concepts, notions and procedures occurs. The application of DEs as a tool to model phenomena (physical, biological, or chemical) is considered as important as learning techniques to solve DEs. Technology also plays an important integrative role highlighting the range of possible multiple representations of the same mathematical object (the DE). The modelling activities design pays particular attention to the pseudo-concrete and physical domains and to moving between key stages of the specialised modelling schema of Rodríguez (2010).

Duan, Wang and Wu claim that the teaching of mathematical modelling at tertiary level would be more stimulating if combined with student-involved research. Based on the Innovation Practice and Internship Base for Mathematical Modelling at the National University for Defence Technology in China, a student innovation training system from the lowest level to senior undergraduates has been built to cultivate modelling competency in students to solve practical problems under supervision. In addition to generic mathematical modelling ability and modelling associated with a particular profession, students are expected to develop an open and cooperative consciousness. An example of mathematical positioning with the Global Navigation Satellite System (GNSS) is introduced to show the whole process in detail: stimulation of a focusing interest (e.g., by considering implications for reliability of satellite signals in today's world communications and the problem of multipath signals from the same source), mathematizing, and solution through modelling of the situation. The focus is on one of the mathematical issues that arise in increasing the processing speed, accuracy and reliability of GNSS – the multipath case. This case is used to demonstrate how students are encouraged to explore solutions, select a method and implement this method, and develop active and critical thinking.

As the last tertiary example, Villareal, Esteley and Smith demonstrate how it is possible for teacher educators to engage pre-service teachers in active modelling projects in a context (Argentina) where such engagement is rare in schools and in universities. What attracted these educators to mathematical modelling was its open and interdisciplinary nature and the opportunity for reflection on mathematics and the social role of mathematics and mathematical modelling. They were thus drawn to the socio-critical modelling perspective (Kaiser and Sriraman 2006; Rosa and Orey, this volume) and this is the point of departure of their work from that of others (e.g., Caron and Bélair 2007; Widjaja 2013) as they focus on developing this perspective in their pre-service teachers. The pre-service teachers in this study developed 12 modelling projects but only those that explored socio-economic and ecological conscience themes focused on authentic social concerns. Differences in how pre-service teachers approached social issues and the modelling of these were attributed to the link they were able to establish with their own socio-environmental contexts and their understanding of what it meant to model from a socio-critical perspective. Their difficulties were related to theme selection and concerns about the sophistication of the mathematics they would use to model. In the process of modelling they came to realise that mathematical modelling gave them a new way of doing mathematics and perceiving its implications for society and the possibility to reflect at a meta-level on their modelling activity and project these forward onto possible opportunities for them to use modelling in their future classrooms.

Luna, Souza and Lima show the applicability of mathematical modelling in a primary school (Years 1–5) in Brazil. Their aim is to analyse, using Bernstein's theory (1990), the kinds of text that are produced in a mathematical modelling activity at this level of education. The participants were the teacher and her Year 4 students (9–10 years old). The authors take a socio-critical perspective (Kaiser and Sriraman 2006) to the teaching of modelling expecting students to begin identifying the presence of mathematics in society and to reflect critically about its presence from the beginning of primary school. The topic of the activity is thus *Virtual Water*, all the water used in the production of foods from cultivation to manufacture. According to Bernstein's theory, the principles governing the pedagogical practice of a given context allow the production of different texts for the instruction of school mathematics. The authors view text as being verbal, written or gestural. What is of interest to them are power and control relations governing communication practices within social relations during modelling activity. As a result of a weakening of these relations during modelling, students have the opportunity of preparing legitimate texts of the instructional discourse by proposing mathematical procedures and concepts and developing their own answers to the problems posed. English (2013) has reported a similar finding previously. *Contextualised texts* were produced by both students and the teacher when a student raised a query about the meaning of a mathematical concept (cubic metre) and both students and teacher discussed this within a familiar context (the capacity of a swimming pool). The students also produced *critical mathematical text* characterised by criticism of the presence of mathematics in a given text used in society (product labelling) and *interdisciplinary texts* as they made their own relations between the instructional discourse of school mathematics and other instructional discourses of schooling (e.g., science) prompted by engagement in the modelling activity. The authors suggest that the specific production of these texts could be a result of the pedagogical practices used during modelling in the primary school setting.

Silva, Santana and Carneiro present a case study in Youth and Adult Education, an area where there is not much previous research in ICTMA or elsewhere. The authors describe the implementation of a modelling activity illustrated by transcripts of student interaction at selected critical incidents in a lesson sequence. They attempt to capture critical pedagogical practices as a teacher works using a multidisciplinary approach to improve students' learning. The aim is to show possible contributions of a mathematical modelling environment for using mathematics to understanding situations studied in other disciplines; whilst taking account of students' prior knowledge and everyday experiences. Students were clearly interested and engaged by the can recycling context for the task but they had difficulties with formulation and mathematizing exhibiting the previously reported over reliance on linear models (De Bock et al. 2007). The authors illustrate how the teacher acted as a mediator between the real situation and the mathematical objects used in representations of essential features of the situation in mathematising to facilitate students refining these mathematisations.

A modelling challenge program for junior high school and high school students in Japan is the topic of the chapter by Yanagimoto, Kawasaki and Yoshimura. In the first year of the program, two tasks with scientific and social content were prepared. The first task was concerned with the phenomenon of a cart with a sail rolling down a slope, and the other with the income and expenditure of an electric power company (see Yoshimura this volume). The authors prepared two different types of tasks as they wanted to see whether there was a preference for one type of context over the other. However, as only eight students participated in the event, the sample size was too small to be able to do this. The project, as an attempt to foster modelling in mathematics classrooms in Japan, is highly relevant and it joins a number of such projects within the ICTMA community (e.g., examples in Kaiser et al. 2011; Stillman and Brown 2014).

Finally, Orey and Rosa provide an interesting insight into the mathematics of an architectural feature. The issue being investigated through ethnomodelling (Rosa and Orey 2013) was how the architects / builders managed to capture the shape of a hanging chain in the wall of a brick building. From the outset it was presumed that this was done by the builders hanging a chain between the support posts and using it as a template. This seems to be confirmed by the mathematics. It is unlikely that any other practical method would have been used by the builders so quadratic or exponential shapes are unlikely. Consequently the purpose for doing the mathematics is to understand how the builders may have managed to obtain such a feature in their wall. Thus by looking at various models students have the opportunity to explore the relationship between mathematical ideas developed in the daily practice of community members, in this case builders, and formal mathematical objects as advocated by D'Ambrosio (1990/1998), amongst others.

1.8 Conclusion

The ICTMA community operates from diverse research paradigms and theoretical underpinnings in its work. However, what all members of this community of researchers share is a genuine desire "to establish a learning culture of mathematical modelling" (Stillman et al. 2013, p. 22). In this volume, in particular the influences of cultural, social, and cognitive perspectives on the shaping of the work in the community become visible. This is meant to stimulate further research in the area of learning and teaching mathematical modelling at all educational levels. That builds on existing foundations or takes the field in new directions. In particular, diversification in theoretical approaches utilized and the pursuit of deeper insights into significant challenges in the teaching, learning and assessing of mathematical modelling and applications should lead to further advancement of our field.

References

Andresen, M., & Petersen, A. (2011). Modelling chemical equilibrium in school mathematics with technology. In G. Kaiser, W. Blum, R. Borromeo Ferri, & G. Stillman (Eds.), *Trends in teaching and learning of mathematical modeling* (pp. 519–528). New York: Springer.

Antonius, S. (2007). Modelling based project examination. In W. Blum, P. Galbraith, H.-W. Henn, & M. Niss (Eds.), *Modelling and applications in mathematics education* (pp. 409–416). New York: Springer.

Antonius, S., Haines, C., Jensen, T. H., & Niss, M., (with H. Burkhardt) (2007). Classroom activities and the teacher. In W. Blum, P. Galbraith, H.-W. Henn, & M. Niss (Eds.), *Modelling and applications in mathematics education* (pp. 295–308). New York: Springer.

Aslaken, H., & Santosa, F. (2013). Technology issues. In A. Damlamian, J. F. Rodrigues, & R. Sträßer (Eds.), *Educational interfaces between mathematics and industry* (pp. 359–364). Cham: Springer.

Barbosa, J. (2007). Teacher-student interactions in mathematical modelling. In C. Haines, P. Galbraith, W. Blum, & S. Khan (Eds.), *Mathematical modelling: Education, engineering and economics* (pp. 232–240). Chichester: Horwood.

Bernstein, B. (1990). *Class, codes and control: Vol. 4. The structuring of pedagogic discourse.* London: Routledge.

Berry, J. S., Burghes, D. N., Huntley, I. D., James, D. J. G., & Moscardini, A. O. (Eds.). (1986). *Mathematical modelling methodology, models and micros.* Chichester: Ellis Horwood.

Blomhøj, M., & Jensen, H. T. (2003). Developing mathematical modelling competence: Conceptual clarification and educational planning. *Teaching Mathematics and Its Applications, 22*(3), 123–139.

Blomhøj, M., & Højgaard Jensen, T. (2007). What's all the fuss about competencies? In W. Blum, P. L. Galbraith, H.-W. Henn, & M. Niss (Eds.), *Modelling and applications in mathematics education: The 14th ICMI study* (pp. 45–56). New York: Springer.

Blum, W. (2011). Can modelling be taught and learnt? Some answers from empirical research. In G. Kaiser, W. Blum, R. Borromeo Ferri, & G. Stillman (Eds.), *Trends in teaching and learning of mathematical modelling* (pp. 15–30). Dordrecht: Springer.

Blum, W., & Leiß, D. (2007). How do students and teachers deal with modelling problems? In C. Haines, P. Galbraith, W. Blum, & S. Khan (Eds.), *Mathematical modelling: Education, engineering & economics* (pp. 222–231). London: Horwood.

Blum, W., & Niss, M. (1991). Applied mathematical problem solving, modelling, applications, and links to other subjects – State, trends and issues in mathematics instruction. *Educational Studies in Mathematics, 22*(1), 37–68.

Blum, W., Galbraith, P. L., Henn, H.-W., & Niss, M. (Eds.). (2007). *Modelling and applications in mathematics education.* New York: Springer.

Bourdieu, P. (1977). *Outline of a theory of practice.* Melbourne: Cambridge University Press.

Briedenbach, W. (1969). *Methodik des Mathematikunterrichts in Grund-und Hauptschulen. Band 1 – Rechnen.* Hannover: Hermann Schroedel Verlag KG.

Brousseau, G. (1997). *Theory of didactical situations in mathematics.* Dordrecht: Kluwer.

Brown, J. P. (2015). Complexities of digital technology use and the teaching and learning of function. *Computers & Education, 87*, 112–122. doi: 10.1016/j.compedu.2015.03.022

Brown, J. S., Collins, A., & Duguid, P. (1989). Situated cognition and the culture of learning. *Educational Researcher, 18*(1), 32–42.

Bruder, R., & Prescott, A. (2013). Research evidence on the benefits of IBL. *ZDM: The International Journal on Mathematics Education, 45*(6), 811–822.

Caron, F., & Bélair, J. (2007). Exploring university students' competencies in modelling. In C. Haines, P. Galbraith, W. Blum, & S. Khan (Eds.), *Mathematical modelling: Education, engineering and economics* (pp. 120–129). Chichester: Horwood.

Chapman, O. (2009). Teachers' conceptions and use of mathematical contextual problems in Canada. In L. Verschaffel, B. Greer, W. Van Dooren, & S. Mukhopadhyay (Eds.), *Words and worlds: Modelling verbal descriptions of situations* (pp. 227–244). Rotterdam: Sense.

Crouch, R., & Haines, C. (2004). Mathematical modelling: Transitions between the real world and the mathematical world. *International Journal of Mathematics Education in Science and Technology, 35*(2), 197–206.

D'Ambrosio, U. (1998). *Ethnomathematics: The art or techniques of explaining and knowing* (P. B. Scott, Trans.). Las Cruces: International Study Group on Ethnomathematics (Original work published 1990).

Davis, P. J. (1991). Applied mathematics as a social instrument. In M. Niss, W. Blum, & I. Huntley (Eds.), *Teaching of mathematical modelling and applications* (pp. 1–9). Chichester: Ellis Horwood.

De Bock, D., Van Dooren, W., Janssens, D., & Verschaffel, L. (2007). *The illusion of linearity: From analysis to improvement*. New York: Springer.

de Oliveira, A. M. P., & Barbosa, J. C. (2013). Mathematical modelling, mathematical content and tensions in discourses. In G. A. Stillman, G. Kaiser, W. Blum, & J. P. Brown (Eds.), *Teaching mathematical modelling: Connecting to research and practice* (pp. 67–76). Dordrecht: Springer.

Dewey, J. (1916). *Democracy and education*. New York: Macmillan.

Dewey, J. (1917). The need for a recovery of philosophy. In J. Dewey (Ed.), *Creative intelligence: Essays in the pragmatic attitude* (pp. 3–69). New York: Holt.

Durkheim, E. (1982). *The rules of sociological method* (W. D. Halls, Trans.). New York: Free Press.

English, L. D. (2002). Setting an agenda for international research in mathematics education. In L. D. English (Ed.), *Handbook of international research in mathematics education* (pp. 3–19). New York: Routledge.

English, L. D. (2013). Complex modelling in the primary and middle school years: An interdisciplinary approach. In G. A. Stillman, G. Kaiser, W. Blum, & J. P. Brown (Eds.), *Teaching mathematical modelling: Connecting to research and practice* (pp. 491–505). Dordrecht: Springer.

Freire, P. (2000). *Pedagogy of the oppressed*. New York: Continuum.

Frejd, P. (2013). Modes of modelling assessment – A literature review. *Educational Studies in Mathematics Education, 84*(3), 413–438.

Galbraith, P. (2007a). Assessment and evaluation – Overview. In W. Blum, P. Galbraith, H.-W. Henn, & M. Niss (Eds.), *Modelling and applications in mathematics education* (pp. 405–408). New York: Springer.

Galbraith, P. (2007b). Dreaming a 'possible dream': More windmills to conquer. In C. Haines, P. Galbraith, W. Blum, & S. Khan (Eds.), *Mathematical modelling: Education, engineering and economics* (pp. 44–62). Chichester: Horwood.

Galbraith, P. (2013). From conference to community: An ICTMA journey–The Ken Houston Inaugural Lecture. In G. A. Stillman, G. Kaiser, W. Blum, & J. P. Brown (Eds.), *Teaching mathematical modelling: Connecting to research and practice* (pp. 27–43). Dordrecht: Springer.

Galbraith, P. L., & Clatworthy, N. J. (1990). Beyond standard models – Meeting the challenge of modelling. *Educational Studies in Mathematics, 21*(2), 137–163.

Galbraith, P., Stillman, G., Brown, J., & Edwards, I. (2007). Facilitating middle secondary modelling competencies. In C. Haines, P. Galbraith, W. Blum, & S. Khan (Eds.), *Mathematical modelling (ICTMA 12): Education, engineering and economics* (pp. 130–140). Chichester: Ellis Horwood.

Gibson, J. J. (1977). The theory of affordance. In R. Shaw & J. Bransford (Eds.), *Perceiving, acting and knowing: Toward an ecological psychology* (pp. 67–82). Hillsdale: Erlbaum.

Giroux, H. (1988). *Teachers as intellectuals: Toward a critical pedagogy of learning*. Westport: Bergin and Garvey Press.

Goos, M. (1998). Technology as a tool for transforming mathematical tasks. In P. Galbraith, W. Blum, G. Booker, & I. D. Huntley (Eds.), *Mathematical modelling: Teaching and assessment in a technology-rich world* (pp. 103–113). Chichester: Horwood.

Greefrath, G., & Leuders, T. (2013). Verbrauch im Haushalt – Schätzen und Überschlagen. In S. Prediger, B. Barzel, S. Hussmann, & T. Leuders (Eds.), *Mathewerkstatt* (pp. 5–22). Berlin: Cornelsen.

Haines, C., & Izard, J. (1995). Assessment in context for mathematical modelling. In C. Sloyer, W. Blum, & I. Huntley (Eds.), *Advances and perspectives in the teaching of mathematical modelling and applications* (pp. 131–149). Yorklyn: Water Street Mathematics.

Haines, C., Izard, J., & Le Masurier, D. (1993). Modelling intensions realised: Assessing the full range of developed skills. In T. Breiteig, I. Huntley, & G. Kaiser-Messmer (Eds.), *Teaching and learning mathematics in context* (pp. 200–211). Chichester: Horwood.

Haines, C., Crouch, R., & Davies, J. (2001). Understanding students' modelling skills. In J. F. Matos, W. Blum, K. Houston, & S. P. Carreira (Eds.), *Modelling and mathematics education: Applications in science and technology* (pp. 366–380). Chichester: Ellis Horwood.

Hein, N., & Biembengut, M. (2006). Modelaje matemático como método de investigación en clases de matemáticas. In M. Murillo (Ed.), *Memorias del V Festival Internacional de Matemática* (pp. 1–25). Puntarenas: Colegio Universitario de Puntarenas.

Henn, W.-H. (1998). The impact of computer algebra systems on modelling activities. In P. Galbraith, W. Blum, G. Booker, & I. D. Huntley (Eds.), *Mathematical modelling: Teaching and assessment in a technology-rich world* (pp. 115–123). Chichester: Horwood.

Henning, H., & Keune, M. (2007). Levels of modelling competencies. In W. Blum, P. L. Galbraith, H.-W. Henn, & M. Niss (Eds.), *Modelling and applications in mathematics education* (pp. 225–232). New York: Springer.

Hestenes, D. (2010). Modeling theory for math and science education. In R. Lesh, P. L. Galbraith, C. R. Haines, & A. Hurford (Eds.), *Modeling students' mathematical modelling competencies* (pp. 13–41). New York: Springer.

Højgaard Jensen, T. (2007). Assessing mathematical modelling competency. In C. P. Haines, P. Galbraith, W. Blum, & S. Khan (Eds.), *Mathematical modelling: Education, engineering and economics* (pp. 155–164). Chichester: Horwood.

Houston, K., & Neil, N. (2003). Assessing modelling skills. In S. Lamon, W. Parker, & K. Houston (Eds.), *Mathematical modelling: A way of life* (pp. 165–178). Chichester: Horwood.

Izard, J., Haines, C., Crouch, R., Houston, K., & Neill, N. (2003). Assessing the impact of teaching mathematical modelling. In S. Lamon, W. Parker, & K. Houston (Eds.), *Mathematical modelling: A way of life* (pp. 165–178). Chichester: Horwood.

Jablonka, E. (2007). The relevance of modelling and applications: Relevant to whom and for what purposes? In W. Blum, P. L. Galbraith, H.-W. Henn, & M. Niss (Eds.), *Modelling and applications in mathematics education* (pp. 193–200). New York: Springer.

Jacobs, H. R. (1994). *Mathematics: A human endeavour* (3rd ed.). New York: W.H. Freeman.

Julie, C. (2002). Making relevance relevant in mathematics teacher education. *Proceedings of the 2nd international conference on the teaching of mathematics (at the undergraduate level).* Hoboken: Wiley.

Kaiser, G. (2007). Modelling and modelling competencies in school. In C. Haines, P. Galbraith, W. Blum, & S. Khan (Eds.), *Mathematical modelling: Education, engineering and economics* (pp. 110–119). Chichester: Ellis Horwood.

Kaiser, G., & Sriraman, B. (2006). A global survey of international perspectives on modelling in mathematics education. *ZDM: The International Journal on Mathematics Education, 38*(3), 302–310.

Kaiser, G., Blum, W., Borromeo Ferri, R., & Stillman, G. (Eds.). (2011). *Trends in teaching and learning mathematical modelling*. New York: Springer.

Kaiser, G., Kooj, v. d. H., & Wake, G. (2013). Educational interfaces between mathematics and industry at school level. In A. Damlamian, J. F. Rodrigues, & R. Sträßer (Eds.), *Educational interfaces between mathematics and industry* (pp. 263–270). Cham: Springer.

Kaufman, D., Moss, D. M., & Osborn, T. A. (Eds.). (2003). *Beyond the boundaries: A transdisciplinary approach to learning and teaching*. Westport: Praeger.

Larson, C. (2010). Modeling and quantitative reasoning: The summer jobs problem. In R. Lesh, P. L. Galbraith, C. R. Haines, & A. Hurford (Eds.), *Modeling students' mathematical modelling competencies* (pp. 111–118). New York: Springer.

Lesh, R., & Doerr, H. (Eds.). (2003). *Beyond constructivism: A models and modeling perspective*. Mahwah: Erlbaum.

Lesh, R., & English, L. (2005). Trends in the evolution of models & modeling perspectives on mathematical learning and problem solving. *ZDM: The International Journal on Mathematics Education, 37*(6), 487–489.

Lesh, R., Galbraith, P. L., Haines, C. R., & Hurford, A. (Eds.). (2010). *Modeling students' mathematical modelling competencies*. New York: Springer.

Li, T. (2013). Mathematical modeling education is the most important educational interface between mathematics and industry. In A. Damlamian, J. F. Rodrigues, & R. Sträßer (Eds.), *Educational interfaces between mathematics and industry* (pp. 51–58). Cham: Springer.

Maaß, K. (2004). *Mathematisches Modellieren im Unterricht-Ergebnisse einer empirischen Studie*. Hildesheim, Berlin: Franzbecker.

Maaß, K. (2006). What do we mean by modeling competencies? *ZDM, 38*(2), 113–142.

Maaß, K. (2007). Modelling in class: What do we want the students to learn. In C. Haines, P. Galbraith, W. Blum, & S. Khan (Eds.), *Mathematical modelling: Education, engineering & economics* (pp. 63–78). London: Horwood.

Marton, F., & Tsui, A. (2004). *Classroom discourse and the space of learning*. Mahwah: Erlbaum.

Matsuzaki, A. (2004). 'Rub' and 'stray' of mathematical modelling. In H.-W. Henn, & W. Blum (Eds.), *ICMI study 14: Applications and modelling in mathematics education* (Pre-conference vol., pp. 181–186). Dortmund: University of Dortmund.

Matsuzaki, A., & Saeki, A. (2013). Evidence of a dual modelling cycle: Through a teaching practice example for undergraduate school student. In G. A. Stillman, G. Kaiser, W. Blum, & J. P. Brown (Eds.), *Teaching mathematical modelling: Connecting to research and practice* (pp. 195–205). New York: Springer.

Mellin-Olsen, S. (1987). *The politics of mathematics education*. Dordrecht: Reidel.

Mercer, N. (2000). *Words & minds*. London: Routledge.

Michelsen, C., & Sriraman, B. (2009). Does interdisciplinary instruction raise students' interest in mathematics and the subjects of the natural sciences? *ZDM: The International Journal on Mathematics education, 41*(1–2), 231–244.

Moscardini, A. O., Prior, D. D., & Cross, M. (1984). On the use of simulation software in higher educational courses. In J. S. Berry, D. N. Burghes, I. D. Huntley, D. J. G. James, & A. O. Moscardini (Eds.), *Teaching and applying mathematical modelling* (pp. 339–355). Chichester: Ellis Horwood.

National Governors Association Center for Best Practices & Council of Chief State School Officers. (2010). *Common core state standards for mathematics*. Washington, DC: Author. http://www.corestandards.org/assets/CCSSI_Math%20Standards.pdf

Niss, M. (1992). *Applications and modelling in school mathematics: Directions for future development*. Roskilde: IMFUFA, Roskilde University Centre.

Niss, M. (2001). Issues and problems of research on the teaching and learning of applications and modelling. In J. F. Matos, W. Blum, S. K. Houston, & S. P. Carreira (Eds.), *Modelling and mathematics education*. Chichester: Horwood.

Niss, M. (2003). Mathematical competencies and the learning of mathematics: The Danish KOM project. In A. Gagatsis & S. Papastavridis (Eds.), *Third Mediterranean conference on mathematical education* (pp. 115–1240). Athens: Hellenic Mathematical Society and Cyprus Mathematical Society.

Niss, M. (2010). Modeling a crucial aspect of students' mathematical modeling. In R. Lesh, P. L. Galbraith, C. R. Haines, & A. Hurford (Eds.), *Modeling students' mathematical modeling competencies* (pp. 43–59). New York: Springer.

Niss, M., & Højgaard, T. (Eds.). (2011). *Competencies and mathematical learning. Ideas and inspiration for the development of mathematics teaching and learning in Denmark* (English translation of original 2002). Roskilde University: IMFUFA. http://pure.au.dk/portal/files/41669781/THJ11_MN_KOM_in_english.pdf

Niss, M., Blum, W., & Galbraith, P. (2007). Introduction. In *Modelling and applications in mathematics education* (pp. 3–32). New York: Springer.

Nunes, T., Schliemann, A. D., & Carraher, D. W. (1993). *Street mathematics and school mathematics*. Cambridge: Cambridge University Press.

Palm, T. (2008). Impact of authenticity on sense making in word problem solving. *Educational Studies in Mathematics, 67*(1), 37–58.

Pead, D., Ralph, B., & Muller, E. (2007). Uses of technologies in learning mathematics through modelling. In W. Blum, P. L. Galbraith, H.-W. Henn, & M. Niss (Eds.), *Modelling and applications in mathematics education* (pp. 309–318). New York: Springer.

Prior, D. E., & Moscardini, A. O. (1984). Modelling growth and development. In J. S. Berry, D. N. Burghes, I. D. Huntley, D. J. G. James, & A. O. Moscardini (Eds.), *Teaching and applying mathematical modelling* (pp. 290–305). Chichester: Ellis Horwood.

Rodríguez, R. (2010). Aprendizaje y enseñanza de la modelación: el caso de las ecuaciones diferenciales. *Revista Latinoamericana de Matemática Educativa, 13*(4–1), 191–210.

Rosa, M., & Orey, D. (2013). Ethnomodelling as a research lens on ethnomathematics and modelling. In G. A. Stillman, G. Kaiser, W. Blum, & J. P. Brown (Eds.), *Teaching mathematical modelling: Connecting to research and practice* (pp. 117–127). Dordrecht: Springer.

Runesson, U. (2005). Beyond discourse and interaction. Variation: A critical aspect for teaching and learning mathematics. *Cambridge Journal of Education, 35*(1), 69–87.

Saeki, A., & Matsuzaki, A. (2013). Dual modelling cycle framework for responding to the diversities of modellers. In G. A. Stillman, G. Kaiser, W. Blum, & J. P. Brown (Eds.), *Teaching mathematical modelling: Connecting to research and practice* (pp. 89–99). Dordrecht: Springer.

Salling Olesen, H. (2008). Workplace learning. In P. Jarvis (Ed.), *The Routledge international handbook of lifelong learning* (pp. 114–128). London: Routledge Mental Health.

Sebeok, T. A. (1991). *Semiotics in the United States*. Bloomington: Indiana University Press.

Shaw, J., & Puckett-Cliatt, M. (1989). Developing measurement sense. In P. Trafton & A. Schule (Eds.), *New directions for elementary school mathematics: 1989 yearbook* (pp. 149–155). Reston: National Council of Teachers of Mathematics.

Skovsmose, O. (1994). *Towards a philosophy of critical mathematics education*. Dordrecht: Kluwer.

Sriraman, B., & English, L. (Eds.). (2010). *Theories of mathematics education: Seeking new frontiers*. New York: Springer.

Stillman, G. (1998). The emperor's new clothes? Teaching and assessment of mathematical applications at the senior secondary level. In P. Galbraith, W. Blum, G. Booker, & I. D. Huntley (Eds.), *Mathematical modelling: Teaching and assessment in a technology-rich world* (pp. 243–253). Chichester: Ellis Horwood.

Stillman, G. (2000). Impact of prior knowledge of task context on approaches to applications tasks. *Journal of Mathematical Behavior, 19*(3), 333–361.

Stillman, G. (2001). Development of a new research tool: The cognitive demand profile. In J. Bobis, M. Mitchelmore, & B. Perry (Eds.), *Proceedings of the twenty fourth annual conference of the Mathematics Education Research Group of Australasia Incorporated* (Vol. 2, pp. 459–467). Sydney: MERGA.

Stillman, G. (2002). *Assessing higher order mathematical thinking through applications*. Ph.D. thesis, The University of Queensland, Brisbane

Stillman, G. (2004). Strategies employed by upper secondary students for overcoming or exploiting conditions affecting accessibility of applications tasks. *Mathematics Education Research Journal, 16*(1), 41–72.

Stillman, G. (2010). Implementing applications and modelling in secondary school: Issues for teaching and learning. In B. Kaur & J. Dindyal (Eds.), *Mathematical applications and modelling* (pp. 300–322). Singapore: World Scientific.
Stillman, G. (2011). Applying metacognitive knowledge and strategies in applications and modelling tasks in secondary school. In G. Kaiser, W. Blum, R. Borromeo Ferri, & G. Stillman (Eds.), *Trends in teaching and learning of mathematical modelling* (pp. 165–180). New York: Springer.
Stillman, G., & Brown, J. P. (2014). Evidence of *Implemented Anticipation* in mathematizing by beginning modellers. *Mathematics Education Research Journal, 26*(4), 763–789.
Stillman, G., Brown, J., & Galbraith, P. (2010). Identifying challenges within phases of mathematical modeling activities in Year 9. In R. Lesh, P. L. Galbraith, C. R. Haines, & A. Hurford (Eds.), *Modeling students' mathematical modeling competencies* (pp. 385–398). New York: Springer.
Stillman, G., & Galbraith, P. (1998). Applying mathematics with real world connections: Metacognitive characteristics of secondary students. *Educational Studies in Mathematics, 36* (2), 157–194.
Stillman, G., & Galbraith, P. (2011). Evolution of applications and modelling in a senior secondary curriculum. In G. Kaiser, W. Blum, R. Borromeo Ferri, & G. Stillman (Eds.), *Trends in teaching and learning of mathematical modelling* (pp. 689–699). New York: Springer.
Stillman, G. A., Kaiser, G., Blum, W., & Brown, J. P. (2013). *Teaching mathematical modelling: Connecting to research and practice*. Dordrecht: Springer.
Sweller, J. (2010). Cognitive load theory: Recent theoretical advances. In J. L. Plass, R. Moreno, & R. Brüken (Eds.), *Cognitive load theory* (pp. 29–47). New York: Cambridge University Press.
Tall, D. (2004). Introducing three worlds of mathematics. *For the Learning of Mathematics, 23*(3), 29–33.
Tikhomirov, O. K. (1981). The psychological consequences of computerization. In J. V. Wertsch (Ed.), *The concept of activity in Soviet psychology* (pp. 256–278). Armonk: M. E. Sharpe.
Törner, G., Schoenfeld, A. H., & Reiss, K. M. (2007). Problem solving around the world: Summing up the state of the art. *ZDM: The International Journal on Mathematics Education, 39*(5/6).
Treilibs, V. (1979). *Formulation processes in mathematical modelling*. Unpublished Master of Philosophy, University of Nottingham.
Valsiner, J. (1997). *Culture and the development of children's action* (2nd ed.). New York: Wiley.
Villa-Ochoa, J. A., & López, C. M. J. (2011). Sense of reality through mathematical modelling. In G. Kaiser, W. Blum, R. Borromeo Ferri, & G. Stillman (Eds.), *Trends in teaching and learning of mathematical modelling* (pp. 701–711). New York: Springer.
Vos, P. (2011). What is 'authentic' in the teaching and learning of mathematical modelling? In G. Kaiser, W. Blum, R. Borromeo Ferri, & G. Stillman (Eds.), *Trends in teaching and learning of mathematical modelling* (pp. 713–722). New York: Springer.
Widjaja, W. (2013). Building awareness of mathematical modelling in teacher education: A case study in Indonesia. In G. A. Stillman, G. Kaiser, W. Blum, & J. P. Brown (Eds.), *Teaching mathematical modelling: Connecting to research and practice* (pp. 583–593). Dordrecht: Springer.
Wijaya, A., Van den Heuvel-Ranhuizen, M., Doorman, M., & Robitzsch, A. (2014). Difficulties in solving context-based PISA mathematics tasks: An Analysis of students' errors. *The Mathematics Enthusiast, 11*(3), 555–584.
Wineburg, S., & Grossman, P. (Eds.). (2000). *Interdisciplinary curriculum: Challenges to implementation*. New York: Teachers College Press.

Part I
Innovative Practices in Modelling Education Research and Teaching

Chapter 2
Mathematical Modelling as a Strategy for Building-Up Systems of Knowledge in Different Cultural Environments

Ubiratan D'Ambrosio

Abstract Knowledge is a cumulative succession of strategies developed by humans living in different natural and cultural environments in response to the pulsions of survival and transcendence. The objective of knowledge is to understand, to explain and to cope with selected facts and phenomena of reality, ideally reality as a whole. Mathematical modelling is such a strategy that deals with facts and phenomena. In this chapter, how knowledge is generated (cognition), how it is individually and socially organised (epistemology) and how it is expropriated by power structure, institutionalised and given back to the people who generated it through filters (politics) is discussed. These steps are treated in an integrated and holistic way.

2.1 Introduction

In the text, I use, many times, the terms *artifacts* and *mentifacts*. These words, together with *sociofacts*, were introduced by biologist Julian Huxley (1887–1975) as the bases for a theory of culture (Huxley 1955). These terms have also been used in cultural semiotics. Whereas artifacts are the elements of the material culture, the mentifacts can be understood as the elements of the mental culture. Mentifacts include the symbols and codes of a culture, the signifier and the artifacts are related to the users of signs, the signified. Mental culture can, therefore, be regarded as a set of symbols and codes. This is studied by several authors and I mention Umberto Eco as a good reference. Particularly relevant are folkloristic studies in which models, that is, artifacts, come from mentifacts (Hale 2013).

I base my arguments in a behavioural hierarchy that leads to individual behaviour, which includes learning, the acquisition of knowledge and strategies for action, and to social behaviour, which results from the encounter of an individual

U. D'Ambrosio (✉)
Post-Graduate Program in Mathematics Education, UNIAN Universidade Anhanguera de São Paulo, São Paulo, Brazil
e-mail: ubi@usp.br

G.A. Stillman et al. (eds.), *Mathematical Modelling in Education Research and Practice*, International Perspectives on the Teaching and Learning of Mathematical Modelling, DOI 10.1007/978-3-319-18272-8_2

and another individual. These behaviours, individual and social, generate the context of cultural behaviour, including the processes of cultural transmission and mutual exposure of diverse cultures, are objects of study of the dynamics of cultural encounters. The transfer of knowledge, particularly of technology, is a crucial issue in the analysis of the process of development, fundamental to understanding the process of globalisation.

Preliminarily, I am interested in understanding the process of learning, acquisition of knowledge and strategies for action, that constitute a hierarchy of behaviours.

2.2 The Generation of Knowledge

Initially, it is the individual behaviour, which implicitly includes the processes of learning and, in particular, of the acquisition of language. Following this, we have social behaviour that develops and evolves within the so-called educational process.

Therefore, social behavior becomes more complex and generates a cultural phenomenon. It is vital to understand how arts and techniques, which incorporate artifacts to reality, develop into ideas, such as religion, values, philosophies, ideologies and sciences, as mentifacts, which are also incorporated to reality in a broad sense, that is artifacts + mentifacts + natural facts and phenomena. Once incorporated to reality, artifacts and mentifacts change it. Thus I conceptualize technology as the synthesis of artifacts (instruments) and mentifacts. That is, technology represents a merger of doing with the knowledge, contributing to the ways that man deals with reality and copes with situations and problems. Not only material instruments, such as tools and practices are responsible for action, but also the substratum of mentifacts, mainly religion and ideology. A very pertinent example is the agricultural use of transgenic. In history, the emergence of the gothic is an example. Instruments, both material and intellectual, such as counting, are responsible for ad hoc solutions.

We may understand the construction of knowledge as a three-step process:

1. How are ad hoc practices and solution of problems developed into methods?
2. How are methods developed into theories?
3. How are theories developed into scientific invention?

While methods are essentially a rational and coordinated use of techniques, theories are impregnated modes of explanation and understanding, based on myths, on spirituality and even religions, on science and mathematics and in ideology, which are all mentifacts.

Let's examine in more detail each step and the dynamics of their evolution. We discuss the learning process as something that creates a context which is the interaction of a genetic program and the environment. This is the subject of an important line of research, usually identified as 'nature versus nurture'.

Since the early philosophers this discussion is central. The psychologists have joined the philosophers over the relative importance of the environment, that is, upbringing, experience, and learning ('nurture'), and heredity, that is, genetic inheritance ('nature'), in determining the make-up of human personality and of intelligence as an organism, as related to behaviour and knowledge. The implications of these discussions for eugenics are obvious, in which differences in the capacities of individuals (and hence their behaviour) can be attributed to inherited differences in their genetic make-up.

Moving into these discussions leads to the theme religion versus science. Recent research leads to what has been called the epic of creation versus the epic of evolution. This is an area that gains in importance.[1]

I will try to avoid going into this discussion by just assuming life as an observable fact and recognizing that body and mind follow parallel and interconnected paths in this process of interaction of the genetic program with the environment. In the process, reality is recognized and analysed, thus originating intentional actions, concepts of meaning, which are response to will and need. Space, time, causality, imitation, the ludic and other categories play important roles in the process of interaction of the genetic program with the environment.

This is well illustrated by child behaviour. The concept of reality changes step by step, and the child, initially reacting only to instincts of survival (breathing, eating $\approx$ need) incorporate decision-making ($\approx$ will) and go from individual behaviour to social behaviour. The action of a child which initially results purely from their perception of situations and objects in their self-centred universe, changes upon reflection on the consequences of the action. Thus proceeds a modification of the action, considering all the information resulting from the complex of the senses, the emotions and memory combined. This action changes reality by adding facts, both artifacts and mentifacts (i.e., objects, things, ideas, and values), to that reality. Such change of reality by the action of the individual immediately provokes new thinking, new behaviour, interaction with new information already stored and newly acquired information. As a result, new action is initiated, with immediate effect in reality and, as a consequence, the addition of new facts. It is the individual as a maker of the reality by the addition of facts produced by the individual.

Man assumes the role of a creator, generating knowledge (mentifacts) and its *thingification* (in the sense of becoming an artifact). I use the term thingification to emphasize the material aspect of the action. Many authors, for example Marx, have used the word as well as *reification*. Both knowledge and their thingification are in the form of arts, sounds, objects, things in general, ideas, images, fantasy, concepts, theories, values and interpretations, in order to cope with, to understand, and to explain reality. They are added to the existing reality, enlarging and remaking it, to best fit the individual needs and will. These remarks are appropriate to discuss knowledge of different cultural systems, as it occurred in the conquests after

[1] See, for example, *ZYGON: Journal of Religion & Science,* Volume 44 Number 1, March 2009, which is entirely devoted to the theme.

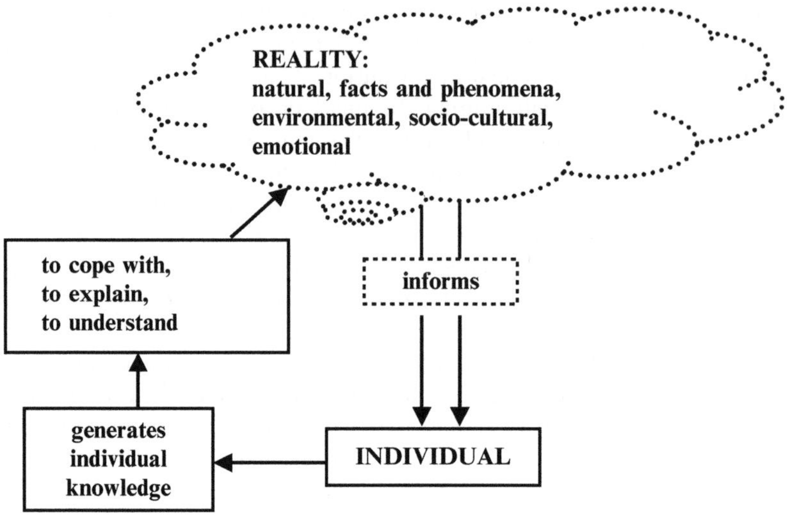

Fig. 2.1 The cycle of individual generated knowledge (After D'Ambrosio 2009, p. 90)

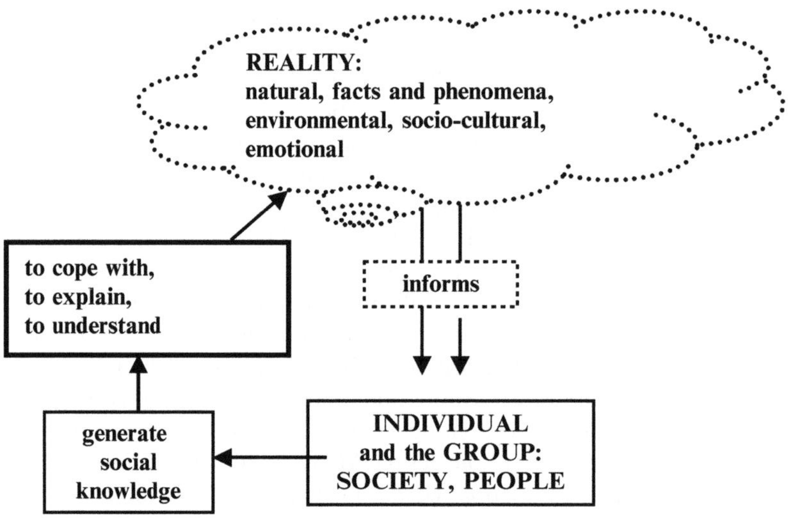

Fig. 2.2 The cycle of social knowledge

Columbus and the emergence of other knowledge systems as a result of the dynamics of cultural encounters (D'Ambrosio 1992). The knowledge cycle introduced in this chapter, served as the basis for elaborating Figs. 2.1, 2.2, and 2.3 used here.

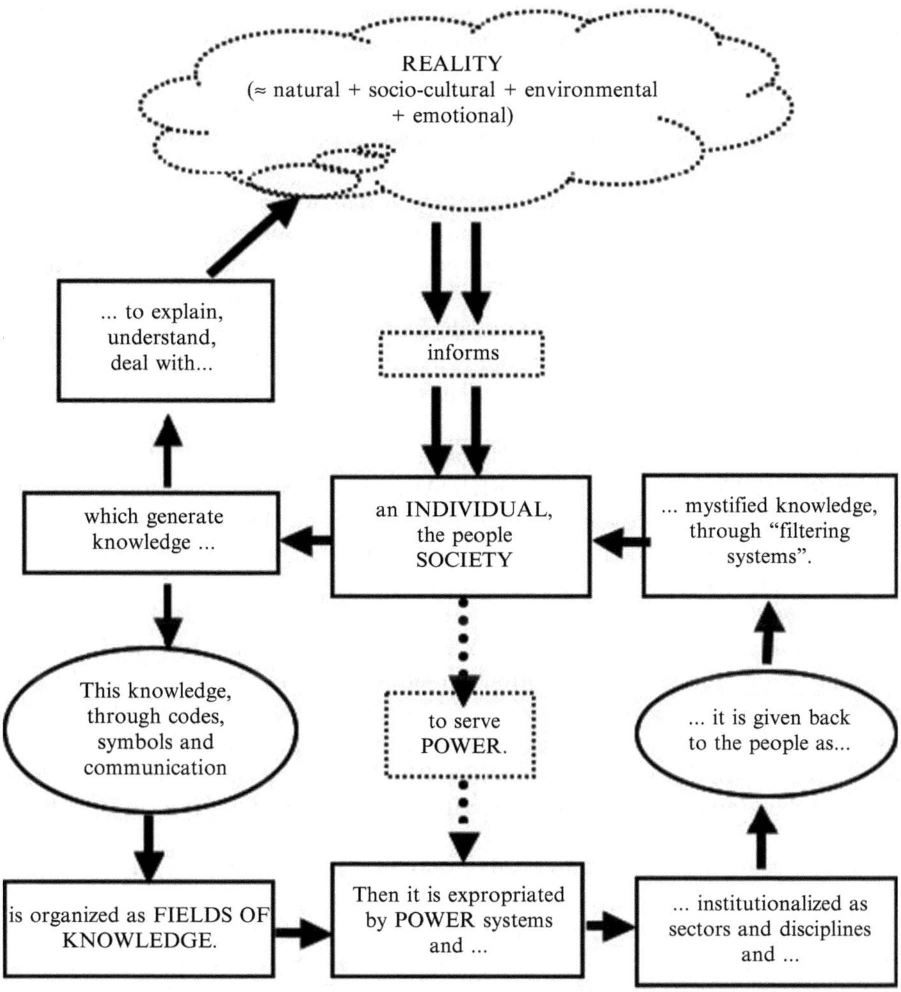

Fig. 2.3 The full cycle of knowledge

The individual is not alone. Gregariousness is a characteristic of animal species. How do the individual and the other interact? Communication plays a fundamental role in the interaction.

Particularly important is the emergence of language. When did utterance of humans become a word (Kenneally 2007)? Language is greatly advantageous in conveying to others individual will and needs. I will simply admit that through communication, even before the emergence of language, individuals interacted with others to produce knowledge and to making their behaviours compatible.

Similarly with 'nature versus nurture', there is a controversy about the 'individual versus social' in building up knowledge. The main question is how social structures impact the cognitive structures of the individual and how structures of individual consciousness and cognition can and do impact the structures of society. Essentially, the question becomes: how do individuals and society interact in cognitive actions, as well as in socio-political actions? How do ideologies, for example, languages, arts, religions, styles of knowing, become established, and how are social actions coordinated, for example as political movements? (See Wu 2007)

Thus, through encounters and interaction of individuals, there is mutual exposure and exchange of ad hoc practices and solutions of problems organized by each individual as knowledge. These are in general different practices and solutions. Through neurophysiological processes, as yet not well understood, which certainly include language and mimicry, the ad hoc practices and solutions for common problems, organized as individual knowledge, are shared and transformed, and result in socially organized knowledge. Thus, the cycle of knowledge is represented as in Fig. 2.2.

These two figures for the cycle of knowledge are understood as: (1) the cycle in which REALITY informs the INDIVIDUAL, who processes the information and exerts an ACTION (bodily and mental $\approx$ individual knowledge) which affects and *modifies* REALITY, which, once modified, informs (now incorporating newer elements) the INDIVIDUAL, who realizes a different ACTION (bodily and mental $\approx$ *modified* individual knowledge), which again modifies REALITY, and so on; (2) the cycle in which REALITY informs the GROUP (individual, society, people), who processes the information and exerts an ACTION (bodily and mental $\approx$ social knowledge) which affects and modifies REALITY, which, once modified, informs (now incorporating newer elements) the GROUP, who realizes a different ACTION, which again modifies REALITY, and so on. Hence, we may consider the individual cycle of knowledge, which is active as far as there is life:

$$\ldots \to \text{reality} \to \text{individual/group} \to \text{action} \to \text{reality} \to \ldots$$

This cycle synthesizes life as a dynamic process, to which every animal is subjected. Action manifests in several ways. Action may be the result of instinct and leads to the satisfaction of the *pulsion* of survival, meaning the permanent drive towards the survival of the individual and of the species, in other words nourishment and mating, which are subordinated to physiology, sociobiology and ecology, as a common characteristic of all living species. I will clarify the use of the word pulsion, rarely used in English. It is widely recognized that the English translation of the texts of Sigmund Freud is problematic, particularly the concept of *trieb*, which has been translated as "drive" and "instinct". Both are not faithful to the concept. Instead the French, Spanish and Portuguese translations use the word "pulsion". This word, which in the English language is used in different contexts, rarely in psychoanalysis, represents very well my understanding of survival.

In the human species, survival is a pulsion, which is loaded with emotions and intensions. Like every animal species, humans satisfy the pulsion of survival developing strategies to work with the most immediate environment, which supplies air, water, food, and with the other of the same species, necessary for procreation. This means, everything that is necessary for the survival of the individuals and of the species. These strategies are modes of behaviour and individual and collective knowledge, which include communication. Through interaction, there is an action of learning how to cope with the pulsion of survival. This gives origin to a form of communication between individuals which results in sharing strategies for individual survival. There is a form of learning, which involves mimicry and other sophisticated forms of interaction. Recent work in primatology shows some rudimentary form of instruction in chimpanzees.

Humans differ from other animal species. The species *homo* subordinates the strategies developed by other animal species for survival, a drive towards satisfying needs, which is usually called instinct, to will. In other words, the *homo* species go beyond survival and the continuation of species. Will leads to choices, preferences and desires, thus to emotions. As a consequence, another pulsion, which I call the *pulsion of transcendence*, is intrinsic to the *homo* species. The pulsion of transcendence is responsible for the needs of explaining, of understanding and of creating or, in other words, for transcending our own existence and projecting ourselves into the past and into the future. This is responsible for the development of instruments and techniques, for codes and a sophisticated communication system which has a cognitive dimension and developed into language. The use of instruments and techniques, of codes and communication, is organized as labor and power. For a more detailed discussion see D'Ambrosio (2012).

In satisfying the pulsion of transcendence, the species *homo* develops the perception of past, present and future, and their linkage, and the explanations of facts and phenomena encountered in their natural and imaginary environment. These are incorporated to the memory, individual and collective, and organized as arts and techniques, which evolve as representations of the real (models), as elaborations about these representations which result in organised systems of explanations of the origins and the creation of myths and mysteries (mentifacts). Some of the representations materialize as objects, concrete representations and sophisticated instruments (artifacts).

All this behaviour encounters support in the memory, where myths, mysteries, history and the traditions are organised, generally as religions and value systems. Explanations of the origins and the creation and of myths and mysteries generate curiosity and will to know the future, and give rise to divination organised and theorised as divinatory arts.

Probably the most basic of all systems of knowledge, present in every culture, are the divinatory arts. The human species, different from any other animal species, developed the concepts of *past*, *present* and *future*, and how they are enchained. Nothing is more characteristic of the human species than the desire to know the future. Thus, divinatory arts, such as astrology, the oracles, logic, the I Ching, numerology and the sciences, in general, through which we may know what will happen, are exemplary systems of knowledge. All these divinatory arts are all based on observing, comparing, classifying, ordering, measuring, quantifying, inferring, which are the quintessence of mathematical ideas.

The cycle of knowledge leads to the explanation of individual behaviour, social behaviour and cultural behaviour as the result of the incessant change of reality, as expressed in the cycle ...→reality→individual/group→action→reality→....

Systems of knowledge reveal not only their convenience to explain reality, facts and phenomena, but also they are important strategies to cope with daily situations and problems not only for the individual, but for society and the people in general.

Societies are organised subjected to different forms of power structure. The power structures recognise the advantage of mastering these strategies for their benefit, hence they proceed in expropriating and controlling these strategies, and consequently the system of knowledge in which they are based. Thus, the knowledge shared by the group is detained and controlled by the power structure and is institutionalized as clergy, as norms and laws, as disciplines, as academies, indeed in many ways, which are controlled by classes subordinated to the power structure. They are given back to the people as mystified systems of knowledge, subjected by filters. The mystification and filters guarantee that the systems of knowledge and the strategies associated with them do not challenge the power structure.

The full cycle of knowledge includes its generation, individually and socially, its organisation, its expropriation, institutionalization, transmission and diffusion, through systems of education and different forms of filters (such as examinations, degrees, certifications). Thus we are led to the full cycle of knowledge, as in Fig. 2.3.

These steps shown in Figs. 2.1, 2.2, and 2.3 are commonly treated as disciplines, respectively cognition, epistemology and politics. A serious limitation to understanding knowledge, as an intrinsic characteristic of the human species in response to the pulsions of survival and transcendence, is to treat it in separate steps, through the academic disciplines just mentioned.

2.3 How About Modelling?

Consider again the cycle ...→reality→individual/group→action→reality→.... In it, selected facts and phenomena of reality inform individuals and groups. Obviously, no one has full access, awareness and knowledge of reality; no one is omniscient. Our natural limitations give us access to selected facts and phenomena. The reason and the form of selection are extremely complex. They go from an uneven capability of individuals and groups to receive information, in some cases related to sensorial qualities or deficiencies, in other cases to the interest in the information received. The interest may be because of needs, or preference or merely by chance. Anyway, the information received is processed, in a way not yet well understood. The individual or group exerts an action of generating artifacts and mentifacts from the selected part of reality. They are incorporated into reality as representations, which inform the individuals or groups and the cycle goes on. The main question is then, how individuals and groups deal with the representations of selected facts and phenomena.

Fig. 2.4 Amazonian *pariko*

In a representation, reality is restricted to selected facts and phenomena and the result is a sort of "isolated individualized reality". To deal with the "isolated individualized reality", individuals attribute codes or parameters to the selected facts and phenomena. These parameters may be of a mathematical nature, such as mathematical forms and mathematical symbols. The isolated individualized reality, with the mathematical symbols attributed to the selected facts and phenomena, is a *mathematical model* of it. As an example, consider a *pariko*, typical of an Amazonian culture (Fig. 2.4). This artifact is a model of a complex social reality. It serves as a form of identity of its owner: indicates age, affiliation, origin and many other components of social life. However, it is impossible to attribute to this model parameters of a mathematical nature. Other examples may be drawn from urbanization.

Through models, humans try to give explanations of myths and mysteries, and these explanations are organized as arts, techniques, theories, as strategies to explain and deal with facts and phenomena. These strategies, have been historically organized, in different groups, in different spatial and temporal contexts, which are the support of cultures, as systems of knowledge.

The result is a sort of "isolated individualised reality", restricted to the representation of selected facts and phenomena. The "isolated individualised reality", dealt with the resource of parameters, is a model. Individuals are informed and elaborate on the model analysing the parameters associated with it.

Intellectual resources allow the individual to deal with the model and the parameters created by the individual are representations of facts and phenomena of the reality in the broad sense. The most common intellectual resources are based on observing, comparing, classifying, ordering, measuring, quantifying, and

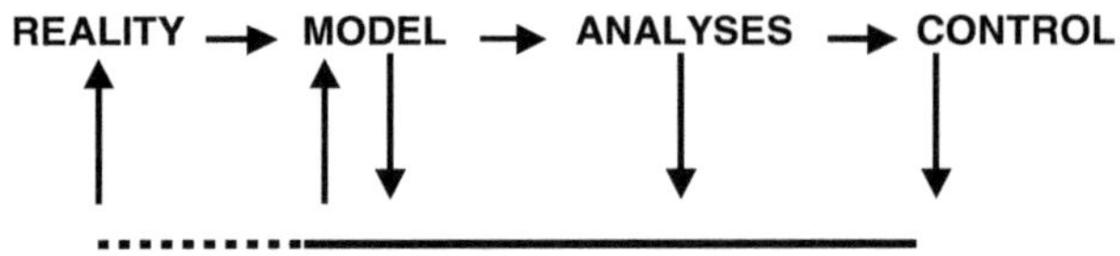

Fig. 2.5 The practice of mathematical modelling

inferring. As mentioned earlier, these intellectual resources are the basic pillars on which mathematics is based.

These parameters may be in terms of formal mathematics. I call *mathematical modelling* the process of dealing with a model in which the parameters associated with it, which is the objective of coping with and explaining selected facts and phenomena of reality, are in terms of formal mathematics.

The practice of mathematical modelling is an iterative method starting with reality, with which we started by selecting parameters, constructing a model, proceeding to its mathematical analysis, verifying results through control procedures and reformulating the model, repeating the analyses and control until we reach a satisfactory perception of the selected facts and phenomena. This is illustrated by the diagram in Fig. 2.5.

In each step, the practitioner reformulates the choice of parameters and resumes the process, which eventually allows a better understanding of the selected facts and phenomena of reality, which is the goal that justifies our practices as scientists.

References

D'Ambrosio, U. (1992). The cultural dynamics of the encounter of two worlds after 1492 as seen in the development of scientific thought. *Impact of Science on Society, No. 167, 42*(3), 205–214.

D'Ambrosio, U. (2009). Mathematical modelling: Cognitive, pedagogical, historical and political dimensions. *Journal of Mathematical Modelling, 1*(1), 89–98.

D'Ambrosio, U. (2012). The program ethnomathematics: The theoretical basis and dynamics of cultural encounters. *Cosmópolis: A Review of Cosmopolitics, 3–4*, 13–41.

Hale, M. (2013). Airship captains, pith helmets, & other assorted brassy bits: Steampunk, personas and material-semiotic production. *New Directions in Folklores, 11*(1), 3–34.

Huxley, J. S. (1955). Evolution, cultural and biological [Guest Editorial]. In W. L. Thomas Jr. (Ed.), *Yearbook of anthropology* (pp. 1–3–25). Chicago: University of Chicago Press.

Kenneally, C. (2007). *The first word: The search for the origin of language*. New York: Viking.

Wu, W. (2007). Vygostsky and Lenin on learning: The parallel structures of individual and social development. *Science & Society, 71*(3), 273–298.

Chapter 3
The Meaning of the Problem in a Mathematical Modelling Activity

Lourdes Maria Werle de Almeida and Karina Alessandra Pessoa da Silva

Abstract In this chapter we present a reflection on the assignment of meaning associated with the problem identified and solved in a mathematical modelling activity. Initially, we present our understanding of the notions of problem and mathematical modelling. To treat the meaning issue about the problem in modelling activities we have based our approach in Peircean Semiotics where the meaning is associated with the generation of interpretants during the development of activities. To illustrate our understanding, we describe briefly, the case of a modelling activity developed by students in a mathematics degree course, presenting the generation of interpretants by one student and indications of meaning assignment for the problems revealed by him.

3.1 Introduction

Discussions about the meaning associated with meaningful activities have been recurrent in literature (see, e.g., Hoffmann 2004; Steinbring 2002). In general, arguments about what is meant by meaning or about how its assignment can be inferred, are rare in research or experience with mathematical modelling reports although not non-existent (see, e.g., Ärlebäck and Frejd 2013; Stillman 2011). Beyond the wariness about what is meaning, different aspects involved in a modelling activity development need to be considered. Thus, one can ask: the meaning would be about what? The meaning could be about the content, which appears in the activity, about the problem, about the relationship established with

L.M.W. de Almeida (✉)
Department of Mathematics, University of Londrina, Antonio Pisicchio, 155, apto 303, Londrina, PR 86050-482, Brazil
e-mail: lourdes.maria@sercomtel.com.br

K.A.P. da Silva
COMAT, Federal University of Technology of Paraná, Campus Cornélio Procópio – Avenida Alberto Carazzai, 1640, Cornélio Procópio, PR 86300-000, Brazil
e-mail: karinasilva@utfpr.edu.br

G.A. Stillman et al. (eds.), *Mathematical Modelling in Education Research and Practice*, International Perspectives on the Teaching and Learning of Mathematical Modelling, DOI 10.1007/978-3-319-18272-8_3

reality through the problem, amongst other factors. In this chapter, particularly, we present considerations about the meaning assigned by students to a problem which has been studied in a mathematical modelling activity. There are many different views taken on the objectives and the traditions of the introduction of problems in mathematical classes (see, e.g., Borasi 1986; Erickson 1999; Poggioli 2001; Zawojewski and Lesh 2003) and we will present some considerations about the use of problems.

With regard to mathematical modelling activities, we argue the development of an activity is guided by the search for a solution to a problem. We therefore believe that the problem is also an object of study in modelling activities and highlight the meaning assignment to this object as an important issue for research into mathematical modelling education.

In this chapter, we base our considerations about meaning on the semiotics of Charles Sanders Peirce (1839–1914), the Peircean Semiotic, analysing the generation of interpretant signs of students when they are solving a problem while developing mathematical modelling activities. (For English readers consult Peirce 1992, 1998.) To support our argument, we treat the involvement of one student studying for a mathematics degree course and his interpretants during the development of a modelling activity about tree pruning.

3.2 Problems, Mathematical Modelling and Meaning

For many centuries, there has been a focus on understanding what could turn into a problem. In addition, in this context, Euclides (between centuries III and IV B.C.) referred to a problem as like something which is a part of an observation imbued with some knowledge, and exploring something yet unknown, treating of a problem as an axiomatic construction. Yet, the mathematician-logician-philosopher, Charles Sanders Peirce, considered that the problem-situation is the base for a question. It would become problematic in its own subjection to a question. George Pólya (1887–1985), author of the book *How to solve it* (Polya 1971), claims a problem only exists as a problem when there is a difficulty which the problem solver wishes to overcome. There is the imperative to want to resolve something with which we have difficulty. Poggioli (2001), on the other hand, argues that a problem is a situation which the subject wishes to do something about, but does not know the necessary actions to materialize this goal and needs to develop the means to this materialization. Zawojewski and Lesh (2003) argue that, in general, for it to be a problem there is not a readily accessible procedure that guarantees or completely determines a solution, but the individual must make a solution as well as they can do it. Considering these interpretations, the term problem is understood in this chapter to be like a situation, which the individual does not have a scheme a priori for its resolution and there is not specific procedures previously known or solutions yet indicated. It is with this understanding we explore the problem in mathematical modelling activities.

In general, the term *mathematical modelling* refers to the search for a mathematical representation for an object or non-mathematical phenomenon. The modelling process creates a complex structural relationship between two entities of different epistemological nature: the situation that is to be modelled and the mathematical system. So, with a mathematical modelling activity we approach, through mathematics, a problem-situation not essentially mathematical (Almeida 2010). A mathematical modelling activity can be described in terms of an initial situation (problematic), of a final desired situation (which represents a solution to the initial situation) and of a set of procedures and concepts which are necessary for one to go from the initial situation to the final one. Literature usually refers to this initial problematic situation as the *problem-situation*; and, in general, a mathematical representation or mathematical model is associated with the final desired situation. Lesh and Doerr (2003) refer to mathematical models as being conceptual systems that are expressed using external notation systems and that are used to describe or explain the behaviors of other systems. So, in a mathematical activity the individuals do not have a scheme a priori that indicates specific procedures but are operating on their own interpretations and have to decide what the problem is. In this sense, the most of their time and their effort tends to be on trying to establish the meaning assignment of the situation, of the problem.

Some examples of research which refers to meaning assignment to procedures of the students in mathematical activities are Blomhøj and Kjeldsen (2011) and Ärlebäck and Frejd (2013). In our research, considering the definition of 'problem' corresponds to the initial stage of a modelling activity development, we sought to understand the meaning assignment by the students for the problem when they try out modelling activities. Therefore, in modelling activities, it is more than defining a problem; it is necessary to solve the problem. In this process, different representations are used constituting signs that point, indicate or represent mathematical objects.

For it to be possible to understand a meaning assignment for a problem and for a mathematical object, it is necessary to analyse the signs that the students use to suggest, indicate or represent the object that they are dealing with. To treat issues related to meaning assignment in mathematical modelling activities we base our approach on that developed by Charles Sanders Peirce who organized a general analytic-philosophical doctrine – Peircean Semiotics. The *sign*, according to Peirce (1972), has the function of representing an object to someone (an *interpreter*), creating in someone's mind another sign, the *interpretant*. This new sign is a rational process that is created in the interpreters' mind (i.e., in the mind of a human being). Thus, the interpretant is another sign that can indicate the meaning of the first. Therefore, the meaning of a sign can be evidenced in another sign: the interpretant. In this way, a method to understand the meaning assignment for the problem is to analyse the interpretants of the students during the resolution of a problem as well, that is, what is said about this problem after the signs were generated. Taking into consideration aspects of Peirce's theory of meaning and interpretants we addressed the meaning assignment for the problem in mathematical modelling activities.

3.3 Design of the Study

In order to achieve our aim of understanding meaning assignment for the problem in mathematical modelling activities we present a case study where we analyse the interpretants of one interpreter (student) when he developed modelling activities. The interpreter, who is the subject of our research, is a student of the 4th year of a mathematics degree course of a Brazilian University involved with modelling activities in a Mathematical Modelling subject.

Students were engaged in some modelling activities during 1 year (2011). The activity that we refer to in this chapter is the concluding activity of the subject and was developed by a group of students at the end of the second half of this year. The development of the activity lasted for 05 h (classes). Although the activity has been developed by a group of students, we analyse the interpretants (signs) of one student (interpreter) of this group to show the utility of the semiotic lens to act as a window on meaning assignment. This student participated in all classes as well as in a semi-structured interview. We analysed the written report, video and audio collected during the development of the activity. The activity that the students engaged in is about the tree pruning and streetlights in Londrina city and was developed by a group of six students. We present interpretants of one of these students, named here as Carlo, a pseudonym.

3.4 The Meaning of the Problem and the Generation of Interpretants in Mathematical Modelling Activities

For the development of the activity students had to define a problem to study in groups in their mathematical modelling classes. One student (Carlo) suggested: "The street from my house is kind of dark – it is not very bright; the neighbors and I think there are problems with the size of the trees and their pruning; could we study something about it?" From this idea, the students collected data from websites and had access to the Manual of Public Lighting of Companhia Paranaense de Energia (Copel 2011). Amongst the information provided in the Manual of Copel, students found a reference to tree pruning presented by an image as shown in Fig. 3.1.

To perform the pruning of the tree, one takes into account its size, the height of the lamp post next to it (h), the distance (D) the tree is away from the lamp post and the type of lamp used. Using this information, the students were interested initially in studying the problem of street lighting on the street where they live. The student, who suggested the theme for the activity, lived on the street Sierra Parecis. Later, they also collected data at the street Bento Munhoz of Rocha Neto, next to the place where people walk at the lakeside. Considering the data collected at the two streets, the students posed the problem: What is the pruning point of trees for these streets according to the type of lamppost?

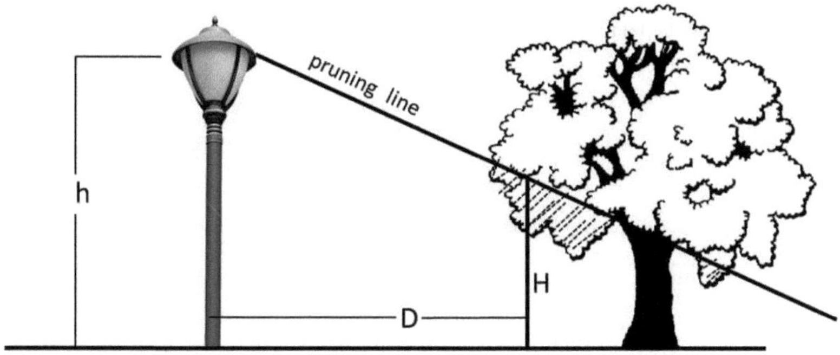

D: distance from the pole to lowest branch of the tree
h: the position that the lamp is mounted
H: height of the pruning

$H = -0.26\,D + h$ (the function that indicates the height of pruning)

Fig. 3.1 Figure representing the height of tree pruning (Handbook of Public Lighting – Copel 2011, p. 23)

This problem is a sign, an interpretant of the student, and indicates that, in a certain way, the meaning for the problem is associated with the fact that the issue that the modelling activity intended to study – the pruning point of trees – is exactly a problem on the street in which the student lives (Sierra Parecis). Moreover, the choice of the street Bento Munhoz of Rocha Neto also indicates that there was a certain meaning in the problem that the activity proposed to study as indicated in the response of Carlo in an interview:

Carlo: The Street Bento Munhoz of Rocha Neto is next to a place of people walking at the lakeside so it must be very well lit!

In this case, the student's assertion reveals a social concern. Then the meaning to the problem can arise also from that concern. To solve the proposed problem, the group of students had to 'produce' the necessary data. The existence of different types of streetlights in the two streets required them to perform measurements in places with regard to the height of the lampposts, lamp height and types of trees.

A mathematical approach to the problem was permeated by the generation of interpretants in order to obtain a solution. The meaning assignment to the problem was revealed during the construction of mathematical models associated with the solution. The relationship between the problem and mathematical objects is shown in the setting out of the development activity. Students organized this into three steps: the position of the lamp on the lamp post, the radius of lighting curve and the lighting, as shown in Fig. 3.2.

To determine the position of the lamppost they used similarity of triangles. This denotes mathematical knowledge that assists in decision-making for developments in solving a problem:

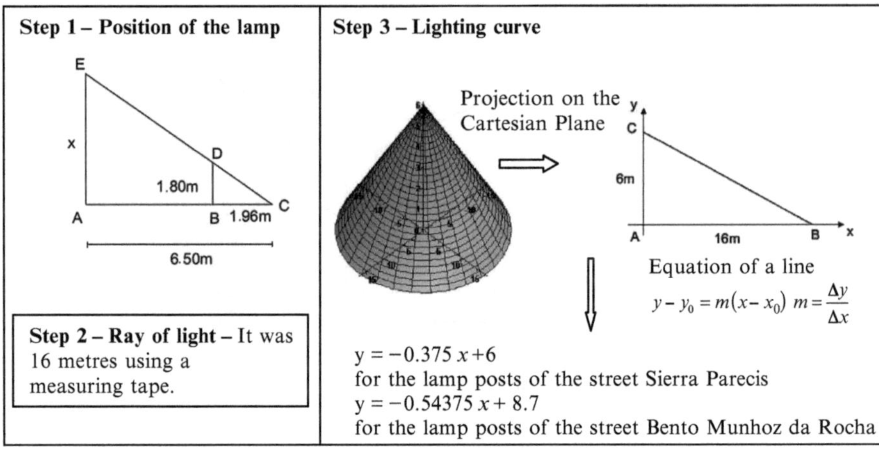

Fig. 3.2 Signs (interpretants) generated by the students in the activity

Carlo: For the deduction of our model, we went to the field, we took measurements. Then we needed the height of the lamp, and then we used similarity of triangles to try to find the height of the lamp. We took the height of [stops] is [stops]. We actually calculated the height of the lamppost. Then it was easier for us to measure the height of a person to see that he/she was away from the edge of the light and we used the similarity of triangles to find the height of the lamppost.

It thus appears that the interpretants can be linked to students' procedures. The procedures used by the students to find a solution also indicate meaning assignment to the problem. The interpretants indicate what the student needs to do and how to do it to find a solution to the problem. Awareness of the consequences of the use of signs to represent the object is a condition to "determine what a concept means" (Peirce 1989, p. IX). While acknowledging that the mathematical object was something 'simple' for the education level, the meaning assignment to this object may be a result of the relationship between the mathematical object and the reality around them.

Interpreting the results beyond the reconsidering of the mathematics in a problem-situation is an aspect that also configures the assignment of meaning to the problem. This is what we can infer from examining Carlo's answer (interpretant) when asked about the use of the graphic resource for representing the problem in Fig. 3.2.

Carlo: Because I [stops] In fact in the following, in the plan it was easy to view; only we found that the tree is not a linear thing, especially the crown of her! The crown of it had a very similar behavior to that of a sphere. So what happens? If I traced the ray of light considering a branch like this [gesturing], I might have a branch, vertically [stops] or I could have a branch horizontally, as I will consider the line for pruning now by looking for one branch and not to another? So, we found that the model in the plane would not be an interesting condition so we had to go to 3D. And then we believed the graphic display gives an abstraction much better of the situation for you to understand and speak [stops] the point of the cone, the ray and of the light. We took into consideration and doubted [stops] will be a cone, will be a sphere but the graphic I think it helps you, allows you to view to see how this is happening the behavior of these different representations. In other words, the intersection of the cone with the sphere, of course had a little more work, but we do not focus on this question of deduction of the equation of the sphere, the equation of the cone but this was not our object of study.

By mentioning the importance of refining the problem situation to try to represent it mathematically through a three-dimensional graph, the argument brought to the fore by Carlo was that "models provide only approximations of the real behavior" (D'Ambrosio 2009, p. 91). What is evident is that the intention of the students was to make an approximation to reality. Although they acknowledged that the model obtained was 'simple' from the mathematical point of view regarding the level of education at which they were at, they proposed a more 'sophisticated' approach to the situation by means of computer resources, but showing a representation only. In this sense, the meaning consists of what Peirce (2005) designates as "an idea of the feeling or predominantly an idea of acting and being acted" (p. 194).

Observing the models obtained from the height of the tree pruning for the two streets as shown in Fig. 3.2, the students realized that there were differences between the coefficients of each of the linear functions. From this observation, they reflected on why Copel uses a 'standard equation' ($H = -0.26\,D + h$) for tree pruning.

Carlo: Then yeah, we found two equations, but how [were] they related with Copel's? So what is the ideal pruning? We plotted the equations into the graph [presenting graphs in the Cartesian plane see Fig. 3.3], [then] started to notice that whenever these red here [pointing to the graphical representation of $f_1(x)$ and $f_2(x)$] are the ones Copel can translate but are the same equation to Copel, the same angular coefficient and this is the h [stops] of Munhoz of Rocha, which is well lit there and the g is the Sierra Parecis.

Thus, in general, meaning assignment seems to be revealed through the familiarity that the student has been building over the issue, indicating that interpretants correspond with the interpretive effect that the sign produces in the mind. Taking into account the statement of Peirce (2005) that the "meaning is the declared interpretant of a sign" (p. 222), we can infer that generation of signs by the student can support his answer that in two different streets the height for tree pruning is different. On the other hand, the equation found for pruning height on the street Sierra Parecis where the student lives, indicating that tree pruning is necessary on this street is also an interpretant that reveals the meaning assignment to the problem.

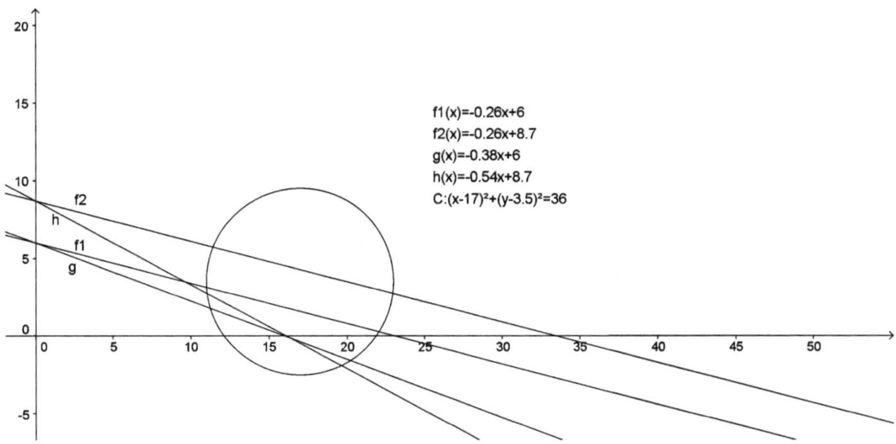

Fig. 3.3 Graphical representation of the lines indicated by Copel's pruning

3.5 Discussion and Implications for Teaching, Learning and Research

The data analysed in this chapter comes from an activity at the end of the year of a Mathematical Modelling subject. Thus the student whose interpretants we analysed, was not a 'newcomer' to mathematical modelling. However for this activity, particularly, the group of students had a need to collect data by going to the streets to take measurements of trees and poles and in this sense, the activity is a challenge even for students who have had previous experience with mathematical modelling.

Thus the student rather than using a fixed interpretation or procedure to process data or to solve the problem, essentially operated on his own interpretation and decisions and so the discussion and the communication with other members of the group were important. What we can infer in this case is that the student's use of procedures to define and to solve the problem also may be an effect of the modelling activities the student has experienced already.

If, on the one hand, the meaning of the definition of the problem 'What is the point of pruning trees for these streets according to the type of lamppost?' can be associated with a particular situation experienced by the student (Carlo who lives in Sierra Parecis Street); on the other hand, the interpretants' generation reveals that the meaning of the solution would be accompanied by the experience the student has had. This is what the statement of the student in the interview indicated: "we found the model in the plane would not be an interesting condition, so we had to go to 3D." This result is similar to a finding by Ärlebäck and Frejd (2013) in which the authors observed that the 'signifiers' of the students were, to some extent, associated with lack of modelling experiences of the students.

According to Steinbring (2002, p. 113) "Mathematical concepts and mathematical knowledge are not a priori given in the external reality, neither as concrete, material objects, nor as independently existing (platonic) ideas". They are mental objects (Changeux and Connes 1992) or social facts (Searle 1997) or cultural objects (Hersh 1997). So the meaning of these theoretical, social or cultural objects has to be generated by the individual in interaction with experience and abstract reference contexts. Mathematical modelling activities have potential to establish these interactions, or at least provide students with indications of how these interactions can be. Furthermore, the activity described in this chapter indicates that it requires the student to combine several tools (procedures, principles, use of concepts) to obtain a large and complex tool that allows them to point out a solution.

When we study mathematical modelling using a semiotic perspective we can infer that the interpretants that Carlo generated reveal what he knows about tree pruning, streetlights and until now about similarity of triangles and projection of three-dimensional figures. In this sense, our argumentation about meaning assignment is aligned to Hoffmann (2004) when he notes that the interpretant might be a spontaneous reaction within a person's mind but it also can be any arbitrarily created meaning within a certain group of persons or the shared standard reaction to a certain sign within a group which may be defined in a general way by certain societal or cultural characteristics.

Blomhøj and Kjeldsen (2011), in addressing the reflections on skills in mathematical modelling by students involved in a project, note the meaning corresponds to internal reflections of students. These authors propose that "they need to be exposed to didactical situations that challenge them to reflect upon and critique the modelling process and the function of models in different contexts" (p. 386).

In our research, when we proposed the concluding activity of the subject in classes, students were faced with a challenging situation that made them reflect on their procedures as related to defining a problem and with respect to the tools used for its resolution. In this case, Carlo used computer software to improve the solution. Although he has used computer software, the interpretants do not clarify whether he accepted the solution proposed by the software or if he engaged in some reflection on its use. In this regard, a survey focused on assigning meaning and reflection to the object with the aid of computers is a future research area to be developed. Stillman (2011), in this context, argues that in modelling, reflection can make sense only when related to mathematical content and the processing decisions by means of which the content is evoked and implemented. The author defends the use of metacognition to attain crucial importance to promote reflection when students are engaging in mathematical modelling.

We hope that reflection on the analysis of interpretants in a mathematical modelling activity can also reach other researchers who, like us, seek to understand the meaning assignment of students involved in modelling activities. A semiotic approach to mathematical modelling activities in the classroom can guide the teacher's work aimed at assigning meaning to their students as well as being a better basis for teachers' decision making and interventions.

References

Almeida, L. M. W. (2010). Um olhar semiótico sobre modelos e modelagem: metáforas como foco de análise. *Zetetiké. FE-Unicamp, 18*, número temático, 387–414.

Ärlebäck, J. B., & Frejd, P. (2013). Modelling from the perspective of commognition – An emerging framework. In G. A. Stillman, G. Kaiser, W. Blum, & J. P. Brown (Eds.), *Teaching mathematical modelling: Connecting to research and practice* (pp. 47–56). Dordrecht: Springer.

Blomhøj, M., & Kjeldsen, T. H. (2011). Students' reflections in mathematical modelling projects. In G. Kaiser, W. Blum, R. Borromeo Ferri, & G. Stillman (Eds.), *Trends in teaching and learning of mathematical modelling (ICTMA 14)* (pp. 385–395). New York: Springer.

Borasi, R. (1986). On the nature of problems. *Educational Studies in Mathematics, 17*(2), 125–141.

Changeux, J. P., & Connes, A. (1992). *Gedankenmaterie*. Berlin: Springer.

Copel. (2011). *Manual de Iluminação Pública. 1998*. http://wwwcopel.com/hpcopel/manual_iluminacao.pdf. Accessed 25 Oct 2011.

D'Ambrosio, U. (2009). Mathematical modeling: Cognitive, pedagogical, historical and political dimensions. *Journal of Mathematical Modelling and Application, 1*(6), 89–98.

Erickson, D. K. (1999). A problem-based approach the mathematics instruction. *Connecting Research to Teaching, 92*(6), 516–521.

Hersh, R. (1997). *What is mathematics, really?* London: Jonathan Cape.

Hoffmann, M. H. G. (2004). Learning by developing knowledge networks. *Zentralblatt für Didaktik der Mathematik, 36*(6), 196–205.

Lesh, R., & Doerr, H. M. (2003). Foundations of a models and modeling perspective on mathematics teaching, learning, and problem solving. In R. Lesh & H. M. Doerr (Eds.), *Beyond constructivism: Models and modeling perspectives on mathematics problem solving, learning, and teaching* (pp. 3–33). Mahwah: Erlbaum.

Peirce, C. S. (1972). *Semiótica e Filosofia: textos escolhidos*. São Paulo: Cultrix.

Peirce, C. S. (1989). *Escritos coligidos*. São Paulo: Nova Cultural.

Peirce, C. S. (1992). In N. Houser & C. Kloesel (Eds.), *The essential Peirce* (Vol. 1). Bloomington: Indiana University Press.

Peirce, C. S. (1998). *The essential Peirce* (Vol. 2). Bloomington: Indiana University Press.

Peirce, C. S. (2005). *Semiótica*. São Paulo: Perspectiva.

Poggioli, L. (2001). *Estrategias de resolución de problemas*. Serie Enseñando a aprender. Caracas: Polar. http://www.paideavirtus.cl/mce/PDFS/Ensenando.pdf. Accessed 22 Apr 2012.

Polya, G. (1971). *How to solve it*. Princeton: Princeton University Press.

Searle, J. R. (1997). Die Konstruktion der gesellschaftlichen Wirklichen. *Zur Ontologie sozialer Tatsachen*. Hamburg: Rowolhlt.

Steinbring, H. (2002). How do mathematical symbols acquire their meaning? – The methodology of the epistemology-based interaction research. In H. G. Weigand, N. Neill, A. Peter-Koop, K. Reiss, G. Törner, & B. Wollring (Eds.), *Developments in mathematics education in German-speaking countries* (pp. 113–123). Franzbecker: Hildesheim.

Stillman, G. (2011). Appling metacognitive knowledge and strategies in applications and modelling tasks at secondary school. In G. Kaiser, W. Blum, R. Borromeo Ferri, & G. Stillman (Eds.), *Trends in teaching and learning of mathematical modeling (ICTMA 14)* (pp. 165–180). New York: Springer.

Zawojewski, J. S., & Lesh, R. (2003). A models and modeling perspective on problem solving. In R. Lesh & H. M. Doerr (Eds.), *Beyond constructivism: Models and modeling perspectives on mathematics problem solving, learning, and teaching* (pp. 317–336). Mahwah: Erlbaum.

Chapter 4
Extending the Reach of the Models and Modelling Perspective: A Course-Sized Research Site

Corey Brady, Richard Lesh, and Serife Sevis

Abstract For over 30 years, researchers have engaged in inquiry within the Models and Modelling Perspective (MMP), taking as a fundamental principle that learners' ideas develop in coherent conceptual systems called *models*. Under appropriate conditions, such as in Model Eliciting Activities (MEAs), this research has shown how learners' models can grow through rapid cycles of development toward solutions involving creative mathematics. These externalized models, and other *thought-revealing artifacts,* can become rich objects for reflection by learners, for formative assessment by teachers, and for analysis of idea-development by researchers. This chapter describes a new research effort to expand the reach of this MMP tradition, engaging questions about the interconnected models and modelling processes of students and teachers at larger, course-length scales.

C. Brady (✉)
School of Education and Social Policy, Northwestern University, 2120 Campus Drive, Evanston, IL 60208, USA
e-mail: cbrady@northwestern.edu

R. Lesh
Mathematics Education (Emeritus), Indiana University, 327 Woodside Avenue, Park City, UT 84060, USA
e-mail: ralesh@me.com

S. Sevis
Middle East Technical University, Universiteler Mahallesi, Dumlupinar Bulvari No.1, 06800 Cankaya/Ankara, Turkey

Department of Curriculum and Instruction, Mathematics Education, School of Education, Indiana University, 201 North Rose Avenue, Office 3002B, Bloomington 47405, USA
e-mail: serisevi@umail.iu.edu

G.A. Stillman et al. (eds.), *Mathematical Modelling in Education Research and Practice*, International Perspectives on the Teaching and Learning of Mathematical Modelling, DOI 10.1007/978-3-319-18272-8_4

4.1 Introduction

In this chapter we discuss ongoing work that is intended to carry forward the agenda of the Models and Modelling Perspective (MMP). We are in the process of assembling and testing a web-based research repository, dedicated to the creation and refinement of a suite of research tools to study the interacting and continually evolving modelling processes of students and teachers, in the context of a course-sized collection of curricular materials dealing with Quantification and Data Modelling. Our motivation and rationale for building this site resonates strongly with currents in the international research community, as seen at ICTMA 16. The site builds on the foundation of MMP research, which in turn was shaped by the perspectives of the American Pragmatists, as well as both cognitive and sociocultural constructivist perspectives. With it, we hope to support the kinds of extended inquiry and collaboration needed to broaden the reach of the MMP community.

We begin our account by reviewing the framing research questions of the MMP tradition. We then describe how these questions connect with claims about the nature of knowledge, and we show how MMP researchers have developed research tools to produce data and evidence to answer these questions. Next, we focus in on our ongoing research, describing the assumptions and conjectures that underlie our development of a course-sized research site to carry forward these lines of inquiry. As with prior MMP research, we show how our work will create and assemble new research tools to generate evidence for the new kinds of question we are asking. In particular, we suggest how it will support investigations into the various relations and interactions between pairs of constructs not always recognized as interdependent, such as: (a) student development and teacher development; (b) conceptual knowledge and procedural knowledge, facts, and skills; and (c) learning and assessment (both formative and summative).

4.2 Research Questions Addressed by the MMP

As a program of research, the MMP was developed explicitly to investigate the following kinds of questions about learning:

- How can we characterize realistic problem-solving situations where solutions demand elementary-but-powerful mathematical constructs and conceptual systems?
- What kinds of "mathematical thinking" are emphasized in such situations?
- What does it mean to "understand" the most important of these ideas and abilities?
- How do such competencies develop, and what can be done to facilitate their development?

- How can we document and assess the most important (deeper, higher-order, more powerful) conceptual achievements that are needed for full participation as citizens in increasingly complex societies and professions? and
- How can we identify students who have exceptional potential that is not adequately measured by standardized tests?

These questions are tightly connected with the portrayal of knowing and learning that has been developed through MMP research. Here, the MMP builds on perspectives originating with Piaget and Vygotsky as well as with the American Pragmatists (c.f., English et al. 2008; Lesh and Doerr 2003). In particular, Mousoulides et al. (2008) summarized key elements of the MMP debt to Dewey, Mead, James and Peirce:

- Conceptual systems are human constructs, and thus also are fundamentally social in nature (Dewey and Mead).
- The "worlds of experience" that humans strive to understand and explain are rarely static. They are most often products of human creativity, continually changing in response to the evolving needs of humans who create and re-create them (James).
- Meaning, in this setting, tends to be distributed across a variety of representational media (ranging from spoken language to written language, to diagrams and graphs, to concrete models, to experience-based metaphors). Each of these representational forms foregrounds some facets of experience and backgrounds others (Peirce).
- Knowledge is organized around concrete experiences at least as much as around abstractions. The ways of thinking needed to make sense of realistically complex situations nearly always must integrate ideas from more than a single discipline, textbook topic area, or theory (Dewey).
- In a world filled with technological tools for expressing and communicating ideas, it is naïve to suppose that all "thinking" goes on inside the minds of isolated individuals (Dewey).

4.3 Claims About the Nature of Knowing and Learning

MMP research, rooted in these perspectives and pursuing questions such as the ones listed above, has illuminated the nature of knowing and learning in authentic problem-solving settings. A key feature of such settings is that they challenge learners to engage in original mathematical work (i.e., to produce mathematics that is *new to them*), rather than merely applying mathematics learned from an authoritative source. There are many dimensions to the image of knowledge that has emerged from this research; in this section we indicate three such dimensions that have been influential in guiding inquiry into both teacher and student knowledge. Then, in the following section we describe how the MMP develops and refines research tools to operationalize these dimensions of knowledge, permitting them to be externally expressed in *thought-revealing artifacts* created by the learners themselves.

Dimension 1 Practical knowledge is understood as an *interpretation system* for making sense of phenomena. In particular, this means that in realistic problem settings, experts distinguish themselves from non-experts not only by what they *do* but also by what they *see* in such situations. Moreover, learners' concrete past experiences and accumulated models also serve as *lenses* through which they can view and interpret new situations. Adopting this view also suggests that many aspects of learner's knowledge will be tacit and instinctive. Learners may be able to use these knowledge resources to guide actions before they can subject them to analysis as hierarchical, logical structures of rules and procedures. That said, interpretation systems have the property that they are *both* structures for action *and* structures that can be reflected upon. That is, knowledge as models and interpretation systems can serve as (a) windows or lenses which one can *look through* to view the world, or (b) objects in themselves which one can *look at* and analyze.

Dimension 2 Knowledge is constituted as much by *connections* forged by the learner among big ideas of the domain and between these ideas and prototype situations, as by an 'intrinsic' understanding of the big ideas themselves. Adopting this view also suggests that many aspects of knowledge may be highly situated, multiply-determined, and bound up with particular concrete experiences. The process of learning may therefore be expected to be multi-dimensional and non-linear, in spite of attempts to rationalize the *teaching* of material in the form of logical, linear sequences. Thus, in articulating learning goals, a distinction must be made between (a) lists of names of topics that should be emphasized in teaching, (b) operational definitions of what it means for students to have learned these big ideas, and (c) over-time accounts of students' growing appreciation of the significance and interrelatedness of these big ideas (see Learning Progress Maps, below).

Dimension 3 As discussed above, important kinds of knowledge are characterized by the ability to see situations in a certain way, or by having the skill or competency to act in a certain manner. Mastery of this kind of knowledge involves knowing *when* and *where* to apply it, as one conceptual tool among a repertoire. Adopting this perspective leads to the practical construct of *problem-solving personae*: stances or roles adopted by problem-solving individuals and groups, in response to the situation at hand. Expertise in this dimension involves recognizing that no single behavior, technique, or heuristic is valuable independent of context. Moreover, research into these personae increasingly suggests that while they may have a logical or technical core, they also involve "soft" aspects of knowledge, including attitudes, feelings, and beliefs (both about oneself and the domain).

4.4 Research Tools and the Data Generated by Inquiry Within the MMP

For each facet of what it means to know and understand, MMP researchers seek means to operationalize that dimension of knowing. Their history of success in doing this has supported the conviction that to conceptualize something is to be able to externalize it in thought-revealing artifacts, given an appropriate research setting to do so. Thus, for each dimension of knowledge we seek to study, MMP researchers produce and refine tools to facilitate the generation of a relevant type of research data and evidence. In this section we illustrate this with reference to the three facets of knowing discussed above.

Dimension 1: Knowledge as Interpretation Systems Increasing awareness of the complexity and diversity of learners' emerging knowledge in this dimension has been a principal driver of the genre of research activities known as Model-Eliciting Activities (MEAs). Research has shown that learners can externalize their local, situated systems of sense-making (which the MMP calls their *models*) given compelling problem settings and the means to represent these systems. Each element of the design of MEAs—both the activities themselves and their implementation in classroom settings—is driven by the goal of optimizing the processes of idea development and improving our view into the models and modelling processes of learners as they emerge in time.

Dimension 2: Making Connections Here we investigate students' work in constructing connections among the big ideas of a domain, as well as teachers' sense of inherent connections, and teachers' sense of the connections that their students are actually making. Researching this dimension of knowledge implies additional design criteria for MEAs (in particular, that these occasions for authentic mathematical production are also sufficiently *open* to permit different learners to incorporate different combinations of big ideas in productive ways). In addition, a focus on connections has also provoked the development of a series of specific research tools. Learning Progress Maps support learners and teachers in recognizing and reporting connections that are made between ideas at different stages in model development. Concept Analysis Wheels support and document emergent thinking about the organization of the discipline as a whole and the relations among its structures.

Dimension 3: Building and Using Problem-Solving Personae To study the broad spectrum of problem-solving behaviors and their deployment as situations shift over time, MMP researchers have developed a family of Reflection Tools (RTs). For instance, RTs have been created to study shifting roles in group interactions; changes in the affective dimension of "flow" (Csikszentmihalyi 1997) in the course of problem-solving; shifts in groups' self-assessment of their progress; and so forth. For each behavior or experiential aspect, a separate research tool is developed to support learners in the generation of thought-revealing artifacts focused on that aspect. The wide range of these tools reflects the complexity of problem-solving personae that have emerged in the context of MEAs.

4.5 Extending the Questions; Expanding the Toolkit

Building on the successes of prior MMP research, we aim to study new levels and dimensions of the models and modelling processes of students and teachers. For example,

- How can learners' models best be extended and expanded in classroom settings where multiple problem solving teams have worked in parallel?
- How can learners connect their models and a domain's big ideas with procedural skills and with a familiarity and facility with tools of the domain?
- How do knowledge and models develop and mature at larger time scales?

While it has been possible to pursue many of the research questions discussed earlier in this chapter through studies involving a single MEA or a sequence of several of them, questions like the ones above require new design structures that can be deployed over a course-sized experience. In particular, these questions require attention to students' and teachers' ways of unpacking the models created in MEAs, incorporating them into normal classroom discourse; integrating them into shared canonical ways of thinking and generating mathematical interpretations; and linking these models and big ideas with a host of techniques, tools, and procedural skills. These activities have not yet been a focus of the majority of MMP research. Studying models and modelling at these scales also demands the development of a series of new research tools. These will be based on our existing assumptions and conjectures but also designed to produce evidence that will test and revise or refine those assumptions and conjectures and support iterative cycles of development—that is, modelling at the researcher level.

4.6 Some Assumptions and Conjectures

In this section, we outline some of the assumptions and conjectures that form a basis for our initial development of a course-sized research site. We focus on a subset that are most relevant to currents in the international research discourse. In particular, we describe our relations to research on (a) learning progressions, and (b) explicitly teaching problem solving and heuristics.

4.6.1 Learning Progressions

Inquiring into the organization of knowledge and learning experiences as they unfold in time, many researchers in the international community have engaged the notion of a *learning progression*. In many manifestations of this approach, the objective is to identify a *trajectory* among the big ideas of a discipline. This may be grounded in mathematical structure, historical development, and/or studies of connections among ideas that are resonant with conceptions that have been identified in learners before instruction has occurred. These documented connections are then used to rationalize a fixed, linear order in which the big ideas are presented.

Our conjecture is that while this notion may be relevant for the study of *teaching* or instructional planning, it does not adequately address the question of possible productive connections among ideas from the viewpoint of the *learner*. For example, in prior work with MEAs in the context of courses focused on data modelling, we have found that the big ideas of a course can be productively connected in a variety of possible ways, and that connections develop in multiple dimensions simultaneously. Moreover, the earliest emergence of big ideas often occurs in modelling solutions, where learners formulate embryonic concepts and procedures by assembling ideas from a variety of textbook topic areas. Thus, facilitating model development may demand a responsive instructional approach, which unpacks these partially-formed insights and ideas, sorting them out and supporting more explicit connections among knowledge elements. In particular, this leads us to the conjecture that there may not be a single, a priori way to establish these connections for all learners independent of their idiosyncratic and situated modelling work.

Our approach on this matter is critical to designing our course-sized research site, affecting the very metaphors we have for learning at larger time scales. In particular, we associate the learning-progression perspective with a *ballistic* model of learning, in which a teach-first, apply-later philosophy is applied to propel classroom learners through course materials, at least in its initial exposure to the big ideas of the curriculum. In contrast, we explore the viability of alternatives to a ballistic model, which we unpack below.

4.6.1.1 Alternative Model #1: Learning as Finding One's Way Around in a Terrain

Instead of envisioning concept development using a ballistic metaphor, where a single point moves along a path in space, we find it useful to substitute a metaphor where the discipline is conceptualized as a multi-dimensional terrain, and where big ideas are conceived as something akin to mountains in a topographic view (see also, Zawojewski et al. 2013). Within this metaphor, we have found that development often proceeds in ways that resemble the formation and passage of interacting weather systems over the region. This metaphor honors the value and specificity of each manner of moving through the domain while also recognizing that different routes and systems are possible, each with its own characteristic strengths and weaknesses. Facilitating classroom inquiry in which different students or groups are exploring different regions in the disciplinary topography, however, requires the support of new instructional principles as well as additional tools and structures. If successful, the classroom group's exploration of a course's terrain would yield a range of possible "views" of the topographically high points (the big ideas), constructing powerful personal models that illustrate connections among them

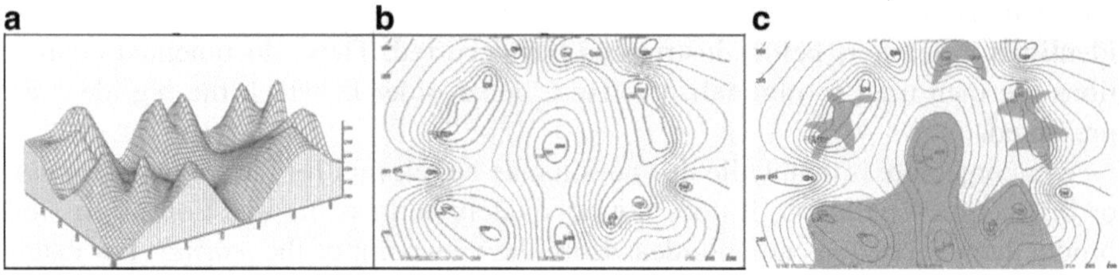

Fig. 4.1 The ideas of a course conceived as a terrain (**a**) (**b**) (c.f., Zawojewski et al. 2013, p. 492). Mapping progress (**c**) (c.f., Zawojewski et al. 2013, p. 498)

and make use of different forms of procedural and conceptual knowledge. Learning Progress Maps (prototypes shown in Fig. 4.1) are intended to support this instructional work and to study the utility of the terrain metaphor.

4.6.1.2 Alternative Model #2: An Evolutionary Model for the Development of Ideas

Prior MMP research also suggests another alternative to the ballistic model: one rooted in evolutionary theory. This metaphor highlights the importance of the *diversity* of ideas within problem-solving groups; of conditions for these diverse ideas to be placed in *communication* with each other; of an immanent mechanism for the *selection* of some ideas over others to be pursued; and of a means for the *survival* and *accumulation* of changes in ideas over iterative cycles. These four factors—diversity, communication, selection, and accumulation—are critical elements of nearly all evolutionary processes that involve living organisms in complex ecosystems. Hence in this view, the classroom is seen as an ecology in which a diversity of ideas evolves, by analogy with the diversity of species in a natural ecosystem. The classroom group is analogous to a team of naturalists, understanding each of these idea-species both in terms of their history and in terms of the relations with the conceptual environment. Like the terrain metaphor, this evolutionary metaphor honors the organic and nonlinear development of knowledge that the MMP has illuminated.

4.6.2 Teaching Problem Solving and Heuristics

Another current in international research involves revived interest in the explicit teaching of problem-solving strategies. Some attempts to teach heuristics or metacognitive strategies are rooted in work by Pólya (1945) and Schoenfeld (1985, 1987), among others. This approach holds that context-free instruction in heuristics may make the actual practices of problem solving more generally learnable. However, doing so by "cleaning up" *descriptive* accounts of real process and presenting them as *prescriptive* guides to follow, can be based on a conception of problem solving that sees the situated "messiness" of actual problems as incidental. Moreover, these problem-solving approaches based on Pólya's and Schoenfeld's work hold that instruction in abstract concepts should precede "application" of these concepts to solve particular problems. In contrast, MMP researchers view messiness and intuition as intrinsic elements of the modelling process. Problem-solving individuals and groups appropriate the ideas they are working with and make them effective tools for problem solving in idiosyncratic ways, relying on dimensions of knowledge and understanding that emerge *along with* their solutions (*because of* their problem-solving activity). This view is far from a situation in which students are taught to recognize mathematical structures in the "givens" of a problem and apply corresponding techniques.

How, then, does the MMP tradition answer the question of how to guide problem-solving practice as it unfolds? At their simplest, models can be seen as representations that capture what the modeller sees as the essence of the situation they are attempting to conceptualize. They can employ a variety of media (like stories, drawings, etc). As with other human attempts at representation, their development tends to involve an iterative series of drafts—some building on prior drafts, others exploring new directions, techniques, or emphases. Even though no single draft may be the "best" in all dimensions, it is still possible to compare drafts against one another, identify their strengths and weaknesses, and isolate the best parts of different drafts, *with reference to a given representational purpose.*

Thus, an answer to the question of guidance comes with the "Self-Evaluation Principle" of MEAs. Critical to the specification of MEAs is the *purpose* for which the model will be used by a client. Given such a description, learners themselves can identify the degree to which their current models are useful or powerful tools for those specific purposes. Thus usefulness and power, with reference to the specifications of the problem, enable learners themselves to assess and regulate their work, rather than appealing to the authority of the teacher or to guidelines in a decontextualized model of problem-solving behavior.

We should note in passing that we believe that the kind of situated knowing described above is also a feature of effective decision making in *teaching* practice. This has strong implications for our models of professional development. A view like ours, which emphasizes contextual understanding, will favor models of teacher professional development organized as *in situ* reflections on practice. A core element in the design of our course-sized research site must therefore be to offer the means for teachers to engage in "inline" modelling of their own students' learning processes and to become "connoisseurs of student work," sharing these emerging skills and sensitivities with colleagues in a low-overhead manner and with an eye to their usefulness and power in actual teaching settings.

4.7 Implications for Design

In recent years, MMP research has helped to lay the foundation for the inquiry we are currently undertaking. For instance, various efforts to assemble MEAs into larger, coherent instructional unities have led to the emergence of Model Development Sequences (MDSs), which support instruction that makes the most of MEAs in practical classroom contexts. In particular, a given MDS may include the following, in addition to one or more MEAs:

- Reflection Tool Activities, in which student groups turn their attention to describing individual and group level processes, functions, roles, conceptions, and beliefs. These include Ways of Thinking Sheets, various surveys and questionnaires, Concept Maps, Observation Sheets, Self-Reflection Guides, and Quality Assurance Guides for the products created in MEA activities.
- Product Classification Activities, in which students categorize the kinds of thinking involved in their solutions to MEAs.
- Model Extension Activities, often involving dynamic mathematics software, in which the class extends and formalizes promising elements of mathematical thinking that have appeared in student solutions.
- Model Adaptation Activities, where students generalize models, transferring them to related but different problems from those they originally were created to solve.

These MDS elements are designed to be highly modular, in order to accommodate (as well as to reveal) the needs and intentions of teachers that engage with them. One example of how these components might be laid out across multiple days in an instructional unit is shown in Fig. 4.2 (in this example, a 4-day sequence).

Importantly, each of the elements of the MDS acts as a research tool as well as an instructional component. At the research level, the data produced by these elements will serve to illuminate processes of learning and idea development that are critical to answering our research questions. Similarly, many of the other research tools described above (including Learning Progress Maps and Concept Analysis Wheels) will generate new research data when developed *iteratively* by students and teachers as the course unfolds.

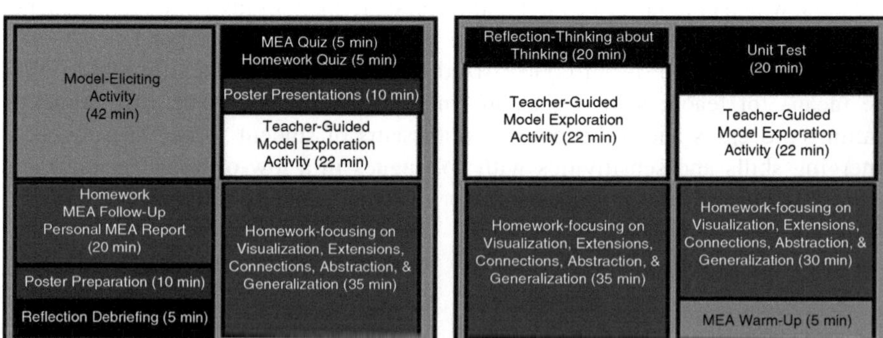

Fig. 4.2 A 4-day segment from a model development sequence

4.8 Conclusion: Contributions of a Course-Sized Research Site

In building our research site, we aim to illuminate new dimensions of knowing and learning, and we have already established initial versions of many of the research tools that will support this work. As suggested in the discussion of the design of MDS units, we are undertaking this work with a commitment to flexibility-in-use that we describe as *designing for scale*. Our course materials are presented in modularized and easily modifiable, reconfigurable, and extendable components.

Thus, teachers and researchers will be able to engage with these materials in a variety of ways, each embodying a unique (teacher or researcher level) model of knowledge development. The data and evidence produced by this diversity of models will help support choices amongst them, to refine our own assumptions and conjectures, and to iteratively shape our own understandings so that more useful and powerful conceptions survive and accumulate. From our description, it should be clear that we are conceiving this work as a modelling process for ourselves. We aim for our course-sized research site to provide value also to the broader teacher and researcher community,

- by facilitating the development, sharing, and testing of new Research Tools for different facets of research into dimensions of learning as they emerge;
- by offering a shared setting for the refinement of the *design principles* for research tools to produce evidence about learning at scales higher than the MEA;
- by fostering the accumulation of knowledge in teacher and researcher communities, exposing our process and inviting broad participation in constructing the site; and
- by encouraging the formation of collaborative communities of teachers and researchers, allowing participants to identify possible collaborators through shared interests in research tools and facets of problems of research or instruction.

This expansion of the MMP perspective builds on a long history of research success, whilst opening the way to have a new level of practical impact on classroom instruction.

References

Csikszentmihalyi, M. (1997). *Finding flow: The psychology of engagement with everyday life*. New York: Basic Books.

English, L. D., Lesh, R., & Fennewald, T. (2008). *Methodologies for investigating relationships between concept development and the development of problem solving abilities*. Paper presented at ICME-11, Monterrey. Retrieved April 20, 2012, from http://tsg.icme11.org/tsg/show/20

Lesh, R., & Doerr, H. (2003). In what ways does a models and modeling perspective move beyond constructivism? In R. Lesh & H. Doerr (Eds.), *Beyond constructivism: A models and modeling perspective* (pp. 519–556). Mahwah: Erlbaum.

Mousoulides, N., Sriraman, B., & Lesh, R. (2008). The philosophy and practicality of modeling involving complex systems. *The Philosophy of Mathematics Education Journal, 23*, 134–157.

Pólya, G. (1945). *How to solve it*. Princeton: Princeton University Press.

Schoenfeld, A. H. (1985). *Mathematical problem solving*. New York: Academic.

Schoenfeld, A. H. (1987). What's all this fuss about metacognition? In A. Schoenfeld (Ed.), *Cognitive science and mathematics education* (pp. 189–215). Hillsdale: Erlbaum.

Zawojewski, J. S., Magiera, M., & Lesh, R. (2013). A proposal for a problem-driven mathematics curriculum framework. *The Mathematics Enthusiast, 10*(1&2), 469–506.

Chapter 5
Prescriptive Modelling – Challenges and Opportunities

Mogens Niss

Abstract This chapter deals with a distinction between two kinds of mathematical modelling purposes and related modelling endeavours, descriptive modelling and prescriptive modelling. Whilst descriptive modelling is usually the focus of attention of practice, research and development in mathematics education, prescriptive modelling – in which the aim is to design, organise or structure certain aspects of extra-mathematical domains – is hardly noticed, let alone investigated in mathematics education. After having presented three concrete examples of prescriptive modelling, this chapter makes a plea for paying attention to its cultivation and investigation in mathematics education contexts. It does so by analysing prescriptive modelling in relation to the so-called modelling cycle and finishes by outlining challenges and opportunities for such an endeavour.

5.1 Introduction

Classically, the purpose of mathematical modelling is to capture, represent, understand, or analyse existing extra-mathematical phenomena, situations or domains, usually as a means of answering practical, intellectual or scientific questions – and solving related problems – pertaining to the domain under consideration. A few examples of such questions include: Which of several internet subscription schemes should I choose? When should the owner of a vineyard harvest his/her grapes so as to balance quantity and quality? What is the expected long-term development of the Danish population, in total and across age groups? How tall is the huge construction crane working across the street? Will the North Pole ever be ice-free during summer prior to 2050? None of these questions can be answered by mathematical means alone. On the other hand, none of them can be answered in a satisfactory manner without the involvement of some kind and amount of mathematics either.

M. Niss (✉)
IMFUFA, Department of Science, Roskilde University, Universitetsvej 1,
Roskilde DK 4000, Denmark
e-mail: mn@ruc.dk

G.A. Stillman et al. (eds.), *Mathematical Modelling in Education Research and Practice*, International Perspectives on the Teaching and Learning of Mathematical Modelling, DOI 10.1007/978-3-319-18272-8_5

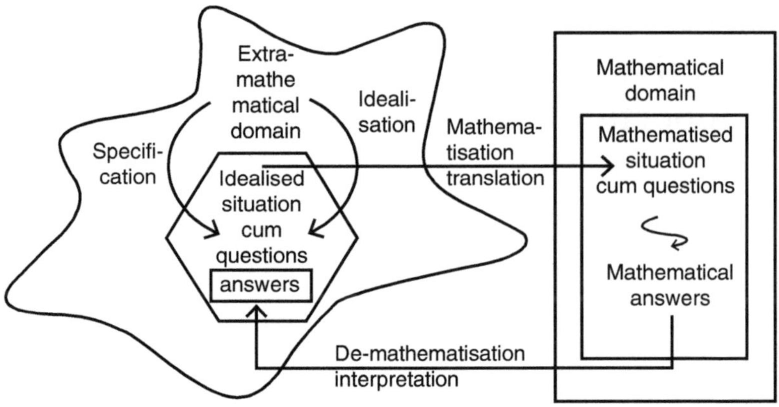

Fig. 5.1 The modelling cycle (Niss 2010, p. 44)

Let us agree to call, in short, modelling for purposes such as the ones just mentioned *descriptive modelling*. In descriptive modelling we attempt to capture and come to grips with, and – if relevant – eventually act in, some extra-mathematical domain in the real world, in a field of practice or in an academic discipline. For very good reasons, educational research and development regarding mathematical models and modelling have traditionally focused on descriptive modelling. The processes of descriptive modelling are typically represented by one of several more or less similar versions of the *modelling cycle*, for example the one shown in Fig. 5.1 (an earlier version was published in Niss 2010).

The focus of much research and development is on the key sub-processes in this cycle:

Preparing the extra-mathematical domain, labelled *D* and depicted as an "amoeba" in Fig. 5.1, for modelling by specifying and idealising the situation considered, making assumptions about it, and by choosing and formulating the questions to be answered. The resulting idealised situation is represented by a hexagon in Fig. 5.1.

Mathematising the idealised situation and questions, by translating (by way of a 'mapping' f) all the objects, phenomena, relationships, assumptions, and questions in *D* into mathematical "representatives" of them in some chosen mathematical domain *M*, depicted as a rectangle on the right-hand side of the figure. The sub-rectangle zooms in on the mathematical questions and the answers to them. The mathematisation gives rise to the *model* (D, M, f), conceived of as a triple in which no component can be left out of consideration.

Dealing with the mathematised situation, that is using mathematical concepts, considerations, theorems, procedures, techniques, and reasoning to *derive and justify answers* to the mathematical questions that resulted from the mathematisation process. This process of mathematical treatment is depicted as the curved arrow in the innermost rectangle within *M*.

De-mathematising the mathematical outcomes of the mathematical treatment undertaken, by translating the outcomes back (interpreting them) into extra-mathematical answers to the initial extra-mathematical questions driving the modelling endeavour in the first place.

Validating the model by confronting the model output with the known reality of the extra-mathematical domain, and by assessing the quality and relevance of the answers obtained in relation to the purpose of the modelling undertaken.

In the literature, most of these different sub-processes of the modelling cycle have been studied more or less extensively from theoretical, empirical and practical perspectives (see, for example, Blum et al. 2007 and Stillman et al. 2010).

In addition to descriptive modelling there is, however, also another kind of purpose of mathematical modelling, namely to *specify or design objects or structures* that are meant to inhabit some extra-mathematical domain whilst possessing (if possible) certain required or desired properties. Here are a very few examples: Where should a new power plant or a huge shopping centre be located? In what way should seats be apportioned amongst parties in parliamentary elections? In what way should the m members of a board of directors be elected by the electorate from p candidates ($m < p$)? What would be a good measure of the degree of acidity of substances? What would be a good measure of the viscosity of a liquid/fluid? How should a loan amortisation scheme be defined so as to fulfil certain requirements? How should plane drawings be constructed so as to faithfully reflect our spatial vision? What dimensions should a box (a right-angled parallelepiped) holding 1 l have so as to minimise the use of material?

The purpose of such modelling is not primarily to come to grips with some existing part of the world, that is an extra-mathematical domain, but *to design, prescribe, organise or structure certain aspects of it*. Inspired by Davis (1991), let us agree to call modelling for such purposes *prescriptive modelling*. In prescriptive modelling the ultimate aim is to pave the way for *taking action based on decisions* resulting from a certain kind of mathematical considerations, in other words 'to change the world' rather than only 'to understand the world'.

The *didactics of prescriptive mathematical modelling* is the focus of attention in this chapter. In order to provide substance to the discussion to follow in subsequent sections, let us analyse some examples in greater detail. They have been deliberately chosen to be simple and authentic.

5.2 Examples

5.2.1 *Example 1: BMI (Body Mass Index)*

In many health contexts it is customary to consider what is called a person's *body mass index*, BMI, originally introduced by Quetelet in the mid-nineteenth century (Wikipedia a). It is defined as

Table 5.1 Labels of BMI intervals

$BMI \leq 18.5$	$18.5 \leq BMI \leq 25$	$25 \leq BMI \leq 30$	$30 \leq BMI$
Underweight	Normal weight	Overweight	Obese

$$BMI = w/h^2,$$

where w is the person's weight (measured in kilograms) and h is the person's height (measured in metres). Thus the unit of BMI is kg/m^2. Along with the index come four consecutive intervals containing the possible values of BMI (all of which are positive). A normative label is attached to each interval as in the following ubiquitous table (Table 5.1).

According to this table any individual whose BMI belongs to a given interval is equipped with a certain label. Recently, in some quarters, the labelling has been modified such that underweight is associated with $BMI \leq 20$, normal weight corresponds to the interval $20 \leq BMI \leq 30$, whereas obesity still corresponds to $30 \leq BMI$. Sometimes the two extreme intervals, "underweight" and "obese" are divided into several subintervals.

In order to understand what is happening here, in terms of modelling, let us make use of the modelling cycle.

First of all the *aim* of the endeavor was clearly to establish an index of '(relative) heaviness' for individuals in a human population. This aim can be seen as being derived from a *meta-question*: how can we create a quantitative measure to capture the evident variation of heaviness within and across human populations? Perhaps an underlying agenda for this question was to establish a seemingly objective platform for providing medical counselling to people whose body mass was likely to cause them trouble sooner or later.

To this end, every human being is strongly idealised inasmuch as the only traits of the person taken into consideration are his/her weight and height. In other words, the person is parametrised (a special way of being represented mathematically) by a non-negative real-valued 2-vector (w, h). This vector is then used as the argument of a function f: $R_+^2 \rightarrow R_+$, given by $f(w, h) = w/h^2$, combining weight and height to define our measure of 'relative heaviness'. It is then decided to mathematise the set of all people – the global population – by putting each member into one of four classes, according to which of four non-negative intervals, [0,18.5], [18.5, 25], [25,30], [30, ∞], their f-value belongs to. The literature does not seem to specify the classification of the borderline cases 18.5, 25 and 30, which introduce a slight ambiguity into the labeling scheme. Apart from this, the *mathematisation* of the situation is now complete. This mathematisation translates real world entities (individuals and populations, scales and yardsticks) into a mathematical domain consisting of real number spaces, and (rational) functions on them.

The *mathematical treatment* of the mathematised entities is next to trivial. It simply consists in calculating for any individual his or her BMI from w and h, and then locating the resulting value in one (or two) of the intervals.

The *de-mathematisation* is equally trivial as it consists in attaching real world labels to the individual, based on his/her BMI location: underweight, normal weight, overweight, or obese.

As to the *validation* of this model, it hardly makes sense to confront the de-mathematised model output with the known reality of the extra-mathematical domain (unless an independent *non-impressionistic* measure of relative heaviness was established). The same is true as regards assessment of the quality and relevance of the answers obtained in relation to the purpose of the model. In other words, in contradistinction to what is the case with descriptive modelling, validation of the BMI model is not really possible. More specifically, in principle the model *cannot be falsified.*

However, it can be *criticised* and *meta-validated.* For example, what would happen if another measure were chosen, for example, $f(w, h) = w/h^{\alpha}$, perhaps with $\alpha = 1$ or $\alpha = 3$? And what would happen if the interval boundaries were changed? That could have a major impact on the attribution of heaviness labels to people. Why does the model not distinguish between men and women, between young and old, and between different ethnic groups? Why is body composition not taken into account? If adopted across populations, what distributions of BMI values and heaviness would we encounter? Questions such as these would help shed light on some of the consequences of putting the model to use, but that is different to validating the model directly.

This examination suggests that the descriptive modelling cycle becomes *very rudimentary* when applied to the BMI model. No initial substantive modelling questions were posed, only a meta-question, "what might a measure of relative heaviness of human beings look like?" No extra-mathematical facts or assumptions were involved in the modelling, except one very significant assumption: only weight and height matters. The specific mathematisation, that is the formula $f(w, h) = w/h^2$ largely came out of the blue, as did the labelling of the intervals. The mathematical treatment of the model proposed boiled down to a mere calculation of the specific BMI value, given w and h, followed by locating the value in an interval (or two). Similarly, the de-mathematisation boiled down to attaching a weight class label to the value obtained. Finally, no direct validation of the model was possible, only a wide spectrum of meta-validation and critique of the use of the model. Thus, only the first couple of sub-processes of the modelling cycle were in play in a non-trivial manner.

5.2.2 Example 2: A-Paper (DIN) Formats

Imagine that you wish to *design* a system of paper sheet formats with the following properties:

- Each sheet of paper is rectangular
- The area of the largest sheet in the system is 1 m^2

Fig. 5.2 A-paper formats

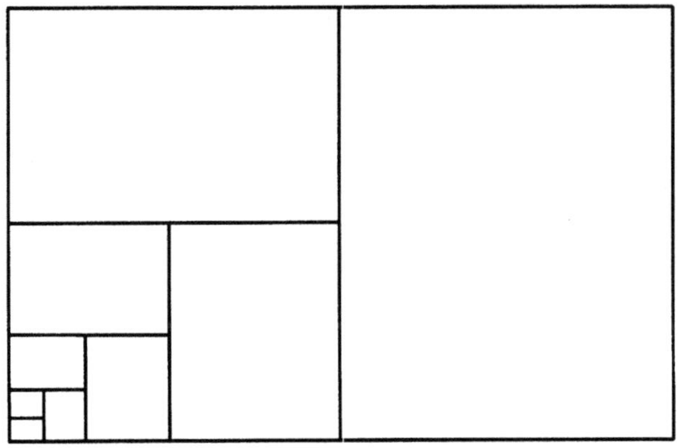

- If any sheet of paper in the system is bisected across a mid-point transversal between the two longer sides, each half sheet is also in the system and is similar to the original one, that is side proportions remain the same.

These three wishes are depicted in Fig. 5.2.

Once again, let us analyse the modelling taking place here in terms of the modelling cycle.

First, the *aim* is simply to design – if we can – the desired system of paper formats by specifying the exact side lengths of each of the different sheets in the system. No assumptions about reality are being made (except that paper exists and can be cut into various shapes). They have been replaced by requirements or wishes. The modelling questions then are "can these requirements be fulfilled?" and "if yes, how?"

As to the *mathematisation*, each sheet of paper, the *n*th sheet ($n \geq 0$) being denoted by A_n, is mathematised as a rectangle, and parametrised in terms of its dimensions $(l_n, s_n) \in R_+^{\ 2}$, where l_n indicates its longer side and s_n its shorter side. The remaining requirements are mathematised as (a) the unit of length is cm, of area cm^2; (b) $l_0 \cdot s_0 = 10{,}000$ cm^2; (c) the sheets are similar, that is, for every $n \geq 0$ we must have $l_{n+1}/s_{n+1} = l_n/s_n$, as well as $l_{n+1} = s_n$ and $s_{n+1} = l_n/2$. The mathematical domain involved consists of real numbers and sequences.

The mathematised questions then become: does there exist a sequence $(\mathrm{l_n}, \mathrm{s_n})$, $\mathrm{n} \geq 0$, composed of positive elements, that satisfies (b) and (c)? And if so, what elements does/can it have?

The *mathematical treatment* needed to answer these questions is as follows:

First, we observe that since for all $n \geq 0, l_{n+1}/s_{n+1} = l_n/s_n$ and $l_{n+1} = s_n$ and $s_{n+1} = l_n/2$ we obtain $l_n/s_n = l_{n+1}/s_{n+1} = s_n/(l_n/2)$, from which we deduce that $2s_n^2 = l_n^2$, that is $l_n = 2^{1/2}s_n$. As this also holds for $n = 0$, we have $l_0 = 2^{1/2}s_0$. Moreover, since $l_0 s_0 = 10^4$, insertion of the previous identity yields, $2^{1/2}s_0^2 = 10^4$, whence $s_0 = 10^2/2^{1/4}$. Furthermore we have that $l_0 = 2^{1/2}\, s_0 = 2^{1/2}\, 10^2/2^{1/4} = 2^{1/4}\, 10^2$. In summary, $l_0 = 2^{1/4}\, 10^2$ and $s_0 = 10^2/2^{1/4}$. Next we have $l_1 = s_0 = 10^2/2^{1/4}$ and

$s_1 = l_0/2 = 2^{1/4}\ 10^2/2 = 10^2/2^{3/4}$. Continuing by recursion and induction we obtain for any $n \geq 0$:

$$l_n = 10^2/2^{(2n-1)/4} \text{ and } s_n = 10^2/2^{(2n+1)/4},$$

all of which are, by the way, irrational numbers. This gives the complete answer to the questions above.

The *de-mathematisation* consists of nothing but translating the numbers obtained back to units in cm to answer the initial questions. Yes, there does exist a uniquely determined (infinite!) sequence of paper sheet formats satisfying the requirements or wishes initially posed. The dimensions of the nth sheet, A_n, are

$$l_n = 100/2^{(2n-1)/4}\text{cm, and } s_n = 100/s^{(2n+1)/4}\text{cm.}$$

For example, the dimensions of the prevalent A4 sheet (A_4) are $l_4 = 100/2^{7/4} \sim 29.73$ cm and $s_4 = 100/2^{9/4} \sim 21.02$ cm.

How about the *validation* of this model? Well, the questions posed were answered in the positive (and uniquely). The resulting design, when implemented, *creates reality*, it does not describe it. So, from this perspective confrontation of the model with reality has no meaning. The situation would be completely different, however, had the initial modelling question been entirely different, for instance as follows. Here we have a bunch of different paper formats. What were the principles according to which they were designed? Hypothesising underlying principles such as the third requirement for the model design above, one might perform descriptive modelling and then validate the model outcomes by confronting them with empirical data from actual paper sheets.

The modelling cycle for this prescriptive modelling is also very rudimentary. Initially there was no existing extra-mathematical domain to model. We wanted to design a new part of reality by specifying some requirements or wishes to be fulfilled by our design, if possible. These requirements are already of a pre-mathematical nature, since we speak about rectangular sheets, folding them along a mid-point transversal and obtaining similar shapes. No assumptions about reality were involved. Mathematisation amounted to translating the design requirements into mathematical requirements amenable to mathematical treatment. This treatment was the centrepiece of the work done. Once this was completed, de-mathematisation was trivial: attach units to the numbers and cut the paper sheets accordingly. No validation was possible beyond the mere observation that a (unique) positive and satisfactory answer to the design questions exists. Confrontation with the properties of an existing extra-mathematical reality prior to the design scheme developed does not make sense. Meta-validation, too, does not make sense here. Again, only a few sub-processes of the modelling cycle were in play in a proper sense.

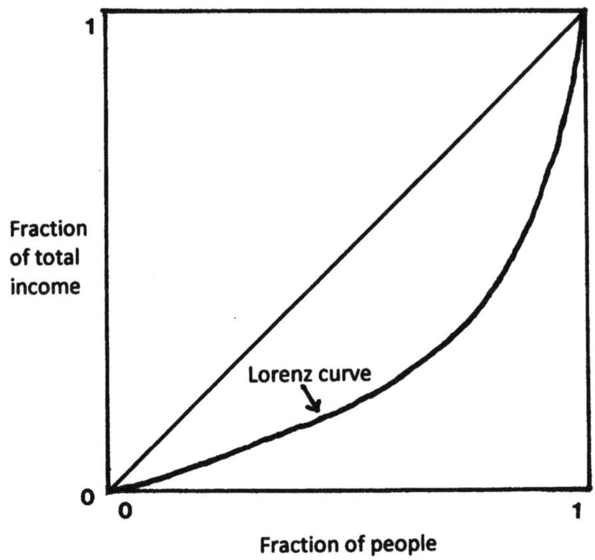

Fig. 5.3 The Lorenz curve of income distribution

5.2.3 *Example 3: The Gini Coefficient of Income Inequality*

Assume that we want to create an index of inequality of income in a population. To this end we begin by specifying the fraction $L(p)\,(\in [0;\ 1])$ of the total income of the population which is owned by the fraction $p(\in [0;\ 1])$ of the population having the lowest income. L is called the *Lorenz function* and its graph in $[0; 1] \times [0; 1]$ the Lorenz curve as shown in Fig. 5.3. In case of complete equality, the population fraction p with the "lowest" income would own exactly the fraction p of the total income, i.e. $L(p) = p$ for every p. In the opposite case of maximal inequality one "person" would own everything, that is $L(p) = 0$ for $p < 1$ and $L(1) = 1$.

In 1912 Italian mathematician Corrado Gini proposed (Gini 1912; Wikipedia b) an index of inequality based on the "actual accumulated inequality", specified as the integrated difference between the Lorenz function corresponding to complete equality and the actual Lorenz function, that is the area between the two curves:

$$I_L = \int_0^1 (p - L(p))dp = \tfrac{1}{2} - \int_0^1 L(p)dp.$$

In the special case of maximal inequality, the accumulated inequality would be $\int_0^1 pdp = \tfrac{1}{2}$.

Now, the so-called *Gini coefficient, G,* is defined as the ratio between the actual accumulated inequality and maximal inequality, that is

$$G = \frac{\frac{1}{2} - \int_0^1 L(p)dp}{1/2} = 1 - 2\int_0^1 L(p)dp,$$

corresponding to twice the area between the Lorenz curve and the "equality line". Clearly, G ∈ [0; 1], where G = 0 corresponds to complete equality and G = 1 to maximal inequality. In summary, G is a measure of the degree of income inequality in a population. The measure allows for comparison of income inequality amongst, say, countries or regions.

The *modelling cycle* for this undertaking has the following features.

First, the *aim* of the endeavour was to construct an index measure which can capture income inequality in a population. This requires the notion of "income" to be specified in such a way that each individual possesses a well-defined income. We want the measure to be constructed to fulfill certain requirements: The index must be defined for any population, once individuals' incomes are known. We want the index to involve the notion of actual accumulated inequality in relation to the maximal possible inequality.

The *mathematisation* undertaken then included four steps. First, all individuals are ranked according to their income in non-decreasing order. Then the accumulated income distribution is mathematised by way of the Lorenz function L: $[0;1] \to [0;1]$, which is a real-valued function. Next, complete equality is mathematised by $L_e\ (p) = p$ for all p, and maximal inequality by $L_{mi}\ (p) = 0$ for $p < 1$, $L_{mi}\ (1) = 1$. Accumulated actual inequality is mathematised by $I_L = \int_0^1 (p - L(p))dp$. Finally, the degree of income inequality in a population is mathematised by the Gini coefficient $G_L = I_L/I_{mi}$.

This mathematisation process is somewhat involved. In contrast, the subsequent *mathematical treatment* is straightforward. For a specific given income distribution L, calculate I_L and I_{mi} to obtain $G_L = I_L/I_{mi}$.

De-mathematisation only consists in stating that the Gini income inequality coefficient of population P is G. As this coefficient is a pure number, no units are involved.

The situation concerning *validation* is much the same as for the BMI model. It hardly makes sense to confront the de-mathematised model output with what we know about the reality of extra-mathematical domain, unless we were in possession of another, independent and non-impressionistic measure of income inequality as a benchmark reference. The same is true with regard to assessing the quality and relevance of the model in relation to its purpose. In other words, direct validation is not really possible.

So, the Gini coefficient model *cannot be falsified*. But it can be *criticised* and *meta-validated* by looking at questions such as: Is the choice of the kinds of incomes taken into consideration reasonable? Which segments of the population are taken into account? How about children? To what extent is comparison of Gini coefficients amongst different populations possible and meaningful?

Again we are faced with a very rudimentary modelling cycle. The extra-mathematical domain consisting of people, populations and incomes does indeed exist and is very complex, as are the features selected for consideration. The mathematisation undertaken was rather intricate and involved, but the remaining sub-processes of the modelling cycle – mathematical treatment, de-mathematisation and direct validation – were trivial or not meaningful, respectively.

5.2.4 Conclusions from the Examples

On the basis of these examples – and hosts of other examples – it seems fair to conclude that prescriptive modelling, even though it shares significant features with descriptive modelling, differs from it in characteristic ways, which become visible when considering the modelling cycle. Whilst the preparation of the extra-mathematical domain for modelling by way of specification and idealisation can be rather similar in descriptive and prescriptive modelling, the mathematisation part may – but need not – be very different, since in prescriptive modelling there may not be any clue whatsoever concerning how to come up with a sensible model to meet the aim of the modelling. The de-mathematisation and the validation parts are normally very different as they are largely absent in prescriptive modelling. Mathematical treatment sometimes is similar and sometimes different in the two sorts of modelling, depending on the context and situation. This suggests that the sub-processes of the modelling cycle may not be able to fully capture what happens in evaluating models arising from prescriptive modelling. Instead, meta-validation becomes crucial.

It should be highlighted that the difference we focus on is between two different kinds of *modelling purposes*, not between two different kinds of models. Thus, the very same model can arise in descriptive and in prescriptive modelling. For example, the model in the A-paper format example above arose as part of a prescriptive modelling endeavour. However, the very same model may arise as part of a descriptive modelling endeavour, if the task were to identify the design principles underlying a collection of real paper sheets belonging to the A-paper system. That is why we distinguish between descriptive and prescriptive *modelling*, not between descriptive and prescriptive *models*.[1]

It is also worth mentioning that undertaking a prescriptive modelling task may well involve undertaking certain descriptive modelling tasks along the way. This is the case, for instance, in the Gini model example above, when the ranked income

[1] In Blum and Niss (1991, p. 39), the authors spoke about *descriptive models* rather than *descriptive modelling*, and also about *normative* models (not prescriptive models). Thus, the terms proposed in the present chapter represent a move from focusing on the *product* (the model) to focusing on the modelling *purposes*, whilst replacing 'normative' with 'prescriptive', following Davis (1991).

distribution of a population and the Lorenz function – results of descriptive modelling – are being used as fuel for the construction of the Gini coefficient. On closer inspection, this phenomenon of prescriptive modelling sometimes involving aspects of descriptive modelling, and vice versa, turns out to be rather widespread. For instance, this happens all the time in optimisation, which is nothing but an instance of prescriptive modelling.

5.3 Teaching and Learning of Prescriptive Modelling

The examples above show that the modelling cycle of prescriptive modelling may well be very rudimentary in key places. Depending on which sub-processes of the modelling cycle are affected, this seems to give rise to significant challenges to the teaching and learning of prescriptive modelling.

In cases in which the prescriptive modelling task is very generally and vaguely defined, as well as *complex*, for example, when the task is "create a measure or a descriptor of…", it is usually not realistic to expect students to be able to undertake the preparation of the situation to be modelled and to come up with a subsequent mathematisation of it by transfer of experiences from descriptive modelling only. The BMI index and the Gini coefficient are points in case. This is not surprising, as the invention of good measures or descriptors (not to be confused with descriptions) may actually prove to be significant scientific work. Just think of notions and measures such as the pH value of acidity, acceleration, elasticity, least squares, and hosts of concepts from descriptive (sic!) statistics. Unless students are guided (or very well trained), the modelling process in such cases is likely to stop before it begins. Let us agree to name such prescriptive modelling cases, in which the focus is on inventing notions, descriptors or measures, *Type I* cases.

If, however, a *Type I* task is less complex, students may in fact be able to put their possible knowledge of, and experiences with, standard descriptor models (e.g., average, ratio, rate of change) to use in new situations. This is where model-eliciting activities (promoted by Lesh, Doerr and others, e.g., Lesh and Doerr 2003, and Amit and Jan 2010) and emergent modelling (Gravemeijer 2007) become helpful and significant.

In the complex *Type I* cases, which put high demands on mathematisation, for students to prepare the situation at issue for modelling and then mathematise it requires, most likely, tight guidance. In contrast, the subsequent mathematical treatment and the de-mathematisation become trivial in principle, and validation of the model usually makes no sense. However, as emphasised above, meta-validation of models generated by such endeavours is essential.

If, on the other hand, the prescriptive modelling endeavour is well defined in the sense that it rests on clearly formulated specific requirements or wishes of a pre-mathematical nature, the first parts of the modelling process may well resemble those of descriptive modelling, even though the entire modelling cycle remains rudimentary. The A-paper format example above is a point in case. Preparing the situation for modelling and subsequently mathematising it are often straightforward and accessible for students undertaking independent work. The mathematical treatment becomes the core activity and may be more or less demanding, but not fundamentally inaccessible to students. De-mathematisation and validation are typically trivial. Meta-validation of the model will focus on the impact of the resulting model on the (degree of) fulfilment of the initial requirements or wishes, or on the impact of these requirements / wishes themselves on the modelling endeavour, especially in cases where not all requirements or wishes could be satisfied or, on the contrary, in cases where several satisfactory solutions exist, amongst which choices can then be made if additional criteria are invoked. In other words, in such cases – which we may agree to call *Type II cases* – prescriptive modelling can well be within reach for students, even though the mathematical treatment may sometimes prove to be a challenge.

Prescriptive modelling is omnipresent and highly important with significant scientific, practical and societal impact. This implies that meta-validation of models arising from prescriptive modelling is of crucial importance. Against the background outlined above, one might think that since the modelling process required in tasks of *Type I* is only partially accessible to students, teaching and learning of prescriptive modelling should be restricted to modelling tasks of *Type II*. However, given the general importance of *Type I* modelling tasks, that would be a wrong conclusion to draw. Rather, we must realise that we have to deal with both kinds of prescriptive modelling in mathematics education practice, research and development, and then try and find ways to overcome the difficulties encountered.

5.4 Challenges and Opportunities

Our overarching task and challenge in this context is to devise teaching and learning environments and activities for prescriptive modelling, also covering situations in which the students are unlikely to be able to complete every part of the modelling process without assistance. I propose two focal points for such activities as ways to meet this challenge.

Firstly, we should engage students in *identifying and analysing the hidden or explicit assumptions, requirements or wishes*, as well as the prerequisites, underlying the modelling (to be) undertaken, in particular as far as the mathematisation part is concerned. Secondly, we should engage students in *meta-validation* of the models arising from prescriptive modelling undertakings with respect to (at least) the following points:

- The consequences of the modelling outcomes for the discourse on the issues addressed by the modelling endeavour;
- Comparing and contrasting the model obtained with actual or potential model alternatives;

- Considering the impact of changed requirements or wishes for the modelling and its outcomes.

Since prescriptive modelling in mathematics education has only been subjected to theoretical and empirical research to a very limited extent, and then primarily at the lower end of the complexity scale as is the case with model-eliciting activities, our second challenge is to design and implement such research. An obvious first focus for such research would be to investigate the similarities, differences and relationships between descriptive and prescriptive modelling, and the ways these (may) play out in mathematics education. As part of this endeavour, it will be important to look into ways to further develop or supplement – or replace? – the modelling cycle when it comes to adequately capturing the processes involved in prescriptive modelling, especially as far as validation and meta-validation are concerned.

The agenda just outlined is an extensive one. There is a long way to go to complete it. It is time to get going.

References

Amit, M., & Jan, I. (2010). Eliciting environments as "nurseries" for modeling probabilistic situations. In R. Lesh, P. L. Galbraith, C. R. Haines, & A. Hurford (Eds.), *Modeling students' mathematical modeling competencies* (pp. 155–166). New York: Springer.

Blum, W., & Niss, M. (1991). Applied mathematical problem solving, modelling, applications, and links to other subjects – State, trends and issues in mathematics instruction. *Educational Studies in Mathematics, 22*(1), 37–68.

Blum, W., Galbraith, P. L., Henn, H.-W., & Niss, M. (Eds.). (2007). *Modelling and applications in mathematics education: The 14th ICMI study*. New York: Springer.

Davis, P. J. (1991). Applied mathematics as a social instrument. In M. Niss, W. Blum, & I. Huntley (Eds.), *Teaching of mathematical modelling and applications* (pp. 1–9). Chichester: Ellis Horwood.

Gini, C. (1912). *Variabilità e mutabilità*. Bologna: C. Cuppini.

Gravemeijer, K. (2007). Emergent modelling as a precursor to mathematical modelling. In W. Blum, P. L. Galbraith, H.-W. Henn, & M. Niss (Eds.), *Modelling and applications in mathematics education. The 14th ICMI study* (pp. 137–144). New York: Springer.

Lesh, R. A., & Doerr, H. M. (Eds.). (2003). *Beyond constructivism: Models and modeling perspectives on mathematics problem solving, learning and teaching*. Mahwah: Erlbaum.

Niss, M. (2010). Modeling a crucial aspect of students' mathematical modeling. In R. Lesh, P. L. Galbraith, C. R. Haines, & A. Hurford (Eds.), *Modeling students' mathematical modeling competencies* (pp. 43–59). New York: Springer.

Stillman, G., Brown, J., & Galbraith, P. (2010). Identifying challenges within transition phases of mathematical modeling activities at year 9. In R. Lesh, P. L. Galbraith, C. R. Haines, & A. Hurford (Eds.), *Modeling students' mathematical modeling competencies* (pp. 385–398). New York: Springer.

Wikipedia a. www.wikipedia.org/wiki/Body_mass_index. Accessed 14 Nov 2013.

Wikipedia b. www.wikipedia.org/wiki/Gini_coefficient. Accessed 14 Nov 2013.

Chapter 6
An Approach to Theory Based Modelling Tasks

Xenia-Rosemarie Reit and Matthias Ludwig

Abstract The MokiMaS project *(Modelling Competency in Mathematics Classes of Secondary Education)* addresses the question how mathematical modelling competencies can be evaluated in a holistic context and points out a theory-based approach to assess students' modelling competency at the end of lower secondary education. In particular this chapter discusses criteria based modelling tasks, constituting the core of the test instrument. The piloting of the modelling tasks gave interesting insights into cognitive and structural differences of tasks. Furthermore, a strategy to classify the difficulty of modelling tasks on the basis of a thought structure model, describing the cognitive load of solution approaches is presented. Finally, the nature of a metrologically reasonable modelling task is discussed.

6.1 Introduction

There is broad consensus that the integration of the concept of mathematical modelling into school must be increased. Of course this awareness is not new and efforts have been set up during the last decade. However, not least PISA and TIMSS reveal wide gaps of not only German students especially in the area of mathematical modelling. Nevertheless, it is reasonable to question to what extent mathematical modelling can be assessed in detail with a test instrument which "aim[s] to evaluate [whole] education systems" (OECD n.d.). Test instruments which have been developed in a smaller setting are almost always based on a treatment developed within a study, such that these tests are restricted to specific sub-competencies of mathematical modelling (e.g., Zöttl et al. 2011). Moreover, Frejd (2012, p. 84) found that every third paper and pencil test consists of multiple choice items of Haines et al. (2000). As a result, sub-competencies can be evaluated but it is a debatable point whether modelling competency as a whole is tested. Within his

---

X.-R. Reit (✉) • M. Ludwig
Institute of Mathematics Education, Goethe-University Frankfurt, IDMI,
Robert-Mayer-Straße 6-8, Frankfurt 60325, Germany
e-mail: reit@math.uni-frankfurt.de; ludwig@math.uni-frankfurt.de

G.A. Stillman et al. (eds.), *Mathematical Modelling in Education Research and Practice*, International Perspectives on the Teaching and Learning of Mathematical Modelling, DOI 10.1007/978-3-319-18272-8_6

large literature review Frejd claims that "not a single item was found in this study that assessed all aspects of modelling (holistic view)" (Frejd 2012, p. 80). In this chapter we show the results of a pilot study in which modelling tasks have been developed following a holistic concept.

6.2 Theoretical Framework and Method

The aim of the MokiMas project *(Modelling Competency in Mathematics Classes of Secondary Education)* is to develop a paper and pencil test to assess mathematical modelling at the end of lower secondary education. This test pursues a holistic[1] approach and will be based on modelling tasks developed according to specific criteria.

6.2.1 Task Criteria

The modelling tasks have been developed according to predefined criteria that form a theoretical framework (Reit and Ludwig 2013, p. 805). Hence, we focused on walking the world with eyes open and discovering mathematics everywhere (Blum 2006, p. 26). In doing so, the item *Taj Mahal,* for example, has been developed during a journey to India. The following criteria serve as a basis for the development of tasks:

- authentic context (Maaß 2007),
- realistic numeric values (Müller et al. 2007),
- problem solving character (Maaß 2007),
- naturalistic format for questions,
- openness relating to the task space.

Authenticity and relation to reality are core elements of modelling tasks. We agree with Palm (2002 cited in Vos 2011) who defines authenticity of a task by having its origin in reality. Furthermore, the task related situation actually occurs or can occur in reality. There is no a priori known solution algorithm for the task that can be applied by students directly. That means that the solution makes itself out to be a problem at the students' level. The questioning is either close to the living environment of the students or takes up a realistic question that could arise in reality. Openness of tasks is reflected by the task space. There has to be more than

[1] In this context "holistic" is said to underline the fact that open modelling tasks are used, in contrast to multiple choice tasks or those which concentrate on subcompetencies.

one solution approach leading to a solution. The solution approaches distinguish themselves by their mathematical model; thereby, students are able to have more options to arrive at a solution. Openness should rather be based on alternative mathematical models to solve the tasks rather than on approximating sizes. Concerning this, we do not deny that making assumptions is an important part of mathematical modelling but we want it to be limited to a degree that ensures an assessable solution interval.

6.2.2 Degree of Difficulty

A common instrument to determine the *degree of difficulty* is the solution rate by applying a dichotomous rating. Since the answer format in the test is very open and compared to others rather extensively, it seems to us not adequate to determine the degree of difficulty only by a dichotomous rating. However, to have a well-founded basis to start from, we firstly rated the student solutions according to whether the task has been solved or not. Of course it has to be defined what is meant by "the task is solved". Incorrect solution approaches which are unrewarding, are considered to be unsolved, as well as those which show an abandoned solution process and thus end up with no solution. Solution approaches drawing defective conclusions which lead to unreasonable solutions are also rated as "unsolved". If the chosen mathematical model leads to an overall reasonable solution, the solution approach is regarded as "solved". This means that a solution approach has also been rated as "solved" even if the chosen numeric values are not quite appropriate. Therefore the solution rate does not necessarily provide an indication of the modelling competency because the solution approaches which are regarded as "solved" have recognisable great differences.

To adjust the method of determining the degree of difficulty to our task format we also applied a method based on the thought structure of the respective solution approach. That means that we classified the student solutions in the main solution approaches and analysed their thought complexity. In doing so, we tried to detect the single thought steps, referring to the *Simplex-Komplex*-idea of Breidenbach (1969). According to Breidenbach, a Simplex is a task consisting of three items and every item can be determined by the two others (p. 180). However, for the study we used the more broadly defined term *thought operation* to be able to include also solution steps which do not fit in the Simplex structure, such as addition of three values or duplication. In other words, we break the solution approaches into their individual components and establish a relation between the order of thought operations and cognitive demand. By analyzing the main solution approaches for thought operations, we can reveal single cognitive steps which can be illustrated in a kind of arithmetic tree. Combining this idea with Cognitive Load Theory (CLT), which describes the load of cognitive resources in the working memory (cf. Van Gerven 2003, p. 490), leads to a promising method to determine the degree of difficulty of holistic mathematical modelling tasks. Especially actions to be processed simultaneously (which are characterised in the following as the *width of the solution*) stress the working memory (element interactivity) (cf. Sweller 2010, p. 41). That means that several aspects in a task which are related to each other and have to be considered and understood at the same time lead to a high cognitive load for the working memory (cf. Sweller 2010). Applying this theory to the arithmetic tree consisting of thought operations, especially the width of a solution approach, appears to be an integral part when investigating the degree of difficulty.

6.3 Design of the Pilot Study

The focus of this pilot study was not to assess the students' competency in mathematical modelling but to evaluate the adequacy of the tasks with regard to the following main questions:

1. Is the working time adequate?
2. Which solution approaches are used?
   (How many distinguishable and adequate solution approaches are used?)
3. What is the degree of difficulty?
4. What is the solution rate?

To come closer to the answer of what is a metrologically reasonable modelling task, we analysed the student solutions in detail in view of what solution approach had been used and whether the solution approach led to a reasonable solution. In addition, we had a closer look at how the solution process was performed by identifying its thought structure.

6.3.1 The Modelling Tasks

The piloting of a first booklet was conducted with grade 9 students from German grammar schools. Due to technical reasons two items (*Toothpaste* and *Taj Mahal*) have been tested by 108 and one item (*Potato*) by 79 students from different schools. The students had 60 min in total to solve the task under seatwork conditions. Additionally the time for one task was restricted to 20 min since we wanted to find out if the working time was adequate.

Toothpaste Item

Even little children learn how to brush their teeth and tooth brushing is part of our daily routine. Can you give a general formula for how many days a toothpaste lasts? Reason mathematically.

In the *Toothpaste* item the students are asked to come up with a general formula for how long a toothpaste tube lasts. We assume that the students have a mental picture of the dimensions of either the toothbrush or the toothpaste tube. With the given picture we try to facilitate making assumptions by indicating that both, toothbrush and toothpaste, are approximately equal in length.

The *Taj Mahal* item is more demanding in terms of reading requirements. The façade of the famous mausoleum, Taj Mahal, has to be cleaned and thus surrounded by scaffolding made of bamboo cane as is usual in India. The question is: How many metres of bamboo cane are necessary to do so?

Taj Mahal Item

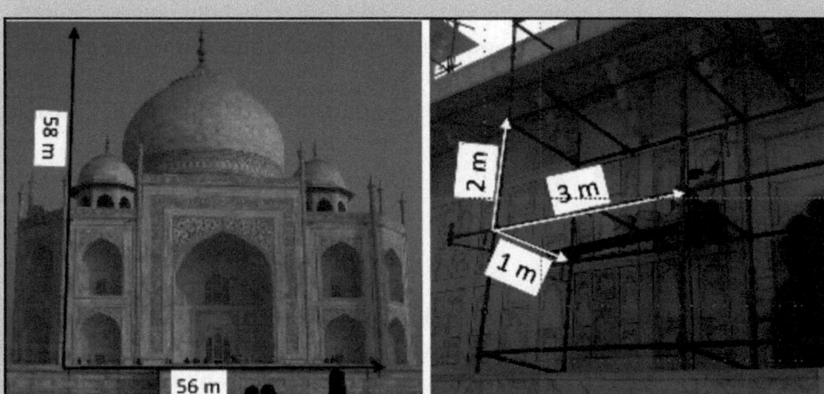

The mausoleum of Taj Mahal in India was built by a Great Mogul in honour of his wife. The base structure is essentially a large cube with chamfered corners. This white marble facade discolours over time due to

(continued)

air pollution. Therefore the fragile outer facade has to be cleaned regularly elaborately by hand. Traditionally Indians use scaffoldings made of bamboo canes for this. In a first step the main building (without dome) needs to be cleaned.

The question is how many metres of bamboo cane are necessary to surround the main building completely? Reason mathematically.

Inspired by a TV documentary about industrial manufactured French fries, the *Potato* item has been developed which asks: How many potato sticks can be obtained from one potato of standardised shape?

Potato Item

Industrial manufactured French fries are supposed to be equal in size and the single sticks are cut out lengthwise. Therefore the whole potato cannot be used. The potato tubers which look similar to the picture above, are regularly formed and approximately 10 cm in length.

How many of these potato sticks can be obtained from one potato? Reason mathematically.

6.4 Results

The focus of our investigations lies in developing suitable modelling tasks which can be used in a test setting. Therefore we analysed the student solutions in detail bearing in mind the issues posed in Sect. 6.3. After a first look through the student solutions, we classified them into several main solution approaches for each item. These main approaches then have been analysed separately and comparatively in detail. To be able to make a statement about the *degree of difficulty* we applied, on the one hand, the Simplex idea of Breidenbach (1969) and additionally the solution rate (See Sect. 6.2). To do so we determined the thought structure of each main solution approach individually and within an item. We then analysed their commonalities or differences to come to a general statement about the *degree of difficulty*. We analysed how many Simplex structures had to be processed one after another (*length of the solution approach*) and whether a simultaneous performing of Simplex structures was necessary (*width of the solution approach*). This method of determining the degree of difficulty has then been compared to the solution rate of the respective item.

6.4.1 Toothpaste Item

Having a look at the *Toothpaste* item, it became obvious that the student solutions can be classified into two main solution approaches (See Figs. 6.1 and 6.2).[2] Figure 6.1 characterises a procedure where the student interprets the toothpaste as a cylinder and estimates radius and height maybe based on the given picture and/or a mental picture. Then the volume of the toothpaste is calculated to be able to divide this result by the estimated daily toothpaste consumption.

In a different solution approach (See Fig. 6.2) the student estimates the toothpaste used either in ml or mg and assumes a number of teeth brushings per day. The total consumption per day is then calculated and divided by the previously estimated toothpaste used.

When analysing the thought structure behind these solutions, it becomes obvious that two thought operations (circles in Figs. 6.1 and 6.2) have to be performed consecutively in both approaches. That means that the approaches seem to be

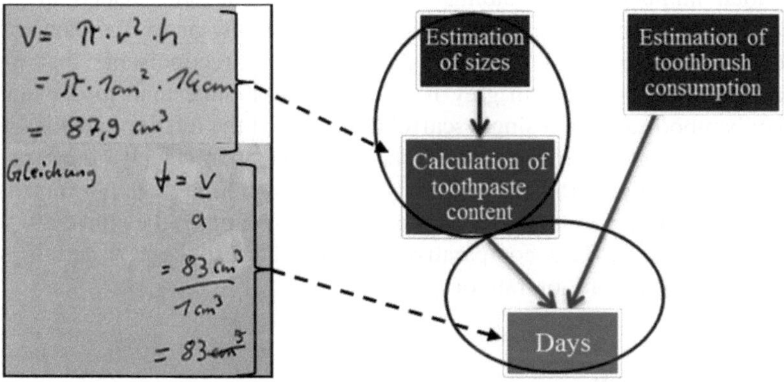

Fig. 6.1 *Item Toothpaste*: abstract of student solution 1 and its thought structure

[2] Solutions of the *Toothpaste* item will be explained more extensively and serve as an example for the setting up of the thought structure model for the other items.

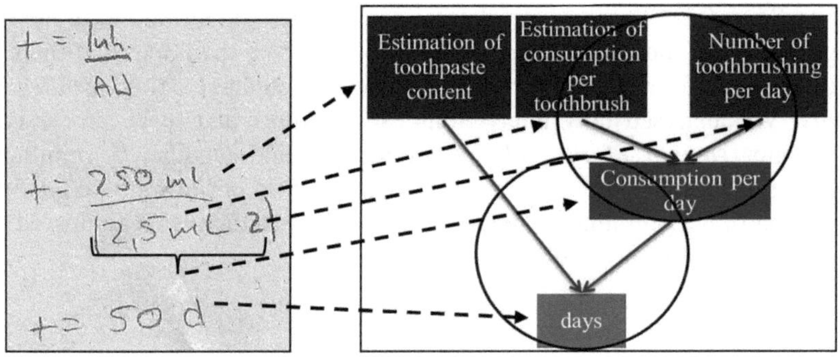

Fig. 6.2 *Toothpaste item*: abstract of student solution 2 and its thought structure

comparable in their degree of difficulty having a solution space with a width of 1 (one thought operation per solution level) and length of 2 (two thought steps have to be conducted consecutively). This suggests a relatively small degree of difficulty which is in line with a rather high solution rate of 76 %.

6.4.2 Taj Mahal Item

The analysis of *the Taj Mahal* item indicates that there are also two main solution approaches. In the following we only briefly describe the solution strategy of the two approaches. The general concept of the two approaches is the calculation of the length of bamboo cane needed to build one scaffolding on the one hand (i.e., unshaded left side of both parts of Fig. 6.3), and on the other hand, the calculation of the total number of scaffoldings needed (i.e., right side shaded in grey in Fig. 6.3). Multiplication leads to an answer of how many metres of bamboo cane are needed to surround the Taj Mahal. It is clearly recognisable that in both approaches the part shaded in grey is a lot more elaborate than to calculate the length of bamboo cane for a single scaffolding. Even if the left approach in Fig. 6.3 requires one thought operation more than the right approach, we can assume a comparable degree of difficulty since this thought operation does not increase the maximal width of the approach. That means that both approaches have a width of 3 and a length of 4. This comparatively complicated thought structure is also reflected by the low solution rate of 24 %.

6.4.3 Potato Item

When analysing the solution approaches of *Potato* item two main approaches became apparent. Both approaches show the same thought structure (See Fig. 6.4). Comparing the thought structure to that of the *Toothpaste* item similarities

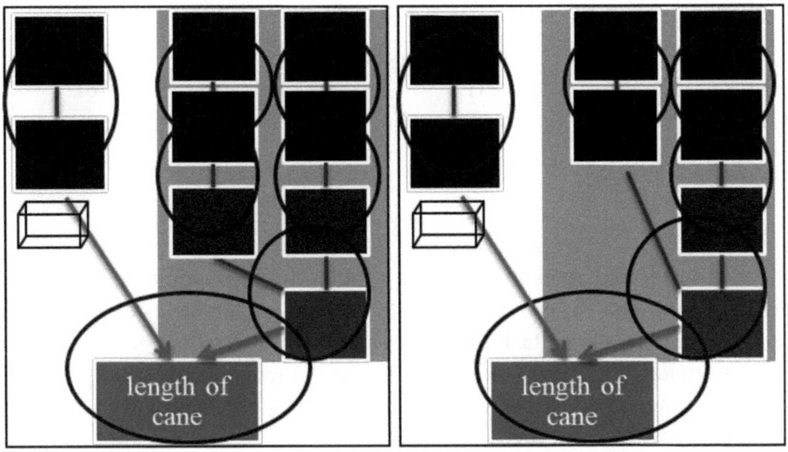

Fig. 6.3 *Item Taj Mahal*: thought structure of the two main solution approaches

Fig. 6.4 *Potato item*: thought structure of the two main solution approaches

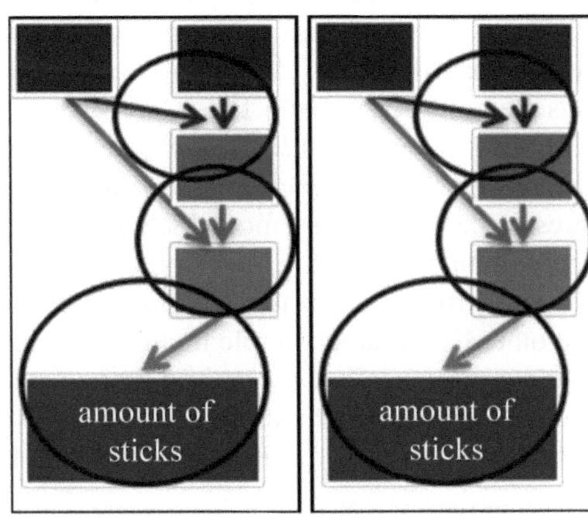

are obvious. This item also shows a width of one but a length of 3. A solution rate of 40 %, which is in between those of *Taj Mahal* and *Toothpaste*, supports the findings of the thought structure method.

6.5 Discussion

In the following we want to interpret the results with reference to the main questions of Sect. 6.2. With regard to the working time it became obvious that 20 min per task is too long which reduces the total working time needed for a booklet. This is in fact

Table 6.1 Thought structure and solution rate of the items

Item	*n*	Thought structure (length, width)	Solution rate (%)
Toothpaste	108	(2,1)	76
Taj Mahal	108	(4,3)	24
Potato	79	(3,1)	40

a positive observation since 60 min for a booklet might be too demanding in terms of fatigue phenomena. It was unclear if the tasks meet the criterion of openness. It was reasonable to assume a task space that allows for more than one distinguishable solution approach but it was not clear if the students would really apply these individually. Since at least two solution approaches per task were found, we can assume that all criteria are met.

One aim is to determine the degree of difficulty in order to be able to weight each solution approach by its difficulty to arrive at an exact assessment. It is reasonable that a dichotomous rating does not account for the true complexity of the tasks; therefore the solution approaches have been represented as an arithmetic tree, illustrating the necessary thought operations. These illustrations enabled us to reveal their degree of difficulty. A solution rate has been determined to support this method (See Table 6.1).

Following Sweller (2010, p. 41), in particular the width of a solution approach is responsible for the *degree of difficulty* since a simultaneous processing of data stresses the working memory (See Sect. 6.2.2). This is also reflected in our results when thought structure and solution rate are compared. A more complicated thought structure as in the *Taj Mahal* item corresponds to a low solution rate. On the other hand, rather simple thought structures, in particular those with a width of one such as the *Toothpaste* and *Potato* items, have a higher solution rate. The thought structure method seems to be a promising method to support the determination of the degree of difficulty, as it is in line with the solution rates.

6.6 Outlook

Although further tests have to be performed, these data suggest that using the thought structure idea on the basis of cognitive load theory may be a suitable method to determine the *degree of difficulty*. A further question now lies in how far the results of length and width can contribute to assessing the student solutions. It would be desirable to use this method to weight the student performance and thus, to not only assess the result but also the solution process in an adequate way.

References

Blum, W. (2006). Die Bildungsstandards Mathematik, Einführung. In W. Blum, R. Drüke-Noe, & O. Köller (Eds.), *Bildungsstandards Mathematik: konkret, Sekundarstufe 1: Aufgabenbeispiele, Unterrichtsanregungen, Fortbildungsideen* (pp. 14–32). Berlin: Cornelsen Verlag Sciptor GmbH & Co. KG.

Breidenbach, W. (1969). *Methodik des Mathematikunterrichts in Grund- und Hauptschulen. Band 1 – Rechnen*. Hannover: Hermann Schroedel Verlag KG.

Frejd, P. (2012). Modelling assessment of mathematical modelling – A literature review. In C. Bergsten, E. Jablonka, & M. Raman (Eds.), *MADIF-8: Evaluation and comparison of mathematical achievement* (pp. 81–90). Linköping: Svensk förening för MatematikDidaktisk Forskning – SMDF.

Haines, C., Crouch, R., & Davis, J. (2000). *Mathematical modelling skills: A research instrument* (Technical report No. 55). Hatfield: University of Hertfordshire, Department of Mathematics.

Maaß, K. (2007). *Mathematisches Modellieren – Aufgaben für die Sekundarstufe*. Berlin: Cornelsen Verlag Scriptor.

Müller, M., Leiß, D., Schukajlow, S., Blum, W., & Messner, R. (2007). Auswendiggelernt – Abgehackt – Abgefragt? In *Beiträge zum Mathematikunterricht* (pp. 723–726). Hildesheim: Franzbecker.

OECD. (n.d.). *Programme for International Student Assessment (PISA)*. Retrieved March 15, 2013, from http://www.oecd.org/pisa/

Reit, X.-R., & Ludwig, M. (2013). Wege zu theoretisch fundierten Testaufgaben zur. In G. Greefrath, F. Käpnick, & M. Stein (Eds.), *Beiträge zum Mathematikunterricht 2013* (pp. 805–808). Münster: WTM-Verlag.

Sweller, J. (2010). Cognitive load theory: Recent theoretical advances. In J. L. Plass, R. Moreno, & R. Brünken (Eds.), *Cognitive load theory* (pp. 29–47). New York: Cambridge University Press.

Van Gerven, P. W. (2003). The efficiency of multimedia learning into old age. *British Journal of Educational Psychology, 73*(4), 489–505.

Vos, P. (2011). What is 'authentic' in the teaching and learning of mathematical modelling? In G. Kaiser, W. Blum, R. Borromeo Ferri, & G. Stillman (Eds.), *Trends in teaching and learning of mathematical modelling (ICTMA 14)* (pp. 713–722). New York: Springer.

Zöttl, L., Ufer, S., & Reiss, K. (2011). Assessing modelling competencies using a multidimensional IRT approach. In G. Kaiser, W. Blum, R. Borromeo Ferri, & G. Stillman (Eds.), *Trends in teaching and learning of mathematical modelling (ICTMA 14)* (pp. 427–437). New York: Springer.

Chapter 7
Facilitating Mathematisation in Modelling by Beginning Modellers in Secondary School

Gloria Ann Stillman, Jill P. Brown, and Vince Geiger

Abstract Based on theoretical considerations, a possible means of gaining a resolution of the long standing issues of problem formulation and specification and their successful mathematisation by relatively naïve modellers is proposed. This is based on empirical evidence having been provided for paradigmatic cases of the construct of implemented anticipation as proposed by Niss, that is, foreshadowing of future action projected back onto decisions about current action during ideal mathematisation. A Foreshadowing and Feedback Framework to Engage Beginning Modellers in Implemented Anticipation and its theoretical underpinnings are outlined and illustrated using empirical data.

7.1 Introduction

At ICTMA 12 in 2005 the beginning of a research agenda focussing on secondary students' learning and applying modelling skills in a Technology-Rich Teaching and Learning Environment (TRTLE) was highlighted (Galbraith et al. 2007). The specific aim was to learn more about the critical points that represent transitions between phases in the solution of a modelling endeavour. A direct outcome of this early work was a blockages framework (Galbraith and Stillman 2006). This has since been extended (Stillman et al. 2010) and validated by ourselves

G.A. Stillman (✉)
Education VIC, Faculty of Education & Arts, Australian Catholic University,
1200 Mair Street, Ballarat, VIC 3350, Australia
e-mail: gloria.stillman@acu.edu.au

J.P. Brown
Education VIC, Faculty of Education & Arts, Australian Catholic University,
Level 1, 250 Victoria Parade, Melbourne, VIC 3065, Australia
e-mail: jill.brown@acu.edu.au

V. Geiger
Education QLD, Faculty of Education & Arts, Australian Catholic University,
1100 Nudgee Road, Brisbane, Banyo, QLD 3014, Australia
e-mail: vincent.geiger@acu.edu.au

G.A. Stillman et al. (eds.), *Mathematical Modelling in Education Research and Practice*, International Perspectives on the Teaching and Learning of Mathematical Modelling, DOI 10.1007/978-3-319-18272-8_7

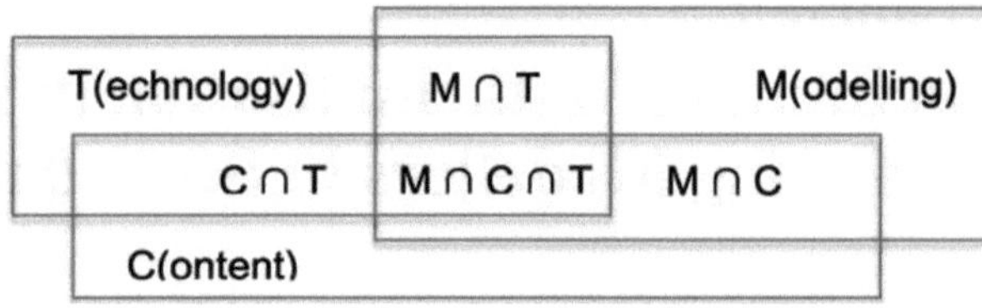

Fig. 7.1 Modelling (M), mathematics content (C), and technology (T) interactions (Galbraith et al. 2007, p. 131)

(Stillman et al. 2007) and others (Huang 2012). If we use a different lens on the same transitions looking at what is possible rather than what blocks progress, we believe that we might gain traction on the long standing issues of problem formulation and specification and their successful mathematisation by relatively naïve modellers.

The notion of *anticipation* as a key form of mathematical thinking is central to resolving these issues. The complexity of the interactions between mathematical content, modelling meta-knowledge and technology that students cope with when successfully modelling in a TRTLE (see Fig. 7.1) emphasise the importance of student perception and judicious enactment of affordances offered in a TRTLE (Brown 2013). In more recent work Stillman and Brown (2012, 2014) have investigated whether there is empirical evidence for the Niss notion of "implemented anticipation" (2010, p. 55). Furthermore we propose that this idea be extended to technology use when modelling in TRTLEs. In this chapter a scaffolding framework that novice modellers in school could use during the transitions associated with problem specification, formulation and mathematisation will be presented. This framework is grounded in the theoretical argument and empirical evidence to date.

7.2 Theoretical Frame

Researchers in the area of applications and mathematical modelling use diagrams of the modelling cycle as a means of communicating what appears to be happening at a task and cognitive level during modelling. These diagrams are also used to scaffold modelling activity of beginner modellers. The Niss (2010) modelling diagram separates the extra-mathematical domain (i.e., the real world situation and the modeller's idealisation of this) and the mathematical domain. Finding or generating a problem from a real world messy situation is a crucial cognitive step (Getzels 1979). Once a problem has been found it needs to be posed. This happens when a problem is formulated in such a way that it is amenable to mathematical analysis (Stillman 2015). Idealisation (formulation of an ideal problem from the real situation) occurs through making assumptions and identifying essential elements or features in the situation which are of interest that are then formulated, that is, specified into a problem statement which may take the form of, or include, question(s). The idealised situation is mathematised through translation into the mathematical domain. The mathematical domain includes the mathematical model made of the situation, mathematical questions posed and mathematical artefacts (e.g., graphs and tables) used in solving the mathematical model. Mathematical outputs (i.e., answers) have then to be interpreted in terms of the idealised situation and the real situation that stimulated the modelling (i.e., back into the extra-mathematical domain). These outputs can then answer questions posed about the real situation or, if they are inadequate for this purpose, stimulate further modelling.

Anticipation as a thought form has a long genesis in educational literature. Dewey (1916, p. 83) writes of "conjectured anticipation" as involving "a tentative interpretation of the given elements, attributing them to a tendency to effect certain consequences." Furthermore he recognised that it has both "projection" and "prospective" aspects (Dewey 1917, p. 13). Both aspects come into play when engaging in mathematical modelling. Within the context of mathematical modelling we define anticipating as the foreseeing of what will be useful mathematically subsequently in transitions between phases of the modelling process. This anticipation involves both foreshadowing and feedback loops between phases informing decision making. In order to produce a theoretical model of the mathematisation process, Niss (2010) used this idea but coined the term, "implemented anticipation" (p. 54). The use of the past tense in "implemented" was deliberate as successful mathematisation, from the Niss perspective, involves not only anticipating what will be useful mathematically in subsequent steps of the cycle but also implementing that anticipation in decision making and carrying through of actions bringing those next steps to fruition. He is thus harnessing the potential for guiding future actions that Dewey saw in anticipation when he wrote: "Imaginative forecast of the future is this forerunning quality of behaviour rendered available for guidance in the present" (Dewey 1917, p. 13).

Figure 7.2 is an interpretation of the *Niss model of ideal mathematisation*. Firstly, the idealisation and specification of the real situation from the extra-mathematical domain involves deciding what elements or features are essential as well as posing any related question or statement of the problem in light of their anticipated usefulness in mathematising. Secondly, when mathematising this formulation of the problem situation the modeller needs to do this by anticipating

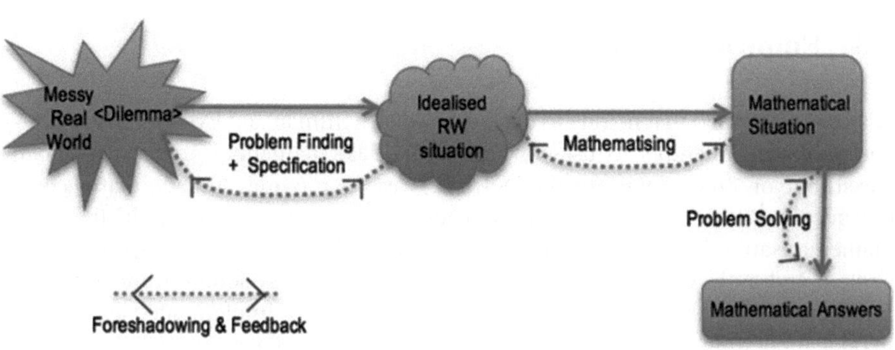

Fig. 7.2 Niss's model of ideal mathematisation

mathematical representations and mathematical questions that, from previous experience, have been successful when put to similar use. Thirdly, when anticipating these mathematical representations, the modeller has to be cognisant of the utility of the selected mathematisation and the resulting model in future problem solving processes to provide mathematical answers to the mathematical questions posed by the mathematisation. This thus involves anticipating mathematical procedures and strategies to be used in problem solving after mathematisation is complete.

It is our contention that "implemented anticipation" applies not only to mathematics and mathematical artefacts but also to use of digital tools. We hypothesize that at any of these transitions, anticipated use of digital tools in effecting use of representations or solving procedures could also come into play (Galbraith et al. 2007) through perception of affordances of such technology (Brown 2013). The dotted double arrowed paths in Fig. 7.2 represent this three step foreshadowing and feedback that are captured in successful implemented anticipation. There is an obvious correspondence between this foreshadowing of the results of future actions being "projected back onto current actions" (Niss 2010, p. 55) and the idea that a "sense of direction" is of crucial importance in modelling (Maaß 2006; Treilibs 1979). Furthermore, we hypothesize that ideal mathematisation occurs when the relationship between the modeller and modelling and technology is that of the artisan where both the modelling and technology become an "Extension-of-Self" (Geiger 2009).

Niss (2010) identifies four enablers to successfully use implemented anticipation in mathematising a real or realistic situation. These are that modellers need to: (1) possess relevant mathematical knowledge, (2) be capable of using this when modelling, (3) believe a valid use of mathematics is modelling real phenomena, and (4) have persistence and confidence in their mathematical capabilities (p. 57). It is reasonable to expect that beginning modellers would experience the challenge of ideal mathematising and have difficulties related to the three aspects of implemented anticipation. Furthermore, these difficulties could possibly be explained by lack of one or more of these four enablers.

7.3 Empirical Evidence

Niss has confirmed that no one to date, other than Stillman and Brown (2012, 2014), has published attempts to use his model of ideal mathematisation in analysing classroom or other data. As the purpose was to provide "paradigmatic cases" (Freudenthal 1981, p. 135) for the existence of the Niss construct in theorising mathematisation in modelling tasks, data examined came from an open modelling context and a classroom context where relative novice modellers were participating. Empirical evidence was found for all three aspects of implemented anticipation in open modelling situations with Year 10 and 11 students and for the last two in the classroom data analysed (Stillman and Brown 2014) where students were not given the freedom to choose their own situation to model. In Stillman and Brown (2014) we claim to have presented a set of exemplars from practice that form the basis for our paradigmatic cases for the existence of the Niss construct in theorising mathematisation in modelling or realistic tasks.

Unsuccessful mathematisation attempts were able to be explained using the Niss enablers (Stillman and Brown 2014). In the open modelling situation the flaw in mathematisation in one exemplar project was due to lack of relevant mathematical knowledge as what was required was beyond the students' current knowledge. In the classroom context, unsuccessful mathematisation attempts were related to inability of the particular students to use relevant mathematical knowledge in modelling rather than lack of the other enablers. This was understandable for Year 9 students being introduced to modelling attempting only their third extended modelling task. In both contexts, successful implemented anticipation coincided with all four enablers being present; however, an alternative interpretation in the case of the third enabler, believing a valid use of mathematics is modelling real phenomena, is that students engage with tasks in classrooms as that is the expected norm. Further research is thus needed regarding the enablers.

7.4 Proposed Framework

Vygotsky raises the paradox that "the method [of enquiry] is simultaneously pre-requisite and product, the tool and the result of the study" (1978, p. 65). This is indeed the case for implemented anticipation. Implemented anticipation is a pre-requisite construct that must be suggested theoretically before we could use it to find empirical evidence for its existence in practice. Holzman (1997, p. 52) takes this idea further commenting that "Tool-and-result come into existence together, their relationship is one of dialectical unity, rather than instrumental duality". Thus, as beginning modellers engage in mathematisation of a messy situation implemented anticipation is manifested simultaneously with the tool by which that implemented anticipation can be evidenced, namely, dialogue.

Our overall goal is for students to be *talking and doing* mathematical modelling – *gaining leverage on the mathematisation processes*. In schooling modelling is usually conducted in groups through collective reasoning and activity. We therefore theorise that during this collective engagement with the problem situation and each other, students are drawn into the articulated thinking, reasoning, manifestation of beliefs about modelling and mathematics and confidence in mathematical knowledge and ability to use this, of others within the group. Mercer (2000) uses the term "interthinking" (p. 141) to refer to this "pulling students into a shared communicative space" (Hunter 2012, p. 3). Mathematising is manifested as the group works on the modelling task engaging in *interthinking* and developing a group mathematisation of the situation. Our focus is on how these modellers can be facilitated into the discourse of modelling and mathematisation. We propose a framework to engage beginning modellers in implemented anticipation (see

PROBLEM FINDING (F) & SPECIFICATION (S) FORESHADOWING & FEEDBACK LOOP
F1 What situation are we interested in?
S1 What features/elements of the situation are we interested in?
S2 Which of these features/elements are relevant {irrelevant} {essential}?
S3 What criteria could we use to determine {relevance} {"essential"}? What could we mathematise?
S4 What questions could we pose? Will these be useful in mathematising?
MATHEMATISING (M) FORESHADOWING & FEEDBACK LOOP
M1 How could we mathematize?
M2 What representations {models} could we use?
M3 What representations have any of us used before in something similar?
M4 What mathematical questions could we ask?
M5 What has been successful in the past?
M6 What technological tools {do we know} {are available}?
M7 What would these tools enable us to do?
PROBLEM SOLVING (P) FORESHADOWING & FEEDBACK LOOP
P1 If we use that representation how will it allow us to work mathematically?
P2 If we use that model what mathematics would we need to know or find out how to use?
P3 If we use this model how does that effect what features we mathematized?
P4 If we use this model how does that effect what features of the situation we thought {relevant} {essential}?

Fig. 7.3 Foreshadowing and feedback framework to engage beginning modellers in implemented anticipation

Fig. 7.3) drawing on our extension of the theoretical ideas of Niss. The framework consists of scaffolding questions related to the foreshadowing and feedback loops (Fig. 7.2) associated with problem finding (F) and specification (S), mathematising (M) and problem solving (P). Group members could ask these of each other in order to develop a sense of direction for their modelling. Words in curled brackets are alternatives that could be asked.

7.5 Illustrative Example

To illustrate how this framework might be applied by students an example is presented that is based on data (video, digital photographs, written scripts, poster and questionnaire responses) from an open modelling context where students chose the task and solution pathways (see Stillman et al. 2013 for further details). Potential links to the framework are shown as alphanumeric codes such as F1 meaning the scaffolding question "What situation are we interested in?" as per Fig. 7.3 could be of use at this point in group decision making. At the 2010 AB Paterson Mathematical Modelling Challenge, the first two authors mentored 46 students in 13 mixed groups of 3–4 students from different schools in Queensland and/or Singapore in one large room. One group comprising Tim[1] (an Australian Year 11 student) and 2 Singaporean students (Becky, Year 11, and Colin, Year 10) was purposefully chosen to illustrate how the framework might be used in a

[1] Student names used throughout this chapter are pseudonyms.

TRTLE. These students used 3 laptops and remained in the room the entire time allocated to working on their modelling situation (i.e., 9 h over two consecutive days).

Group members were interested in "the feasibility of Hollywood stunts and whether it is possible...to perform this [sic] stunts in reality" [Questionnaire, Video Day 2] [F1]. One stunt they chose to model was "a scene from a movie, *The Man with the Golden Gun*, and [they were] looking at the car scene in which the car took off from the bridge, spins 360° ... before landing on the other half of the bridge" [Becky, Video Day 2]. The students had visited Dreamworld theme park earlier in the week to gather and analyse data from various rides using technological tools such as 3D accelerometers and real world interface software like Logger Pro [M6, M7]. All three had some knowledge of these tools and their affordances but only Tim was expert in their application as Becky and Colin had not used them before this event. The question the group posed was: "*What is the speed of a car needed to jump across a broken bridge?*" [Questionnaire] [S4, M4]. The features deemed relevant were launch angle, car length, and width of gap in the bridge [S1, S2, S3].

Analysis of still images from the movie DVD was used to calculate their launch angle [M5, M6, M7]. Tim demonstrated he was fully aware of the technical conditions needed for the Logger Pro software to be used reliably to determine angles and distances. He was also aware of the mathematics required and that arbitrary units could be used (see Fig. 7.4a).

Tim: Fortunately enough, the video camera at one of the points is perpendicular to the motion in the car, which is imperative when you are analysing these kind of things. So we are able to take the angle of launch straight off the video (see Fig. 7.4a). Just your simple trigonometry to find out that your angle is 23.8°. [Video, Day 2]

Logger Pro was also used to find the gap in the bridge but this time they used the length of the car to form a reference scale [M7].

Tim: So fortunately enough we had accessed the Logger Pro software, went on-line and found that the length of the car is 4.71 m, set that as our scale and then measured that from two other points on the video which gave us a length of 17.06 m for the car to jump across (see Fig. 7.4b). [Video Day 2]

a **b**

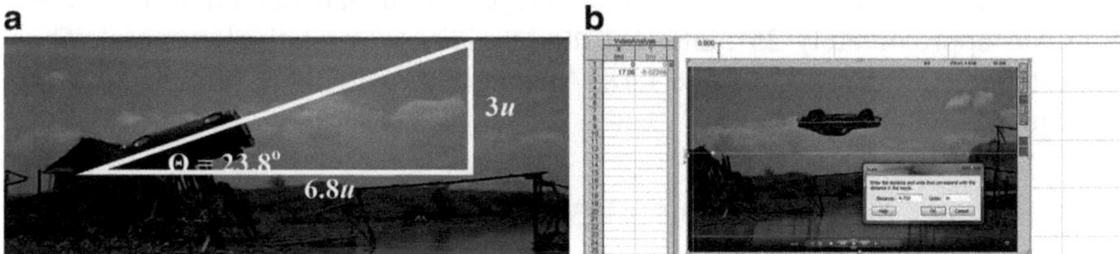

Fig. 7.4 Using Logger Pro software on DVD image to (**a**) calculate angle of launch and (**b**) find length of gap

The group made use of physics concepts related to projectile motion to construct their model of the motion of the car in flight across the gap [M1, M2, M3, M5, P1, P2]. The rotation of the car was ignored [P4].

Tim: Of course, it is easiest to do projectile motion through parametric equations. We see a car jumping off at an angle of say, 24°. Its motion can then be split into x and y components so we have done that by combining the velocity with the angle it jumped off at. . . . And also we have assumed. . .that gravity is equal to 9.8 m per sec [sic] so we have integrated that to see how it affects the displacement, which is -4.9 t^2; then combined that into a parametric equation. [Video, Day 2]

Using a parametric equation the motion of the car in both the x and y components as a function of time is thus modelled as follows: $x(t) = (vcos\theta)t$ and $y(t) = (vsin\theta)t - 4.9t^2$.

However, rather than ignore air resistance the group decided to incorporate this so there were other relevant features of the situation they now needed to include in their problem specification and its mathematisation [P3, P4].

Tim: We have also decided to include air resistance because that is a really large factor that affects projectiles. It is usually negligible but we decided to include it for a far more accurate representation [Video, Day 2].

Their reason for including it was "in reality, the motion of objects is impeded by the drag force" [Poster]. This can be quantified through the drag equation, $F_{drag} = \frac{1}{2} C_d A\rho v^2$ where v is the velocity of the car, ρ is the density of the air which the car is flying through, and A is the reference area, that is, the surface area of the car which is directly travelling through the air [M2, M3]. The AMC driver's manual gave the coefficient of drag of the 1975 Hornet, the car used in the movie, as 0.32. To determine the reference area the front design of the car was modelled by a trapezium and a rectangle (Fig. 7.5). The width was sourced from the Driver's Manual on the internet as 1.793 m. All other measurements were taken using this scale (see Fig. 7.5) in Logger Pro [M7]. The reference area was therefore calculated as

$$(b \times w) + h\left(\frac{a+b}{2}\right) = (1.793 \times 0.448) + 0.583\left(\frac{1.793 + 1.255}{2}\right) = 1.69m^2$$

Basing their argumentation on the film being shot in Thailand and there being "mild cloud in the sky" on the DVD, the group suggested this indicated that the temperature was approximately 25°, and that the humidity was relatively normal. Thus, they estimated air density, ρ, to be 1.18. All constants were then combined into one constant, k.

$$k = \frac{1}{2}C_d A\rho = 0.5 \times 0.32 \times 1.18 \times 1.69 = 0.32$$

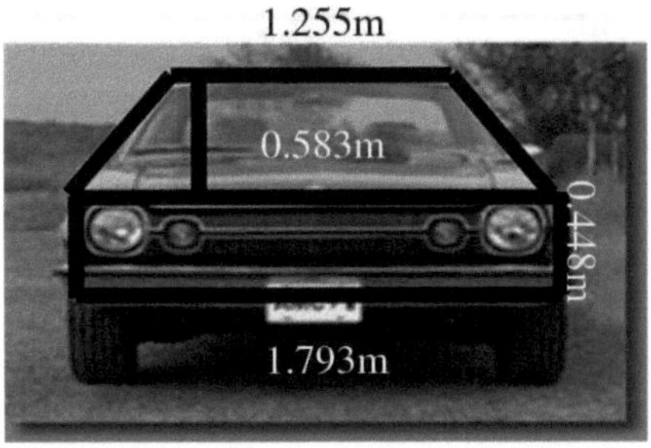

Fig. 7.5 Group's modelling of front design of car for reference area A in drag equation

This gave a drag force, $F_{drag} = kv^2 = 0.32v^2$, which was then incorporated into their parametric equation, but for this to occur acceleration has to be included [S1, S2]. As air resistance impedes the motion in all directions, it is considered in both x and y components. Using Newton's second law, $a = \frac{F}{m}$. The mass of the car with driver was specified by the Special Features section of the DVD as 1,460 kg. Thus, $a = \frac{0.32v^2}{1,460}$. This was then incorporated into their parametric equation as: $x(t) = (v\ cos\theta)t - \frac{0.32(v\cos\theta)^2}{1,460}t^2$ and $y(t) = (v\ sin\theta)t - \left(4.9 - \frac{0.32(v\sin\theta)^2}{1,460}\right)t^2$

As determined earlier using Logger Pro, $x = 17.06$ m, and $\theta = 23.8\ ^{\circ}$ giving: $17.06 = (v\ cos23.8)t - \frac{0.32(v\cos 23.8)^2}{1,460}t^2$. To solve this equation they had anticipated the utility of the affordances of the TI-Nspire computer program for this purpose [M7] (see Fig. 7.6).

Tim: We eventually incorporated the deceleration due to air resistance into the parametric model as well which eventually allowed us to draw this graph (Fig. 7.6). The TI-Nspire is a really good program to allow you to do things like this because I could include sliders which would easily allow me to modify the velocity, angle and all the different components that affect drag.

The output from the computer program was then demathematised by the group as "travelling at 15.3 ms^{-1} which I think equates to 56 or 57 kmh^{-1}, so that is certainly achievable" [Tim, Video 2]. They concluded that the stunt was possible.

7.6 Discussion and Conclusion

Our purpose in this chapter has been to propose a framework of scaffolding questions for beginning modellers related theoretically to the foreshadowing and feedback loops associated with operationalising the construct, implemented anticipation, during ideal mathematisation (Niss 2010). As Willemain and Powell (2007) have documented in researching the differences between novice and expert modellers, it is not that novices do not discuss during modelling, rather it is the focus of that discussion. Novices in their study "spent more time talking about the

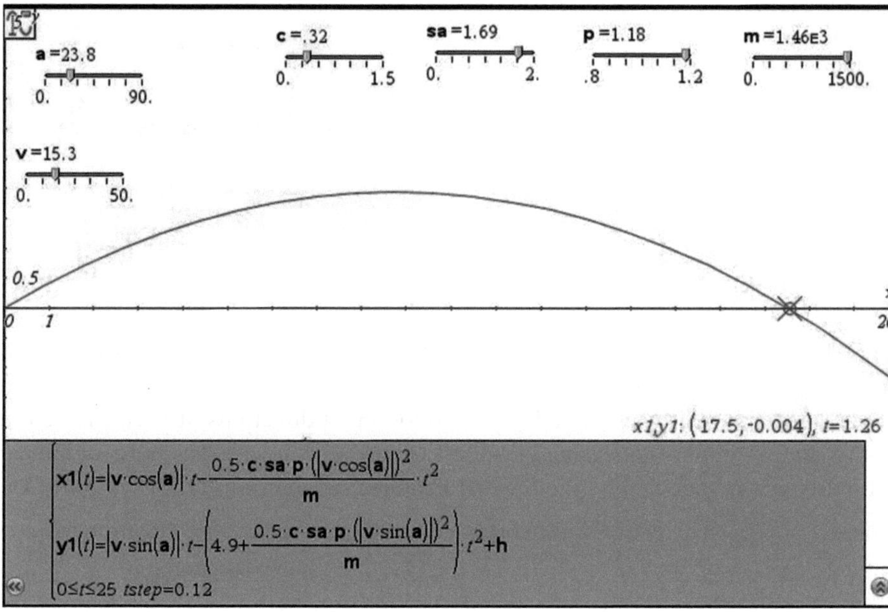

Fig. 7.6 Using sliders in TI-Nspire software to calculate velocity from their final parametric model

problem than building a model and little time assessing their progress" (p. 1280). Framework questions are not meant just to generate rich group discussion in interthinking (Mercer 2000) but rather to facilitate productive discussion that moves novices forward from a problem context focus to actualising model construction. Our claim is that each question can be used at a decision making point to facilitate the discussion moving towards this goal by activating foreshadowing and feedback loops. The feedback, for example, from a foreshadowed refinement to the model of the flight of the car to include air resistance highlights this framework aspect. Once this refinement has been suggested its impact on the relevance of situation features needs to be discussed. Questions P3 and P4 come into play raising group awareness of this feedback effect and providing a means to progress the model refinement rather than it remain at the level of suggestion.

As we believe modelling is enhanced for beginning modellers in a TRTLE we chose an open modelling example where the group perceived and enacted affordances of digital tools to conduct their modelling. They were operating in the space where the interactions shown in Fig. 7.1 were coming to the fore but at their own instigation. Although choosing a situation to model that would allow use of the analysis software they had used recently seems an obvious choice, this was the only group who did so. Real world interface software affords novice modellers the concrete experience of measuring angles and distances directly from still images of real objects frozen in motion. Optimum use of such tools requires technical knowledge as Tim indicated but tool availability makes modelling accessible to a broader spectrum of students. Thus questions M6 and M7 are necessary in the framework if schooling is to keep pace with technological advances in society and workforce changes where model creation and use are no longer left to experts (Willemain and Powell 2007).

The main contribution of this chapter for the potential resolution of issues related to problem formulation and specification and their successful mathematisation by novice modellers is that it transforms and extends Niss's implemented anticipation within his model of ideal mathematisation into an operationalized scaffold so that it can be used by such modellers. The framework related to the foreshadowing and feedback loops shown in Fig. 7.2 facilitates group members asking each other scaffolding questions to develop a sense of direction for their modelling (Treilibs 1979). The framework both facilitates and exposes interthinking (Hunter 2012; Mercer 2000) but with the added purpose of enhancing the effectiveness of discussion in leading to model construction through appropriate mathematisation processes.

References

Brown, J. (2013). *Perceiving and enacting the affordances of Technology-Rich Teaching and Learning Environments (TRTLE's) for student understanding of function*. Unpublished doctor of philosophy thesis, University of Melbourne.

Dewey, J. (1916). *Democracy and education*. New York: Macmillan.

Dewey, J. (1917). The need for a recovery of philosophy. In J. Dewey (Ed.), *Creative intelligence: Essays in the pragmatic attitude* (pp. 3–69). New York: Holt.

Freudenthal, H. (1981). Major problems of mathematics education. *Educational Studies in Mathematics, 12*(2), 133–150.

Galbraith, P., & Stillman, G. (2006). A framework for identifying student blockages during transitions in the modelling process. *Zentralblatt für Didaktik der Mathematik, 38*(2), 143–162.

Galbraith, P., Stillman, G., Brown, J., & Edwards, I. (2007). Facilitating middle secondary modelling competencies. In C. Haines, P. Galbraith, W. Blum, & S. Khan (Eds.), *Mathematical modelling: Education, engineering and economics* (pp. 130–140). Chichester: Horwood Press.

Geiger, V. (2009). The master, servant, partner, extension-of-self framework in individual, small group and whole class contexts. In R. Hunter, B. Bicknell, & T. Burgess (Eds.), *Crossing divides* (pp. 201–208). Adelaide: MERGA.

Getzels, J. W. (1979). Problem finding: A theoretical note. *Cognitive Science, 3*(2), 167–179.

Holzman, L. (1997). *Schools for growth: Radical alternatives to current educational models*. Mahwah: Lawrence Erlbaum.

Huang, C.-H. (2012). Investigating engineering students' mathematical modelling competency. *World Transactions on Engineering and Technology Education, 10*(2), 99–104.

Hunter, R. (2012). Coming to "know' mathematics through being scaffolded to 'talk and do' mathematics. *International Journal for Mathematics Teaching and Learning, 14*. http://www.cimt.plymouth.ac.uk/journal/hunter2.pdf

Maaβ, K. (2006). What are modelling competencies? *ZDM, 38*(2), 113–142.

Mercer, N. (2000). *Words & minds*. London: Routledge.

Niss, M. (2010). Modeling a crucial aspect of students' mathematical modeling. In R. Lesh, P. Galbraith, C. R. Haines, & A. Hurford (Eds.), *Modelling students' mathematical competencies* (pp. 43–59). New York: Springer.
Stillman, G. (2015). Problem finding and problem posing for mathematical modelling. In N. H. Lee & K. E. D. Ng (Eds.), *Mathematical modelling: From theory to practice* (pp. 41–56). Singapore: World Scientific.
Stillman, G., & Brown, J. (2012). Empirical evidence for Niss' implemented anticipation in mathematizing realistic situations. In J. Dindyal, L. P. Cheng, & S. F. Ng (Eds.), *Mathematics education: Expanding horizons* (Vol. 2, pp. 682–689). Adelaide: MERGA.
Stillman, G., & Brown, J. (2014). Evidence of *Implemented Anticipation* in mathematising by beginning modellers. *Mathematics Education Research Journal, 26*(4), 763–789.
Stillman, G., Brown, J., & Galbraith, P. (2010). Identifying challenges within transition phases of mathematical modeling activities at year 9. In R. Lesh, P. Galbraith, C. R. Haines, & A. Hurford (Eds.), *Modeling students' mathematical competencies* (pp. 385–398). New York: Springer.
Stillman, G., Brown, J., & Galbraith, P. (2013). Challenges in modelling challenges: Intents and purposes. In G. A. Stillman, G. Kaiser, W. Blum, & J. P. Brown (Eds.), *Teaching mathematical modelling: Connecting to research and practice* (pp. 217–227). Dordrecht: Springer.
Stillman, G., Galbraith, P., Brown, J., & Edwards, I. (2007). A framework for success in implementing mathematical modelling in the secondary classroom in mathematics. In J. Watson & K. Beswick (Eds.), *Mathematics: Essential research, essential practice* (Vol. 2, pp. 688–707). Adelaide: MERGA.
Treilibs, V. (1979). *Formulation processes in mathematical modelling*. Unpublished Master of Philosophy, University of Nottingham.
Vygotsky, L. S. (1978). *Mind in society*. Cambridge, MA: MIT Press.
Willemain, T. R., & Powell, S. G. (2007). How novices formulate models. Part II: A quantitative description of behaviour. *Journal of the Operational Research Society, 58*, 1271–1283.

Chapter 8
Authenticity in Extra-curricular Mathematics Activities: Researching Authenticity as a Social Construct

Pauline Vos

Abstract In this chapter I study authentic aspects in mathematics education, in particular with respect to mathematical modelling. I dcfinc 'authenticity' as a social construct, building on the French sociologist Émile Durkheim. For an aspect to be authentic, it needs to have: (1) an out-of-school origin and (2) a certification of originality. The study validates this definition, asking: what authentic aspects can be identified within mathematics education? Data were collected from the excursion *Railway Timetable Dynamics*. During the excursion secondary school students were exposed to research carried out by university mathematicians on behalf of the National Railway Company. The authentic aspects were mathematical or non-mathematical. Often the certification was a testimony by an expert.

8.1 Introduction

In mathematics education, we encounter tasks in which mathematics is connected to the real world. Many of these tasks are *word problems*, in which mathematical exercises are 'dressed-up' (Blum and Niss 1991) into a story that contextualizes the mathematics. For example, the division exercise $3\frac{1}{2} \div \frac{1}{4} = \ldots$ may be concealed in a pizza situation: *How many quarter pizza slices can you get out of three and a half pizzas?* Or the same division may be concealed in a money situation: *How many quarter dollars ($0.25) make $3.50 ?*

Generally, in word problems the degree of reality is far fetched: the problems are completely inauthentic. In the above described word problems the division activity remains isolated and the answer is a number and not a solution to a problem in real life. The only need to carry out the division is pedagogical: to do a division of

P. Vos (✉)
Department of Mathematical Sciences, University of Agder, Gimlemoen 25, Kristiansand S 4630, Norway
e-mail: fpvos@hotmail.com

G.A. Stillman et al. (eds.), *Mathematical Modelling in Education Research and Practice*, International Perspectives on the Teaching and Learning of Mathematical Modelling, DOI 10.1007/978-3-319-18272-8_8

fractions. When set into a pizza situation or a money situation, the task may offer learners an idea of how division is useful in out-of-school situations, but research on adult consumers' calculations has shown that school skills are poorly applied in real life situations (Lave 1988). The lack of a link from mathematical word problems to genuine out-of-school life has been criticized (e.g., Boaler 1993; Gerofsky 1996; Verschaffel et al. 2000), pointing at problems such as: word problems ask for skills that are hardly ever needed in real life (e.g., finding the numbers concealed in a text and applying the right operation on these numbers), word problems do not offer students a real experience of the usefulness of mathematics and word problems are not appreciated by students either. Therefore, it remains a challenge of how authentic out-of-school aspects can be integrated into learning environments. In a previous study, I have problematized authenticity (Vos 2011). In this chapter, I will define *authenticity* and study the operationalization and validation of this definition.

8.2 Theoretical Background

8.2.1 Authenticity as Simulation or Imitation

In defining *authenticity,* a number of researchers have put up lists of components. For example, Heck (2010) refers to *authentic student research projects* if: (1) students work on self-selected, challenging, messy, ill-defined, open-ended problems rooted in a real life situation; (2) students do not follow some standard recipes, but examine their problem from different perspectives, using a variety of resources and high-order skills; (3) a broad range of competencies is required to make the project a success, and one of these abilities is making use of ICT for information gathering, data processing and reporting; (4) students' work is open-ended in the sense that there are multiple methods to obtain possible or even competing answers; (5) it offers students the opportunity to be in contact with contemporary, cross-disciplinary research and to learn about the nature of mathematics and science; and (6) students disclose their own understanding through a portfolio, a report or a presentation (Heck 2010, p. 116). Palm (2002, 2007) analyzes authenticity as the extent to which a task simulates a real-life performance. He offers a list to analyze tasks according to: the aspects of the event (might the event occur in real life?), the question (might the question be posed in that event?), the purpose, the information/data (are the data corresponding to real life data?), the presentation, the solution strategies, the circumstances, and the solution requirements.

In the above cases, the researchers designed activities that *imitate* or *simulate* real world activities. However, imitations and simulations are *copies*. The lists of criteria make the imitations or simulations to be near perfect copies of out-of-school activities, but the fact remains: copies are different from the real thing! In the study by Heck (2010) on the modelling of bungee jumping, students experiment with a Barbie doll in the gym, and not with real-life bungee-jumping, where a calculation error may cause death. In the study by Palm (2002, 2007), students work on a task to organize a school excursion. However, they fill in an order form from a fictitious bus company and after the number of necessary buses is filled in, the task ends and there is not a real excursion. So, while these tasks may meet a certain set of criteria set by the researchers, the tasks are still essentially inauthentic because they only *simulate* or *imitate* real life. This fact complicates the definition of authenticity in education, in particular in mathematical modelling education.

8.2.2 Authenticity as a Social Construct

According to dictionary definitions (e.g., Webster's), authenticity means that an object is authentic if it has a true origin and hasn't been copied (or forged). The term is illustrated through the discipline of Archaeology, where found artifacts are considered authentic when they truthfully originate from human activities in the past. The word *authentic* is used as a contrast to 'being a copy', such as imitations and forgeries. The qualification of being authentic is dichotomous: an artifact is either authentic or not; there is no artifact that is 'more or less authentic' or 'more authentic' than another. It is possible that an artifact receives the tag: 'authenticity cannot be established', but this un-decidedness does not render the artifact 'more or less authentic'. To establish authenticity, archaeologists have a number of methods to justify the authenticity (e.g., similarity to other authentic artifacts, the site of finding, the route that the object has travelled from the origin to its present location). In Archaeology, only acknowledged experts can give an artifact the classification of authenticity; if an outsider finds a Roman coin in his/her backyard, she/he will need to consult an expert before the artifact can be confirmed as authentic.

In this exemplary discipline of Archeology, we see that the term *authenticity* is defined in an interplay between objects and actors. The objects are (1) an artifact, that is: the object of study that is to be qualified as 'authentic' or not; (2) an origin, that is: a reality that has produced the object. The actors are: (3) the experts who decide on the qualification of authenticity; and (4) the outsiders who observe the artifact and trust the expert's authority.

The sociological analysis of the use of the term authenticity in disciplines such as Archaeology, Law and Art, shows us that authenticity is a *social construct*: it is an agreement reached through a social process. Social constructs were first studied by the French sociologist Émile Durkheim (Berger and Luckmann 1966). He studied the common language and knowledge of a community. Social constructs are, for example, 'money value', 'democracy', or 'science'. Generally, social constructs become institutionalized and they are not negotiable by individuals. Thus, a Roman coin found in someone's backyard will not be taken as authentic until an institutionalized procedure has been applied, that is until an expert has confirmed its originality, eventually with a counter-expertise.

Considering authenticity as a social construct, for an aspect in mathematics education to be considered as authentic it requires:

1. an out-of-school origin
2. a certification.

These requirements align with the definition of authenticity by Niss (1992, cited in Palm 2002, p. I-20): "We define an authentic extra-mathematical situation as one which is embedded in a true existing practice or subject area outside mathematics, and which deals with objects, phenomena, issues, or problems that are genuine to that area and recognized as such by people working in it." In this definition, the given situation has a true origin and this origin is attested to by experts. My definition of authenticity as a social construct is wider: (1) the origin can be mathematical, as long as it is from out-of-school, and (2) a certification can also be offered by other experts than "people working in it", such as stakeholders linked to other problems than workplace problems, such as consumer or environmental problems.

The study described in this chapter will validate the above definition by scrutinizing a series of mathematics activities on the presence of authentic aspects and whether the authenticity can be related to an origin and a certification. The research question is: *what authentic aspects can be identified within mathematics education*, and in particular: are authentic aspects always related to non-mathematical objects or can they also be mathematical?

8.3 Methods

For this study I observed systematically a series of mathematics activities on the presence of authentic aspects. The mathematical activities were part of a school excursion for mathematics. I selected an excursion instead of standard classroom activities, because I assumed to identify (1) more authentic aspects and (2) clear authentic aspects in extra-curricular, out-of-school mathematics activities than in regular mathematics classes. The selected excursion is one out of a collection of excursions that are organised by *ITS Academy*, a project in which four tertiary institutions in the city of Amsterdam collaborate with secondary schools (see www.ITSAcademy.nl). The goal is to motivate secondary school students for STEM studies (science, technology, engineering and mathematics) by organising motivating experiences for senior secondary school students (grades 10, 11 and 12). Many activities are organised within the schools, for example university researchers come to the schools to deliver guest lectures. Also, laboratory equipment can be borrowed by schools, and there are video lectures and on-line applications to demonstrate the work of researchers in schools. There are also excursions that take the students out of their school environment. For the natural sciences, the excursions are focused around the university's laboratory equipment that schools cannot afford. To raise more interest for Astronomy, there is an excursion to the university's telescope. For mathematics there are several excursions, for example on the modelling in Biodiversity (dynamic systems) and in Business Studies (the Black-Scholes-formula for investments). The study presented in this chapter focuses on the excursion *Railway Timetable Dynamics*, in which Graph Theory is used to model the routing of passenger trains. In this excursion, the students physically visit *Science Park,* the campus of University of Amsterdam where the Faculty of Science and Mathematics can be found.

The excursion was studied by two independent coders on aspects that could be coded as authentic: the author and a research colleague. The coders familiarised themselves with the setting of the excursion, the sequence of activities, and in the spring of 2012 they joined visits by groups of secondary school students from two secondary schools. Students' activities were videotaped, the accompanying teachers and three randomly selected students were interviewed. Thereafter, the coders created a table, listing authentic aspects; for each authentic aspect they applied the criteria for authenticity (origin and certification), so they noted a true non-educational origin and the way the justification was offered to the students. Comparing between coders, it was noted that their observations overlapped largely, albeit differing in wording. Not all aspects were observed by both coders, but they reached consensus.

8.4 Results

In the excursion described here, *Railway Timetable Dynamics*, the physical environment of the excursions was the first authentic aspect. The university premises with professors, researchers and university students, and with university halls and laboratories were all authentic. The buildings and all people present were *real.* Clearly, the secondary school students were not taken to a theatre, in which a play was performed by actors. The physical presence at the university scenes, being able to see and speak with researchers and sit in university lecture halls served to certify the authenticity of these aspects.

In the excursion *Railway Timetable Dynamics* the students are introduced to a problem from mathematical Graph Theory. In 2007 a new railway timetable was introduced by the National Dutch Railway Company (Nederlandse Spoorwegen, NS), based on research on bottlenecks in the timetable by Prof Alexander Schrijvers and his team. The research was carried out on behalf of the NS. In the excursion, the students meet with a NS-manager through live video conferencing. They see him against the background of the NS control room, while he explains the urge of the railway company to collaborate with research mathematicians, see Fig. 8.1. The live connection serves to certify the origins of the timetable problem of NS.

After meeting live with the NS-spokesman, the students' task is to create a timetable for the railway system in the northern part of the Province of North-Holland. This region was chosen because it is in the vicinity of Amsterdam and thus, it is a familiar region to most students. Also, this region is a peninsula and thus the graph for its railway system is mathematically not too complex, see Fig. 8.2.

The activities in the excursion are scaffolded: students are introduced to the mathematical symbols and techniques and they learn how a cyclic timetable is repeated every hour. Deliberately, the start of the activities is kept simple, so that

Fig. 8.1 Live video conferencing with the railway control room

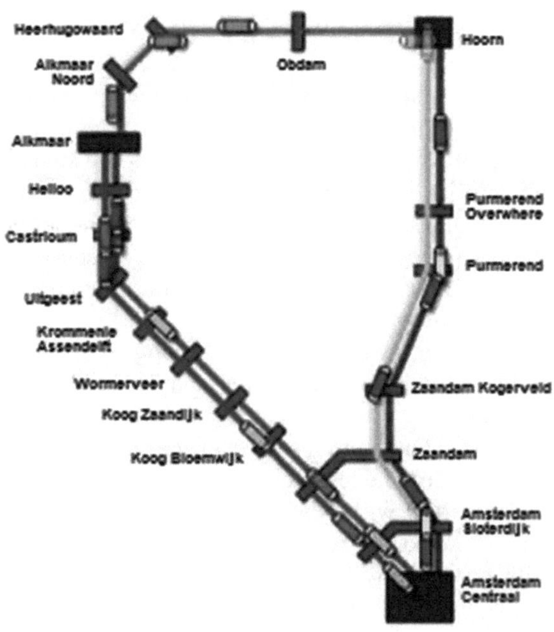

Fig. 8.2 Picture from the task in the excursion *Railway Timetable Dynamics*

all students can step in, even if they are somewhat weaker in mathematics. They start to formulate assumptions that need to be included into the model, such as the limitations caused by faster trains that do not stop at all stations, the maximal waiting time for a train (e.g., if a train waits for 20 min at a platform, this will cause passenger dissatisfaction), the transit time for passengers, and the safety distance between

Fig. 8.3 Students working with the modelling software for the excursion *Railway Timetable Dynamics*

departing and arriving trains. After lunch, the students used modelling software to build their own timetable, see Fig. 8.3. This software was specially created for the excursion and simulates the software that the mathematical researchers used originally. It is explained to the students, that constructing software was part of the mathematicians' job when working on the bottle-necks in the NS timetable.

The original problem, to create a new timetable, was observed as authentic, as it originated from the NS and it was certified in the live video conferencing. This live connection turned out to be pivotal in convincing students that the mathematical activities really served to solve practical problems. However, the problem given to the students to find a timetable for the province of North Holland was observed as inauthentic, because it was a reduced problem for pedagogical reasons and the students merely simulated the researchers' work. The original work was far more complex and could not be covered in the time span of the excursion. Also, students did not need to report back solutions to the NS in such a way that their solution would be implemented. If they made errors, it did not have consequences for the company nor for the passengers. Also, the software used was inauthentic, as well as the 'research experience' (the excursion simulated the work of research mathematicians; the students only worked for a few hours on one problem, not for weeks or longer, not under time pressure). However, an authentic aspect in this excursion was the 'researcher's frustration', that is: students worked for a long time on one single problem without finding a reasonable solution, and thereby discovering that the problem was far more complex than anticipated. The authenticity of this experience was certified by the university researcher who recounted his own frustrations and his need for perseverance while doing research.

Another authentic aspect found was the applicability of mathematics, that is: the excursion showed the students how real mathematics researchers use mathematics to improve real timetabling problems of railway companies. This was certified by the expert in the video-conferencing meeting.

8.5 Conclusion and Discussion

In this small-scale study we can observe a range of authentic aspects in mathematical modelling education. The authentic aspects can be distinguished as mathematical or non-mathematical. Observed authentic aspects that are within the field of mathematics: authentic symbols (identical to the ones used by the researchers), authentic research questions (to resolve the bottlenecks in the railway timetable), authentic research experiences (working for a long time and not finding an answer). As non-mathematical aspects, we observed authentic buildings, authentic professionals, authentic applicability of mathematics to authentic problems (graph theory and its applicability to the railway company's timetable problem) and authentic problem settings (stations, platforms, waiting time for passengers).

In this excursion, *Railway Timetable Dynamics*, we also observed the reduction of authentic aspects into inauthentic aspects: we observed how activities from the real world were downsized, simplified and reduced to match the educational setting, which is framed by time constraints and limited skills of the participants. This reduction serves the educational purposes, which were: to offer students experiences and give them a 'feel' of what real professionals do with real mathematics in the real world. Within the learning environment for modelling education not everything can or needs to be authentic. There are so many different aspects within a learning environment, such as goals, resources, media, activities and so forth, that it is unnecessary to have all aspects be authentic. If all aspects of a learning environment would be authentic, the students would take up the full task of a professional. This would mean that mathematical errors would have serious consequences, such as errors in the railway timetable or loss of money for the company.

The extra-curricular experience for students offers an epistemological insight into the learning of mathematics. Often, we discern in the learning of mathematics different strands needed for becoming mathematically proficient, such as: conceptual knowledge, procedural fluency, strategic problem solving competencies, and so forth. The present study demonstrates the necessity for students to construct knowledge *about* mathematics, in particular knowledge about the utility of mathematics (e.g., for solving complex timetable problems). This knowledge may be attained without fully understanding the underlying mathematical concepts or procedures. Whether this knowledge affects students' mathematical disposition is a point for further studies.

In this present study, authenticity is strongly related to the work of professional mathematical researchers. Of course, this is due to the object of study: an excursion to a university will mirror the work of university mathematicians. If the focus had been on the modelling of consumer activities, the authentic aspects would relate to consumer aspects.

Taking authenticity as a social construct, we verified whether each aspect had an out-of-school origin and whether this origin could be certified. Often the certification consisted of a testimony by an expert, in particular the video conferencing with the NS control room played a pivotal role in the certification. We noted that for a number of authentic aspects the origins and certifications were not always explicitly given, while this could have been done. For example, the symbols used in the mathematical model were authentic, being the original symbols that the university researchers had introduced in their research process. However, this fact was not made explicit and the students complained of the symbols' awkwardness. Probably the excursion trainers were not aware of the importance of the certification, and the visiting students did not ask. Here, we obviously see points for improvement in the design of learning environments.

In this chapter I defined authenticity as a social construct, whereby each authentic aspect has an out-of-school origin and a certification of this. This definition was practical for separate aspects in a learning environment. This definition differs from those definitions with lists of criteria, in which cause and effect are reversed. When a modelling activity contains authentic aspects, this causes the activity to be fuzzy, open-ended, requiring higher-order thinking and so forth. It means that the lists of features offered by a number of authors (e.g., Heck 2010; Palm 2002, 2007) are *implications* of authentic aspects and not *indicators* or *descriptors* of authenticity.

References

Berger, P. L., & Luckmann, T. (1966). *The social construction of reality: A treatise in the sociology of knowledge*. Garden City: Anchor Books.

Blum, W., & Niss, M. (1991). Applied mathematical problem solving, modelling, applications, and links to other subjects – State, trends and issues in mathematics instruction. *Educational Studies in Mathematics, 22*(1), 37–68.

Boaler, J. (1993). The role of contexts in the mathematics classroom: Do they make mathematics more "real"? *For the Learning of Mathematics, 13*(2), 12–17.

Gerofsky, S. (1996). A linguistic and narrative view of word problems in mathematics education. *For the Learning of Mathematics, 16*(2), 36–45.

Heck, A. (2010). Modelling in cross-disciplinary authentic student research projects. *International Journal for Technology in Mathematics Education, 17*(3), 115–120.

Lave, J. (1988). *Cognition in practice: Mind, mathematics and culture in everyday life*. Cambridge, MA: Cambridge University Press.

Palm, T. (2002). *The realism of mathematical schools tasks – Features and consequences*. Ph.D. thesis, Umeå University, Umeå.

Palm, T. (2007). Features and impact of the authenticity of applied mathematical school tasks. In W. Blum, P. L. Galbraith, H.-W. Henn, & M. Niss (Eds.), *Applications and modelling in mathematics education* (pp. 201–208). New York: Springer.

Verschaffel, L., Greer, B., & De Corte, E. (2000). *Making sense of word problems*. Lisse: Swets & Zeitlinger.

Vos, P. (2011). What is 'authentic' in the teaching and learning of mathematical modelling? In G. Kaiser, W. Blum, R. Borromeo Ferri, & G. Stillman (Eds.), *Trends in teaching and learning of mathematical modelling* (pp. 713–722). New York: Springer.

Chapter 9
The Teaching Goal and Oriented Learning of Mathematical Modelling Courses

Mengda Wu, Dan Wang, and Xiaojun Duan

Abstract By analyzing the characteristics of mathematical modelling courses in Chinese universities, we present a viewpoint on how to teach this course, including what the instructor should teach and what would be expected to be learnt by students. It is worth emphasising that "teaching motivated by mathematical modelling thought" should be the main goal of mathematical modelling courses. The mathematical methods and modelling cases are the carriers of thought transmission and should serve the goal so that the student can better comprehend the mathematical thought.

9.1 Introduction

In China, mathematical modelling courses are included in the curriculum schedule in most universities (Xie 2010). Most Chinese teachers also believe that mathematical modelling activities can improve students' ability to solve problems. The findings of Dan and Xie (2011) from a small experimental study at a Chinese engineering university with average-level students "provide strong evidence to support the mathematical education reform in China with regards to mathematical modelling courses and related activities as a vehicle to improve the students' innovation ability" (p. 465). Ban (2001) surveyed Chinese high school students and found "the high school students in China spent lots of time in learning math [ematics], but they don't feel the effect of the math[ematics], the mathematical modelling activities can improve the interest in learning math[ematics]". However, mathematical modelling courses have not yet imposed a normalized teaching system as have the traditional mathematical courses, such as Calculus, Linear Algebra, Probability and Statistics (Ye et al. 2003). In Chinese universities, several ways of teaching mathematical modelling courses have been paid attention by

M. Wu (✉) • D. Wang • X. Duan
College of Science, National University of Defence Technology, Yanwachi Street 137, Changsha, Hunan 410073, P.R. China
e-mail: mdwu@nudtwmd.com; casablanman@hotmail.com; xj_duan@163.com

G.A. Stillman et al. (eds.), *Mathematical Modelling in Education Research and Practice*, International Perspectives on the Teaching and Learning of Mathematical Modelling, DOI 10.1007/978-3-319-18272-8_9

Chinese teachers, such as mathematical modelling teaching by mathematical method or modelling case, also including teaching modelling method by case researching (cf Clements 1984). Although these mathematical methods are important parts in mathematical modelling courses, it is impossible for students to become proficient at all of these mathematical methods through the teaching of mathematical modelling courses since the course time is limited. So, a mathematical modelling course should provide a way of teaching in its own systematic thinking pattern (Li 2013) and not be only a collection of various methods.

This chapter will mainly focus on collecting together some of the wisdom of practice about the essential nature of mathematical modelling teaching at tertiary level from our experience in China and relating it to similar ideas expressed in other contexts. By analyzing the characteristics of mathematical modelling cases, two viewpoints have been presented on teaching this course in this chapter, namely what should the instructor teach (i.e., the *teaching goal*) and what should be learnt by the students (i.e., the *orientation of the learning*). It is emphasized that summarizing and spreading the message about the thinking underpinning mathematical modelling should be the main goal of teaching a mathematical modelling course, whilst the mathematical methods and modelling cases are the carriers of thought transmission and should serve the goal to help students to comprehend and learn the mathematical thinking. To this end, Blomhøj and Kjeldsen (2013), from their course at Roskilde University, as well as others (e.g., Li 2013), have pointed out "that the modelling context provides a window to students' understanding and their images of the mathematical concepts they work with as well as to their understanding of a mathematical model and modelling" (p. 151). We think that the teaching goal of the mathematical modelling courses should be to develop the thinking underpinning mathematical modelling. The main goal should not be that the mathematical modelling course is just to help students become proficient in the application of mathematical methods; the more important goal is to help students understand the general principles and thinking of mathematical modelling through the process of constructing the mathematical model and solving it (i.e., to develop meta-knowledge about modelling). This process will improve the thinking ability of students with respect to quantitative thinking.

In this chapter, we advocate several means of teaching such thinking during a mathematical modelling course, including intuitive insights into how the problem should be modelled, making connections, innovation through critique, choice based on common sense and cultivating inductive thinking. In fact, these thoughts also appear in many applied mathematics courses such as applied differential equations, and mathematical biology; but such courses in China that emphasize the application and teaching of methods, lack the summary of thinking. Next, these viewpoints will be illustrated respectively.

9.2 Intuitions: Insights into the Nature of the Problem

To some extent, innovation relies on intuition. The intuitive ability to detect the rules hidden in a complicated real-world phenomenon often leads to innovative ideas. In fact, there are many mathematical modelling teaching cases where the final conclusions are contrary to "common sense". Deep analysis of such cases can continuously correct students' first impression to a more profound and accurate degree. Therefore, through this process, the student's quantitative thinking ability could be enhanced (Borromeo Ferri 2011). Furthermore, the author of Sciencenet blog posting (2013) points out that the college students without this self-criticism ability for intuitive knowledge will be destined to become a passive watcher of society rather than a participant.

For example, people easily form an intuition that the scientific reasoning ability of a person is dependent on the knowledge owned; but in fact, it is not necessarily true. In January 2009, Bao et al. (2009) published an article on learning and scientific reasoning. The article revealed that: overall knowledge and scientific reasoning ability are unrelated. The relationship between knowledge and scientific reasoning of students is similar to that of lens imaging and the image's resolution. The amount of knowledge has been increased for students who graduated from university; however the improvement for their scientific reasoning may not be much. Similarly, the resolution does not become better along with the size of the image increasing. Figure 9.1 (Bao et al. 2009) shows comparative results for China and USA first year college students on various measures. Force Concept Inventory (FCI), Brief Electricity and Magnetism Assessment (BEMA) are used to test student STEM (Science, Technology, Engineering, and Mathematics) content knowledge and Lawson's classroom test is used to test scientific reasoning ability. From these comparison data, Bao et al. (2009) drew the following conclusion: the magnitude of physical knowledge of American students is 50 % less than that of Chinese students, but their scientific thinking abilities are almost the same.

The same statement could be made from the following two cases:

Case 1: Lanchester Battle Model (Lucas 1983)

In the Lanchester battle model, only two factors influencing the outcome of the battle are considered: numerical strength and fighting capability. The variables x_0 and y_0 are denoted as the initial numerical strengths of side X and side Y; a and b are denoted as the effective fighting capability coefficients of side X and side Y. Under some simplified hypothesis, using the Lanchester battle model, the requirement of side Y winning is:

$$\left(\frac{y_0}{x_0}\right)^2 > \frac{b}{a}$$

Before giving the final conclusion of this model, students are asked to solve this problem: If the initial numerical strength of side Y is just one half less than that of side X, then side Y must improve its fighting capability coefficient to win the battle. How much should the fighting capability of side Y be improved to guarantee winning the battle?

Most students answer that only when the fighting capability of side Y is two times higher than that of side X, can side Y win the battle. However, the Lanchester

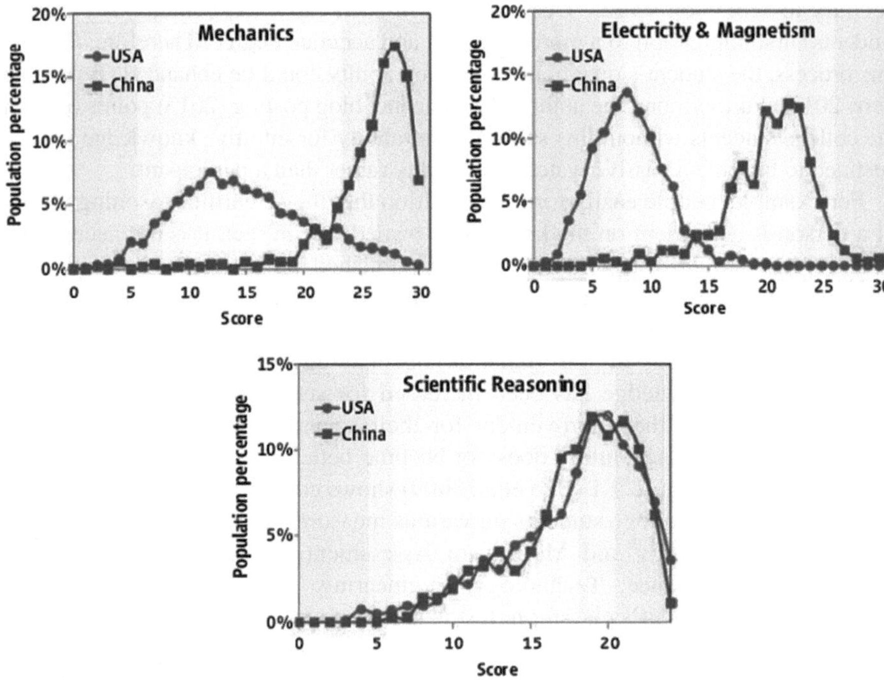

Fig. 9.1 Knowledge and ability relation between Chinese students and American students

battle model shows that only when the fighting capability of side *Y* is four times higher than that of side *X*, can side *Y* win the battle!

This case shows that most students are using a "linear thinking style", which is a habitual initial thinking pattern. The linear thinking mode is available and helpful on many occasions, but when facing a nonlinear problem, we may fail to solve the problems correctly (De Bock et al. 2007; Van Dooren et al. 2013). In mathematical modelling cases, students should learn to rethink the initial thinking style, correcting the frustrated initiation and reducing the mistake-making possibility.

Case 2: More Means the Smoother Journey? (Braess's Paradox 2008; Mind Your Decisions 2009)

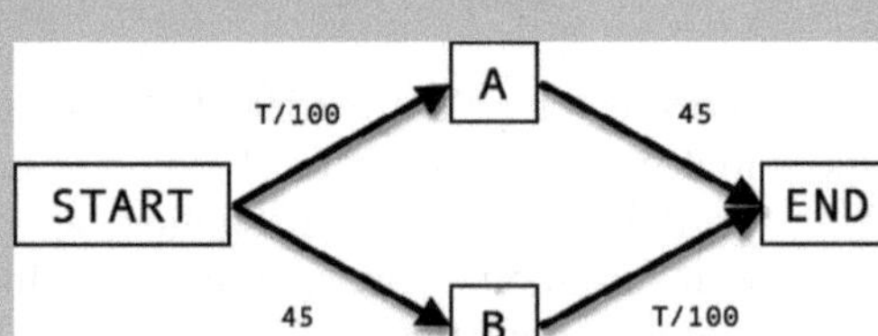

In this case, due to poor transportation conditions, the time *t* spent on the road from the "start" to A is linked to the total amount of vehicles *T*, which

(continued)

satisfies the condition $t = \frac{T}{100}$. While the road from "start" to point B has good transportation conditions, the time t spent on this road is fixed at 45 min. In a certain time, suppose that 4,000 vehicles are required to pass from the "start" to the "end", how do you allocate these vehicles to the different roads?

This is another case that is contrary to "common sense". According to Nash non-cooperation game equilibrium theory (Nash 1951), the Nash equilibrium point is: each path carries 2,000 vehicles. Then all drivers of vehicles will think their choice is optimum, the Nash equilibrium point is reached at this point.

By the above vehicle allocation, the time required from "start" to "end" will be 65 min. Now, suppose that a new road between A and B was built due to the long time cost, which is shown as follows in Fig. 9.2. Additionally, suppose that the time required between A and B can be ignored when compared with the whole time cost.

Generally, people think that more roads will lead to less time cost from the "start" to the "end". However, according to Nash equilibrium theory (Nash 1951), all vehicles will choose the same path: START → A → B → END. Under the non-cooperation condition, all participants believe that their choice was optimum. In this case, every vehicle would take 80 min to pass from "start" to "end".

The research of the German mathematician Dietrich Braess comes to the conclusion: if the journey time of each road is linearly related with the vehicle number of this road in a transportation network, there must exist a Nash equilibrium point. It may cause the whole network to be under adverse conditions, which is called the Braess paradox phenomenon (Braess 1969; Braess et al. 2005).

9.3 Connections: Reconstruction of Personal Knowledge System

From the viewpoint of solving the real-world problem (Brown 2015), the construction of a personal knowledge system should not only involve the amount of mathematical knowledge, but also encompass the "soft ability" (if the amount of knowledge is called hard ability, here soft ability is the ability to analyze the problem, modelling, design the experiment, etc.) for the application of the knowledge (Brown this volume; Greefrath et al. 2011). This is like the performance of a

Fig. 9.2 Tracking model of transportation

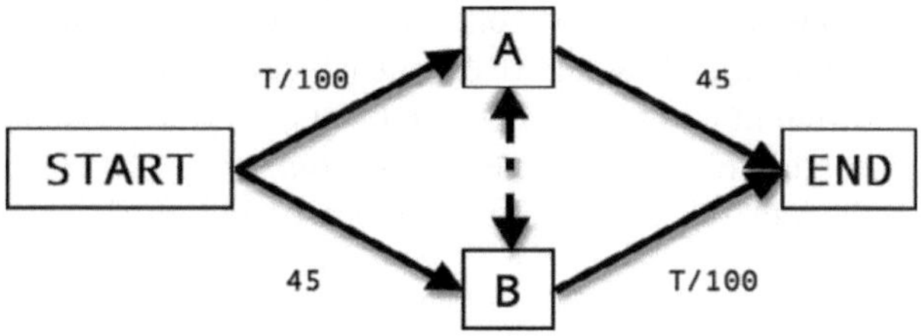

database system depending not only on storage size, but also on input, output and data retrieval functions. More importantly, mathematical modelling teaching should focus on forming and strengthening the knowledge application ability of learners (Spandaw 2011).

In the teaching process, "connections" should be emphasized, including the connection between "knowledge" and "problem", and the connection between "knowledge" and "knowledge". The famous mathematician Stefan Banach has said "A mathematician is a person who can find analogies between theorems; a better mathematician is one who can see analogies between proofs and the best mathematician can notice analogies between theories. One can imagine that the ultimate mathematician is one who can see analogies between analogies" (BrainyQuote.com 2014). Higher level thinking not only relies on the amount of knowledge, but also depends on the connections among knowledge. The knowledge system we advocate is a system constructed with knowledge and connections. A successful modelling process will not be prefect without deep understanding of the problem and a thorough understanding of the model. Only when the two parts are clearly understood, will the optimal connection between knowledge and problem be "linked" and, based on that, a perfect modelling could have a good foundation.

9.4 Innovation: Critical Thinking Ability

Innovation usually comes from critical thinking. Innovation usually cannot come forth without critique and accurately understanding a problem. A critical thinking ability is an essential characteristic of innovation thinking. In the modern information age, information content and information spreading have a rapid growth; people will be more likely to suffer from an "information indigestion problem". So the critical thinking habits should be guiding ideas in teaching of mathematical modelling courses. The teaching based on the real problem and case, which is a major characteristic of mathematical modelling courses, can help the students to achieve the training of critical thinking habits as pointed out also by Blomhøj and Kjeldsen (2011). Let us elaborate this idea through a further case.

Case 3: Locks Packing Problem – Complaint Degree Tolerance

A certain factory produces locks, the key of each lock has 5 slots, the height of each slot was chosen among the set {1,2,3,4,5,6}. Under the current industrial technique condition, the experimental results to test whether the key of a lock would open other locks are as follows: for 2 keys matched to different locks, if there exist 4 slots with the same height in all 5 slots and the height difference of the rest of slots is only 1, then these two keys may open the lock of each other. On the other hand, the locks are often packed together

(continued)

with 60 locks to sell, while customers often buy several boxes or more. Then they may find different keys that may open the same lock. This situation is called inter-opened locks. Now your group must try to solve the following question:

(1)–(3) Design the way for packing the locks into a box.
(4) According to this packing method, how do you measure the customer's complaint degree (dissatisfaction with inter-opened locks) qualitatively?

The following conclusion can be gained by computing the mathematical expectation of the number of inter-opened locks: average 2.33 pairs of inter-opened locks can be found in a box; 9.41 pairs of inter-opened locks can be found in two boxes. Then many modellers easily form the following reasoning process: the more boxes the customers buy, the more inter-opened locks, then the higher degree of complaint will be gained.

This reasoning process may seem reasonable. However, other modellers query this. They criticize the viewpoint: the more inter-opened locks, the higher the complaint degree. This assumption is reasonable only when the customer tests all the locks to judge whether they can open each other. However, will the customer really do all the tests? If a customer buys 50 boxes of locks, then the cumulative test time to carry out all tests would reach 4.5 million. It is impossible! So it is more likely for those customers to do inter-open lock tests randomly. Then the problem reaches an essential point: which is the probability for finding inter-opened locks does not linearly increase as the number of boxes purchased increases. So those modellers come to the opposite conclusion that the more lock boxes the customers buy, the less complaint they will have. This case requires students to pay attention to engaging critical thinking when it is considered.

9.5 Choice: the Ability of Macro-coordination and Direction Control

If we say students master the skill of designing the machine parts by classroom study, the mathematical modelling process could be regarded as constructing a machine, which involves the determination of the modelling tasks and a trade-off among all constraints. In the mathematical modelling process, many situations and different schemes need to be continuously selected and compromised. It is related to macro-control ability for coordinating all these aspects. This ability could be trained only in the process of solving practical problems (Stillman 2011).

Case 4: Cable Car Operation Plan – Realistic Alternative

A certain scenic spot attracts many tourists due to its beautiful scenery. Because of many tourists, there is a waiting queue for tourists to ride the cable car to reach the summit of the mountain. Now the problem is: how to solve this waiting time problem for tourists?

This is a test problem for our mathematical modelling thinking. Most students' answers are: speeding up the cable car. It is pity that it is a wrong answer. We denote by Q the carrying capacity of a cable car (which means the number of tourists being carried in unit time), V as the cable car's speed, P as the maximum number of tourists that can be carried for each cable car, and L as the distance between two neighbouring cable cars. Then the tourist carrying capacity for the whole cable system can be expressed by the following formula:

$$Q = (V \times P)/L$$

Considering safety, cable car arrival interval is fixed. It means that V/L is a constant determined by the safety factor. So when accelerating the cable car's rate, the distance between two neighbouring cars should also be increased, this change cannot help in increasing the whole carrying ability. The correct answer to this problem is slowing down the cable car's speed. This is unexpected. In fact, slowing down the cable car's speed also cannot improve the carrying capacity; however, since the cable car's speed is slow, tourists will have more time on the cable, which decreases the waiting time on the ground, the degree of complaint is naturally reduced.

In fact, there are many interesting mathematical modelling cases that are not consistent with common sense. In brief, students not only learn the mathematical modelling method and computation skills through studying these cases, but also include judgment ability and trade-off ability to the real-world problem.

9.6 Inductions: Improve Thinking Mode

In the modern information world, data modelling has become an important approach to mathematical modelling (Engel and Kuntze 2011) since large datasets have become available (see Kutz 2013). The mathematical teaching generally trains students' deductive reasoning ability in an inferential thinking mode, while mathematical modelling courses trend to train students' inductive reasoning ability. In the mathematical modelling teaching process, we should consciously emphasize the training of students' inductive reasoning ability. This will help to cultivate students' thinking mode to adapt to the information world's requirements.

Case 5: Forecasting Popularity – Induction of the Impact Factors (Roja et al. 2012)
News articles are very dynamic due to their relation to continuously developing events that typically have short life spans. For a news article to be popular, it is essential for it to propagate to a large number of readers within a short time. Hence predicting the popularity of news items prior to their release is an interesting and challenging task.

This is a complex problem for students because there are so many factors influencing the popularity of the new articles. We also provide a lot of data to students. Students need to induce the most influential factors from data. In this case, the students will finish the following steps:

- Extract the influential factors by statistical analysis;
- Quantify the influential weight of the factor;
- Build the model between popularity and the influential factors by a statistical model, and verify the model;
- Predict the popularity of an article.

In Roja et al. (2012), three factors are extracted as influential factors on the popularity (T). They are the score (C) of the type of information, the score (S) of the source of information, and the max score (Ent_{max}) of the named entities (such as a place, a person or a organization, etc.). Raoja et al. built a model as:

$$\ln(T) = 1.24\ln(S) + 0.45\ln(C) + 0.1Ent_{\max} - 3$$

Students extract different influential factors and build various models to get the description of popularity. Although students cannot build as perfect a model as Roja et al. do, students try to understand how to induce the information from lots of data. Further, the teacher can guide the students in training their inductive reasoning ability.

9.7 Conclusions

The chapter discusses two main aspects about the teaching of mathematical modelling courses: what should the instructor teach and what should be learnt by the students. Some viewpoints are proposed: (1) We should attach importance to the training of the thinking of students in mathematical modeling course. In the real world, students will face unfamiliar problems; then, the training of thinking will help students analyze and respond to the problem effectively rather than recall of methods from memory. (2) The teacher should pay more attention to training of students' systematic thinking modes, not just the mathematical methods. (3) Summarizing and spreading the message about the thinking underpinning mathematical modelling should be the main goal of teaching a mathematical modelling course, whilst the mathematical methods and modelling cases are the carriers of thought transmission. (4) Students learn and understand the process of the mathematical modelling through the training of the thinking. At the same time, students also learn and understand the mathematical methods. This teaching mode will help students improve their quantitative thinking ability to deal with real-world problems in mathematical modelling courses. Only then might we reverse Bao et al.'s finding (2009) and creativity be fully enhanced and discovered during students' university life.

Acknowledgment This work is supported by the Course Construction Program, and the Innovation Practice and Internship Base for Mathematical Modelling, of NUDT in China.

References

Ban, C. (2001). *Theory and experimental study of mathematical modelling to raise creative thinking of high-school students*. Master's degree thesis, Tianjin Normal University (In Chinese).

Bao, L., et al. (2009). Learning and scientific reasoning: Comparisons of Chinese and U.S. students show that content knowledge and reasoning skills diverge. *Science, 323*(5914), 586–587.

Blomhøj, M., & Kjeldsen, T. (2011). Students' reflections in mathematical modelling projects. In G. Kaiser, W. Blum, R. Borromeo Ferri, & G. Stillman (Eds.), *Trends in teaching and learning mathematical modelling* (pp. 385–395). New York: Springer.

Blomhøj, M., & Kjeldsen, T. (2013). Students' mathematical learning in modelling activities. In G. A. Stillman, G. Kaiser, W. Blum, & J. P. Brown (Eds.), *Teaching mathematical modelling: Connecting to research and practice* (pp. 141–152). Dordrecht: Springer.

Borromeo Ferri, R. (2011). Effective mathematical modelling without blockages – A commentary. In G. Kaiser, R. W. Blum, R. Borromeo Ferri, & G. Stillman (Eds.), *Trends in the teaching and learning of mathematical modelling* (pp. 181–185). New York: Springer.

Braess, D. (1969). Uber ein paradoxon der verkehrsplanung. *Unternehmensforschung, 12*, 258–268.

Braess, D., Nagurney, A., & Wakolbinger, T. (2005). On a paradox of traffic planning. *Transportation Science, 39*(4), 446–450.

Braess's Paradox Roads Example. (2008). *Braess's paradox*. Retrieved October 11, 2013, from Wikipedia: http://en.wikipedia.org/wiki/Braess%27s_paradox

BrainyQuote.com. (2014, October 15). *Stefan Banach*. http://www.brainyquote.com/quotes/quotes/s/stefanbana311401.html

Brown, J. P. (2015). Visualization tactics for solving real world tasks for solving real world tasks. In G. A. Stillman, W. Blum, & M. S. Biembengut (Eds.), *Mathematical modelling in education research and practice: Cultural, social and cognitive influences* (pp. 431–442). Cham: Springer.

Clements, R. R. (1984). The use of the simulation and case study techniques in the teaching of mathematical modeling. In J. S. Berry, D. N. Burghes, I. D. Huntley, D. J. G. James, & A. O. Moscardini (Eds.), *Teaching and applying mathematical modeling* (pp. 316–324). Chichester: Ellis Horwood.

Dan, Q., & Xie, J. (2011). Mathematical modelling skills and creative thinking levels: An experimental study. In G. Kaiser, R. W. Blum, R. Borromeo Ferri, & G. Stillman (Eds.), *Trends in the teaching and learning of mathematical modelling* (pp. 457–466). New York: Springer.

De Bock, D., Van Dooren, W., Janssens, D., & Verschaffel, L. (2007). *The illusion of linearity: From analysis to improvement*. New York: Springer.

Engel, J., & Kuntze, S. (2011). From data to functions: Connecting modelling competencies and statistical literacy. In G. Kaiser, R. W. Blum, R. Borromeo Ferri, & G. Stillman (Eds.), *Trends in the teaching and learning of mathematical modelling* (pp. 397–406). New York: Springer.

Greefrath, G., Siller, H.-S., & Weitendorf, J. (2011). Modelling considering the influence of technology. In G. Kaiser, R. W. Blum, R. Borromeo Ferri, & G. Stillman (Eds.), *Trends in the teaching and learning of mathematical modelling* (pp. 315–329). New York: Springer.

Kutz, J. N. (2013). *Data-driven modeling & scientific computation: Methods for complex systems & big data*. Oxford: Oxford University Press.

Li, T. (2013). Mathematical modeling education is the most important educational interface between mathematics and industry. In A. Damlamian, J. F. Rodrigues, & R. Sträßer (Eds.), *Educational interfaces between mathematics and industry* (pp. 51–58). Cham: Springer.

Lucas, W. F. (1983). *Modules in applied mathematics: Differential equation models*. New York: Springer.

Mind Your Decisions. (2009, June 1). *Why the secret to speedier highways might be closing some roads: The Braess paradox*. http://mindyourdecisions.com/blog/2009/01/06/why-the-secret-to-speedier-highways-might-be-closing-some-roads-the-braess-paradox/

Nash, J. (1951). Non-cooperative games. *Annals of Mathematics, 54*(2), 286–295.

Roja, B., Sitaram, A., & Bernardo, H. (2012). The pulse of news in social media: Forecasting popularity. *Proceedings of the sixth international AAAI conference on weblogs and social media*, Dublin, pp. 26–33.

Sciencenet. (2013, February 28). *College students should build up the self-criticism ability on intuition*. http://blog.sciencenet.cn/blog-3377-665760.html

Spandaw, J. (2011). Practical knowledge of research mathematicians, scientists, and engineers about the teaching of modeling. In G. Kaiser, R. W. Blum, R. Borromeo Ferri, & G. Stillman (Eds.), *Trends in the teaching and learning of mathematical modelling* (pp. 679–688). New York: Springer.

Stillman, G. (2011). Applying metacognitive knowledge and strategies in applications and modelling tasks at secondary school. In G. Kaiser, R. W. Blum, R. Borromeo Ferri, & G. Stillman (Eds.), *Trends in the teaching and learning of mathematical modelling* (pp. 165–180). New York: Springer.

Van Dooren, W., De Bock, D., & Verschaffel, L. (2013). How students connect descriptions of real-world situations to mathematical models in different representational modes. In G. A. Stillman, G. Kaiser, W. Blum, & J. P. Brown (Eds.), *Teaching mathematical modelling: Connecting to research and practice* (pp. 385–394). Dordrecht: Springer.

Xie, J. (2010). Mathematical modeling courses and related activities in China universities. In A. Araújo, A. Fernamdes, A. Azevedo, & J. -F. Rodrigues (Eds.), *Proceedings of EIMI 2010 conference (educational interfaces between mathematics and industry)* (pp. 597–606). Bedford: COMAP.

Ye, Q.-X., Blum, W., Houston, S. K., & Jiang, Q.-Y. (2003). *Mathematical modeling in education and culture*. Chichester: Horwood.

Part II
Research into, or Evaluation of, Teaching and Learning

Chapter 10
Modelling Competencies: Past Development and Further Perspectives

Gabriele Kaiser and Susanne Brand

Abstract This chapter aims to describe the understanding of modelling competencies and how it has developed since the establishment of the conference series "International Conference on the Teaching of Mathematical Modelling and Applications" in the early eighties of the last century. Based on these descriptions and the distinction of different strands of the modelling competency debate, the chapter will point out, how the discussion has evolved. Using as example an empirical study on the structure of modelling competency, namely holistic versus analytic, and a comparison of two approaches to foster modelling competencies, namely a holistic versus an atomistic approach, it will be pointed out that possible future developments will be based on more sophisticated psychometrical methods for the measurement of modelling competencies leading to comprehensive descriptions of 'modelling competency'.

10.1 Development of Modelling Competencies from the Past Until Today

In recent years modelling competencies and their promotion have been included in many curricula all over the world, many innovative projects aim to foster modelling competencies and there exist many studies on their promotion. However, there still does not exist a joint understanding of modelling competencies. So, the question arises, how modelling competencies were understood at the beginning of the discussion on teaching and learning of mathematical modelling, how this understanding has developed in the past and in which directions future developments might go.

G. Kaiser (✉) • S. Brand
Faculty of Education, University of Hamburg, Von-Melle-Park 8, Hamburg 20146, Germany
e-mail: Gabriele.Kaiser@uni-hamburg.de; susanne.brand@uni-hamburg.de

G.A. Stillman et al. (eds.), *Mathematical Modelling in Education Research and Practice*, International Perspectives on the Teaching and Learning of Mathematical Modelling, DOI 10.1007/978-3-319-18272-8_10

In the following first part of this chapter we focus on the understanding and conceptualisation of the construct modelling competency in the past, its understanding and occurrence at the beginning of the more recent modelling discussion starting with the implementation of the conference series "International Conference on the Teaching of Mathematical Modelling and Applications" in the early eighties of the last century. Within this development fostered by this conference series and the contributing group of researchers, differing descriptions and conceptualisations of modelling competency and methods of measurement of modelling competencies were proposed, which allow us to identify four strands of modelling competencies. Besides many commonalities these four strands have developed their own understanding of modelling competencies, namely either analytic or holistic. Based on analyses of these different conceptualisations and measurement methods of modelling competencies an empirical study will be described exemplarily, which incorporates aspects of these strands of the discussion and which has the potential to show possible directions for future developments.

10.1.1 Modelling Competencies at the Beginning of the Modelling Discussion

At the first conference of the international conference series on the teaching of mathematical modelling and applications in 1983 at the University of Exeter, later called ICTMA, David Burghes (1984), the conference chair, described the overarching goal of the promotion of mathematical modelling as follows:

> The basic philosophy behind the approach ... of the modelling workshop for higher education is that to become proficient in modelling, you must fully experience it – it is no good just watching somebody else do it, or repeat what somebody else has done – you must experience it yourself. I would liken it to the activity of swimming. You can watch others swim, you can practice exercises, but to swim, you must be in the water doing it yourself. (1984, p. xiii).

However, despite this widespread consensus on the overall goal of teaching and learning mathematical modelling, it remained in the publications of this time quite vague, what it means to be proficient in modelling. No explicit discussion of the definition or description of the construct 'modelling skills' or 'modelling abilities' took place at this time, no publication can be found in the proceedings of the respective conferences on the teaching of mathematical modelling and applications. The emphasis in the proceedings of ICTMA1 to ICTMA3 lay on the development of courses and their evaluation, the creation of new examples and the widening of mathematical themes to be tackled with a modelling spirit, especially the discussion about new ways to design modelling courses was important (see Berry et al. 1984, 1986, 1987; Blum et al. 1989). Ways of assessment became important soon within the context of big curriculum projects developing new modelling courses and accompanying assessment materials (for a detailed analysis of the development of the modes of assessment within ICTMA see Frejd 2013). For example the *Enterprising Mathematics Course* from the Centre of Innovation in Mathematics Teaching at the University of Exeter, one of the most important projects at this time, presented materials from this context-based course with accompanying assessment materials at ICTMA4 (Francis and Hobbs 1991), in which for the first time achievement tests for modelling were developed based on a definition of different achievement levels. Especially the evaluation of coursework, in which modelling tasks had to be tackled independently, created intensive discussions at this time. Another important project at this time from the Shell Centre for Mathematical Education at the University of Nottingham, the *Numeracy through Problem Solving Project,* aimed to implement mathematical modelling for students of all abilities (Swan 1991). Based on ground-breaking work from Treilibs, already developed in 1979, the project refined the distinction of different kinds of abilities necessary within modelling processes distinguishing technical and strategic skills from other ability components.

Characteristic of this time was the reference and connection to applied problem solving, which became apparent in a joint publication by two theme groups from the 6th International Congress on Mathematical Education (ICME-6) hold in Budapest in 1988 on "Problem Solving, Modelling and Applications" and on "Mathematics and Other Subjects". In their overview Blum and Niss (1989) devote one paragraph on assessment and tests and summarise the state-of-the-art at this time at a general level as follows:

> In conclusion we could say that currently the *role* of assessment and tests is to *inform* students and teachers rather than to provide bases for decisions or measures, and that the *character* of assessment and testing is mostly *qualitative* and *absolute* with no reference to well-defined standards. (p. 19)

The usage of the construct of abilities or skills is characteristic for this time which is not only typical for the modelling tradition, but reflects world-wide trends in psychology and pedagogy. Although no clear definition of the concept of modelling abilities or modelling skills has been developed at this time at an international level, significant studies on the identification of modelling abilities have taken place at national level. The already mentioned ground-breaking study by Treilibs (1979), in which he explored the relation of modelling to conventional ability for school mathematics, is one important example. He distinguished different component skills of modelling, which discriminate able from less able modellers, namely the ability to generate pertinent variables, to select important variables, to identify the question to be posed, to generate relationships between the variables and to select adequate relationships. In her empirical study on the teaching and learning of mathematical modelling Kaiser-Meßmer (1986) described different abilities within her theoretical framework and distinguished abilities to apply known mathematics in order to solve real world problems from modelling abilities, which referred to the abilities to carry out a modelling process, called model building skills. These model building skills referred to the different phases of a modelling process, which was described as a cycle starting from the real world going to the mathematical world and coming back to the real world. In addition general pedagogical abilities such as problem solving abilities, ability to develop heuristic strategies and to communicate were formulated and examined in an accompanying empirical study. These two quite different studies separate for the first time within the modelling debate explicitly sub-abilities as components of the modelling ability and connect them to the various phases and steps of a modelling process. These two studies reflect the state-of-the-art at this time.

Elaborate discussions on the scope and the focus on the assessment of mathematical applications promoted the modelling discussion significantly, the question, what and why to assess, how to assess modelling without destroying the spirit of modelling shaped the development based amongst others by the broad overview given by Niss (1993) at ICTMA5. New assessment schemes were developed in order to evaluate modelling courses continuing work already presented at earlier conferences (Gillespie 1993).

At this time a new pioneering approach to assess mathematical modelling was developed in the frame of a British and Australian Assessment Research Group using rating scales in order to evaluate the efficiency of mathematical modelling and in broader terms to assess the quality of students' modelling activities. Based on experts' rating of the level of quality of the modelling processes and their products, test instruments were developed, which allowed assessing students' progress during the modelling courses, and referred to early versions of Item Response Theory. The group aimed at developing a comprehensive assessment strategy which includes oral presentations too. This work was widely acknowledged and discussed, not only for the first time at ICTMA6 and many following ICTMAs (Haines and Izard 1995; Haines et al. 2001), but also, for example, at the 7th International Congress on Mathematical Education (ICME-7) (Haines et al. 1993).

Another important theme, which arose at this time, was metacognition, its importance for modelling and possibilities to promote modelling skills. According to Tanner and Jones (1995), who referred to the problem solving discussion, "metacognition involves awareness and control of one's own thinking" (p. 61) and can therefore be promoted via peer and self assessment (Tanner and Jones 1993). They emphasise that distinct elements of metacognition are not only knowledge of one's own thought processes and the control and application of that knowledge, important are "the learners' beliefs about the nature of mathematics and themselves as learners" (Tanner and Jones 1995, p. 61). Related to this work was the approach by Matos and Carreira (1995), which was connected to applied problem solving as well and analysed the cognitive processes involved in applied problem solving activities. They had already distinguished empirically different phases of modelling activities and different cognitive activities such as the identification of variables based on intuitive perceptions referring to different phases of the modelling process. Finally, these processes lead to the development of mental models, which are underlying the modelling activities and guiding the modelling process.

The aspect of assessment, that is, assessing modelling activities, to which the issue of modelling abilities or skills or competencies belong, received growing attention in the discussion on the teaching and learning of mathematical modelling as it was displayed in the ICTMA proceedings. At the end of the last century contributions – for example by Stillman (1998) – mentioned explicitly mathematical competence and their relation to meta-knowledge, with the latter aspect continuing a discussion already started by Matos and Carreira (1995).

Summarising the current knowledge of this early phase of the modelling discussion we can state that attempts to clarify students' modelling skills and/or modelling abilities were generated by several approaches, innovative ways to assess these abilities or skills were developed, not restricted to simple summative scores, but using more ambitious psychometrical models. Especially the relevance of meta-cognition and the analysis of cognitive processes were emphasised already at this stage and were continued in the more recent modelling discussion.

10.1.2 Role of Modelling Competencies in the Recent International Modelling Discussion

The more recent international modelling discussion can be described via the 14th ICMI-Study on Applications and Modelling, which provided a survey of the state-of-the-art on the teaching and learning of mathematical modelling embracing a broad community (Blum et al. 2007). Modelling competencies played a strong role at the study conference and within the ICMI study book and showed the growing importance of mathematical modelling competencies, a construct now widely used, and the high relevance of measuring modelling competencies.

Blomhøj and Højgaard Jensen (2007) introduced the construct competence referring to the Danish KOM project as follows: "*Competence* is someone's insightful readiness to act in response to the challenges of a given situation" (p. 47). With this definition, competence is "headed for action, based on but identical to neither knowledge nor skills" (p. 47). Based on the distinction of various phases of the modelling process, Blomhøj and Højgaard Jensen (2007) (similarly Blomhøj and Højgaard Jensen 2003) described two different approaches to support modelling competencies, namely the holistic and the atomistic approaches. The holistic approach requires a full-scale modelling process, where the students work through all phases of a modelling process. That approach is seen as being contrary to the atomistic approach, where students concentrate on the processes of mathematising and analysing models mathematically, because these are seen as especially demanding. They conclude:

> we believe .... a balance between the holistic approach and the atomistic approach is necessary when considering the design of an entire educational programme aiming at (among other things) developing the students' mathematical modelling competence. Neither of the two approaches alone is adequate (p. 137).

Apart from this very broad and unique approach to modelling competencies the ICMI study on modelling and applications in mathematics education contained several more papers – within a chapter devoted to modelling competencies – which are classified by Greer and Verschaffel (2007) according to three levels of mathematical modelling:

> implicit (in which the student is essentially modelling without being aware of it), explicit modelling (in which attention is drawn to the modelling process), and critical modelling (whereby the roles of modelling within mathematics and science, and within society, are critically examined). (p. 219)

Contributions based on the implicit usage of mathematical models dealt with arithmetic operations as mental models (Usiskin 2007) or focused on students' modelling behaviour from a psychological perspective (De Bock et al. 2007), papers belonging to the group of explicit modelling proposed a scheme with levels of modelling (Henning and Keune 2007), assessed the phases of mathematical modelling (Houston 2007) or elaborated on the differences between novices and experts (Haines and Crouch 2007). To the group of critical modelling belongs amongst others the already mentioned paper by Blomhøj and Højgaard Jensen (2007).

Connected to this discussion further elaborations took place at ICTMA12, where a whole chapter in the proceedings are devoted to recognising modelling competencies (Haines et al. 2007) including a plenary on this theme (Maaß 2007). There the already mentioned projects reported further results of their studies such as Crouch and Haines (2007) on the differences between experts and novices modelling behaviours, or Izard (2007) on assessing progress in mathematical modelling via special rating scales, or Højgaard Jensen (2007) on the assessment of modelling competency bearing in mind various facets of this construct. In addition, further chapters in the proceedings allowed an overview on the international discussion on modelling competencies and its worldwide understanding for example from an Australian perspective (Galbraith et al. 2007), or a Chinese perspective (Jiang et al. 2007) or a Japanese point of view (Ikeda et al. 2007).

At the same time the European Society for Research in Mathematics Education (ERME) established a working group devoted to the teaching and learning of applications and modelling for their biennial congresses. At the Fourth Congress of the European Society for Research in Mathematics Education (CERME4) in 2005 a rich variety of papers on competency were presented, which led subsequently to the publishing of two issues of *ZDM – The International Journal on Mathematics Education* (2006). Maaß (2006) proposed a comprehensive understanding of modelling competencies referring to work by Blum and Kaiser, the latter describing modelling activities with future teachers and students, which aimed to foster modelling competencies (Kaiser and Schwarz 2006). At

CERME4 already mentioned researchers presented their work, for example Haines and Crouch (2006) or Henning and Keune (2006). Unfortunately the subsequent conferences of this series did not focus specifically on modelling competencies.

A deepening of the understanding of modelling competencies took place at ICTMA14, where a comprehensive chapter in the proceedings is devoted to modelling competency (Kaiser et al. 2011). The protagonists of the discussion on modelling competencies were reporting on the progress of their projects such as Blomhøj and Kjeldsen (2011) who emphasised the role of students' reflection within the promotion of modelling competencies or Henn (2011), who analysed obstacles or support of modelling competencies. Furthermore, Engel and Kuntze (2011) related modelling competencies to statistical literacy, Frejd and Ärlebäck (2011) described results of the testing of modelling competencies in Sweden, and Zöttl et al. (2011) reported on the assessment of modelling competencies based on a multidimensional IRT-approach.

Summarising the earlier and more recent discussion on modelling competencies, it can be concluded that modelling competencies can be seen as a settled topic in the current modelling discussion, being promoted in various projects all over the world. Analysing this discussion one can identify four important strands, which have shaped the debate on modelling competencies:

- The introduction of *modelling competencies* in an *overall comprehensive concept of competencies* within the Danish KOM project (Niss, Blomhøj, and Højgaard Jensen)
- The assessment of modelling skills and the *development of assessment instruments* within a British-Australian group (mainly Haines, Houston, and Izard)
- The development of a *comprehensive concept* of modelling competencies based on *sub-competencies* and its evaluation by the German group on modelling (Blum, Kaiser, Maaß)
- The integration of *metacognition* into modelling competencies by the Australian modelling group (Galbraith, Stillman, Brown), based on earlier work by Carreira and Matos and Tanner and Jones.

This list is of course not comprehensive, there exist highly important work on modelling competencies outside these strands, which cannot be described here due to space restrictions, but can be found in the mentioned publications. In the following we will briefly analyse the essential characteristics of these four approaches in order to identify characteristics for future developments.

10.1.3 Four Strands of the Earlier and the Recent International Discussion on Modelling Competencies

As first strand the Danish KOM project can be identified, which serves as the central source for the clarification of the concept of modelling competencies. Already as early as 2002 the project developed a comprehensive approach on the definition of mathematical competencies. Niss and Højgaard (2011) described mathematical competency as "a well-informed readiness to act appropriately in situations involving a certain type of mathematical challenge" (p. 49) comprising factual knowledge and concrete skills to carry out certain types of mathematical activities. They distinguished eight competencies and defined modelling competency as one of the eight competencies, which are not seen as independent sub-competencies, but seen as aspects of competencies describing mathematical competency holistically in the sense of Shavelson (2010). Modelling competency is defined by Niss and Højgaard (2011) as follows:

> This competency involves, on the one hand, being able to *analyse* the foundations and properties of existing models and being able to assess their range and validity. Belonging to this is the ability to "*de-mathematise*" (traits of) existing mathematical models, i.e. being able to decode and interpret model elements and results in terms of the real area or situation which they are supposed to model. On the other hand, the competency involves being able to *perform active modelling* in given contexts, i.e. mathematising and applying it to situations beyond mathematics itself. (p. 58)

In addition Niss and Højgaard (2011) distinguished three dimensions of a "person's mastery of a competency", which has been taken up in the modelling discussion and differentiated for modelling by Blomhøj and Højgaard Jensen (2007) as follows: namely:

- degree of coverage relates to the part of the modelling process the students work with and the level of their reflection;
- technical level refers to the kind of mathematics students use;
- radius of action describes the domain of situations in which students are able to perform modelling activities.

This distinction has been discussed widely and serves in many projects as classification schema for modelling problems.

As a second strand of the modelling discussion on competencies, the continuing work of the already mentioned British and Australian Assessment Research Group on the development of assessment instruments for modelling skills and modelling competencies can be reconstructed. The aim of this group can be identified as development of rating scales, assessment schemas and modelling tests in order to evaluate the efficiency of modelling courses and to explore more theoretical questions such as expert-novice student behaviour. Of special interest is the procedure, which was originally used by the group already in the 1990s in order to generate these scales. They approached modelling experts, who developed descriptors for assessing students' modelling achievements and who proposed "indicators of competence" (Haines and Izard 1995, p. 135) as descriptors of students' modelling behaviour. Based on early versions of item-response theory rating scales were developed describing students' performance and descriptor effectiveness at related scales (Haines et al. 1993). As a result of their work the group generated a modelling test with multiple-choice items using a partial credit system. Based on an analytic understanding of modelling competency they distinguished various sub-competencies along the modelling cycle and developed one item per sub-competency. The total score achieved served as description of the students' modelling competencies and aimed to contribute to an overall rating scale on achievements in modelling. The developed multiple-choice items allowed a pre- and post-design and were used in many studies as a robust assessment instrument (e.g., Kaiser 2007).

As the third strand of the debate on modelling competencies, the work of German modellers can be identified, who refer to an analytic understanding of competency and base the concept of modelling competencies on different sub-competencies connected to various phases of the modelling cycle including process-oriented sub-competencies. Departing from the widely quoted definition by Weinert (2001) on competency, Kaiser (2007) defined modelling competencies analytically in distinction to modelling abilities as follows: "Modelling competencies include, in contrast to modelling abilities, not only the ability but also the willingness to work out problems, with mathematical aspects taken from reality, through mathematical modelling" (p. 110).

This approach can be characterised by the detailed description of modelling competencies by sub-competencies developed along the various phases of the modelling cycle, namely:

- competencies to understand real-world problems and to construct a reality model;
- competencies to create a mathematical model out of a real-world model;
- competencies to solve mathematical problems within a mathematical model;
- competency to interpret mathematical results in a real-world model or a real situation;
- competency to challenge solutions and, if necessary, to carry out another modelling process. (Kaiser 2007, p. 111)

Maaß (2006, 2007) emphasised that metacognition is an important influencing factor on the development of modelling competencies and defined the role of metacognition within modelling as follows: "metacognition in this study describes thinking about one's own thinking and controlling one's own thought processes" (2006, p. 118). In addition further competencies were required for an overall comprehension of modelling competencies, amongst others to model at least in parts independently, social competencies such as the ability to work in a group and to communicate about and via mathematics, to critically reflect about the modelling process and to judge its results (Maaß 2006, 2007; Kaiser 2007). Departing from this or similar definitions of modelling competencies a wealth of studies has been carried out, developing amongst others various competence descriptions and evaluating the efficiency of modelling interventions (amongst others Blum 2011; Henning and Keune 2006, 2007; Ludwig and Reit 2013; Zöttl et al. 2011).

As the fourth strand of the discussion, the integration of metacognition into the concept of modelling competencies by the Australian modelling group can be identified. This approach emphasised the importance of reflective metacognitive activity during modelling activities, especially in the transition phases between the different stages in the modelling process. Going back to the ground breaking definition of metacognition and cognitive monitoring by Flavell (1976), Stillman (2011) emphasised that a person's ability of cognitive control relies on metacognitive knowledge, metacognitive experiences, goals and strategies. Concerning modelling that approach implied that metacognitive activities such as controlling or monitoring cognitive activities were needed strongly during all phases of the modelling process, although not all activities were productive. Overall, Stillman (2011) called for the

> orchestration by teachers of the optimal use by their students modellers of metacognitive knowledge and strategies so as to develop students' competencies in their productive use of these within the modelling context to obtain not only a satisfactory outcome to the current modelling activity but also to further their long-term reflective activity for modelling purposes. (p. 179)

The work of this group (cf. Galbraith et al. 2007; Stillman and Galbraith 1998) has many commonalities with the approaches described as the third strand concerning the role of modelling activities along the modelling cycle, especially concerning the distinction of similar sub-competencies based on an analytic understanding of competency. Both strands emphasised the high importance of metacognition. As already described in the previous section, there existed several earlier studies on metacognition in modelling processes such as the studies by Matos and Carreira (1995) and Tanner and Jones (1993, 1995), which have already been mentioned in the description of the earlier modelling traditions.

Having identified four important strands of the discussion, the question arises, in which direction the discussion on modelling competencies may develop. Summarising the characteristics of these four strands commonalities can be identified, especially the connection to the modelling process, maybe described with slight differences. In any case, one important distinction becomes obvious referring to the general discussion on competence (competences) versus competency (competencies). In the general discussion on competence and its measurement two different approaches on conceptualising modelling competency are important, namely either holistically or analytically (see Shavelson 2010). Within holistic approaches competences are seen as a whole, they describe an underlying complex ability that can be reached to a certain extent. A criterion or performance level is set, those who fail to reach this level are not competent and those reaching this level are described as being competent, sometimes several hierarchical levels are distinguished. The first strand on modelling competence shaped by Niss, Blomhøj, and Højgaard Jensen can be assigned to this holistic approach. In contrast to this holistic description analytic approaches separate different facets or sub-competencies of competency, in our case the modelling competency. The overall achievement is determined based on the achievements within the different sub-competencies. The other three strands can be assigned to this analytic approach. This fundamental distinction is sometimes differently phrased as the question of whether competence can be described as competence levels with criterion or performance levels set or as competence structures exploring the relation of performance and underlying abilities. These two different kinds of approaches are not seen as mutually exclusive, but ideally as complementary. These different theoretical approaches on competency have consequences for the description of competence development, which are either seen as continuous progression shifting successively from the lowest to the highest level or as a non-continuous process with qualitative leaps (for details see Klieme et al. 2008). Although the distinction is theoretically clear, the usage of the terminology competency and competence is not clear-cut neither in the general competence debate nor in the current modelling discussion.

Returning to the construct modelling competency, these principally different approaches described as holistic versus analytic approach or as competence levels versus competence structure pose several fundamental questions for modelling traditions: How do we describe and conceptualise modelling competency, as an underlying general ability or composed of different sub-abilities? How shall modelling competency be measured, with an overall score or by a score composed of different sub-competencies? How do modelling competencies develop and how can they be fostered most efficiently, holistically or along the various sub-competencies, the latter being called atomistic by Blomhøj and Højgaard Jensen (2007)? Additionally, apart from this central distinction, which modes of assessment methods are appropriate for the evaluation of modelling competencies? Are probabilistic test methods already used by the supporters of the second strand of the international discussion adequate in order to evaluate the achievements of students in modelling courses?

In the following we will describe one study as an exemplar that provides first answers to these questions and seems to have the potential to point out future directions of research on modelling competencies. This empirical study departs from the question on the structure of the construct, modelling competency, and asks which description – holistic or analytic – is more appropriate. In addition the study goes back to the related question already posed by Blomhøj and Højgaard Jensen (2003), namely: Which kind of promotion of modelling competency is more effective, holistic or atomistic?

10.2 Further Perspectives on Modelling Competencies: Evaluation of the Structure of Modelling Competencies and Their Promotion

10.2.1 Aims and Design of the Study

The study is based on complex modelling problems, which are dealt with by students in co-operative, self-directed learning environments and foster autonomous learning via different modelling approaches. As already mentioned the study departs from the questions on the structure of modelling competency, holistic or analytic and the adequate approach of the promotion of modelling competency, holistic or atomistic. While Blomhøj and Højgaard Jensen (2003) discussed that an integration of both modelling approaches might be adequate for comprehensive modelling problems, it remains an open – not empirically evaluated – question, if one of the two approaches is more efficient in the promotion of modelling competencies. To answer these research questions, an empirical comparative study was carried out considering various aspects of the four strands of the discussion on modelling competency (for a detailed description of the study see Brand 2014a, where a description of the instrument in German can be found, for a short description in English see Brand 2014b).

The study took place in 2012 in various schools in Hamburg, the second biggest city of Germany. The participating classes were divided into two groups: one group tackled modelling problems according to the holistic approach, another group worked on modelling problems according to the atomistic approach. The intervention period started in February 2012 with a teacher-training course for each group. During the modelling project six 90-min modelling activities as well as a modelling test in a pre-, post- and follow-up-design was integrated into their ordinary mathematics lessons (for an overview see Fig. 10.1).

The modelling activities of the holistic group consisted of complete modelling problems with an increasing complexity while the modelling activities of the atomistic group included sub-tasks of modelling. The sub-tasks covered especially the transitions between the real world and mathematics and contained active and passive parts, that is, tasks that require, for example, the building of own real and mathematical models respectively and tasks that include the assessing and validating of given models or solutions. To ensure the comparability of the treatment, around 80 % of the modelling activities were observed by future teachers. In addition to this, all teachers filled in a short questionnaire after each activity about the course of the modelling lessons.

All in all, 15 classes of grade 9 of 4 secondary higher-track schools and 2 comprehensive schools participated in the project. Eight classes were assigned randomly to the holistic approach, seven classes to the atomistic approach. An overview of the sample is given in Table 10.1. Unfortunately two groups of the atomistic approach dropped out and did not participate in all three measurements.

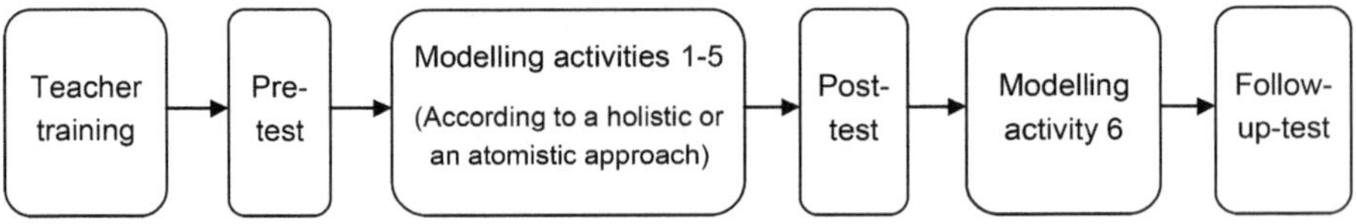

Fig. 10.1 Design of the study

Table 10.1 Sample – number of participating students

	MP 1	MP 2	MP 3	Panel
Holistic approach	168	164	169	132
Atomistic approach	159	152	97	72
Total	327	316	266	204

Altogether, 204 students participated in all 3 measurement points (MP), 132 students of the holistic group and 72 students of the atomistic group. The results presented are based on this longitudinal study (called panel). The initial mathematical abilities of the students as well as their social background were different within the two comparative groups but comparable between both groups.

To evaluate the progress of the students' modelling competencies, a modelling test in a pre-, post- and follow-up-design was developed and conducted. The structure of the modelling test refers to the work described above in the second strand, amongst others by Haines and Crouch (2001) and others, who developed items in a multiple choice format which tested different sub-competencies of mathematical modelling. A strong advantage of the approach chosen is the possibility to measure different sub-competencies of students independently from strengths and weaknesses in specific phases of the modelling process. With this test design the overall question on the structure of modelling competency can be tackled as well. This test design is similar to the design by Zöttl et al. (2011) and differentiates between sub-competencies – called in the following *dimensions* – covering the three sub-processes of mathematical modelling and an overall modelling competency to solve complete modelling tasks as follows:

Dimension 1: Simplifying/Mathematising
This sub-process focuses on the transition between the real world and mathematics and contains items referring to simplifying and mathematising of given modelling problems.

Dimension 2: Working Mathematically
The items of this dimension relate to working mathematically in a mathematical model.

Dimension 3: Interpreting/Validating
This sub-process covers the transition between mathematics and the real world and includes items that require interpreting and validating of given solutions and solving approaches.

Dimension 4: Overall Modelling Competency
This dimension includes in addition to the ability to carry out complete modelling tasks metacognitive abilities, mainly monitoring the modelling process.

The modelling test consisted of two test booklets for each measurement point, the booklets as well as the different measurement points were connected with a sufficient number of anchor items. The number of items used per dimension of the modelling competency varied between 15 and 24 per measurement point and is shown in Table 10.2 (for details see Brand 2014a).

For scaling purposes the methods of multidimensional item response theory (IRT) were used (Rost 2004) as was done previously with earlier versions of IRT methods in the second strand. In detail, the scaling was carried out with Conquest (Wu et al. 2007). The different dimensions of the modelling competency are assumed as being the latent variables that can be estimated as a multivariate function of the items solved. In order to answer the question of overall structure

Table 10.2 Number of items in the modelling test by dimension

	Number of items			
Dimension	MP 1 (booklet 1/ booklet 2)	MP 2 (booklet 1/ booklet 2)	MP 3 (booklet 1/ booklet 2)	Anchor items
Simplifying/ mathematising	16 (15/11)	22 (17/16)	19 (13/17)	11
Working mathematically	19 (17/14)	23 (17/18)	24 (19/21)	12
Interpreting/ validating	20 (15/16)	18 (18/15)	19 (14/16)	11
Overall modelling competency	15 (15/15)	15 (15/15)	15 (15/15)	14

Table 10.3 Four different psychometrical models of the structure of modelling competency

Model A: 1-dimensional model

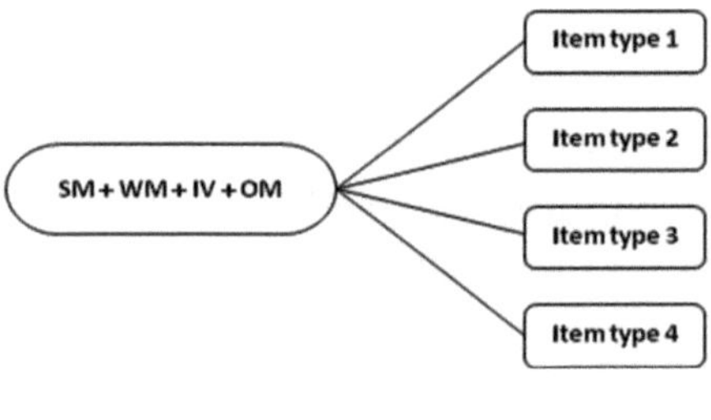

Model B: 4-dimensional between-item model

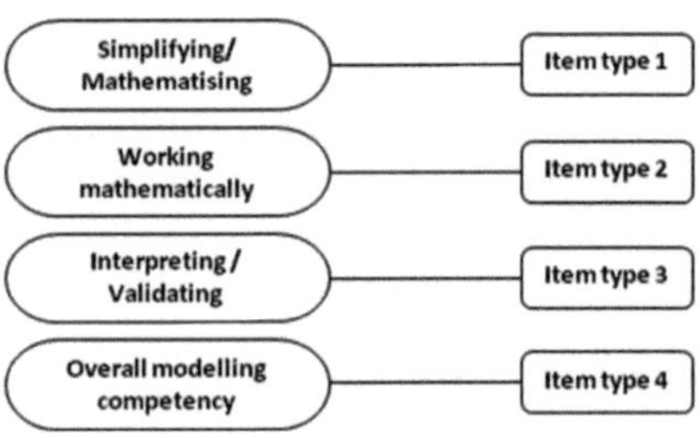

Model C: 3-dimensional within-item model

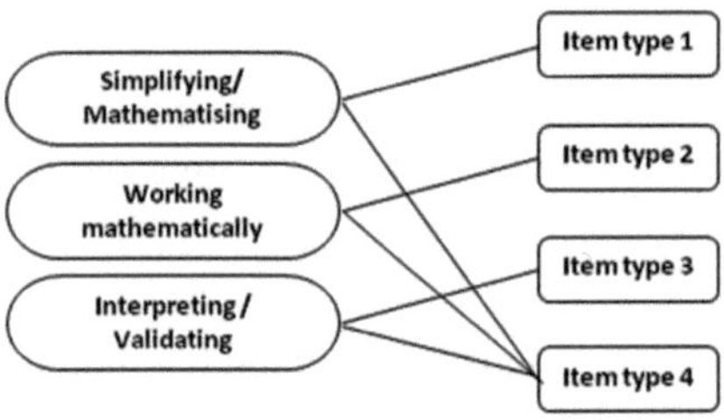

Model D: 4-dimensional within-item model

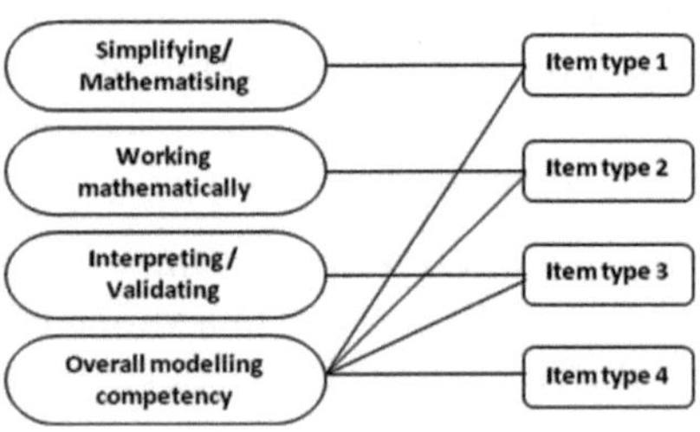

of modelling competency, different psychometrical models of the structure of the modelling competency (see Table 10.3) were scaled and compared using the psychometrical measures Akaike Information Criterion (AIC), Bayes Information Criterion (BIC) and Consistent AIC (CAIC), in order to select the relatively best model for the collected data (Rost 2004). The one-dimensional model A assumes a domain-bridging, total modelling competency – reflecting a holistic approach on modelling competence – while the other models describe various forms of analytic approaches on modelling competency. The four-dimensional between-item model

B postulates that each item loads only on one of the four dimensions. The three-dimensional within-item model C starts from the premise that all answers to the items are explicable by the three sub-processes of mathematical modelling while the four-dimensional within-item model D assumes that an overall modelling competency loads on every type of item.

The data were scaled using an approach of so-called virtual persons for all items of the three measurement points. Following the selection of the best model on the general structure of the modelling competency, weighted likelihood estimates (WLE) were estimated as individual ability parameters. The students' ability parameters for each measurement they participated in were then converted to an average value of $M = 50$ and a standard deviation of $SD = 10$. In order to answer the question on the efficiency of the holistic versus the atomistic approach to foster modelling competency within the two groups, in a first step the average test performances of the students were tested for significance that were corrected by the Bonferroni method. Additionally, the effect sizes of the performance differences were calculated. To analyse the effect of the modelling approach on the performance increase, a two-factor ANOVA with repeated measures was also carried out.

10.2.2 Results

The comparison of the different models of scaling points to the four-dimensional between-item model (see Table 10.4, model B). The indicators of a global model fit (AIC, BIC, CAIC) are the lowest for this model, so this model describes the data the best compared to the others (Rost 2004). The indicators AIC, BIC and CAIC of model D are very close to the results of model B, but the reliabilities of the different dimensions of the modelling competency of model B are higher than of model D. Model A has the best reliability, but in this case the indicators AIC, BIC and CAIC are higher than the ones of model B and therefore fit worse. To conclude, the

Table 10.4 Model selection

	Model A	Model B	Model C	Model D
Number of items	95	95	95	95
Deviance	61,721.53	60,454.06	61,911.25	60,458.39
AIC	61,911.53	60,644.06	62,101.25	60,648.39
BIC	62,002.59	60,735.12	62,192.31	60,739.46
CAIC	62,097.59	60,830.12	62,287.31	60,834.46
EAP-/PV-reliability	0.899	0.767	0.702	0.633
Dimension 1 (SM)		0.817	0.783	0.661
Dimension 2 (WM)		0.796	0.751	0.643
Dimension 3 (IV)		0.821		0.852
Dimension 4 (OM)				

data of this study suggest an analytic description of modelling competency with different sub-competencies.

Concerning the question of the efficiency of the atomistic versus the holistic approach to promote modelling competency the data are varied: concerning the different dimensions of modelling competency the data show highly significant increases in both groups between the first and the second as well as between the first and the third measurement (see Table 10.5). Concerning the first dimension, *simplifying and mathematizing*, the data showed with 0.88 a larger effect size increase for the holistic group between the first and the second measurement compared to 0.72 for the atomistic group. Between the first and the third measurement there was a higher effect size for the atomistic group (0.68) than for the holistic group (0.59). Between the second and the third measurement there are much stronger oblivion effects in the holistic group (see Table 10.5 and for a detailed documentation of all results see Brand 2014a).

The effect sizes in increase in the dimension of *working mathematically* were higher in the atomistic group between the pre- and the post-test (0.57 compared to 0.47) as well as between the post- and the follow-up-test (0.46 instead of 0.32). The effect sizes in the dimension of *interpreting and validating* were larger in the holistic group than in the atomistic group between measurement point one and measurement point two (0.77 versus 0.69) and between measurement point one and

Table 10.5 Means and performance increases of the different dimensions of the modelling competency

	Mean (SD)			Difference (Cohen's *d*)		
Group	MP 1	MP 2	MP 3	MP1 → MP2	MP1 → MP3	MP2 → MP3
Simplifying/mathematising						
Holistic group	48.26 (11.29)	57.60 (9.87)	54.90 (11.14)	+9.33*** (0.88)	+6.64*** (0.59)	−2.69* (−0.26)
Atomistic group	51.21 (7.80)	57.62 (9.84)	57.08 (9.39)	+6.41*** (0.72)	+5.87*** (0.68)	−0.54 (−0.06)
Working mathematically						
Holistic group	49.94 (10.18)	54.92 (10.83)	53.10 (9.29)	+4.98*** (0.47)	+3.16*** (0.32)	−1.82* (−0.18)
Atomistic group	48.85 (9.16)	54.76 (11.35)	53.81 (12.33)	+5.91*** (0.57)	+4.95*** (0.46)	−0.96 (−0.08)
Interpreting/validating						
Holistic group	47.93 (9.42)	55.83 (11.07)	54.38 (10.50)	+7.90*** (0.77)	+6.45*** (0.65)	−1.45 (−0.13)
Atomistic group	50.55 (8.86)	56.73 (8.95)	55.79 (9.59)	+6.19*** (0.69)	+5.24*** (0.57)	−0.95 (−0.10)
Overall modelling competency						
Holistic group	49.78 (9.34)	58.08 (9.20)	55.55 (9.62)	+8.30*** (0.90)	+5.78*** (0.61)	−2.53** (−0.27)
Atomistic group	50.75 (9.74)	57.28 (9.46)	54.21 (9.81)	+6.52*** (0.68)	+3.46* (0.35)	−3.07* (−0.32)

***p < 0.000, **p < 0.01, *p < 0.05

measurement point three (0.65 compared to 0.57). In the *overall modelling competency* there are larger effect sizes in increase in the holistic than in the atomistic group between the first and the second (0.90 instead of 0.68) as well as between the first and the third measurement (0.61 versus 0.35).

The two-factor ANOVA with repeated measures showed no significant influence of the modelling approach in the two dimensions *working mathematically* and *interpreting and validating*. However, there was a significant effect of the holistic approach on the performance increase in the competence facet *simplifying and mathematising*. In the *overall modelling competency* also a significant effect ($p < 0.1$) of the holistic approach was found (Brand 2014a).

10.2.3 Discussion of the Results and Looking Ahead

From an overall point of view the results on the construct of modelling competency confirm empirically the existence and possibility to distinguish different facets of the modelling competency along the phases of the modelling process with an overall modelling competency as an own overall component. These results on the structure of the modelling competency differ concerning the role of overall modelling competency from the results by Zöttl et al. (2011), which can be explained by various factors such as the design of the intervention and the kind of modelling examples as well as the definition of the overall modelling competency, which include metacognitive aspects in this study. To summarise, the results of the study suggests an analytic description of the construct modelling competency.

Concerning the efficiency of the different approaches on the promotion of modelling competency the study shows highly significant increases in the holistic and the atomistic group in all four dimensions of the modelling competency between the pre- and the post-test as well as between the pre- and the follow-up-test. Comparing these overall results of the holistic and the atomistic group more deeply, both approaches show strengths and weaknesses. The holistic group showed larger effect sizes in the dimension of interpreting and validating as well as in the overall modelling competency while the atomistic group achieved larger effect sizes in the dimension of working mathematically. The dimension of simplifying and mathematising revealed mixed results: while there were higher effect sizes between the pre- and the post-test in the holistic group, there were larger effect sizes in the atomistic group between the pre- and the follow-up-test. Because the activities of the classes were not controlled after the project had taken place, the results between the second and the third measurement points have to be interpreted carefully. To come to a conclusion it can be stated that both approaches foster students' modelling competency in all dimensions and have strengths and weaknesses. The assumption that the holistic approach promotes the overall modelling competency more effectively was confirmed by the data while the hypothesis that the sub-modelling competencies connected to the sub-processes of the modelling cycle can be fostered more efficiently by tackling different modelling sub-tasks via the atomistic approach was not confirmed by the data. Instead, there were higher short-term effects in the holistic than in the atomistic group in the two sub-competencies referring to the transitions between real world and mathematics.

The study shows that within a natural setting using real classes in their ordinary environment complex evaluation studies on the effectiveness of various modelling approaches can be analysed. However, these kinds of analysis demand larger samples, the inclusion of only one or two classes is not sufficient. The study shows that with complex evaluation methods using a careful classroom implementation of the study and advanced psychometrical methods important descriptions of the structure of the ambitious construct modelling competency can be developed. In addition, the study suggests that this complex construct can be described as consisting of a global, overarching modelling competency and several sub-competencies. Further studies are needed in order to examine the dependency of the results from the examples, the intervention approach used and the test instruments. These continuation studies might lead to better grounded results on the structure of the construct modelling competency and long-term effects of modelling projects and the possibility of various approaches to support modelling competencies in an overall sense.

References

Berry, J. S., Burghes, D. N., Huntley, I. D., James, D. J. G., & Moscardini, A. O. (Eds.). (1984). *Teaching and applying mathematical modelling*. Chichester: Ellis Horwood.

Berry, J. S., Burghes, D. N., Huntley, I. D., James, D. J. G., & Moscardini, A. O. (Eds.). (1986). *Mathematical modelling methodology, models and micros*. Chichester: Ellis Horwood.

Berry, J. S., Burghes, D. N., Huntley, I. D., James, D. J. G., & Moscardini, A. O. (Eds.). (1987). *Mathematical modelling courses*. Chichester: Ellis Horwood.

Blomhøj, M., & Højgaard Jensen, T. (2003). Developing mathematical modelling competence: Conceptual clarification and educational planning. *Teaching Mathematics and Its Applications, 22*(3), 123–139.

Blomhøj, M., & Højgaard Jensen, T. (2007). What's all the fuss about competencies? In W. Blum, P. L. Galbraith, H.-W. Henn, & M. Niss (Eds.), *Modelling and applications in mathematics education: The 14th ICMI study* (pp. 45–56). New York: Springer.

Blomhøj, M., & Kjeldsen, T. (2011). Students' reflections in mathematical modelling projects. In G. Kaiser, W. Blum, R. Borromeo Ferri, & G. Stillman (Eds.), *Trends in teaching and learning of mathematical modelling (ICTMA 14)* (pp. 385–395). Dordrecht: Springer.

Blum, W. (2011). Can modelling be taught and learnt? Some answers from empirical research. In G. Kaiser, W. Blum, R. Borromeo Ferri, & G. Stillman (Eds.), *Trends in teaching and learning of mathematical modelling (ICTMA 14)* (pp. 15–30). Dordrecht: Springer.

Blum, W., & Niss, M. (1989). Mathematical problem solving, modelling, applications, and links to other subjects: State, trends and issues in mathematics instruction. In W. Blum, M. Niss, & I. Huntley (Eds.), *Modelling, applications and applied problem solving* (pp. 1–21). Chichester: Ellis Horwood.

Blum, W., Berry, J. S., Biehler, R., Huntley, I. D., Kaiser-Messmer, G., & Profke, L. (Eds.). (1989). *Applications and modelling in learning and teaching mathematics*. Chichester: Ellis Horwood.

Blum, W., Galbraith, P. L., Henn, H.-W., & Niss, M. (Eds.). (2007). *Modelling and applications in mathematics education: The 14th ICMI study*. New York: Springer.

Brand, S. (2014a). *Erwerb von Modellierungskompetenzen. Empirischer Vergleich eines holistischen und eines atomistischen Ansatzes zur Förderung von Modellierungskompetenzen*. Wiesbaden: Springer Spektrum.

Brand, S. (2014b). Effects of a holistic versus an atomistic modelling approach on students' mathematical modelling competencies. In P. Liljedahl, C. Nicol, S. Oesterle & D. Allan (Eds.), *Proceedings of the Joint Meeting of PME 38 and PME-NA 36, Vol. 2* (pp. 185–192). Vancouver: PME.

Burghes, D. N. (1984). Prologue. In J. S. Berry, D. N. Burghes, I. D. Huntley, D. J. G. James, & A. O. Moscardini (Eds.), *Teaching and applying mathematical modelling* (pp. xi–xvi). Chichester: Ellis Horwood.

Crouch, R., & Haines, C. (2007). Exemplar models: Expert-novice student behaviours. In C. Haines, P. Galbraith, W. Blum, & S. Khan (Eds.), *Mathematical modelling (ICTMA 12): Education, engineering and economics* (pp. 90–109). Chichester: Ellis Horwood.

De Bock, D., Van Dooren, W., & Janssens, D. (2007). Studying and remedying students' modelling competencies: Routine behaviour or adaptive expertise. In W. Blum, P. L. Galbraith, H.-W. Henn, & M. Niss (Eds.), *Modelling and applications in mathematics education: The 14th ICMI study* (pp. 241–248). New York: Springer.

Engel, J., & Kuntze, S. (2011). From data to functions: Connecting modelling competencies and statistical literacy. In G. Kaiser, W. Blum, R. Borromeo Ferri, & G. Stillman (Eds.), *Trends in teaching and learning of mathematical modelling* (pp. 397–406). Dordrecht: Springer.

Flavell, J. H. (1976). Metacognitive aspects of problem solving. In L. B. Resnick (Ed.), *The nature of intelligence* (pp. 231–235). Hillsdale: Erlbaum.

Francis, B., & Hobbs, D. (1991). Enterprising mathematics: A context-based course with context-based assessment. In M. Niss, W. Blum, & I. Huntley (Eds.), *Teaching of mathematical modelling and applications* (pp. 147–157). Chichester: Ellis Horwood.

Frejd, P. (2013). Modes of modelling assessment – A literature review. *Educational Studies in Mathematics, 84*(3), 413–438.

Frejd, P., & Ärlebäck, J. B. (2011). First results from a study investigating Swedish upper secondary students' mathematical modelling competencies. In G. Kaiser, W. Blum, R. Borromeo Ferri, & G. Stillman (Eds.), *Trends in teaching and learning of mathematical modelling* (pp. 407–416). Dordrecht: Springer.

Galbraith, P., Stillman, G., Brown, J., & Edwards, I. (2007). Facilitating middle secondary modelling competencies. In C. Haines, P. Galbraith, W. Blum, & S. Khan (Eds.), *Mathematical modelling (ICTMA 12): Education, engineering and economics* (pp. 130–140). Chichester: Ellis Horwood.

Gillespie, J. (1993). Numeracy to mathematics. In J. de Lange, C. Keitel, I. Huntley, & M. Niss (Eds.), *Innovation in maths education by modelling and applications* (pp. 311–321). Chichester: Ellis Horwood.

Greer, B., & Verschaffel, L. (2007). Modelling competencies – Overview. In W. Blum, P. L. Galbraith, H.-W. Henn, & M. Niss (Eds.), *Modelling and applications in mathematics education: The 14th ICMI study* (pp. 219–224). New York: Springer.

Haines, C. R., & Crouch, R. M. (2001). Recognising constructs within mathematical modelling. *Teaching Mathematics and Its Applications, 20*(3), 129–138.

Haines, C., & Crouch, R. (2006). Getting to grips with real world contexts: developing research in mathematical modeling. In M. Bosch (Ed.), *CERME 4 – Conference of European research in mathematics education* (pp. 1655–1665). Guixols: European Society for Research in Mathematics Education.

Haines, C., & Crouch, R. (2007). Mathematical modelling and applications: Ability and competence frameworks. In W. Blum, P. L. Galbraith, H.-W. Henn, & M. Niss (Eds.), *Modelling and applications in mathematics education: The 14th ICMI study* (pp. 417–424). New York: Springer.

Haines, C., & Izard, J. (1995). Assessment in context for mathematical modelling. In C. Sloyer, W. Blum, & I. Huntley (Eds.), *Advances and perspectives in the teaching of mathematical modelling and applications* (pp. 131–149). Yorklyn: Water Street Mathematics.

Haines, C., Izard, J., & Le Masurier, D. (1993). Modelling intensions realised: Assessing the full range of developed skills. In T. Breiteig, I. Huntley, & G. Kaiser-Messmer (Eds.), *Teaching and learning mathematics in context* (pp. 200–211). Chichester: Horwood.

Haines, C., Crouch, R., & Davies, J. (2001). Understanding students' modelling skills. In J. F. Matos, W. Blum, K. Houston, & S. P. Carreira (Eds.), *Modelling and mathematics education (ICTMA 9): Applications in science and technology* (pp. 366–380). Chichester: Ellis Horwood.

Haines, C., Galbraith, P., Blum, W., & Khan, S. (Eds.). (2007). *Mathematical modelling (ICTMA 12): Education, engineering and economics*. Chichester: Ellis Horwood.

Henn, H.-W. (2011). Why cats happen to fall from the sky or on good and bad models. In G. Kaiser, W. Blum, R. Borromeo Ferri, & G. Stillman (Eds.), *Trends in teaching and learning of mathematical modelling (ICTMA 14)* (pp. 417–426). Dordrecht: Springer.

Henning, H., & Keune, M. (2006). Levels of modelling competence. In M. Bosch (Ed.), *CERME 4. Proceedings of the fourth congress on the European society for research in mathematics education* (pp. 1666–1674). San Feliude Guixols: European Society for Research in Mathematics Education.

Henning, H., & Keune, M. (2007). Levels of modelling competencies. In W. Blum, P. L. Galbraith, H.-W. Henn, & M. Niss (Eds.), *Modelling and applications in mathematics education: The 14th ICMI study* (pp. 225–232). New York: Springer.

Højgaard Jensen, T. (2007). Assessing mathematical modelling competency. In C. P. Haines, P. Galbraith, W. Blum, & S. Khan (Eds.), *Mathematical modelling (ICTMA 12): Education, engineering and economics* (pp. 141–148). Chichester: Horwood.

Houston, K. (2007). Assessing the "phase" of mathematical modelling. In W. Blum, P. L. Galbraith, H.-W. Henn, & M. Niss (Eds.), *Modelling and applications in mathematics education: The 14th ICMI study* (pp. 249–256). New York: Springer.

Ikeda, T., Stephens, M., & Matsuzaki, A. (2007). A teaching experiment in mathematical modelling. In C. Haines, P. Galbraith, W. Blum, & S. Khan (Eds.), *Mathematical modelling (ICTMA 12): Education, engineering and economics* (pp. 101–109). Chichester: Ellis Horwood.

Izard, J. (2007). Assessing progress in mathematical modelling. In C. Haines, P. Galbraith, W. Blum, & S. Khan (Eds.), *Mathematical modelling (ICTMA 12): Education, engineering and economics* (pp. 158–167). Chichester: Ellis Horwood.

Jiang, Q., Xie, J., & Ye, Q. (2007). An introduction to CUMCM. In C. Haines, P. Galbraith, W. Blum, & S. Khan (Eds.), *Mathematical modelling (ICTMA 12): Education, engineering and economics* (pp. 168–175). Chichester: Ellis Horwood.

Kaiser, G. (2007). Modelling and modelling competencies in school. In C. Haines, P. Galbraith, W. Blum, & S. Khan (Eds.), *Mathematical modelling (ICTMA 12): Education, engineering and economics* (pp. 110–119). Chichester: Ellis Horwood.

Kaiser, G., & Schwarz, B. (2006). Mathematical modelling as bridge between school and university. *Zentralblatt für Didaktik der Mathematik, 38*(2), 196–208.

Kaiser, G., Blum, W., Borromeo Ferri, R., & Stillman, G. (2011). *Trends in teaching and learning of mathematical modelling (ICTMA 14)*. Dordrecht: Springer.

Kaiser-Meßmer, G. (1986). *Anwendungen im Mathematikunterricht* (Empirische Untersuchungen, Vol. 2). Bad Salzdetfurth: Franzbecker.

Klieme, E., Hartig, J., & Rauch, D. (2008). The concept of competence in educational contexts. In J. Hartig, E. Klieme, & D. Leutner (Eds.), *Assessment of competencies in educational contexts: State of the art and future prospects* (pp. 3–22). Göttingen: Hogrefe & Huber.

Ludwig, M., & Reit, X.-R. (2013). A cross-sectional study about modelling competency in secondary school. In G. A. Stillman, G. Kaiser, W. Blum, & J. P. Brown (Eds.), *Teaching mathematical modelling: Connecting to research and practice* (pp. 327–337). Dordrecht: Springer.

Maaß, K. (2006). What are modelling competencies? *ZDM – The International Journal on Mathematics Education, 38*(2), 113–142.

Maaß, K. (2007). Modelling in class: What do we want the students to learn? In C. Haines, P. Galbraith, W. Blum, & S. Khan (Eds.), *Mathematical modelling (ICTMA 12): Education, engineering and economics* (pp. 63–78). Chichester: Ellis Horwood.

Matos, J., & Carreira, S. (1995). Cognitive processes and representations involved in applied problem solving. In C. Sloyer, W. Blum, & I. Huntley (Eds.), *Advances and perspectives in the teaching of mathematical modelling and applications* (pp. 71–80). Yorklyn: Water Street Mathematics.

Niss, M. (1993). Assessment of mathematical modelling and applications in mathematics teaching. In J. de Lange, C. Keitel, I. Huntley, & M. Niss (Eds.), *Innovation in mathematics education by modelling and applications* (pp. 41–51). Chichester: Horwood.

Niss, M., & Højgaard, T. (Eds.) (2011). *Competencies and mathematical learning. Ideas and inspiration for the development of mathematics teaching and learning in Denmark.* Roskilde University: IMFUFA. English translation of Danish original (2002). http://pure.au.dk/portal/files/41669781/THJ11_MN_KOM_in_english.pdf

Rost, J. (2004). *Lehrbuch Testtheorie – Testkonstruktion* (2nd ed.). Bern: Huber.

Shavelson, R. (2010). On the measurement of competency. *Empirical Research in Vocational Education and Training, 2*(1), 41–63.

Stillman, G. (1998). The emperor's new clothes? Teaching and assessment of mathematical applications at the senior secondary level. In P. Galbraith, W. Blum, G. Booker, & I. D. Huntley (Eds.), *Mathematical modelling: Teaching and assessment in a technology-rich world* (pp. 243–253). Chichester: Ellis Horwood.

Stillman, G. (2011). Applying metacognitive knowledge and strategies in applications and modelling tasks at secondary school. In G. Kaiser, W. Blum, R. Borromeo Ferri, & G. Stillman (Eds.), *Trends in teaching and learning of mathematical modelling* (pp. 165–180). Dordrecht: Springer.

Stillman, G., & Galbraith, P. (1998). Applying mathematics with real world connections: Metacognitive characteristics of secondary students. *Educational Studies in Mathematics, 36* (2), 157–195.

Swan, M. (1991). Mathematical modelling for all abilities. In M. Niss, W. Blum, & I. Huntley (Eds.), *Teaching of mathematical modelling and applications* (pp. 137–146). Chichester: Ellis Horwood.

Tanner, H., & Jones, S. (1993). Developing metacognition through peer and self assessment. In T. Breiteig, I. Huntley, & G. Kaiser-Messmer (Eds.), *Teaching and learning mathematics in context* (pp. 228–240). Chichester: Ellis Horwood.

Tanner, H., & Jones, S. (1995). Developing metacognitive skills in mathematical modelling – A socio-constructivist interpretation. In C. Sloyer, W. Blum, & I. Huntley (Eds.), *Advances and perspectives in the teaching of mathematical modelling and applications* (pp. 61–70). Yorklyn: Water Street Mathematics.

Treilibs, V. (1979). *Formulation processes in mathematical modelling.* Unpublished thesis, University of Nottingham for the degree of Master of Philosophy.

Usiskin, Z. (2007). The arithmetic operations as mathematical models. In W. Blum, P. L. Galbraith, H.-W. Henn, & M. Niss (Eds.), *Modelling and applications in mathematics education: The 14th ICMI study* (pp. 257–264). New York: Springer.

Weinert, F. E. (2001). Concepts of competence: A conceptual clarification. In D. Rychen & L. Salgnik (Eds.), *Defining and selecting key competencies* (pp. 45–66). Göttingen: Hogrefe & Huber.

Wu, M. L., Adams, R. J., Wilson, M. R., & Haldane, S. A. (2007). *ACER conquest version 2.0. Generalised item response modelling software*. Melbourne: ACER Press.

Zöttl, L., Ufer, S., & Reiss, K. (2011). Assessing modelling competencies using a multidimensional IRT-approach. In G. Kaiser, W. Blum, R. Borromeo Ferri, & G. Stillman (Eds.), *Trends in teaching and learning of mathematical modelling (ICTMA 14)* (pp. 427–437). Dordrecht: Springer.

Chapter 11
How to Support Teachers to Give Feedback to Modelling Tasks Effectively? Results from a Teacher-Training-Study in the Co^{2}CA Project

Michael Besser, Werner Blum, and Dominik Leiss

Abstract Assessing and reporting students' performance is an important part of every-day teaching. Doing so without focusing on marks but instead on supporting students in further learning by giving individual, process-oriented and task-related feedback to them is a central idea of formative assessment. Within a teacher-training-study as part of the research project *Conditions and Consequences of Classroom Assessment* (Co^{2}CA), a special in-service teacher training has been developed to foster teachers' knowledge about formative assessment when dealing with modelling tasks. A test on teachers' pedagogical content knowledge (PCK-test) has been used to evaluate the trainings. Quantitative results point out that teachers having taken part in the trainings outperform teachers not having been trained in formative assessment if dealing with modelling tasks within the PCK-test.

11.1 Introduction

As part of general discussions about competency-oriented teaching of mathematics and about how to improve it (see e.g., Blomhoj and Jensen 2007; Niss 2003), the question of how to assess and report students' performances in such a way that students' learning is supported as much as possible is of central interest. Based on

M. Besser (✉) • D. Leiss
Institute of Mathematics and its Didactic, Leuphana University of Lüneburg, Lüneburg 21335, Germany
e-mail: michael.besser@leuphana.de; dominik.leiss@leuphana.de

W. Blum
Institute of Mathematics, University of Kassel, Kassel 34109, Germany
e-mail: blum@mathematik.uni-kassel.de

G.A. Stillman et al. (eds.), *Mathematical Modelling in Education Research and Practice*, International Perspectives on the Teaching and Learning of Mathematical Modelling, DOI 10.1007/978-3-319-18272-8_11

this question the interdisciplinary research project Co^2CA[1] investigates how assessing and reporting students' performance can successfully be implemented in everyday teaching of competency-oriented mathematics. One main finding of previous work within Co^2CA as a result of a laboratory-study (2009/2010) and a field-study in school (2010/2011) is (see e.g., Besser and Blum 2012): The quality of implementation of assessing and reporting students' performance differs extremely between classes and teachers. Therefore, an in-service teacher training has been developed within a teacher-training-study to foster teachers' knowledge about formative assessment in competency-oriented mathematics, in particular concerning mathematical modelling.

11.2 The Idea of Formative Assessment

While in school students' performance is quite often assessed only once at the end of a course and a summarized report is given to the students which "does not normally have immediate impact in learning" (Sadler 1989, p. 120), theoretical and empirical discussions hint at the importance of a more formative assessment within classrooms, that is (Baker 2007; Black and William 2009; Shepard 2000): Students' performance should be assessed in short intervals, the assessment should be close to students' learning processes, and diagnoses of students' achievement should be used for supporting further learning. Additionally, as a special element of such a formative assessment, appropriate feedback should be given to the students which answers three central questions: "Where am I going? How am I going? and Where to next?" (Hattie and Timperley 2007, p. 88). Moreover, some meta-analyses hint at the following aspects of how feedback "with which a learner can confirm, add to, overwrite, tune, or restructure information in memory" (Butler and Winne 1995, p. 275) as part of formative assessment should look like:

- Kluger and DeNise (1996) point out that feedback does particularly have an impact on students' learning if it is close to the tasks the students had to work on and the students' solution processes for these tasks. "Effects on performance are augmented by (a) cues that direct attention to task-motivation processes and (b) cues that direct attention to task-learning processes" (p. 268).
- Deci et al. (1999) emphasise that feedback should not only tell students whether a solution is right or wrong but also offer information about how to continue learning without any kind of pressure (see also Bangert-Drowns et al. 1991; Pittman et al. 1980).

[1] *Conditions and Consequences of Classroom Assessment.* Research project supported by the German Research Society (DFG) as part of the current priority programme "Kompetenzmodelle zur Erfassung individueller Lernergebnisse und zur Bilanzierung von Bildungsprozessen" (SPP 1293); principal researchers: E. Klieme, K. Rakoczy (both Frankfurt), W. Blum (Kassel), D. Leiss (Lüeneburg). KL 1057/10–3, BL 275/16–3, LE 2619/1–3.

11.3 Mathematical Modelling as Part of Competency-Oriented Mathematics

Within the last few years, several countries have tried to implement national standards for mathematics teaching and learning that should improve the overall quality of teaching and learning at school (see e.g., NCTM 2000). A crucial aspect of these standards is to no longer only tell teachers which mathematical content should be taught at school but to prescribe which mathematical competencies students should acquire. Besides other competencies (e.g., problem solving, reasoning, communicating – see Blomhoj and Jensen 2007), the concept of mathematical modelling can be understood as one central part of competency-oriented mathematics. Working on real world problems, students do not only have to "solve" purely mathematical tasks but also have to cope with everyday problems by using mathematics. In detail, some central elements of the modelling competency can be described as follows (see also Blum 2011; Maaß 2010):

- Given a real world problem, one has to be able to simplify the problem and transfer it into mathematics by creating a so-called mathematical model of the real world problem.
- Being able to find a mathematical solution of the corresponding mathematical problem, to transfer the mathematical result back to reality, and to validate this result (does it make sense in the real world?).

11.4 Teacher Knowledge as Predictor for the Quality of Teaching and Students' Achievement

For nearly the whole twentieth century researchers have tried to explain students' achievements by analysing the role of the teacher in the classroom – sometimes first with a focus on the teachers' personality, then with a focus on learning processes and products, and more recently also with a focus on teachers' expertise. Especially since the work of Shulman (1986, 1987), the idea of explaining successful teaching by teachers' content knowledge (CK), pedagogical content knowledge (PCK) and general pedagogical knowledge (PK) as part of teachers' expertise helps to understand the teachers' role in the classroom. CK, PCK and PK can be said to be "three core dimensions of teacher knowledge" (Baumert et al. 2010, p. 135). By describing, discussing and assessing teacher knowledge, several studies hark back to these dimensions and stress the importance of these aspects of teacher knowledge for teaching and for student learning. With a special focus on mathematics teachers, especially the COACTIV-project (Baumert et al. 2010), the Michigan-Group (Ball et al. 2005) and the TEDS-project (Döhrmann et al. 2012) point out the importance of CK, PCK and PK for the quality of teaching and students' achievement.

Furthermore with a special focus on teachers' pedagogical content knowledge, Baumert et al. (2010, p. 168) point out:

> PCK – the area of knowledge relating specifically to the main activity of teachers, namely, communicating subject matter to students – makes the greatest contribution to explaining student progress. This knowledge cannot be picked up incidentally, but as our finding on different teacher-training programs show, it can be acquired in structured learning environments. One of the next great challenges for teacher research will be to determine how this knowledge can best be conveyed to both preservice and inservice teachers.

11.5 A Teacher-Training-Study for In-service Teachers: Formative Assessment in Competency-Oriented Mathematics

As a result of the importance of implementing formative assessment in everyday teaching of mathematics, the idea that students should be able to work on real world problems and the importance of teacher knowledge (especially CK, PCK and PK) for students' learning, the main research questions of the teacher-training-study are:

1. Is it possible to develop and conduct a teacher training that fosters teachers' pedagogical content knowledge and general pedagogical knowledge about formative assessment in competency-oriented mathematics if dealing with modelling tasks?
2. Is it possible to develop tests (PCK and PK) which can be reliably used to evaluate this teacher training?
3. Does a successful teacher training have any influence on the teachers' way of teaching and/or on students' performance?

11.5.1 Design of the Teacher-Training-Study

In order to answer these questions, the study is conceived as follows (see also Fig. 11.1): Overall 67 teachers took part in the teacher-training-study which started in February 2013 and ended in December 2013. Due to organizational restrictions there are four smaller experimental-groups (EG A1/EG A2 and EG B1/EG B2), each consisting of a maximum of 20 teachers. Over a period of 10 weeks, the teachers of every group had 3 days of training at the beginning and 3 days of training at the end. Furthermore, the teachers had to implement central ideas of the teacher training in their classroom-teaching within the 10 weeks of the teacher-training-study. To control for teachers' prior knowledge, the teachers were tested on PCK at the beginning of the teacher-training-study (the PCK-test was taken from the COACTIV-study; see e.g., Krauss et al. 2008). For being able to discuss differences between the four experimental-groups there were additional tests for teachers on PCK and PK at the end of the teacher-training-study as well as

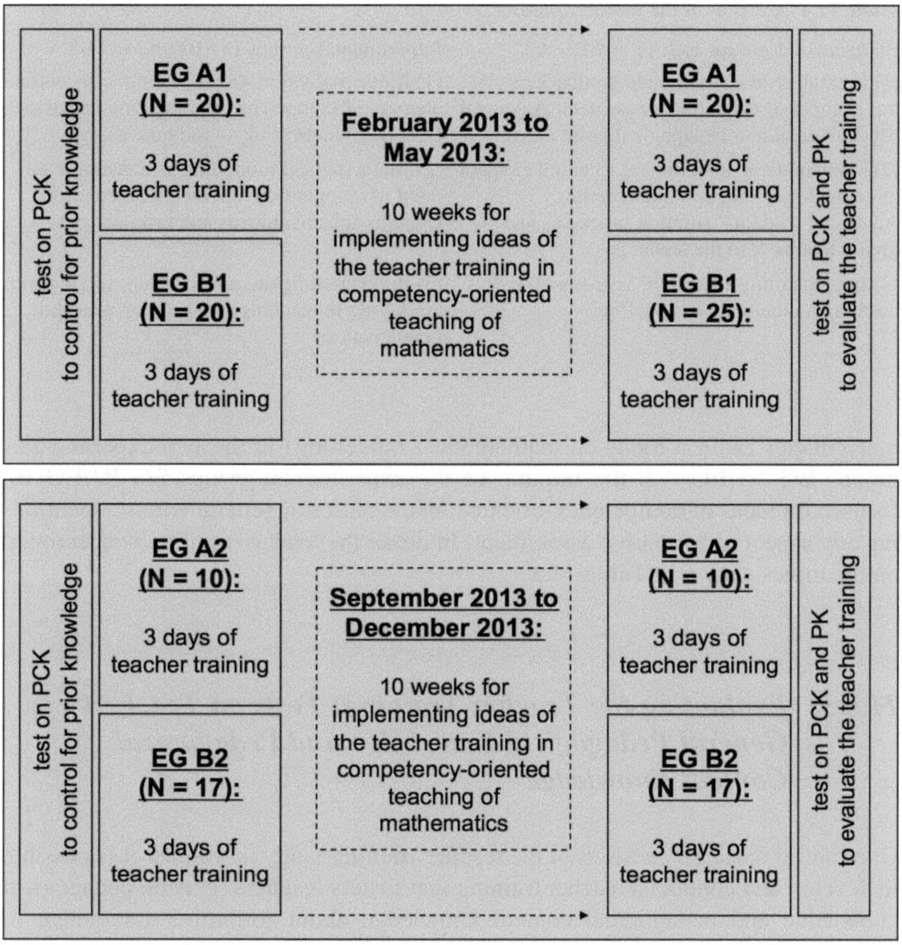

Fig. 11.1 Design of the teacher-training-study

questionnaires for teachers about the perceived usefulness of the training and questionnaires for students about the perceived quality of teaching within the 10 weeks of the study.

11.5.2 Content of the Teacher Training

The content of the teacher training differed between the two experimental-groups EG A1/EG A2 and the two experimental groups EG B1/EG B2 and was identical within EG A1/EG A2 respectively EG B1/EG B2. While there was a training dealing mainly with ideas of formative assessment in competency-oriented

Table 11.1 Content of the teacher-training

Experimental-groups EG A1 and EG A2	Experimental-groups EG B1 and EG B2
(1) Formative assessment and feedback as central element of formative assessment: A general psychological and pedagogical point of view	(1) Mathematical problem solving as a central element of competency-oriented mathematics: General didactical ideas and task-analyses
(2) Mathematical modelling as a central element of competency-oriented mathematics: Analysing students' solution processes and giving feedback to the students	(2) Mathematical modelling as a central element of competency-oriented mathematics: General didactical ideas and task-analyses
(3) Implementing formative assessment in teaching mathematical modelling	(3) Implementing tasks on problem solving and modelling in teaching competency-oriented mathematics

mathematics (with a focus on mathematical modelling) in the two experimental-groups EG A1/EG A2, the training of the experimental-groups EG B1/EG B2 focused on ideas of competency-oriented mathematics in general without mentioning any aspect of formative assessment. In detail the teacher training concentrated on the topics shown in Table 11.1.

11.5.3 Evaluating the Teacher Training: Tests on Teachers' General Pedagogical Knowledge and Pedagogical Content Knowledge

One central research question of the teacher-training-study is whether it is possible to develop and conduct a teacher training that fosters teachers' general pedagogical knowledge and pedagogical content knowledge about formative assessment in competency-oriented mathematics (with a focus on mathematical modelling). For being able to answer this question, special tests on teachers' knowledge are needed which assess PK and PCK concerning formative assessment (FA) in competency-oriented mathematics and which indicate whether teachers in the experimental-groups EG A1/EG A2 outperform their counterparts in the experimental-groups EG B1/EG B2. Since such tests did not exist before, items testing teachers' knowledge concerning the following topics were developed:

- Subtest on general pedagogical knowledge (PK-FA): Items assessing theoretical pedagogical and psychological knowledge (1) about assessment in classroom, (2) about ways to diagnose students' strengths and weaknesses, and (3) about how to give feedback to students' strengths and weaknesses.
- Subtest on pedagogical content knowledge (PCK-FA): Items assessing subject-specific knowledge (1) about modelling processes in mathematics, (2) about how to analyse students' solution processes to modelling tasks, (3) about how to implement general ideas about formative assessment in competency-oriented

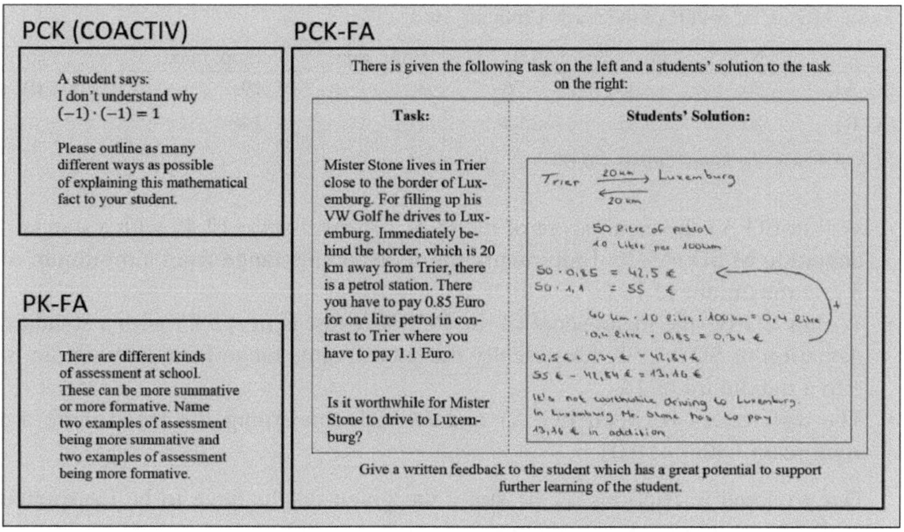

Fig. 11.2 Examples of items; the "PCK-COACTIV-item" is taken from Krauss et al. (2008)

teaching, that is especially about how to give feedback to students if working on modelling tasks. (Examples of items assessing PCK at the beginning of the study by using the COACTIV-test as well as items assessing PK-FA and PCK-FA are given in Fig. 11.2.)

11.5.4 First Results of the Teacher-Training-Study

By August 2013, the first part of the teacher-training-study from February 2013 to March 2013 (EG A1 and EG B1) was conducted and the subtests on teachers' pedagogical content knowledge (administered at the end of the teacher training) focusing on central ideas of formative assessment when dealing with modelling tasks (PCK-FA) have been coded. Therefore results on teachers' performance on the PCK-FA-test comparing EG A1 and EG B1 can be reported here, offering first insights into answers to the research questions (1) and (2). The subtest consists of 10 items (8 items with open responses). Depending on the item up to 1 score-point (2 items), up to 2 score-points (5 items) or up to 3 score-points (3 items) can be reached, a wrong answer is scored by 0. Based on a theoretical maximum of 21 score-points the results of the PCK-FA-test are as follows (see also Table 11.2):

- Altogether 40 teachers participated in the PCK-FA-test, 20 of these teachers participated in the teacher training of EG A1 and 20 in the training of EG B1.
- The reliability of the PCK-FA-test – consisting of 10 items – can be said to be quite good (alpha = 0.80).

Table 11.2 First results of the teacher-training-study

	N	m	SD	Emp. min.	Emp. max.	
EG A1	20	12.45	3.76	3	19	p < .01
EG B1	20	5.85	2.98	1	14	

PCK-FA-test: 10 items; alpha = 0.80

- Within EG A1 the mean-score of the PCK-FA-test is m = 12.45 with a standard deviation of SD = 3.76. Empirically the score-points range from a minimum of 3 to a maximum of 19.
- Within EG B1 the mean-score of the PCK-FA-test is m = 5.85 with a standard deviation of SD = 2.98. Empirically the score-points range from a minimum of 1 to a maximum of 14.
- The differences between EG A1 and EG B1 concerning the mean-score are significant with $p < 0.01$.

Due to some restrictions of the study the given results have to be interpreted carefully concerning the effectiveness of teacher-trainings on teachers' expertise: (1) It is not really a longitudinal study but a study assessing teachers' PCK being sensitive to the trainings once at the end of the trainings. (2) There does not really exist a typical control-group. The study investigates the effectiveness of teacher-trainings by comparing two different experimental groups. (3) The teachers have been assigned to the experimental groups by interest and not randomly. However these preliminary results of a subgroup of the teacher-training-study already point out big differences in teachers' expertise concerning formative assessment if dealing with modelling tasks in competency-oriented mathematics at the end of teacher trainings dealing with different topics.

11.6 Summary and Outlook

According to theory and empirical studies, the idea of implementing formative assessment in competency-oriented mathematics is crucial. Furthermore it has been pointed out that teachers' knowledge (CK, PCK and PK) is essential for students' achievements. Within the teacher-training-study as part of the Co²CA-project, teachers were trained in implementing formative assessment in competency-oriented mathematics. For being able to evaluate the effect of the teacher-training-study, special tests on teachers' general pedagogical knowledge and pedagogical content knowledge have been developed for evaluating and comparing the teachers' performance at the end of the training. Using these tests, teachers trained especially in ideas of formative assessment when dealing with modelling tasks (EG A1 and EG A2) can be contrasted with teachers trained in general didactical ideas of modelling and problem solving in competency-oriented mathematics (EG B1 and EG B2). First results show that the teachers of EG A1 outperform their counterparts of EG B1 at the end of the teacher training. These results are encouraging and hint at the importance of teacher trainings to foster teachers' knowledge about teaching competency-oriented mathematics. Whether these results are really caused by the teacher training – especially if controlling for teachers' general pedagogical content knowledge and teachers' content knowledge – are open questions that have to be answered at the end of the whole teacher-training-study, as well as questions about the effect of teacher training on teachers' general pedagogical knowledge and the quality of teaching.

References

Baker, E. L. (2007). The end(s) of testing. *Educational Researcher, 36*, 309–317.

Ball, D. L., Hill, H. C., & Bass, H. (2005). Knowing mathematics for teaching: Who knows mathematics well enough to teach third grade, and how can we decide? *American Educator, 29* (1), 14–17, 20–22, 43–46.

Bangert-Drowns, R. L., Kulik, C. C., Kulik, J. A., & Morgan, M. (1991). The instructional effect of feedback in test-like events. *Review of Educational Research, 61*, 213–238.

Baumert, J., Kunter, M., Blum, W., Brunner, M., Voss, T., Jordan, A., Klusmann, U., Krauss, S., Neubrand, M., & Tsai, Y.-M. (2010). Teachers' mathematical knowledge, cognitive activation in the classroom, and student progress. *American Educational Research Journal, 47*(1), 133–180.

Besser, M., & Blum, W. (2012). Competency-oriented written feedback in every-day mathematics teaching: How to report on students' solutions or modelling tasks and how to assess the quality of these reports? In *The 12th international congress on mathematical education (Pre-Proceedings)* (pp. 3188–3195). Retrieved from http://www.icme12.org

Black, P., & Wiliam, D. (2009). Developing the theory of formative assessment. *Educational Assessment, Evaluation and Accountability, 21*(1), 5–31.

Blomhoj, M., & Jensen, T. H. (2007). What's all the fuss about competencies? Experiences with using a competence perspective on mathematics education to develop the teaching of mathematical modelling. In W. Blum, P. L. Galbraith, H. W. Henn, & M. Niss (Eds.), *Modelling and applications in mathematics education. The 14th ICMI study* (pp. 45–56). New York: Springer.

Blum, W. (2011). Can modelling be taught and learnt? Some answers from empirical research. In G. Kaiser, W. Blum, R. Borromeo Ferri, & G. Stillman (Eds.), *Trends in teaching and learning mathematical modelling* (pp. 15–30). Dordrecht: Springer.

Butler, D. L., & Winne, P. H. (1995). Feedback and self-regulated learning: A theoretical synthesis. *Review of Educational Research, 65*(3), 245–281.

Deci, E. L., Koestner, R., & Ryan, R. M. (1999). A meta-analytic review of experiments examining the effects of extrinsic rewards on intrinsic motivation. *Psychological Bulletin, 125*(6), 627–668.

Döhrmann, M., Kaiser, G., & Blömeke, S. (2012). The conceptualisation of mathematics competencies in the international teacher education study TEDS-M. *ZDM: The International Journal on Mathematics Education, 44*, 325–340.

Hattie, J., & Timperley, H. (2007). The power of feedback. *Review of Educational Research, 77* (1), 81–112.

Kluger, A. N., & DeNisi, A. (1996). The effects of feedback interventions on performance: A historical review, a meta-analysis, and a preliminary feedback intervention theory. *Psychological Bulletin, 119*(2), 254–284.

Krauss, S., Brunner, M., Kunter, M., Baumert, J., Blum, W., Neubrand, M., & Jordan, A. (2008). Pedagogical content knowledge and content knowledge of secondary mathematics teachers. *Journal of Educational Psychology, 100*(3), 716–725.

Maaß, K. (2010). Classification scheme for modelling tasks. *Journal für Mathematik-Didaktik, 31*, 285–311.
National Council of Teachers of Mathematics. (2000). *Principles and standards for school mathematics*. Reston: NCTM.
Niss, M. (2003). Mathematical competencies and the learning of mathematics: The Danish KOM project. In A. Gagatsis & S. Papastavridis (Eds.), *Mediterranean conference on mathematical education* (pp. 115–124). Athens: 3rd Hellenic Mathematical Society and Cyprus Mathematical Society.
Pittman, T. S., Davey, M. E., Alafat, K. A., Wetherill, K. V., & Kramer, N. (1980). Informational versus controlling verbal rewards. *Personality and Social Psychology Bulletin, 6*, 228–233.
Sadler, D. R. (1989). Formative assessment and the design of instructional systems. *Instructional Science, 18*, 119–144.
Shepard, L. A. (2000). The role of assessment in a learning culture. *Educational Researcher, 29*, 4–14.
Shulman, L. S. (1986). Those who understand: Knowledge growth in teaching. *Educational Researcher, 15*, 4–14.
Shulman, L. S. (1987). Knowledge and teaching: Foundations of the new reform. *Harvard Educational Review, 57*, 1–22.

Chapter 12
A Reflection on Mathematical Modelling and Applications as a Field of Research: Theoretical Orientation and Diversity

Vince Geiger and Peter Frejd

Abstract This chapter explores the nature of theoretical approaches used in mathematical applications and modelling research literature focusing on the orientation and diversity of these approaches within this field. The study is based on the document analysis of a sample of book chapters from significant scholarly volumes dedicated to applications and modelling published between 2002 and 2011. Our analysis reveals that: research is oriented towards learners and teachers rather than to contexts for learning; the number of chapters that tied theory to practice increased whilst chapters focused on purely professional issues decreased; the diversity of theoretical approaches utilised increased over time; and, use of local theories specific to mathematical modelling coalesce around two approaches.

12.1 Introduction

Research within mathematics education known as mathematical modelling and applications has received increasing international recognition over the last decade. This recognition has included: the adoption of the International Community of Teachers of Mathematical Modelling and Applications (ICTMA) as an Affiliated Study Group of the International Commission on Mathematical Instruction (ICMI) in 2004; the conduct of the 14th ICMI Study on Modelling and Applications in Mathematics Education; and a plenary lecture by Werner Blum at the International Congress on Mathematics Education in Seoul during 2012. Such recognition is

V. Geiger (✉)
Education QLD, Faculty of Education & Arts, Australian Catholic University,
1100 Nudgee Road, Brisbane, Banyo, QLD 3014, Australia
e-mail: vincent.geiger@acu.edu.au

P. Frejd
Department of Mathematics, Linköping University, SE-581, 81 Linköping, Sweden
e-mail: peter.frejd@liu.se

G.A. Stillman et al. (eds.), *Mathematical Modelling in Education Research and Practice*, International Perspectives on the Teaching and Learning of Mathematical Modelling, DOI 10.1007/978-3-319-18272-8_12

generally bestowed on a distinctive, rich and coherent field of educational research. The standard by which any field is judged, however, lies in the quality of its publications as these are indicators of the theoretical advances being made, as well as the milieu researchers are attempting to influence within a discipline.

As is the case with mathematics education in general, research literature in mathematical modelling and applications includes an abundance of theories that systematise and provide insight into different aspects of teaching and learning mathematics (Kaiser and Sriraman 2006). The development of theory is evidence of the vitality and vigour of a particular field of research (English 2002; Jablonka et al. 2013). This view is succinctly captured by Sriraman and Nardi (2013) in stating that "The development of theory is absolutely essential in order for significant advances to be made in the thinking of communities (or individuals within them)" (p. 310). Further, the diversity of theoretical approaches can be an indicator of the maturity of a field. Jablonka et al. (2013), for example, argue that the utilisation of theoretical constructs such as single concepts or hybrids from other less known theories can lead to new theoretical stances that provide original insight into research problems. This chapter explores the nature of theoretical approaches used in research literature within mathematical modelling and applications by addressing the following research questions: What is the orientation and diversity of theory used within the field of mathematical modelling and how has this changed over recent time?

The exploration was conducted by applying two different analytical lenses to a sample of relevant book chapters from volumes that offer international perspectives on teaching and learning of mathematical modelling. The first lens is used to determine the current orientation of mathematical modelling and applications in relation to learners, teachers, and learning contexts (English 2002) and its connection to both research and practice (Kilpatrick 2008). A second lens is used to ascertain the diversity of theory and how theories have been used in research literature within the field (Tsatsaroni et al. 2003). The list of theoretical approaches used as anchor points for this analysis is based on both local theories that are specific to mathematical modelling and applications (Kaiser and Sriraman 2006) and general theories used across lines of enquiry within the field of mathematics education (Sriraman and English 2010). These two lenses will be outlined and described as part of the analytical approach used in this chapter.

12.2 Method

In order to make judgements about the direction and diversity of theory within mathematical modelling and applications, a document analysis was conducted on a sample of relevant publications made available through the period 2003–2011. This sample consisted of the books that draw together discussions that took place during the biennial conferences of the International Community of Teachers of Mathematical Modelling and Applications (ICTMA) and the publication that was the outcome of the 14th Study commissioned by the International Commission of

Table 12.1 ICTMA books 2003–2011 and 14th ICMI Study volume included in the document analysis

1. *ICTMA 10* – Ye, Q., Blum, W., Houston, S. K., & Jiang, Q. (Eds.). (2003). *Mathematical modelling in education and culture*. Chichester: Horwood. [ISBN: 1-904275-05-2]
2. *ICTMA 11* – Lamon, S., Parker, W., & Houston, K. (Eds.). (2003). *Mathematical modelling: A way of life*. Chichester: Horwood. [ISBN: 1-904275-03-6]
3. *ICTMA 12* – Haines, C., Galbraith, P., Blum, W., & Khan S. (Eds.). (2007). *Mathematical modelling: Education, engineering and economics*. Chichester: Horwood Publishing. [ISBN: 978-1-904275-20-6]
4. *ICTMA 13* – Lesh, R., Galbraith, P., Haines, C., & Hurford, A. (Eds.). (2010). *Modeling students' mathematical modeling competencies*. New York: Springer. [ISBN 978-1-4419-0560-4]
5. *ICTMA 14* – Kaiser, G., Blum, W., Borromeo Ferri, R., & Stillman, G. (Eds.). (2011). *Trends in teaching and learning of mathematical modelling*. New York: Springer. [ISBN 978-94-007-0909-6]
6. *ICMI Study* – Blum, W., Galbraith, P., Henn, H.-W., & Niss, M. (Eds.). (2007). *Applications and modelling in mathematics education: The 14th ICMI study*. New York: Springer. [ISBN: 978-0-387-29820-7]

Mathematics Instruction (ICMI). The ICTMA books were selected because these represent the perspectives of researchers and practitioners engaged with the work of the peak international body devoted to mathematical modelling and applications. The volume that covers the deliberations of the 14th ICMI Study is included because the *Study* publications of ICMI are widely accepted as representative of the state of the art in a field of research and practice within mathematics education. The publications assessed in this analysis are listed in Table 12.1.

The approach to analysis was co-constructed and implemented by both authors of this chapter. After negotiating the major thrust of the investigation and the development of the two theoretical lenses for the analysis of data (outlined in detail below), the authors generated decision criteria for the categorisation of chapters within the sample. These criteria and elaborations for categorization were then trialed by the authors on a selected section of the ICMI Study 14 volume independently. This was followed by a discussion (over Skype) where the initial categorisations of chapters were discussed, differences of opinion identified, and the final placement of chapters into categories negotiated and agreed upon. The discussion also served to assist in refining the decision criteria. This process was repeated using further chapters from the ICMI Study 14 volume until both authors were confident consistent judgments were being made using both theoretical lenses – greater than 90% agreement on independent assessments. From this point, remaining chapters in volumes from the sample were assessed independently, the process supported by regular meetings, (again via Skype), to discuss any uncertain decisions. *Overviews* and *Section Introductions* were excluded from the process as these are concerned with multiple issues and varying theoretical approaches and so could not be categorised. In a small number of cases, chapters were too difficult to categorise and were recorded as *other*. An additional list category labelled as *no approaches* was created for chapters where authors did not attempt to align with any theoretical approach. Descriptions of the two theoretical lenses and associated decision criteria that informed the categorisation of articles are now presented.

12.2.1 Lens 1: The Orientation of the Field

To establish a framework for determining the orientation of the field of mathematical modelling and applications, we drew on English's (2002) matrix of priorities for mathematics education research which divides advances in theory related to mathematics education into dimensions devoted to learners, teachers, and learning contexts and applied these to the selected publications. At the same time, mathematics education can be viewed from the perspectives of both research and practice (Kilpatrick 2008), a perspective reflected in ICTMA's mission statement.

> The International Community of Teachers of Mathematical Modelling and Applications (the 'Community') is a membership organisation that exists to promote Applications and Modelling (A&M) in all areas of mathematics education – primary and secondary schools, colleges and universities.
>
> (retrieved from http://www.ictma.net/)

Thus, in analysing a chapter, we also considered if the focus of a chapter was: purely theoretical; related to applying theory; or professional – directed at informing practice. The inclusion of these dimensions and perspectives results in the two dimensional framework presented in Table 12.2.

In order to make valid and comparable judgements, the authors developed decision criteria for the placement of chapters into categories that represent a dimension/perspective. In the case of the *dimensions of* English's (2002) priorities for mathematics education, decisions were relatively clear cut as judgements were made on the basis of the principal focus of a chapter on: the activity of a learner (e.g., nature of peer to peer discussion when attempting to validate a result); the actions of a teacher (e.g., role in promoting mathematisation during an application task); and, the nature and role of the learning environments (e.g., use of manipulatives, influence of a student management system).

In the case of the orientation of a chapter, it was necessary to develop elaborations for the associated categories as presented in Table 12.3. In general, these elaborations facilitated clear distinctions between the orientations of a chapter. Cases where an assessment was not clear, or there was disagreement between the authors, were resolved in the manner outlined earlier in the Method section.

Table 12.2 Dimensions and theoretical perspective of chapters

Dimension/perspective	Purely theoretical	Applied theory	Professional
Learners and learning			
Teachers and teaching			
Learning contexts			

Table 12.3 Decision criteria for orientation of chapters

Purely Theoretical: Non-empirical research studies that aim to build or develop new theory, new theoretical framework, etc. This includes chapters focusing on critical reflections on a theory or a theoretical perspective. Position chapters as they relate to the development of theory. Authors may engage in polemic in attempting to discuss the strengths of one theoretic perspective over another – but these must be scholarly with arguments based on research literature – not just author's opinion. The emphasis is on theory development not on data, if data are used these should only be to emphasize or illustrate the theory.
Applied Theory: These are empirical studies that *explicitly* use a theory to analyse data and/or use a theory to discuss the result. The analysis of data might be used to generate new theory or may be descriptive in nature. Data can be of any acceptable format. These chapters must be scholarly and cite the work of relevant authors in the field and it should be evident in these chapters that a theory has been used not just cited.
Professional: These are empirical, descriptive or polemic/discussion chapters that aim to inform teaching and learning practice at any level of education. The chapters may make use of data to inform the reader but not to build or confirm theory.

12.2.2 Lens 2: The Diversity of Theory

To identify the diversity of theoretical approaches used within the sample (Tsatsaroni et al. 2003), we drew on the notions of *local theories* as described by Kaiser and Sriraman (2006) and *general theories* as identified by Sriraman and English (2010). Local theories are those specific to research within a field of scientific endeavour. In the case of mathematical modelling and applications, there are theoretical approaches that are associated specifically with this field of research. In our analysis these include those identified by Kaiser and Sriraman (2006) as: *modelling cycle, modelling competency* and *emergent modelling, models and modelling perspective;* and *approaches for 'authenticity' and defining modelling tasks.* A small group of chapters used a variety of other approaches but these were amalgamated into a single category labelled *other theoretical approaches* because of low frequency counts. General theories are established approaches to viewing and working in the world that are drawn from the field of mathematics education more generally. Examples of these theories include: *radical constructivism, embodied cognition, socio-culturalism, semiotic approaches*, and *neuroscientific approaches.* The full list of general theories included as part of this analysis appears in Table 12.6.

According to Niss (2007) there are researchers in mathematics education that "*do not explicitly invoke or employ any theory* at all in their work" (p. 101). This was evident in a number of chapters and these cases were categorised as *no approaches*. There were also authors who mentioned a theoretical approach in the beginning of a chapter but who did not make use of the theory in any discernable way in the development of a theoretical or conceptual framework used for the analysis of data or as part of a discussion of findings. In this situation we took the decision to categorise these chapters as *no approaches* because the theory, while mentioned, did not appear to influence the argument within a chapter in any way.

12.3 Results

12.3.1 The Orientation of Publications

Chapters in each volume were categorised according to the dimensions and perspectives presented in Table 12.2. Frequencies against volume and dimension, and volume against perspective, are recorded in Tables 12.4 and 12.5. These results are represented as column graphs in Figs. 12.1 and 12.2 respectively.

Figure 12.1 indicates that for most of the sampled period, *Learners and Learning (L)* were the primary focus of activity in mathematical modelling and applications, although this emphasis is less prominent in the most recent volume sampled (ICTMA 14). During the same period, chapters with a focus on *Teachers and Teaching (T)* have waxed and waned, however, there is an increase in the number of chapters with this focus in the last two volumes within the sample (ICTMA 13 and ICTMA 14). Interestingly, there was a balance between *Learners and Learning* and *Teachers and Teaching* in the last ICTMA book – published in 2011. There have been a consistently limited number of publications related to *Learning Contexts (C)* in the sample period.

Figure 12.2 indicates that there is a clear increase in chapters related to applied theory. This increase corresponds with a parallel decrease in chapters devoted to professional concerns, that is, chapters that focus specifically on providing advice to practitioners about ways of teaching or improving learning in mathematical modelling and applications. There were relatively few chapters from authors who sought to address purely theoretical aspects of mathematical modelling and applications, that is, chapters that attempt to critique current theory or seek to generate new theory.

12.3.2 Diversity of Theory

The 252 chapters included in the sample were categorised into the list of theoretical approaches presented in Table 12.6. There were a small number of chapters that fell into more than one theoretical approach (four chapters in ICTMA 12, two in ICTMA 13, and four in ICTMA 14 are categorised into two approaches, and one

Table 12.4 Publications against dimensions learners, teachers, context and other

Publication	Dimensions			
	Learners	Teachers	Context	Other
ICTMA 10	46	46	8	
ICTMA 11	60	20	15	5
ICTMA 12	59	37	4	
ICMI 14	63	17	17	6
ICTMA 13	68	21	11	
ICTMA 14	49	51	0	

Table 12.5 Publication perspectives pure theory, applied theory, and professional and other

Publication	Perspectives			
	Pure theory	Applied theory	Professional	Other
ICTMA 10	0	31	69	
ICTMA 11	5	35	55	5
ICTMA 12	6	43	51	
ICMI 14	16	44	36	6
ICTMA 13	9	53	38	
ICTMA 14	8	59	33	

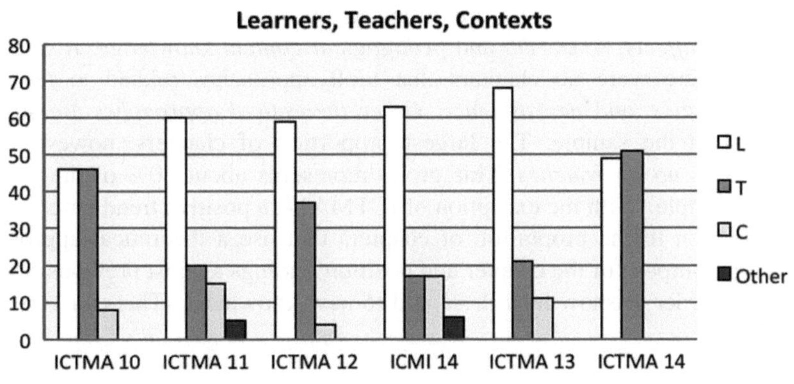

Fig. 12.1 Frequency of chapters assigned to learners (L), teachers (T), and contexts (C)

chapter in each of ICTMA13 and ICMI 14 study are categorised into three approaches).

The categorisation of chapters within this framework demonstrates the diversity of theoretical approaches used in ICTMA proceedings has expanded over the sample period. For example, the ICTMA 10 proceedings included seven different categories of theoretical approaches while the ICTMA 13 and 14 proceedings include 12 and 11 categories, respectively. An anomalous result to this trend is recorded for the 14th ICMI study (2007) where there are 14 categories of theory.

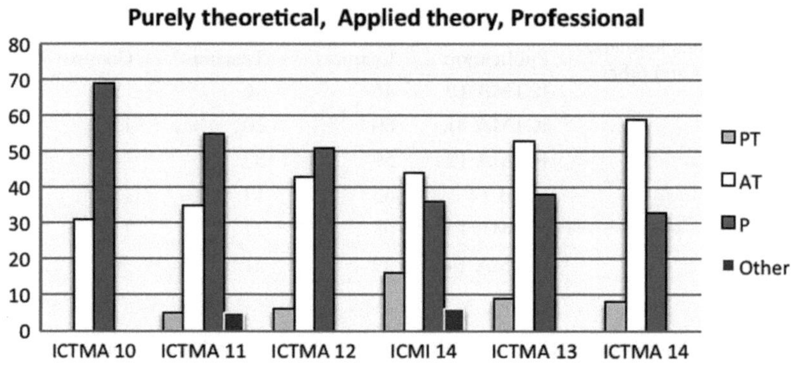

Fig. 12.2 Frequency of chapters assigned to pure theory, applied theory, and professional chapters

This should not be considered surprising as the ICMI Study conference most likely attracted a different group of attendees to ICTMA events even though a degree of overlap is likely. An examination of these data in relation to local theories reveals that the categories of *modelling cycle* and *modelling competencies* were the most frequent theoretical approaches used in this sample with other theories receiving less attention. General theoretical approaches appeared less frequently than local theoretical approaches. Of this group, *socio-cultural* and *instrumental approaches* were the most frequent. These are followed by chapters that focus on *beliefs, attitudes and affects*, *ATD/TDS* and *pedagogical content knowledge*. It should be noted that there were no chapters that took approaches related to *feminism, ethnomathematics*, and *neuroscience*. *Other theoretical approaches* are strongly represented in the sample. The largest proportion of chapters, however, were categorised as *no approaches*. This group represents about 50% of the chapters within the sample. With the exception of ICTMA14, a positive trend emerges over time in relation to the proportion of chapters that use a theoretical approach to establish the purpose of the chapter and position findings against previous research literature in order to show how these add to new knowledge. This can be seen in Table 12.7.

12.4 Conclusions, Discussion and Implications

The data presented in this chapter demonstrate that there has been an increase over time in the use of theory to underpin publications related to mathematical modelling and applications – indication of the developing maturity of this research field. Closer inspection reveals that local theoretical approaches are more frequently used than general theoretical approaches within the selected sample of publications. This may be another marker of maturity as theoretical stances and approaches specific to the field of modelling are now available. This view, however, must be

Table 12.6 The frequencies of theoretical approaches used in the ICTMA 10 to ICTMA 14 proceedings and in the 14th ICMI study

Local theoretical approaches	14th ICMI s	ICTMA 10	ICTMA 11	ICTMA 12	ICTMA 13	ICTMA 14	General approaches	14th ICMI s	ICTMA 10	ICTMA 11	ICTMA 12	ICTMA 13	ICTMA 14
Modelling cycles	1	2		8	5	8	Radical constructivism						
Modelling competencies	6	1	1	3	1	4	Social constructivism						
Emergent modelling	2	1	1	1	1	1	Embodied cognition	1		1		1	1
Models and modelling perspective	2		1		10	4	Pedagogical content knowledge					1	4
Approaches for 'authenticity' and defining modelling tasks	2			1	2	1	Discursive, linguistics, social linguistics, and semiotics approaches	1					
							Sociological approaches	1			1		
Other theoretical approaches	4	1	4	10	8	6	Critical mathematics education			1	1	1	
							Feminist approaches						
No approaches	23	18	13	25	19	25	Beliefs, attitudes and affect	1				2	2
							Ethnomathematics						
							Neuroscience approaches						
							ATD/TDS	1					3
							Instrumental approaches	2	2		1		1
							Pragmatic approaches	1				1	
							Socio-cultural approaches	2			2	2	

Table 12.7 Chronological order against percentage of theory based chapters (2003–2011)

Publication (publication date)	Percentage of theory based chapters
ICTMA 10 (2003)	28%
ICTMA 11 (2003)	41%
ICTMA 12 (2007)	53%
14th ICMI study (2007)	54%
ICTMA 13 (2010)	65%
ICTMA 14 (2011)	58%

tempered by the fact that the use of local theories strongly coalesces around only two approaches – the *modelling cycle* and *modelling competencies*. These are used more frequently than all general theoretical approaches combined. While this might be an indication of the coherence of the research field, it could also indicate that there is too strong a dependence on these particular theoretical approaches. Thus, there is a danger that richness and diversity of thought within the field is too constrained.

While there is clear growth across time in the number of theory oriented chapters included in both ICTMA and ICMI publications, there are subtleties under this general trend that are worthy of closer examination. This change is principally due to an increase in chapters classified as applied research; an encouraging development given that the nature of the field is related to the application of theory to the activities associated with teaching and learning. At the same time, purely theoretical papers were a relative scarcity, which raises concerns about the generation of new theory. An additional issue is the decline in chapters that focus on the concerns of practitioners. This change may be an indication of the increased "academisation" of the field, for example, greater numbers of PhD students have completed studies within the field of modelling and applications and are now contributing to its growth through new publications, or of a decrease in the number of practitioners participating in and lending their voices to ICTMA events. The trend could also indicate the increasing pressure from universities for academics to publish research orientated articles and papers rather than those that seek to inform and support practitioners. All of these issues require further research if we are to understand changes in the field as these develop.

Our analysis also identifies a number of *white spots* that offer potential for further theory development. *White spots* are theoretical approaches within our framework that have not yet been used to explain or theorise phenomena within mathematical modelling and applications. For instance, *sociological approaches, linguistics, social linguistics, and semiotics approaches*, *critical mathematics education*, *pragmatic approaches*, *feminist approaches*, *ethnomathematics*, *and neuroscience* all remain potential aspects of theory within mathematics education which were not utilised in our sample. The view of Sriraman and English (2010), that advancement in the field of mathematics education has often been achieved by adapting theoretical approaches used within and outside mathematics education, means that the *white spots* offer potential for greater theoretical diversity and richness and the resulting advancement of mathematical modelling and applications as a field of research.

References

English, L. D. (2002). Priority themes and issues in international mathematics education research. In L. D. English (Ed.), *Handbook of international research in mathematics education* (pp. 3–16). Mahwah: Erlbaum.

Jablonka, E., Wagner, D., & Walshaw, M. (2013). Theories for studying social, political and cultural dimensions of mathematics education. In M. A. Clements et al. (Eds.), *Third international handbook of mathematics education* (pp. 41–67). New York: Springer.

Kaiser, G., & Sriraman, B. (2006). A global survey of international perspectives on modelling in mathematics education. *ZDM: The International Journal on Mathematics Education, 38*(3), 302–310.

Kilpatrick, J. (2008). The development of mathematics education as an academic field. In M. Menghini, F. Furinghetti, L. Giacardi, & F. Arzarello (Eds.), *The first century of the international commission of mathematical instruction (1908–2008): Reflecting and shaping the world of mathematics education* (pp. 25–39). Roma: Istituto della Enciclopedia Italiana fondata da Giovanni Treccani.

Niss, M. (2007). The concept and role of theory in mathematics education. In C. Bergsten, B. Grevholm, M. Måsøval, & F. Røönning (Eds.), *Relating practice and research in mathematics education: Proceedings of Norma 05* (pp. 97–110). Trondheim: Tapir.

Sriraman, B., & English, L. (Eds.). (2010). *Theories of mathematics education: Seeking new frontiers*. New York: Springer.

Sriraman, B., & Nardi, E. (2013). Theories in mathematics education: Some developments and ways forward. In M. A. Clements et al. (Eds.), *Third international handbook of mathematics education* (pp. 303–325). New York: Springer.

Tsatsaroni, A., Lerman, S., & Xu, G. (2003). *A sociological description of changes in the intellectual field of mathematics education research: Implications for the identities of academics*. Paper presented at annual meeting of the American Educational Research Association, Chicago.

Chapter 13
Problem Solving Methods for Mathematical Modelling

Gilbert Greefrath

Abstract Providing a method for problem solving can support students working on modelling tasks. A few candidate methods are presented here. In a qualitative study, one of these problem solving methods was introduced to students in grades 4 and 6 (Germany), to be used in their work on modelling tasks. The students were observed as they worked and were subsequently interviewed. The results reveal differences between grades, and widely varying problem solving processes. The differences in written final solutions are considerably less pronounced.

13.1 Sub-competencies of Modelling Competencies

The scope of mathematics is twofold: it deals in abstract structures and ideas, and produces models which serve to describe the environment. In order to achieve this second objective, it is essential to integrate realistic problem settings and applications into mathematical teaching. Mathematical modelling has therefore been identified as one of the most crucial general-mathematical skills by the German educational standards in mathematics. Modelling competency is described by NISS et al. (2007, p. 12) as "the ability to identify relevant questions, variables, relations or assumptions in a real world situation, to translate these into mathematics and to interpret and validate the solution to the given situation, as well as the ability to analyse or compare given models by investigating the assumptions", checking properties etcetera. Motivated by the great importance of this competency for mathematical learning, this chapter aims to present and discuss problem solving resources which can help students in their work on modelling problems in mathematics lessons.

Problem solving resources for modelling tasks are usually based on prototypical solutions of modelling problems, as illustrated by modelling cycles. These

G. Greefrath (✉)
Institute of Education in Mathematics and Information Technology, University of Münster, Fliednerstr. 21, 48149 Münster, Germany
e-mail: greefrath@uni-muenster.de

G.A. Stillman et al. (eds.), *Mathematical Modelling in Education Research and Practice*, International Perspectives on the Teaching and Learning of Mathematical Modelling, DOI 10.1007/978-3-319-18272-8_13

modelling cycles describe the persistent dialogue between reality and mathematics. Some cycles are less detailed, regrouping the process of constructing a model from a given situation into one single step. This step is, for example, labelled "modelling/ mathematising" by Ortlieb (2004). Today, modelling usually refers to the cycle as a whole. Mathematising, like in Ortlieb's work, usually refers to the stage which starts on a real problem and ends with the completion of the mathematical model.

A model of the mathematical modelling cycle, elaborated by Blum and his colleagues in the DISUM[1] project, tackles the process from a cognitive perspective, aiming to describe as precisely as possible how students approach a modelling task (see Fig. 13.1). The main departure from the above-mentioned model crafted by Ortlieb is the development of the so-called situation model (Blum and Leiß 2007). The process of model creation is considered in more detail, representing in one initial, additional step the process by which the model's author extracts a mental representation from an initial situation. This expansion was mainly motivated by research into reading comprehension (see, e.g., Kintsch and Greeno 1985).

The capacity to execute each individual step in the modelling cycle can be considered a sub-competency of modelling (cf. Maaß 2006). Blum's modelling cycle provides one way of characterising the individual sub-competencies involved in modelling, as in Table 13.1. Working mathematically (step 4 in Fig. 13.1) is not included as a sub-competency, as it is not specific to modelling competency. Working from other cyclic models of modelling (e.g., Greefrath et al. 2013: Kaiser and Sriraman 2006), other sets of sub-competencies with differing emphasis could conceivably be used. The model of Blum (Fig. 13.1) differentiates the sub-competencies in a clear and detailed way.

Explicitly dividing the modelling process into separate stages is one possible way to reduce complexity for both teachers and students, making it possible to produce more suitable tasks. In particular, this approach to modelling allows individual sub-competencies to be taught separately, enabling a comprehensive modelling competency to develop over time. The insight afforded by the division of the modelling process into several independent sections can be used to create problem solving methods for students, and thus to provide students with problem solving resources during their modelling tasks.

13.2 Problem Solving Methods

Problem solving methods can support students' work on tasks. They are one possible form of problem solving resources. These resources exist in different types and forms. Resources can, for example, help to motivate students to further investigation, providing them with feedback which enables them to judge whether

[1] The DISUM Project (Didactical intervention modes for mathematics teaching oriented towards self-regulation and directed by tasks) led by W. Blum, R. Messner and R. Pekrun.

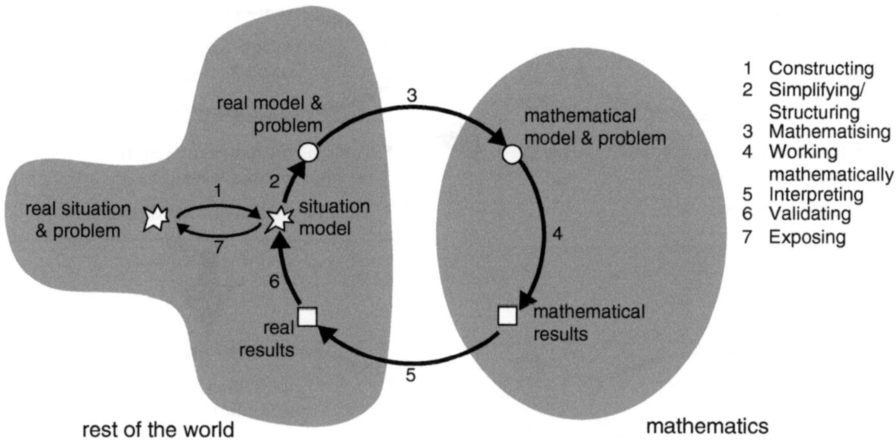

Fig. 13.1 Modelling cycle (Blum and Borromeo Ferri 2009, p. 46)

Table 13.1 Sub-competencies involved in modelling

Sub-competency	Indicator
Constructing	Students construct their own mental model from a given problem, and thus formulate an understanding of their task.
Simplifying	Students identify relevant and irrelevant information from a physical problem.
Mathematizing	Students translate specific, simplified physical situations into mathematical models (e.g., terms, equations, figures, diagrams, functions).
Interpreting	Students relate results obtained from manipulation within the model to the physical situation, and thus obtain physical results.
Validating	Students judge the physical results obtained in terms of plausibility.
Exposing	Students relate the results obtained in the situational model to the real situation, and thus obtain an answer to the problem.

their solution or solution strategy was correct or successful. They can also provide content related hints or information which helps to solve the task. The following categories of resources exist: motivational resources, feedback resources, general strategic resources, content-related strategic resources and content-related resources. Within each individual category, a further distinction between direct and indirect resources can be made. Direct resources explicitly address the student, a specific section of the problem setting, or a concrete mathematical concept. Indirect resources, on the other hand, address the whole class, the task as a whole, or less concrete mathematical concepts (Zech 1998, pp. 315ff.). The problem solving methods presented hereafter are indirect general strategic resources, as although they do refer to general technical problem solving and modelling methods, they do not provide any concrete and contextual indications relating to the content of the task.

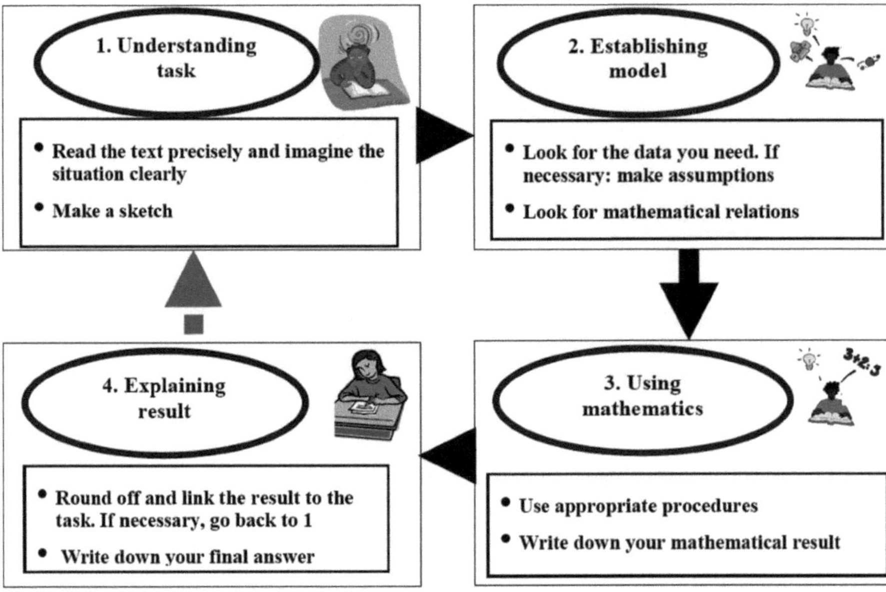

Fig. 13.2 Problem solving method as described by Blum (Blum and Borromeo Ferri, 2009, p. 54)

In the DISUM project, Blum developed a problem solving method for students based on a simplified modelling cycle (see Fig. 13.2). This method comprises four stages, named *understanding the task, establishing the model, using mathematics* and *explaining the results*. Each stage is explained to students with two explicative bullet points. The introduction of the method to students can be supplemented with additional example tasks, allowing them to practise its application.

Böhm (2010) developed a problem solving method in three stages. The first stage aims to understand the problem (*"Problem verstehen"*) to obtain a mental representation of the task at hand. In the second stage, (*"Mathematik ins Spiel bringen und rechnen"*), the problem is mathematised and manipulated mathematically. In the third stage, (*"Resultat auf das Problem beziehen und prüfen"*), the results obtained by mathematical manipulation are related to the problem, and validated. Zöttl et al. (2010) use a similar problem solving method in the KOMMA[2] project. Their method consists of the phases *understanding the task, calculating*, and *explaining the result.*

An alternative problem solving method was presented by Greefrath and Leuders (2013). In the development of this method, the authors considered the stages of problem solving as set out by Polya. In his book *How to solve it*, Polya compiled a catalogue of heuristic questions designed to help manipulation in problem solving tasks. The process of problem solving is divided into the following stages:

[2] KOMMA (KOMpendium MAthematik) is a project aiming at the implementation and evaluation of a learning environment for mathematical modelling.

Understanding the problem, devising a plan, carrying out the plan and *looking back* (Polya 1973). Schoenfeld (1985) added to this by describing some of the stages in more detail. He introduced a distinction at the end of the problem solving process between verification and transition. His suggested problem solving method for students beginning secondary-level education therefore contains five steps, and can be used for both modelling and problem solving tasks:

- Understanding the problem: rephrase it in the students' own words
- Choosing an approach: describe assumptions and plan future calculations
- Performing: perform calculations and manipulations
- Explaining the results
- Checking: check results, calculations and approach

It has been demonstrated that the problem solving methods presented by Blum (Blum and Borromeo Ferri 2009) and Böhm (2010) are structured like a cycle, whereas the problem solving methods described by Zöttl et al. (2010) and Greefrath and Leuders (2013) are structured more linearly, because there is no explicit link to the starting point. This reflects that the first methods are more closely fitted to the modelling cycle, whereas the second set of methods were more closely modelled on the process of problem solving.

There exist some interesting empirical results on students working with problem solving methods. Maaß (2004) used a simplified modelling cycle as a problem solving method. In a qualitative study of grade 7 and 8 school students in Germany, she examined classroom modelling competency, and was able to identify adequate modelling competency in a large subset of the students at the end of the study. Concerning the implementation of the problem solving method, she writes that the students felt they were aided by knowledge about the modelling process and the cycle representation. Böhm (2010) views the three-stage problem solving method as an intermediate tool for the classroom. It should be developed further at a later point, including additional steps, so as not to prevent additional learning later on. The problem solving method is part of a comprehensive programme which aims at fostering modelling-related competencies.

Also in two big research projects in Germany, problem solving methods are used. In the KOMMA project, Zöttl et al. (2010) used structured examples (Reiss and Renkl 2002) of problems solved, consisting of the problem statement and a set of solution steps. Positive results were achieved in the domain of mathematical reasoning and proof by implementing these structured solved examples in the classroom. In a study by Schukajlow et al. (2010) during the DISUM project, significant differences in students' modelling performance could be demonstrated when using the Blum problem solving method (Blum and Borromeo Ferri 2009) applied to Pythagoras' theorem. Lessons that included the problem solving method were found to be the more effective form of teaching and learning. Also, students exposed to the method were more acutely aware of the cognitive strategies they were using, that is the method itself.

However, problem solving methods of this kind are also suspected to bear disadvantages. Meyer and Voigt (2010), for example, have argued that problem solving methods are aimed towards understanding final solutions quickly rather than towards understanding the actual problem solving process, and are little more than additional subject matter for students to learn. Franke and Ruwisch (2010) point out that it can be too much to ask students with little experience in problem solving to search for solutions while simultaneously maintaining a general overview of the process as a whole when using a method.

13.3 Study Design

In a qualitative study, three pairs of German school students in each of grades 4 and 6 (10 and 12 years old), in Grundschule (primary education) and Realschule (secondary education) respectively, were observed while working with the problem solving method set out by Greefrath and Leuders (2013), and subsequently interviewed. The first task (*Teeth Brushing Task*) aimed to achieve an understanding of the method by asking students to arrange the described steps into the correct causal order. This task was given identically to all participating students.

Teeth Brushing Task
Pia is attempting to answer the following question:

How much water could I save in a year while brushing my teeth?

To find an answer, Pia went through several steps:

- I should measure the amount, and add it up.
- Five cupfuls of water flowed through the tap. Five cups are about 1 l. There are 365 days in a year, so that makes 365 l.
- When I brush my teeth, I need water to wash out my mouth. But if I let the water run while I'm brushing, I'm wasting water. How much water runs through the tap over a whole year?
- 365 l is equal to 36 buckets of fresh water that could be saved in a year. That seems about right, because over the course of a year the tap will have been running for many minutes in total.
- So, over a whole year, I could save 365 l of fresh water by turning the tap off while I brush.

Put the steps of Pia's solution in the correct order. Label each step with the name that fits best: understanding the problem, choosing an approach, performing, explaining the results, checking.

The second task involved using the problem solving method. The secondary school students were presented with the method set out by Greefrath and Leuders (2013) as described above, along with the following problem setting:

Use this method to answer the following question: How much fluid do I drink in a week?

The primary school students were given a photograph of a student their age standing next to an oversized flip-flop, and asked to determine the height of a (fictional) person whom the flip-flop might fit.

Flip-Flop Task

Determine the height of a (fictional) person whom the flip-flop might fit.

The observations and interviews were filmed. The videos were transcribed in full and analysed based on Grounded Theory (Strauss and Corbin 1990). The transcripts were analysed line-by-line, and each individual text passage was assigned one of the steps from the problem solving method (understanding, choosing, performing, explaining, checking). The different stages were assigned by two people working independently in order to achieve the highest possible levels of inter-rater reliability in coding.

13.4 Results

Although the first 6th grade pair followed the given problem solving method closely, their attention was particularly focused on the concrete example from the first task. The students mixed up the phases of choosing and performing, and referred alternatively to the problem statement, example, and method. The method did stimulate metacognitive processes, but did not help to optimise the process of answering the question. Even though observation showed that the two students were not successful in following the given method closely, upon being asked whether the method had helped them, they responded as follows: "I thought it was good that we could always look at it. If we had had to do it from nothing, I think it would have been harder. Instead, we had some sort of reference."

The second 6th grade pair followed the given method closely. Some difficulties were only revealed when they were asked to label each phase. A content-related error occurred in the last part of the manipulations. Plausibility considerations did not suffice to disclose the error. This pair of students demonstrated much more clearly the procedure which is hoped to be achieved by providing the problem solving method. Interestingly, one of the two students remarked in the final interview that he saw a similarity to the advice usually given for word problems: question, calculations, results. Overall, the use of this kind of method was met with a very positive reception from the students. The written final solution of the pair reproduced clearly the sequence and structure of the method.

Understanding: How much do you drink in a week?
Choosing: I drink about 2 l every day, so I would calculate 2 L · 7.
Performing: 2 L · 7 = 14 L
Explaining: Each week I drink about 14 L, and each day about 2 L.
Checking: That's equal to 9 · 1.5 l bottles and one 0.5 l bottle.

The third 6th grade pair followed the individual phases of the method. They completed all of the phases, except the last one. However, they followed the example given in the first task more closely than the method itself. This led to difficulties in abstracting from the example in places. The *choosing* and *performing* steps were performed according to the method. The next two steps were, however, mixed up and cut short. This pair also made positive statements about the method, and described the explanations included as important:

Interviewer:	Did you find the method helpful?
Student 1:	So, I think that if the explanations weren't there, with only 'understanding', 'choosing', I wouldn't have been able to do it. It says here also 'write in your own words', 'describe assumptions' and 'plan calculations'. That helped.
Interviewer:	What about if you had had to do the question without a method? Would you have done it in the same way?
Student 2:	I would have found it harder, and maybe needed more time to do it.

This pair also provided a written final solution that was structured according to the given method. Details concerning the interview can be found in Hilmer (2012).

The grade 4 student pairs managed to solve the first task quite easily, putting the steps in order. In the second task, no pairs used the method all the way through to the last step. Two pairs managed nevertheless to provide a good answer. A graphical representation (Fig. 13.3) of the typical answering process of a grade 4 pair shows that the students very quickly abandoned the structure of the method, jumping back and forth between the different stages. Still, the method helped a few

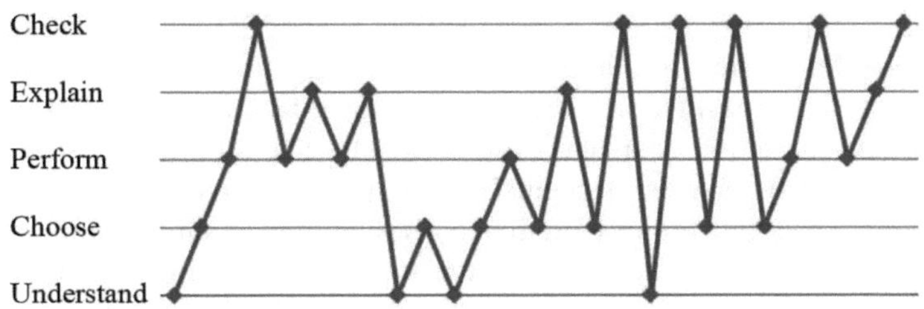

Fig. 13.3 Visualisation of an answering process typical of a grade 4 pair

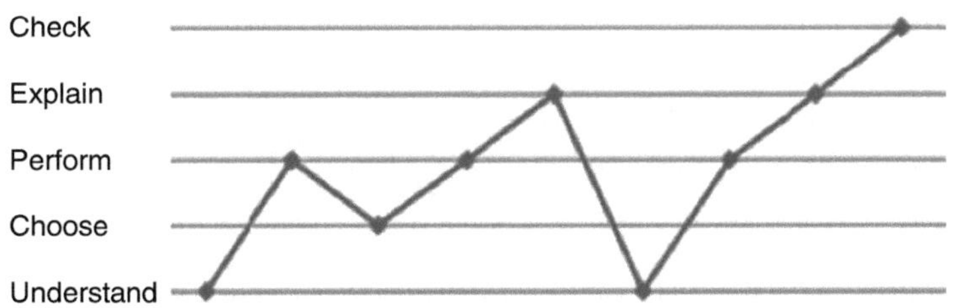

Fig. 13.4 Visualisation of an answering process typical of a grade 6 pair

of the students to make a good start to their answer, and these students worked more independently thereafter. (For details see Förderer 2013.) A typical answering process of a grade 6 pair (Fig. 13.4) shows that the students used the structure of the method, not often jumping back and forth between the different stages.

13.5 Discussion and Conclusion

The results of this study show that students' final solutions can be influenced by providing a method for problem solving. The influence on the written solution is more pronounced than on the solving processes that each pair of students went through, which present considerable differences – from each other and from the model solving process. This is reminiscent of the individual modelling routes that were described by Borromeo Ferri (2007). One aspect that is clearly important is the way that the method is established in lessons. In this study, the method was only presented rapidly, and then used directly afterwards. Clearly, there are students who have difficulty manipulating this type of method, and others who already benefit from a short introduction, helping them to use the method themselves. The results obtained from the students could plausibly be influenced by a more extensive introduction to the method. Whether or not the method had any lasting effect on the problem solving process, all pairs of 6th grade students interviewed recorded positive impressions, and felt supported by its presence. This reinforces some aspects of the findings recorded by Schukajlow et al. (2010). Providing full examples of solved problems, like in the first task of this study, seems to engage some students in a clearly positive way. But there are more factors to control, for example to investigate the effect of metacognitive beliefs (Garofalo and Lester 1985). A more detailed study of school students' problem solving processes seems essential, as providing a method certainly produced a visible impact on the problem solving process. The written final solutions of the students seem, however, to bear little relation to the way the method was in fact used as the problem was being answered.

References

Blum, W., & Borromeo Ferri, R. (2009). Mathematical modelling: Can it be taught and learnt? *Journal of Mathematical Modelling and Application, 1*(1), 45–58.

Blum, W., & Leiß, D. (2007). How do students' and teachers deal with modelling problems? In C. Haines, P. Galbraith, W. Blum, & S. Khan (Eds.), *Mathematical modelling: Education, engineering and economics* (pp. 222–231). Chichester: Horwood.

Böhm, U. (2010). "Aller Anfang ist schwer, Modellieren lernen umso mehr!?"– Erste Schritte auf dem Weg zur langfristigen Förderung von Modellierungskompetenzen im Mathematikunterricht. In R. Bruder & A. Eichler (Eds.), *Materialien für einen realitätsbezogenen Mathematikunterricht Band 15* (pp. 1–14). Hildesheim: Franzbecker.

Borromeo Ferri, R. (2007). Personal experiences and extra-mathematical knowledge as an influence factor on modelling routes of students. *Proceedings of the fifth congress of the European society for research in mathematics education* (pp. 2080–2089), Larnaca.

Förderer, M. (2013). Modellieren mit Hilfe eines Lösungsplans – Analyse der Bearbeitung einer Fermi-Aufgabe in der 4. Klasse. Unpublished master's thesis, University of Münster.

Franke, M., & Ruwisch, S. (2010). *Didaktik Sachrechnens in der Grundschule*. Heidelberg: Spektrum Akademischer Verlag.

Garofalo, J., & Lester, F. (1985). Metacognition, cognitive monitoring, and mathematical performance. *Journal for Research in Mathematics Education, 16*(3), 163–176.

Greefrath, G., & Leuders, T. (2013). Verbrauch im Haushalt – Schätzen und Überschlagen. In S. Prediger, B. Barzel, S. Hussmann, & T. Leuders (Eds.), *mathewerkstatt* (pp. 5–22). Berlin: Cornelsen.

Greefrath, G., Kaiser, G., Blum, W., & Borromeo Ferri, R. (2013). Mathematisches Modellieren – Eine Einführung in theoretische und didaktische Hintergründe. In R. Borromeo Ferri, G. Greefrath, & G. Kaiser (Eds.), *Mathematisches Modellieren für Schule und Hochschule. Theoretische und didaktische Hintergründe* (pp. 11–37). Wiesbaden: Springer Spektrum.

Hilmer, A. (2012). *Modellieren mithilfe eines Lösungsplans – Analyse der Bearbeitung eines Fermiproblems in der 6. Klasse*. Unpublished master's thesis, University of Münster.

Kaiser, G., & Sriraman, B. (2006). A global survey of international perspectives on modelling in mathematics education. *ZDM, 38*(3), 302–310.

Kintsch, W., & Greeno, J. G. (1985). Understanding and solving word arithmetic problems. *Psychological Review, 92*(1), 109–129.

Maaß, K. (2004). *Mathematisches Modellieren im Unterricht. Ergebnisse einer empirischen Studie*. Hildesheim: Franzbecker.

Maaß, K. (2006). What are modelling competencies? *ZDM, 38*(2), 113–142.

Meyer, M., & Voigt, J. (2010). Rationale Modellierungsprozesse. In B. Brandt, M. Fetzer, & M. Schütte (Eds.), *Auf den Spuren Interpretativer Unterrichtsforschung in der Mathematikdidaktik* (pp. 117–148). Münster: Waxmann.

Niss, M., Blum, W., & Galbraith, P. (2007). Introduction. In W. Blum, P. Galbraith, H.-W. Henn, & M. Niss (Eds.), *Modelling and applications in mathematics education* (pp. 3–32). New York: Springer.

Ortlieb, C. P. (2004). Mathematische Modelle und Naturerkenntnis. *Mathematica Didactica, 27* (1), 23–40.

Polya, G. (1973). *How to solve it. A new aspect of mathematical method.* Princeton: Princeton University Press.

Reiss, K., & Renkl, A. (2002). Learning to prove: The idea of heuristic examples. *Zentralblatt für Didaktik der Mathematik, 34*(1), 29–35.

Schoenfeld, A. H. (1985). *Mathematical problem solving*. Orlando: Academic.

Schukajlow, S., Krämer, J., Blum, W., Besser, M., Brode, R., Leiss, D., & Messner, R. (2010). Lösungsplan in Schülerhand: zusätzliche Hürde oder Schlüssel zum Erfolg? *Beiträge zum Mathematikunterricht*, Bd. 2, 771–774.

Strauss, A., & Corbin, J. (1990). *Basics of qualitative research: Grounded theory procedures and techniques*. Newbury Park: Sage.

Zech, F. (1998). *Grundkurs Mathematikdidaktik*. Weinheim und Basel: Beltz.

Zöttl, L., Ufer, S., & Reiss, K. (2010). Modelling with heuristic worked examples in the KOMMA learning environment. *Journal für Mathematik-Didaktik, 31*(1), 143–165.

Chapter 14
Improving Mathematical Modelling by Fostering Measurement Sense: An Intervention Study with Pre-service Mathematics Teachers

Maike Hagena

Abstract Working on mathematical modelling tasks often requires knowledge about different inner-mathematical content of everyday mathematics. It is especially the ability to handle quantities – that is mainly: having a feeling for units of measurement (measurement sense) – being needed to work on modelling tasks successfully. There is nothing known about how modelling competency and measurement sense align. To remedy this, an intervention study with pre-post test design has been conducted focusing on: "Is it possible to improve pre-service teachers' modelling performances by fostering measurement sense?" Forty-eight pre-service teachers out of an experimental group have been trained in measurement sense regarding the quantities, length and area. Another 45 pre-service teachers have been part of a control group. Analysis of pre-service teachers' performances on a modelling-test at the end of the intervention study revealed that pre-service teachers in the experimental group outperformed their counterparts in the control group.

14.1 Introduction

Mathematical modelling as part of competency-oriented mathematics as well as measurement sense are central parts of everyday teaching and learning of mathematics (NCTM 2000). Even though there exists a lot of studies discussing teaching and learning of mathematical modelling (Blum et al. 2007) and many studies about how to teach measurement sense (Clements and Bright 2003), there is a lack of knowledge about how modelling competency and measurement sense combine.

M. Hagena (✉)
Leuphana Universität Lüneburg, Scharnhorststr. 1, Lüneburg 21335, Germany
e-mail: hagena@leuphana.de

G.A. Stillman et al. (eds.), *Mathematical Modelling in Education Research and Practice*, International Perspectives on the Teaching and Learning of Mathematical Modelling, DOI 10.1007/978-3-319-18272-8_14

To date nothing is known about how measurement sense influences modelling competency although Larson (2010) identified "quantitative reasoning as a useful lens to apply to a Models and Modeling Perspective" (p. 117). Larson's findings from students working on a MEA indicated that operations chosen in modelling are "reflective of the relationships they perceive among those quantities". Furthermore Larson (2010, p. 117) stated:

> quantitative reasoning is a central mechanism for iterative refinement of solution processes to real-world problems ... As students work to develop a solution to this problem, they repeatedly reason with and about quantities which they then operate on to create new derived quantities.

These derived quantities are in turn used "to reason with and about the problem situation". For this reason, an intervention study has been conducted investigating the influence of measurement sense on mathematical modelling.

14.2 Measurement Sense as Central Aspect of Handling Quantities

Handling quantities is said to be a weakness around the world (Thompson and Preston 2004); but what are the main difficulties of learners thinking about and working with quantities? Having a closer look at the quantities length and area, empirical research results point out that learners often do not understand the quantities being measured or the units that are used for measurement (Outhred et al. 2003). They also frequently confuse perimeter and area as well as surface area and volume (Martin and Strutchens 2000) and need help to develop a repertoire of benchmarks that can be easily represented and used for estimation (Joram 2003). These mathematical weaknesses and misconceptions concerning handling quantities cannot only be identified when diagnosing learners at school but also when asking pre-service teachers (Kellogg 2010). For this reason it becomes clear that there is a lack of basic understanding of handling quantities, which prevents learners – children as well as adults – from being able to reason about measurement in complex situations (Baturo and Nason 1996). That is why several studies suggest that it is important for learners to understand how and why measurement works (Vasilyeva et al. 2009) as well as to have a feeling for units of measurement (Joram 2003) for building up so called measurement sense successfully. In detail having a good *measurement sense* (*MS*) as a central aspect of handling quantities means:

MS 1 Being able to identify and distinguish different quantities of measurement.
MS 2 Being able to measure, to estimate and to round.
MS 3 Being able to decide whether to estimate, to measure and to round.
MS 4 Having knowledge about units of measurement which includes calculating and converting units.
MS 5 Having a set of meaningful benchmarks for these units and being able to use these benchmarks (Grund 1992; Shaw and Puckett-Cliatt 1989).

14.3 Measurement Sense in the Context of Mathematical Modelling

Translating between reality and the world of mathematics while dealing with real-world problems is at the heart of mathematical modelling (Maaß 2010). From a theoretical point of view if working on modelling problems one has to simplify and structure a real world situation, transfer it into the world of mathematics, work mathematically, interpret a mathematical result and validate a real result (Blum and Leiß 2006). While several studies point out that all these steps of mathematical modelling can cause problems if working on a real world problem (Blum 2011), only little is known about how to support learners to deal with these steps of mathematical modelling successfully. One possible approach for discussing difficulties if thinking about mathematical modelling could be to examine the influence of learners' measurement sense on *mathematical modelling (MM) performances* – since handling quantities is quite often needed to work on modelling problems successfully (for a concrete example see Fig. 14.1):

MM 1 Having to *understand* a given real world problem one firstly has to identify and/or distinguish different quantities of measurement (*MS 1*).

MM 2 *Structuring* and *simplifying* a complex situation often means measuring/ estimating/rounding missing data (*MS 2*). For being able to do so, adequate benchmarks are needed (*MS 5*).

MM 3 *Working mathematically* on a structured, simplified mathematical model requires being able to calculate and convert units of measurement (*MS 4*).

Modelling task "Brickhouse":

How many stones have been used for the outside walls of this house?

Measurement sense (*MS*) needed for mathematical modelling (*MM*) if working on the modelling task successfully:

MM 1 One has to *understand* that the area of the outside walls and not the volume of the house is asked for (*MS 1*).

MM 2 One has to *structure* and *simplify* the situation by estimating length, height and width of the house as well as of one single brick (*MS 2*).

MM 3 One has to *work mathematically* by multiplying/ adding lengths, height and width correctly and by converting m^2 to cm^2 and vice versa (*MS 4*).

MM 4 One has to *validate* a mathematical/ real result by thinking critically whether the calculated amount of bricks is possible/ acceptable/ adequate or not. For doing so measurement sense for all is needed (*MS 1 to MS 5*).

Fig. 14.1 Measurement sense needed for working on the modelling task *Brickhouse*

MM 4 For being able to *validate* a mathematical/real result critically distinct measurement sense for all is needed (*MS 1* to *MS 5*).

Based on these ideas one possible approach for supporting mathematical modelling could be to enable learners to build up measurement sense.

14.4 Improving Mathematical Modelling by Fostering Measurement Sense: An Intervention Study with Pre-service Mathematical Teachers

Based on (1) learners often having difficulties dealing with mathematical modelling tasks, due to the fact that (2) measurement sense is needed to work on many modelling tasks successfully and in line with the circumstance that (3) nothing is known about how learners' measurement sense influences mathematical modelling performances an intervention study with mathematical pre-service teachers has been conducted to answer the following research question:

- Is it possible to improve pre-service teachers' modelling performances by fostering measurement sense?

To be able to answer this question, the following sub-questions are dealt with within the intervention study:

- Is it possible to foster mathematical pre-service teachers' measurement sense within a short intervention? (Sub-RQ 1)
- Does mathematical pre-service teachers measurement sense influence mathematical modelling performances? (Sub-RQ 2)

14.4.1 Design of the Intervention Study

The intervention study took place in January 2013. Overall $N = 93$ pre-service mathematical teachers took part in the study. Before starting the study all participating pre-service teachers had taken part in mathematical modelling courses at university and therefore were familiar with mathematical modelling in general. The pre-service teachers were divided into two groups: $n_{EG} = 48$ pre-service teachers took part in a special intervention fostering measurement sense (i.e., the experimental group—EG). Another $n_{CG} = 45$ pre-service teachers were trained in dealing with fractions (this group can be considered as a control group since measurement sense had not been dealt with—CG). The pre-service teachers were assigned to either experimental group or control group by controlling for performance (marks within mathematical university courses) and gender. The intervention itself lasted for four hours in the experimental group as well as in the control group. To be able

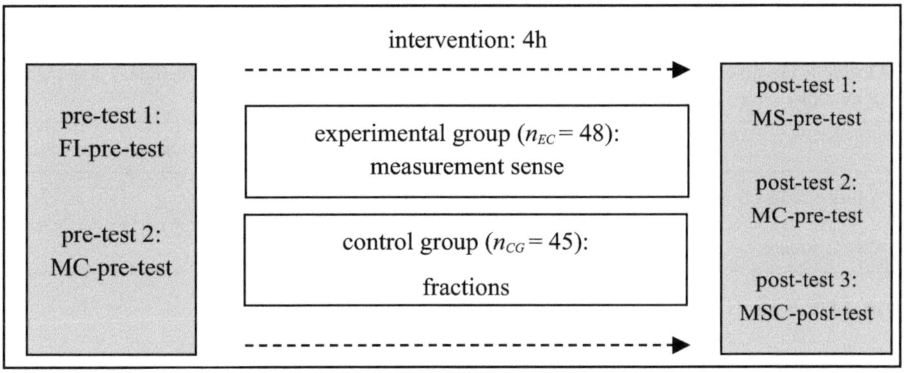

Fig. 14.2 Design of the intervention study

to evaluate the effectiveness of the intervention special tests on pre-service teachers' performances immediately before and after the intervention (for details see Fig. 14.2) were conducted.

14.4.2 Content of the Intervention

Based on the idea of fostering pre-service teachers' measurement sense (Sub-RQ 1), and paying attention to the question as to whether measurement sense influences mathematical modelling (Sub-RQ 2), the following topics were dealt with within the experimental and control groups, respectively: Within the experimental group the pre-service teachers were supported to build up the five central aspects of measurement sense as mentioned in section 2 (*MS 1* to *MS 5*). Within the control group basic concepts about dealing with fractions were trained (calculating with fractions). Neither in the experimental group nor in the control group was mathematical modelling dealt with.

14.4.3 Evaluating the Intervention Study: Tests on Mathematical Pre-service Teachers' Measurement Sense and Modelling Competency

For evaluating the study the following tests were used to answer the research questions mentioned above:

Pre-test 1 – Test on Figural Intelligence (FI-Pre-test) Immediately before starting the study all participating pre-service teachers had to work on a paper-pencil-test

Bus			
Decide if the given quantity is adequate or not. If not, please fill in your own estimate. Take care of the units.			
	Too less	Ok	Too much
A **bus** is about **8 m long**.			
Own estimate:			

Fig. 14.3 Sample item out of the MS-post-test (asking for *MS 2*)

assessing figural intelligence. The test consists of 25 items with closed responses. A tested person has to continue a sequence of given figures logically. The test lasts for 15 min. Results of this test are used to control for intelligence in general.

Pre-test 2 – Test on Modelling Competency (MC-Pre-test) Following the test on figural intelligence the participating pre-service teachers had to work on two modelling tasks with open responses for another 15 min (the tasks are similar to the *Brickhouse Task* in Fig. 14.1). The test was administered as a paper-pencil-test as well and assesses mathematical modelling competency in general.

Post-test 1 – Test on Measurement Sense (MS-Post-test) At the end of the intervention a test on measurement sense concerning length and area was administered to the pre-service teachers. The paper-pencil-test consists of 27 items (20 items with open responses, 7 items with closed responses) and lasted for 20 min. The test requires the ability to estimate (*MS 2*), knowledge about units of measurement (*MS 4*) and the existence of a set of meaningful benchmarks (*MS 5*) as selected parts of measurement sense (for a sample item see Fig. 14.3).

Post-test 2 – Test on Modelling Competency (MC-Post-test) Similar to the MC-pre-test there was a short post-test on mathematical modelling competency in general consisting of another two modelling tasks with open responses – once again these tasks were similar to the *Brickhouse Task* given in Fig. 14.1. The test lasted for 15 min.

Post-test 3 – Test on Modelling Sub-competencies (MSC-Post-test) In addition to the post-test on mathematical competency in general a newly developed paper-pencil test on mathematical sub-competencies was used to evaluate the effectiveness of the intervention. The test lasted for 30 min and consisted of 21 items (10 items with open responses; 11 items with closed responses) requires the abilities to understand (*MM 1*), to structure and simplify (*MM 2*) and to work mathematically (*MM 3*) on a given real world problem as selected parts of mathematical modelling (for a sample item see Fig. 14.4).

Hamster

Hamsters are nocturnal animals which sometimes develop a proper addiction to exercise wheels. In this case they cover a distance of 500 rounds at a stretch.

Which of the following information do you need to calculate the length of the distance a hamster covers while running 500 rounds in his exercise wheel? Mark the correct response(s).

	length of a hamster
	perimeter of the exercise wheel
	time the hamster needs for one rotation of the wheel
	weight of the hamster
	length and width of the exercise wheel
	distance between the paws

Fig. 14.4 Sample item out of the MSC-posttest (asking for *MM 2*)

14.4.4 Selected Results of the Intervention Study

All 93 participants worked on the two pre-tests and the three post-tests. While the FI-pre-test, the MS-post-test and the MSC-post-test can be used for quantitative analyses, the MC-pre-test and MC-post-test will be analysed qualitatively. To date quantitative results concerning the reliability of the FI-, MS- and MSC-tests and some descriptive results as well as some analyses of variance concerning the effectiveness of the treatment can be reported.

Reliability of the Tests The FI-pre-test, the MS-post-test and the MSC-post-test have all been scaled 1-dimensional. The answers to the test items have been scored dichotomously, a correct answer has been scored with 1, an incorrect answer has been scored with 0. The reliability (i.e., Cronbach's alpha) of the FI-pre-test (.74), the MS-post-test (.81) and the MSC-post-test (.81) can be said to be good. This allows these tests to be used for further analyses describing the mathematical pre-service teachers' figural intelligence (FI-pre-test) at the beginning of the intervention as well as the pre-service teachers measurement sense performances (MS-post-test) respectively modelling sub-competency performances (MSC-post-test) at the end of the intervention.

Descriptive Results and Differences Between the Groups As shown in Table 14.1 the pre-service teachers in the experimental (EG) and control (CG) groups scored similarly within the FI-test. This means the two groups can be said to be comparable concerning general figural intelligence. Within the MS-post-test on measurement sense and the MSC-post-test on modelling sub-competencies at the end of the intervention pre-service teachers in the experimental group outperform their counterparts in the control group. The pre-service teachers in EG scored significantly higher than those in CG.

Table 14.1 Descriptive results and differences between the groups

Group	*n*	Min	Max	*M*	*SD*	*t*-test for equality of means
FI-pre-test						
EG	48	9	24	17.52	3.61	$t(91) = 1.58$
CG	45	8	24	16.33	3.62	$p = .117$
MS-post-test						
EG	48	11	26	19.71	4.09	$t(91) = 4.59$
CG	45	5	25	15.58	4.58	$p = .000$
MSC-post-test						
EG	48	5	21	13.67	3.20	$t(91) = 3.44$
CG	45	2	19	11.09	4.00	$p = .001$

Table 14.2 Analyses of variance (ANOVA)

Dependent variable	Fixed factor	Co-variate	Adjusted R^2
MS-posttest	treatment (CG = 0; EG = 1) $F(1, 93) = 18.32, p < .000, =$ .169	FI-pretest $F(1, 93) = 3.90, p = .051, =$ .042	.205
MSC-posttest	MS-posttest $F(18,93) = 1.86, p = .034, =$ .314	FI-pretest $F(1, 93) = 8.05, p = .006, =$ .099	.302

Analyses of Variance (ANOVA) To explain these differences concerning the MS-post-test and the MSC-post-test analyses of variance were conducted (see Table 14.2). These ANOVAs point out: (Sub-RQ 1) If controlling for general intelligence the differences of pre-service teachers' performances within the MS-post-test can partly be explained by the treatment – about 17 % of the variance of the MS-posttest are explained by the treatment. (Sub-RQ 2) Furthermore, about 31 % of the variance of the MSC-post-test can be explained by the pre-service teachers' performances within the MS-post-test if controlling for general intelligence once again.

Referring to the research questions these results point out: Pre-service teachers' measurement sense has been fostered within a short intervention successfully (Sub-RQ 1). Furthermore, differences between pre-service teachers' performances concerning mathematical modelling are partly caused by pre-service teachers' measurement sense (Sub-RQ 2). Overall, it seems to be possible to improve pre-service teachers' modelling performances by fostering measurement sense.

14.5 Summary and Conclusion

As has been shown above measurement sense seems to be a central element of mathematical modelling of particular situations that afford use of measurement skills. Surprisingly, by now nothing is known about the influence of measurement sense on mathematical modelling. Within the intervention study presented above first quantitative insights into the interplay between measurement sense and mathematical modelling have been illustrated: Pre-service teachers' mathematical modelling competencies can be influenced positively by fostering pre-service teachers' measurement sense. As next steps these quantitative results need to be confirmed by conducting further qualitative analyses examining more closely how pre-service teachers' performances develop when working on the open modelling tasks. The tests on mathematical modelling administered to the pre-service teachers before and after the intervention (MC-pre-test, MC-post-test) can be used for doing so. Nevertheless, the quantitative results are encouraging suggesting: Fostering measurement sense in a short intervention seems to be possible. Furthermore, measurement sense seems to influence mathematical modelling that affords use of measurement skills.

References

Baturo, A., & Nason, R. (1996). Students teachers' subject matter knowledge within the domain of area measurement. *Educational Studies in Mathematics, 31*(3), 235–268.

Blum, W. (2011). Can modelling be taught and learnt? Some answers from empirical research. In G. Kaiser, W. Blum, R. Borromeo Ferri, & G. Stillman (Eds.), *Trends in teaching and learning mathematical modelling* (pp. 15–30). Dordrecht: Springer.

Blum, W., & Leiss, D. (2006). "Filling up" – The problem of independence-preserving teacher interventions in lessons with demanding modelling tasks. In M. Bosch (Ed.), *CERME-4 – Proceedings of the fourth congress of the European society for research in mathematics education* (pp. 1623–1633). Sant Feliu de Guixois: IQS Fundemi Business Institute.

Blum, W., Galbraith, P. L., Henn, H.-W., & Niss, M. (Eds.). (2007). *Modelling and applications in mathematics education.* New York: Springer.

Clements, D. H., & Bright, G. W. (Eds.). (2003). *Learning and teaching measurement (Yearbook of the National Council of Teachers of Mathematics).* Reston: National Council of Teachers of Mathematics.

Grund, K.-H. (1992). Größenvorstellungen – eine wesentliche Voraussetzung beim Anwenden von Mathematik. *Grundschule, 12*, 42–44.

Joram, E. (2003). Benchmarks for measurement units. In D. H. Clements & G. W. Bright (Eds.), *Learning and teaching measurement (Yearbook of the National Council of Teachers of Mathematics)* (pp. 195–207). Reston: NCTM.

Kellogg, M. (2010). Preservice elementary teachers' pedagogical knowledge related to area and perimeter. A teacher development experiment investigating anchored instruction with web-based microworlds. Theses and dissertations. Paper 1679. http://scholarcommons.usf.edu/etd/1679. Accessed 11 Sept 2013.

Larson, C. (2010). Modeling and quantitative reasoning: The summer jobs problem. In R. Lesh, P. L. Galbraith, C. R. Haines, & A. Hurford (Eds.), *Modeling students' mathematical modelling competencies: ICTMA13* (pp. 111–118). New York: Springer.

Maaß, K. (2010). Classification scheme for modelling tasks. *Journal für Mathematik-Didaktik, 31* (2), 285–311.

Martin, W. G., & Strutchens, M. E. (2000). Geometry and measurement. In E. A. Silver & P. A. Kenney (Eds.), *Results from the seventh mathematics assessment of the National Assessment of Educational Progress* (pp. 193–234). Reston: NCTM.

National Council of Teachers of Mathematics. (2000). *Principles and standards for school mathematics*. Reston: NCTM.

Outhred, L., Mitchelmore, M., McPhail, D., & Gould, P. (2003). Count me into measurement. A program for the early elementary school. In D. H. Clements & G. W. Bright (Eds.), *Learning and teaching measurement (Yearbook of the National Council of Teachers of Mathematics)* (pp. 195–207). NCTM: Reston.

Shaw, J., & Puckett-Cliatt, M. (1989). Developing measurement sense. In P. Trafton & A. Schule (Eds.), *New directions for elementary school mathematics: 1989 yearbook* (pp. 149–155). Reston: National Council of Teachers of Mathematics.

Thompson, T., & Preston, R. V. (2004). Measurement in the middle grades. Insights from NAEP and TIMSS. *Mathematics Teaching in the Middle School, 9*, 514–519.

Vasilyeva, M., Casey, B., Dearing, E., & Ganley, C. (2009). Measurement skills in low-income elementary school students: Exploring the nature of gender differences. *Cognition and Instruction, 27*(4), 401–428.

Chapter 15
How Do Students Share and Refine Models Through Dual Modelling Teaching: The Case of Students Who Do Not Solve Independently

Takashi Kawakami, Akihiko Saeki, and Akio Matsuzaki

Abstract The purpose of this chapter is firstly to show how students who could not solve an initial task by themselves shared and refined models through dual modelling teaching, and secondly to derive suggestions for dual modelling teaching. Through examining students' worksheets and protocols of video and audio records of the lessons, it was shown that unsuccessful modellers were able to change/modify their own models and classmates' ones and progress their dual modelling cycle by sharing different models. One crucial point for progression in the dual modelling cycle is the sharing of various models, which the modellers could not interpret or find independently. As an intervention strategy for the progression, teachers need to encourage students to share models that are related with both the initial task and their similar one, and to ensure a variety of ways to progress.

T. Kawakami (✉)
Faculty of Children's Studies, Nishikyushu University, Kamizono 3-18-15, Saga-shi, Saga, Japan
e-mail: kawakamita@nisikyu-u.ac.jp

A. Saeki
Graduate School of Education, Naruto University of Education, Nakashima 748, Takashima, Naruto-cho, Naruto-shi, Tokushima, Japan
e-mail: asaeki@naruto-u.ac.jp

A. Matsuzaki
Faculty of Education, Saitama University, Simo-Okubo 255, Sakura-ku, Saitama-shi, Saitama, Japan
e-mail: makio@mail.saitama-u.ac.jp

G.A. Stillman et al. (eds.), *Mathematical Modelling in Education Research and Practice*, International Perspectives on the Teaching and Learning of Mathematical Modelling, DOI 10.1007/978-3-319-18272-8_15

15.1 Dual Modelling Cycle Framework (DMCF)

For responding to the diversity of modellers' modelling processes (Borromeo Ferri 2007) and variation between tasks in their individual modelling progress (e.g., Matsuzaki 2011; Stillman and Galbraith 1998), Saeki and Matsuzaki (2013) elaborated a *dual modelling cycle framework (DMCF)* theoretically (Fig. 15.1). DMCF is constructed using the modelling cycle of Blum and Leiß (2007) and is focused on switching between a first modelling cycle of an initial task and a second modelling cycle of a similar task. In DMCF, three types of cycles (single, double and dual) are distinguished with corresponding progression of the solution (Saeki and Matsuzaki 2013). In the case of using a dual modelling cycle, modellers progress the first modelling cycle by applying or referring to results from the second modelling cycle. DMCF provides opportunities for tackling, reflecting and applying tasks, where previously known or more accessible models are relevant in addressing the tasks.

Matsuzaki and Saeki (2013) illustrated a typical dual modelling cycle through a case study of undergraduate school students. Kawakami et al. (2012) implemented the experimental class based on DMCF for Year 5 students and reported typical successful cases of how students who could solve the initial task by themselves progressed their dual modelling cycle; however the study did not focus on the other students who were unsuccessful independently. The question remains: How does the teacher facilitate the students' progress of their dual modelling cycle? The purpose of this chapter is (a) to show how the students who could not solve the initial task by themselves share and refine models through the dual modelling teaching (Kawakami et al. 2012) and (b) to derive some suggestions for dual modelling teaching corresponding to a diversity of modellers.

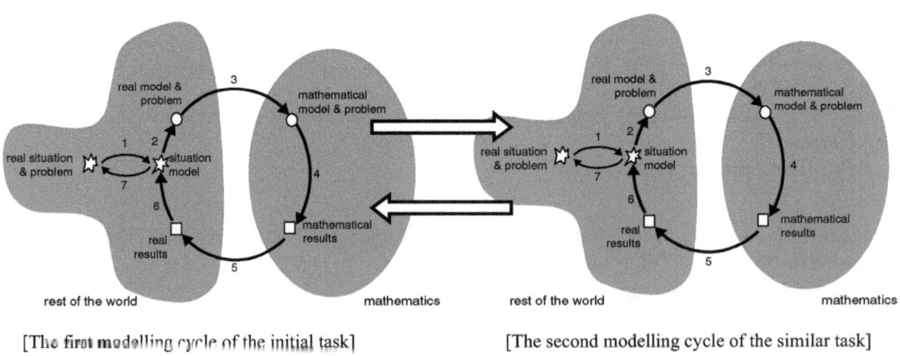

Fig. 15.1 Dual modelling cycle framework (Saeki and Matsuzaki 2013, p. 91)

15.2 Diversity of Student Models in Modelling Teaching Based on DMCF

15.2.1 The Teaching Material Based on DMCF

Oil Tank Task (Initial Task)
There are several types of oil tanks. Their heights are equal but their lengths of diameters are different. Are their lengths of spiral banisters equal or not? As conditions, angles to go up spiral banisters are all the same.

Toilet Paper Tube Task (Similar Task)
It is impossible to open an oil tank along the actual spiral banister. We can instead take a toilet paper tube as a similar shape to an oil tank, which is to be opened along its slit. Consider what the shape of an opened toilet paper tube would be.

The teaching material consists of an *Oil Tank Task* [OT task] (initial task) and a *Toilet Paper Tube Task* [TP task] (similar task). On the first modelling cycle of the OT task, the authors set up 3D models of oil tanks such as in Fig. 15.2. One of the 2D models of oil tanks is a rectangle model (Fig. 15.3). In this case, spirals of both tanks are the same by using mathematical ideas of equivalent transformation and translation of spiral. On the second modelling cycle of the TP task, we perceive that the 2D models of oil tanks are not only rectangle models but also parallelogram models (Fig. 15.4). In the case of a parallelogram model, spirals of both tanks are the same, which is supported by the mathematical idea of translation of the oblique side. Furthermore, we recognised that there was structural or relational correspondence between the rectangle model and the parallelogram model by using mathematical ideas of equivalent transformation and translation of spiral (Fig. 15.5).

15.2.2 Outline of the Modelling Teaching Based on DMCF

The experimental class (see also Kawakami et al. 2012) consisted of three lessons (45 min $\times$ 3). This class consisted of 33 Year 5 students (22 male, 11 female) aged 10–11 years old in a private elementary school in Japan. The data collection comprised video and audio recordings and artefacts of the lessons. In the first and second lesson, the students tackled the *Oil Tank Task* and the *Toilet Paper Tube Task*. In the OT task, the teacher showed 3D models (Fig. 15.2). Through discussion with the students, they drew 2D rectangle models of oil tanks to solve the OT task. However, they could not solve it because an oil tank is too big to open directly. So the teacher provided toilet paper tubes as similar shapes to the oil tank and students

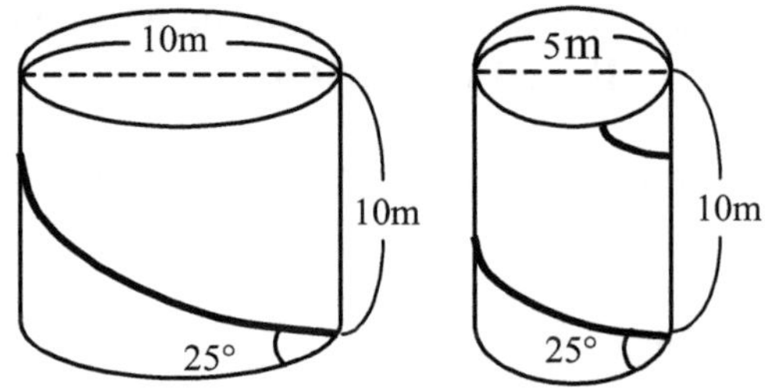

Fig. 15.2 3D models of oil tanks

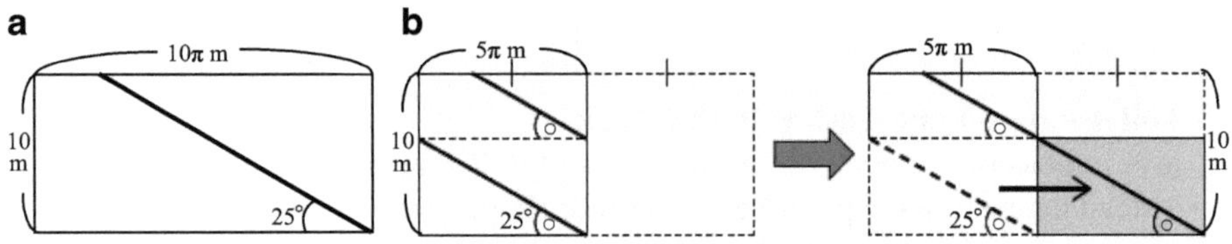

Fig. 15.3 The rectangle model: (**a**) original models, (**b**) half circumference models

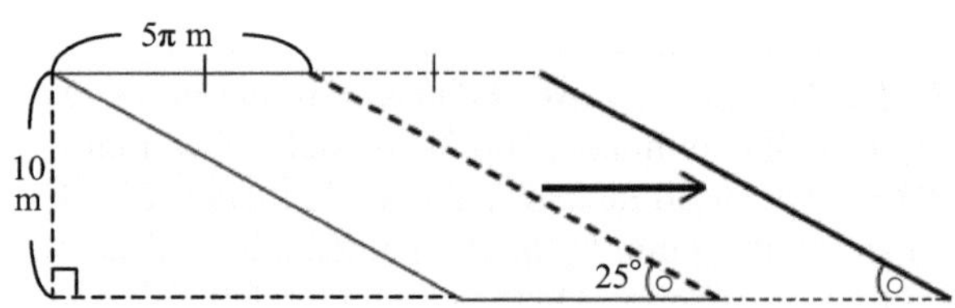

Fig. 15.4 The parallelogram model

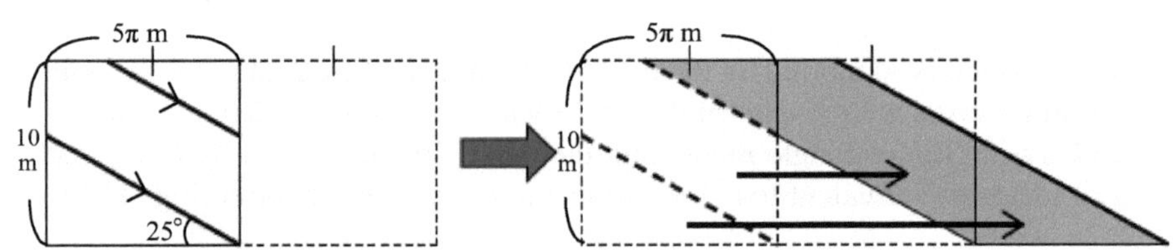

Fig. 15.5 Relationship between the rectangle model and the parallelogram model

made 2D parallelogram models in the TP task. At the end of the second lesson, they tackled the OT task again, with much referring to the TP task.

In the third lesson, the students shared three types of classmates' solutions. At first, the teacher illustrated the solution by focusing on the area of the parallelogram model and calculating the length of the spiral. However, they could not calculate the height of the parallelogram model using their own mathematical skills. Next, the teacher illustrated the solution by measurement of the spiral in rectangle models and parallelogram ones. The teacher said, "We found the approach by calculating is difficult, but we can give an answer by measuring the reduced drawing". Finally, the teacher highlighted Kob's solution by translation of the spiral where he applied equivalent transformations in the rectangle model (Fig. 15.6a) and Matsu's

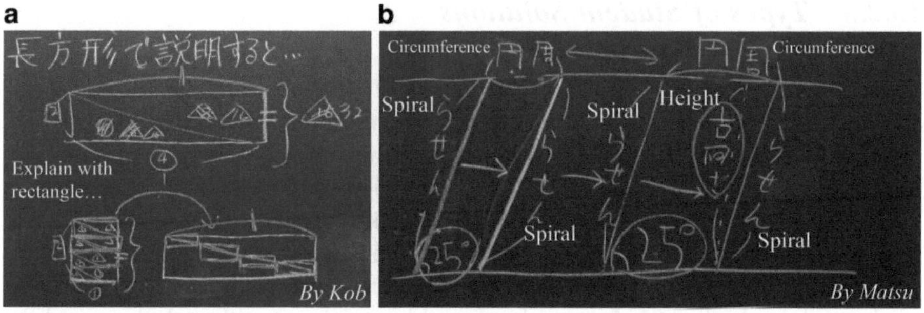

Fig. 15.6 (**a**) The rectangle models (**b**) The parallelogram models

approach of translation of the spiral in the parallelogram model (Fig. 15.6b). By illustrating two kinds of models, the teacher intended students to find the relationship between the first modelling cycle and the second modelling cycle.

Kob drew his idea on the blackboard (see Fig. 15.6a) and tried to explain that the length of the spiral was equal by applying the rate of area and equivalent transformation in the case of a quarter-length of the diameter. However, it was difficult for most of his classmates to understand, as the compound idea in Kob's explanation confused them. Then the teacher complemented his idea by using Tomi's idea in the case of half-length diameter as the following protocols show:

Teacher: Tomi cut the rectangle in half along this dotted line and translated the upper half [of the] rectangle to the bottom. Then the spiral is connected.

Students: I see!

Teacher: Tomi thought about the case as the half-length of the diameter. You have to think about the case of one third and the quarter-length of diameter too.

On the other hand, it was easy for the class to understand the parallelogram approach as the following protocols show:

Teacher: OK, we will look at the approach by parallelogram. Matsu, explain your idea. [Matsu drew his idea on the blackboard as in Fig. 15.6b.]

Matsu: It is probable that these spirals are all the same ... and the length and angle are all the same ... so I think oblique sides of the parallelogram are all the same.

Teacher: So how is this spiral [oblique side of parallelogram] moved?

Students: Translation.

Teacher: That's right. Matsu said, "The length of the spiral is equal because the oblique side of the parallelogram is translated".

15.2.3 Types of Student Solutions

Table 15.1 shows how 33 students tackled the *Oil Tank Task* again at the end of the second lesson. Sixteen students solved it by themselves. In tackling the OT task again, ten students answered by translation of the spiral in the 2D parallelogram model of the oil tank (Fig. 15.6a) (*Type A*), and 5 students answered by translation of spiral in the 2D rectangle model of the oil tank (Fig. 15.6b) (*Type B*). Only one student answered by measurement of spiral in both rectangle models and parallelogram models (*Type C*). On the other hand, 17 students could not solve the OT task by themselves. In *Type A*, four students could not solve it, although they used the parallelogram model. In *Type B*, nine students could not solve it, although they used the rectangle model. In *Type D*, four students could not solve it, although they used other models (e.g., rotation rate of spiral in cylinder model).

Kawakami et al. (2012) analysed the solutions of *Type A* and *Type B* with typical successful cases of how the students who could solve the OT task by themselves progressed the dual modelling cycle. These successful cases give us suggestions to consider what is ideal modelling teaching based on DMCF. However, there are other students who could not solve the OT task by themselves as indicated in Table 15.1. How do they progress the dual modelling cycle through shared classmates' models? How does the teacher facilitate the students' progress of the dual modelling cycle? In the next section, we focus on the solutions of two female students (Kawa's case in *Type A* and Kato's case in *Type B*) and analyse how their original models in the first and second lessons were changed or modified by examining their worksheets and protocols.

15.3 Analysis of Responses of Students who Could not Solve Independently

15.3.1 Kawa's Case

15.3.1.1 Lesson 1: First Trial of *Oil Tank Task* and *Toilet Paper Tube Task*

In the OT task, Kawa drew two rectangle models and put a spiral only in the left rectangle model (Fig. 15.7a), however she did not identify which is the case of the long diameter or of the short diameter. The left rectangle model seems to be mathematically correct. In the TP task, she drew a parallelogram model, however she made a mistake in identifying the length of the spiral as the length of the circumference (Fig. 15.7b).

Table 15.1 Types of student solutions to the *Oil Tank Task* ($N = 33$)

Solution type	Frequency solved	
	Independently	Not independently
A Parallelogram model	10	4
B Rectangular model	5	9
C Parallelogram and rectangular model	1	0
D Other models	0	4

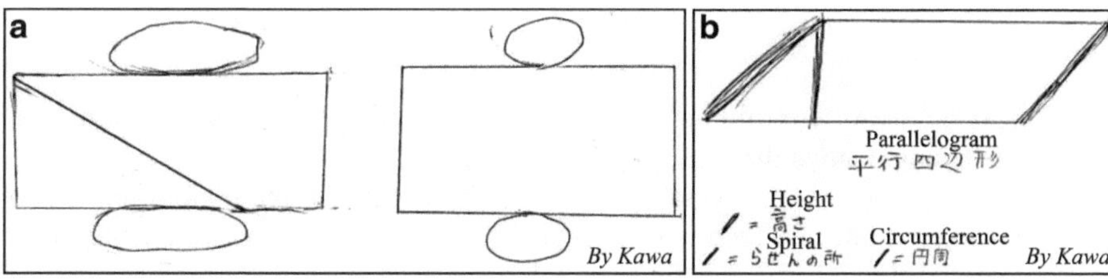

Fig. 15.7 (**a**) The rectangle models (**b**) The parallelogram model

15.3.1.2 Lesson 2: Second Trial of *Oil Tank Task* Based on the *Toilet Paper Tube Task*

Returning to the OT task, Kawa tried to solve it by the parallelogram model in which a pair of adjacent sides are 10 cm and 5 cm; however she could not solve the task. Her parallelogram model of the oil tanks did not correspond to Fig. 15.7b and to Fig. 15.2. She did not understand the structure of the parallelogram model enough.

15.3.1.3 Lesson 3: Final Trial of *Oil Tank Task* Through a Class Presentation

After the class presentation, Kawa wrote Kob's explanation by the rectangle model in the case of a quarter-length of the diameter (Fig. 15.8a). She described the equivalent transformation from a quarter diameter oil tank to the original diameter oil tank by using arrows and the phrase, "Cut and paste here ... spiral is just connected!" So she showed that the spiral was connected as the result of equivalent transformation. She accepted Kob's explanation to be applicable to other diameter length oil tanks by her phrase, "We can vary the length of diameter!" She accepted Kob's rectangle model by referring to Tomi's idea in the case of half-length diameter. Additionally, she wrote a parallelogram model after Matsu's explanation (Fig. 15.8b). She did not copy Matsu's statement from the blackboard, but interpreted Matsu's statement and her classmates' statement of translation by using her own expressions "Circumference is only extended" and "Stretch". In the end, she concluded that changing the length of the diameter of the oil tank has

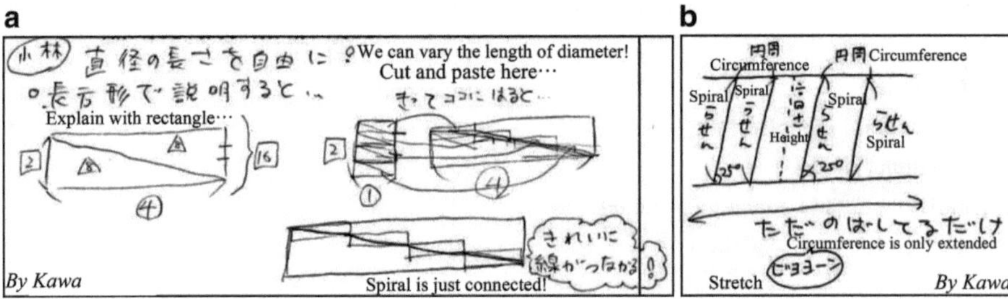

Fig. 15.8 (**a**) The accepted rectangle model (**b**) The accepted parallelogram model

nothing to do with determining the length of the spiral banister of it even if its height or its angle stays the same.

Kawa accepted Kob's rectangle model and Matsu's parallelogram model because she could accept the mathematical structure of both models. Furthermore, she concluded the OT task by the rectangle model and parallelogram model, and generalized it by the latter.

15.3.2 *Kato's Case*

15.3.2.1 Lesson 1: First Trial of *Oil Tank Task* and *Toilet Paper Tube Task*

In the OT task, Kato drew two rectangle models and put spirals on both models. The left rectangle model (Fig. 15.9a) is for the long diameter and right one is for another. However, the latter model is not mathematically correct. In the TP task, she indicated the relationships between the spiral and circumference correctly in a parallelogram model (Fig. 15.9b).

15.3.2.2 Lesson 2: Second Trial of *Oil Tank Task* Based on the *Toilet Paper Tube Task*

Returning to the OT task, Kato used a rectangle model and calculated the length of the circumference and the spiral by using a reduced drawing in which the diameter is 4 cm. This gave a circumference of $4 \times 3.14 = 12.56$ cm. However, she could not solve the task. Her model is not correct mathematically because the spiral is disconnected.

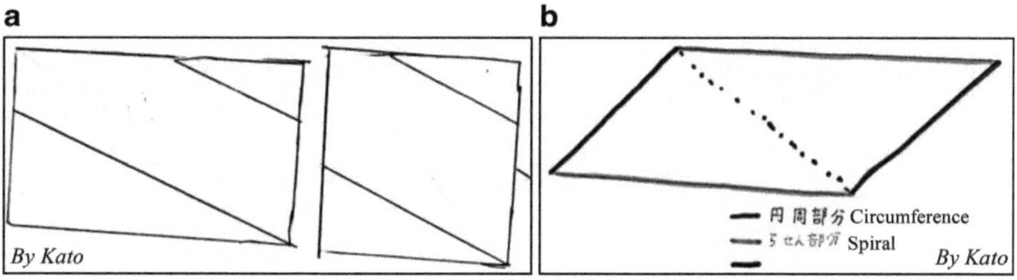

Fig. 15.9 (**a**) The rectangle models (**b**) The parallelogram model

15.3.2.3 Lesson 3: Final Trial of *Oil Tank Task* Through a Class Presentation

After the class presentation, Kato wrote Kob's explanation using the rectangle model in the case of the quarter-length of the diameter (Fig. 15.10a). It seems that she did not accept the relationship of the spiral in the rectangle model and/or apply the mathematical idea of equivalent transformation. Additionally, she wrote a parallelogram model after Matsu's explanation (Fig. 15.10b). She did not copy Matsu's statement from the blackboard, but interpreted Matsu's statement and her classmates' statement of transformation by her expression "You can give a value to the circumference". In the end, she concluded, as did Kawa, that changing the length of diameter of the oil tank has nothing to do with determining the length of the spiral banister of it.

Kato did not adopt Kob's rectangle model because she could not accept the mathematical structure of the rectangle model and/or apply the mathematical idea of equivalent transformation, however, she accepted Matsu's parallelogram model and the relation between circumference and spiral in it. Furthermore, she concluded the OT task and generalised it by using the parallelogram model.

15.4 Discussion

In the first two lessons, Kawa made rectangle models of the oil tank in the first modelling cycle (Fig. 15.7a) and the parallelogram model of the toilet paper tube in the second modelling cycle of DMCF (Fig. 15.7b). However, her parallelogram model was not correct mathematically. Returning to the first cycle from the second cycle, she used the parallelogram model and tried to solve the *Oil Tank Task*, however she could not solve it. From the viewpoint of DMCF, her modelling process is a "double" modelling cycle because she could not progress the first modelling cycle by findings from the second modelling cycle. In other words, she could not transition from the second modelling cycle to the first modelling cycle well. One of the factors of unsuccessful modelling is that she did not accept the relationship between the circumference and the spiral in the parallelogram model.

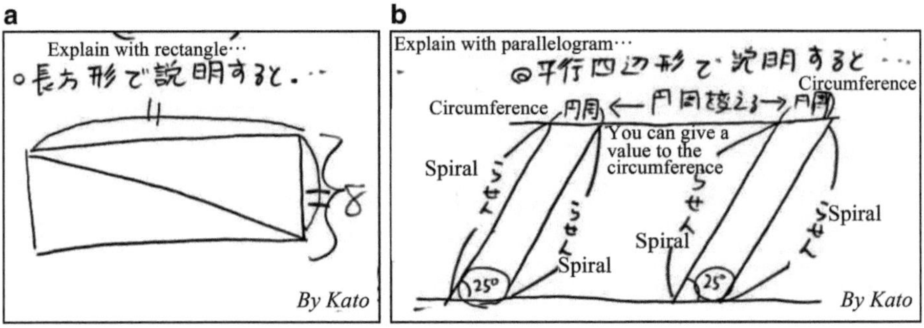

Fig. 15.10 (**a**) The unaccepted rectangle model (**b**) The accepted parallelogram model

In this dual modelling teaching, students confirmed the correspondence between each part in the toilet paper tube and the parallelogram as its development (Kawakami et al. 2012). If the parallelogram model is put on a different orientation, the relationship between the circumference and slit (spiral in oil tank) is changed (Saeki and Matsuzaki 2013). This might influence Kawa's "reconstruction of the figurative context" (Busse and Kaiser 2003, p. 4) offered in the OT task. Through the third lesson, she modified the parallelogram model which she had mistaken by objectifying and interpreting her shared classmates' parallelogram model in her own ways (Fig. 15.8b). She accepted her classmates' rectangle model too (Fig. 15.8a). Additionally, she concluded the OT task and generalized it by using both her accepted parallelogram model and the rectangle one. Her modelling process is a "dual" modelling cycle.

On the other hand, in the first and second lessons, Kato made rectangle models of the oil tank in the first modelling cycle (Fig. 15.9a) and a parallelogram model of the toilet paper tube in the second modelling cycle (Fig. 15.9b), however her rectangle models were not correct mathematically. Returning to the first cycle from the second cycle, she used the rectangle model and tried to solve the OT task, however she could not solve it. Her modelling process is a "double" modelling cycle as was Kawa's in the first and second lessons. Through the third lesson, she concluded and generalised the OT task in objectifying and interpreting only accepted classmates' parallelogram model in her own ways (Fig. 15.10b). When she meant to solve it, she changed the rectangle model to a parallelogram model. Her modelling process is also a "dual" modelling cycle. The reason she did not accept her classmates' rectangle model might be her misunderstanding of the relationship of the spiral to a rectangle and/or the analogy of rectangle models and parallelogram models. In this teaching experiment, the teacher could not spend time confirming the mathematical structure of the rectangle model (Kawakami et al. 2012).

Some unsuccessful modellers were able to change/modify their own models and classmates' ones and progress their dual modelling cycle by sharing different models such as Kawa's case and Kato's case. Kawa and Kato interpreted classmates' models in their own ways and solved them again by applying these models.

Similarly to Brown and Edwards (2013), students' communication of their solutions to the modelling task enhanced their mathematical understandings and modelling. On the other hand, Kawa and Kato sometimes could not transition from the second modelling cycle to the first cycle well because they did not accept the mathematical structure of the models and/or they could not apply mathematical ideas. This also illustrates that individual modeller's prior knowledge related to task context including the mathematical content of a task (e.g., Matsuzaki 2011; Stillman 2000) influences which models are able to be shared.

15.5 Conclusion

One crucial point for progression in the dual modelling cycle is to share various models, which modellers could not interpret or find. As an intervention strategy to stimulate the transition between modelling cycles, teachers need to encourage students to share and use the models that are related with both the initial task and the similar task. For example, teachers need to confirm with students structural or relational correspondences between initial task models and similar task ones and common mathematical ideas in both models. If they do so, unsuccessful modellers can reinterpret and, if necessary, modify the solution of the similar task as a model to reach the solution of the initial task or might integrate both solutions to generalise the conclusion. Thus, teachers have to ensure there are various types of modelling progress for each modeller through dual modelling teaching.

References

Blum, W., & Leiß, D. (2007). How do students and teachers deal with modelling problems? In C. Haines, P. Galbraith, W. Blum, & S. Khan (Eds.), *Mathematical modelling (ICTMA12): Education, engineering and economics* (pp. 222–231). Chichester: Horwood.

Borromeo Ferri, R. (2007). Modelling problems from a cognitive perspective. In C. Haines, P. Galbraith, W. Blum, & S. Khan (Eds.), *Mathematical modelling (ICTMA12): Education, engineering and economics* (pp. 260–270). Chichester: Horwood.

Brown, J. P., & Edwards, I. (2013). Modelling tasks: Insight into mathematical understanding. In G. Kaiser, W. Blum, R. Borromeo Ferri, & G. Stillman (Eds.), *Trends in teaching and learning of mathematical modelling* (pp. 187–197). New York: Springer.

Busse, A., & Kaiser, G. (2003). Context in application and modelling: An empirical approach. In Q.-X. Ye, W. Blum, K. Houston, & Q.-J. Jiang (Eds.), *Mathematical modeling in education and culture: ICTMA 10* (pp. 3–15). Chichester: Horwood.

Kawakami, T., Saeki, A., & Matsuzaki, A. (2012). Necessity for modelling teaching corresponding to diversities: Experimental lessons based on dual modelling cycle framework for the 5th grade pupils. *ICME-12 pre-proceedings* (pp. 3291–3300). Seoul: ICME.

Matsuzaki, A. (2011). Using response analysis mapping to display modellers' mathematical modelling progress. In G. Kaiser, W. Blum, R. Borromeo Ferri, & G. Stillman (Eds.), *Trends in teaching and learning of mathematical modelling: ICTMA 14* (pp. 499–508). New York: Springer.

Matsuzaki, A., & Saeki, A. (2013). Evidence of a dual modelling cycle: Through a teaching practice example for pre-service teachers. In G. A. Stillman, G. Kaiser, W. Blum, & J. P. Brown (Eds.), *Teaching mathematical modelling: Connecting to research and practice* (pp. 195–205). Dordrecht: Springer.

Saeki, A., & Matsuzaki, A. (2013). Dual modelling cycle framework for responding to the diversities of modellers. In G. A. Stillman, G. Kaiser, W. Blum, & J. P. Brown (Eds.), *Teaching mathematical modelling: Connecting to research and practice* (pp. 89–99). Dordrecht: Springer.

Stillman, G. (2000). Impact of prior knowledge of task context on approaches to applications tasks. *Journal of Mathematical Behavior, 19*(3), 333–361.

Stillman, G., & Galbraith, P. (1998). Applying mathematics with real world connections: Metacognitive characteristics of secondary students. *Educational Studies in Mathematics, 36*(2), 157–195.

Chapter 16
Exploring Interconnections Between Real-World and Application Tasks: Case Study from Singapore

Kit Ee Dawn Ng and Gloria Ann Stillman

Abstract Some findings from an interdisciplinary project work (PW) implemented with Year 7 and 8 students (13–14 years old) from three Singapore schools are reported. These are part of a study examining the impact of PW in terms of its learning outcomes (LO). Of interest are findings associated with LO: the extent to which the PW brings about student-perceived "interconnections" between school disciplines, within mathematics, and between school-based mathematics and real-world problem solving. There was an overall increase in mean scores on the scales measuring perception of interconnectedness of mathematics and inter-subject learning (ISL) and beliefs and efforts at making connections (BEC) after PW. ANOVA showed a significant impact of the PW on ISL but not BEC scores. Qualitative results revealed that these seemingly positive results disguised issues with students' ability to make the desired interconnections in a meaningful manner.

16.1 Introduction

Interdisciplinary Project Work (PW) was incorporated into the Singapore curriculum in 1999 (Curriculum Planning and Development Division [CPDD] 1999) to (a) enhance holistic learning (Ng 2008) by highlighting interconnections within and among school subjects (Chan 2001) and (b) demonstrate the relevance between school-based learning and real-world problem solving (CPDD 2012). PW is a platform to foster the development of twenty-first century competencies

K.E.D. Ng (✉)
Mathematics and Mathematics Education Academic Group, National Institute of Education, Nanyang Technological University, 1 Nanyang Walk, Singapore 637616, Singapore
e-mail: dawn.ng@nie.edu.sg

G.A. Stillman
Education VIC, Faculty of Education & Arts, Australian Catholic University, 1200 Mair Street, Ballarat, VIC 3350, Australia
e-mail: gloria.stillman@acu.edu.au

G.A. Stillman et al. (eds.), *Mathematical Modelling in Education Research and Practice*, International Perspectives on the Teaching and Learning of Mathematical Modelling, DOI 10.1007/978-3-319-18272-8_16

(e.g., thinking skills) among Singapore students (Chang 2004). PW became an entry requirement to university in 2005 (Ministry of Education [MOE] 2001). To date, PW has been implemented in various forms (e.g., as an alternative mode of assessment compared to paper-and-pencil tests, as a performance task) across primary, secondary, and pre-university levels in Singapore schools. A PW involving the use of mathematics is perceived as one of the many ways the Singapore mathematics curriculum prepares students in line with current expectations on mathematical literacy which requires flexible, appropriate use of mathematical knowledge and skills for application of mathematics in various contexts (OECD 2013).

PW is situated within real-world contexts. A PW with mathematics as one of its anchor subject-disciplines is an applications task. One learning outcome in PW is knowledge application. This requires students to display an *integrated* but critical use of knowledge from various subject-disciplines to solve problems situated in real-world contexts. In addition, students will be provided with opportunities to perceive the *relevance* and *inter-relatedness* of what is learnt during knowledge application in PW (CPDD 1999). Nonetheless, there appears to be a lack of evidence from local research about whether the integrated use of subject-specific knowledge as well as relevance and inter-relatedness are achieved through PW.

For this research, the term "interconnectedness" was theoretically defined by Ng (2009, p. 167) as comprised of three components arising from critical analysis of the work of Jacobs (1989; 1991) and Boix Mansilla et al. (2000) in line with the PW learning outcome of knowledge application prescribed by the Singapore MOE: "perceptions about (a) possible links between mathematics and other school-based disciplines, (b) the usefulness of mathematics in understanding and learning other subjects, and (c) the complementary relationship between mathematics and other subjects in problem solving." A major aspect of problem solving espoused by the Singapore mathematics curriculum framework (CPDD 2012) is making connections. Connections refer to drawing links between school-based disciplines, within mathematical topics, and between school-based mathematics and the real-world. The complementary relationship described in component (c) above can be viewed in terms of student efforts in drawing upon critical connections during problem solving.

A broader literature analysis of existing research from other contexts on extended real world projects at school level reveals mixed results with respect to students' engagement in learning being enhanced through mathematical tasks embedded within meaningful real-world experiences of students. In tracking the progress of Australian students participating in a variety of interdisciplinary real-world problems over 2 years, Fielding-Wells and Makar (2008) claim that "mathematics problems that crossed curricular boundaries supported both deeper understandings and greater engagement" (p. 1). They concluded that such problems added value to students' cognitive engagement in mathematical learning as the problems prompted more rigorous mathematical representations, reasoning and argumentation processes. On the other hand, Venville, Rennie, and Wallace (2004) reported that students perceived a lack of usefulness of school mathematics in a real-world interdisciplinary mathematics, science and technology task. Wineburg and Grossman (2000) also raised the issue of mathematical learning being placed as a subservient priority during interdisciplinary real-world tasks that usually involved purely mathematical application and direct calculations without necessarily advancing student mathematical content knowledge. Thus, it is unclear whether and how, if so, interdisciplinary real-world projects involving mathematics such as a PW in the form of an applications task has a desired impact on students' engagement in mathematical learning in terms of how they perceive and actually make use of interconnections.

Hence, the research question to be addressed in this chapter is: *To what extent does student participation in a PW involving mathematics implemented over a span of several weeks successfully demonstrate the desired outcome of perceived and displayed "interconnectedness" as part of the knowledge application learning outcome in PW?* This question originates from a larger study in Singapore secondary schools that investigated the impact of a mathematical PW on students' engagement in mathematical learning in view of the stated learning outcomes of PW, one of which is knowledge application (Ng 2009). Knowledge application was partly examined with the lens of "interconnectedness" for the purpose of the larger study. Initial reports on the process of scale development to quantitatively measure the theoretical construct of "interconnectedness" during the larger study can be found in Ng and Stillman (2007) and Ng, Stillman, and Stacey (2007). The theoretical construct of "interconnectedness" became an operational definition resulting from scale development. This definition is summarised in two scales purporting to measure the perception of "interconnectedness": (a) interconnectedness of mathematics and inter-subject learning (ISL) and (b) beliefs and efforts at making connections (BEC). This chapter focuses on presenting qualitative findings from student interviews and work artefacts from a mathematical PW. Brief quantitative results from scale measurement are provided for a more comprehensive profile for understanding the impact of the PW on perception and display of "interconnectedness".

16.2 The Study

As part of the larger case study of PW in Singapore, the participation of Years 7 and 8 Singapore students (aged 13–14) in a PW involving mathematics, science, and geography was investigated to gauge the extent this participation successfully demonstrated the desired outcome of "*interconnectedness*". A multi-site, multi-case based approach (Yin 2009) was taken to glean a deep understanding of the extent and nature of the perception of interconnectedness of mathematics for a selection of PW groups across participating schools. Case study was chosen because of the insights it could offer about students' thinking and behaviours during PW within the naturalistic environment in which it occurred. Mixed methods (Creswell 2007) were used to examine the perception of interconnectedness of mathematics by students. The extent and nature of this perception was analysed and represented using both quantitative and qualitative measures. However, in this report, we will focus on presenting the qualitative results. Pertinent findings from the quantitative analysis will be provided as a background profile for interpretation of the qualitative results.

16.2.1 Sample and Sampling Process

Three government secondary schools that had been implementing PW since its incorporation into the mathematics curriculum agreed to participate in the study. They ran annual PW tasks for year levels not involved in national examinations. They consented to using the PW task designed by the first author (see Sect. 16.2.3) during their routine PW sessions with students in secondary one and two classes (aged 13–14). There were 16 classes altogether (11 from high-stream and 5 from average-stream) of students attempting the PW task. Of these, 398 students (201 males and 197 females) agreed to participate in a questionnaire to gauge their attitudes in interdisciplinary learning. Ten student-groups of four members each agreed to participate in qualitative data collection. Group formation was at the teacher's discretion as the schools already had PW procedures in place.

16.2.2 Data Collection and Analysis

The students took part in the PW through 14–15 weekly one-hour face-to-face sessions facilitated by class teachers. They participated in a pre- and post-PW researcher-designed student questionnaire entitled "Mathematical Attitudes in Interdisciplinary Learning" (MAILL) which included six constructs representing the three attitudinal domains measuring students' perceptions of (a) mathematical confidence, (b) value of mathematics, and (c) interconnectedness of mathematics. The third domain on the perception of the interconnectedness of mathematics, which is the focus here, included two scales: the perception of *mathematics and inter-subject learning* (ISL) and *beliefs and efforts at making connections* (BEC). The two scales were developed by the first author to represent all three components of "interconnectedness" outlined above (see Ng and Stillman 2007; Ng et al. 2007). ANOVA was applied to compare the pre- and post-PW results from these scales.

Video and audio recordings were collected during three student-group discussion sessions for each of the ten cases. These sessions were expected to have more mathematical content in focus. Each student in the group also engaged in three video-stimulated recall interviews (Lyle 2003) which the first author conducted after every discussion session. Field notes on key teacher facilitating episodes within each PW session were recorded. Facilitating teachers were also interviewed at two stages of the PW implementation to understand their pedagogical decisions during the sessions. Raw data were coded using grounded theory (Strauss and Corbin 1998).

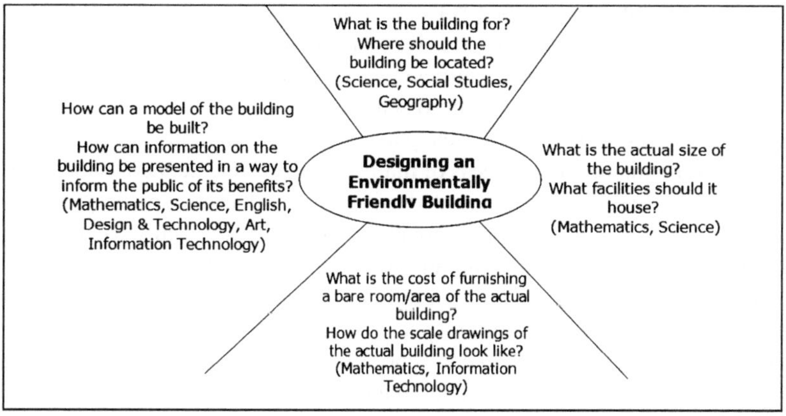

Fig. 16.1 Designing an environmentally friendly task

16.2.3 PW: Designing an Environmentally Friendly Building

The researcher-designed PW task required student-groups to design an environmentally friendly building in a location of their choice within Singapore. It was expected that through this task students would draw upon the use of mathematics, science, and geography in their decision making process (see Fig. 16.1). The mathematical affordances of the task included applications of arithmetic and measurement concepts and skills such as simple calculations, proportional reasoning, area, perimeter, and scale measurement. Student-groups were to construct a physical scale model out of recycled materials according to scale drawings prepared, complete a written report about their project, and lastly make a presentation to their classes.

16.3 Findings

Quantitative results showed an overall increase in mean scores on the two scales measuring the perception of interconnectedness of mathematics and inter-subject learning (ISL) and beliefs and efforts and making connections (BEC) after PW. Post-PW results revealed that ISL and BEC means also increased in the comparisons among subgroups (e.g., gender, stream, and school). ANOVA tests showed a significant impact of the PW on ISL scores. This implied that students' perception about using mathematics for inter-subject learning had been positively enhanced during PW although the partial eta squared value indicated a small effect size. However, no significant difference was observed for BEC indicating PW had little or no impact on students' beliefs and efforts at making connections between mathematics and other subjects during learning. It cannot be determined clearly within the scope of the study why there was a lack of impact of PW on BEC scores.

One reason could be because the prior experience of PW in primary schools by the student sample had already shaped largely positive beliefs about connections between mathematics and other subjects as observed by the mean score of 4.01 (on a 5 point scale) at the start of the research.

Further investigations into ISL as part of the extent and nature of students' perceptions of the interconnectedness of mathematics were extended to the qualitative data. The qualitative results are illustrated using four student-groups. Video recordings of three PW sessions on decision-making about the environmentally friendly building, calculations of the cost of furnishing an area within the building, and scale drawings were analysed. This was done in conjunction with audio recordings of video-stimulated recall interviews with respect to three aspects of interconnectedness: students drawing connections between school-based disciplines, within mathematical topics, and between school-based mathematics and the real-world.

16.3.1 Connections Between School-Based Disciplines

Most of the students when interviewed were conscious of possible connections between mathematics and science in the PW task. Others named specific or general connections between mathematics, art, geography, and design and technology when interviewed:

Researcher: Can you see links between math topics and other subjects? How?

Alice: Yes, because like math before we need to. . .before we draw. . .we are using art to draw then after that we have to use math for scale and then after that. . .I have to use geography research on those EF {eco-friendly} features and everything. . .

Lenny: Yes. . .for example. . .like Math and Design & Technology. . . Geography. . .they interlink.

Although there was a significant increase in ISL scores after PW, actual discussions of inter-subject connections occurred less frequently than expected during the three video-recorded sessions. The students were engaged mainly with direct applications of specific knowledge from one school subject (e.g., science in deciding a water tank design) without drawing upon related mathematical calculations to support their arguments. In other words, the students might not have perceived the integrated use of mathematics within the content of another school subject. This can be observed from the excerpt below which shows William from Group 4 discussing the design of the water tank on the roof of their eco-friendly house:

William: [Nodding with approval.]. . .in the park. . .to prevent the mosquitoes to actually grow, right, we can have something like a block here. . .when the water actually flow{s} through. . .the block will actually open like this. . .and then. . .when the water stops right. . .it will turn back like this. . . {We could have pipes linking between the water tank and the roof. We could also have a net to block the mosquitoes from residing in the pipes.}

William and his group had wanted to recycle rain water collected using a water tank but were concerned with the possibility of the water tank becoming a breeding ground for *Aedes* mosquitoes, a rampant problem in Singapore. The group decided to add a net to the design of the tank not only to filter the water before usage but also to block mosquitoes from coming into the tank. However, the entire discussion was about the net and pipes without drawing in the necessary mathematical calculations on the shape, dimensions, and capacity of the tank.

16.3.2 Connections Within Mathematics Topics

Video and interview data showed that some students perceived the connections between the various mathematical concepts during the PW task whilst others did not. For example, a few student-groups were observed making connections between measurement and geometric concepts during the cost of furnishing discussion session. Figure 16.2 shows Group 8 considered the shape of the toilet floor in their eco-friendly house to be a square and subsequently embarked on the calculations for floor area using the correct measurement algorithms. Nonetheless, these are rather unrealistic for the size and dimensions of the toilet area within a house.

In addition, 7 out of 10 student-groups sampled made connections between the scale used and proportional reasoning during the scale drawing session. For instance, the floor plan from Group 7 representing their eco-friendly mansion had various parts of the building in reasonable proportion with one another. However, Group 1 did not consider realistically the actual floor area of their eco-friendly school hall which typically houses about 800 students when they first decided on the scale of 1 cm on their drawing to represent 1,000 m for the actual building. This lack of connection between scale and spatial visualisation was obvious from this exchange between Chi and Yang and others in their group:

Chi: The scale? {What do you mean by scale drawings?}

Yang: 1 cm to 1,000 ah? Then the school is 1,000 to...
{How about 1 cm on the model representing 1,000 m in the real-life building?}

Kath: [Interrupting Yang's explanations.] That big meh?
{Are you sure this scale is appropriate? Our building will be very big if that's the case!}

Hasyim: What about 1 cm to 100...{How about 1 cm representing 100 m in the real building then?}

Chi: 1 m is here to here [gesturing with his arms] 1,000 m is....[trying to estimate the size of 1,000 m].

Kath: Okay...okay 1,000 m...{Okay, we can have this scale of 1 cm to 1,000 m.}

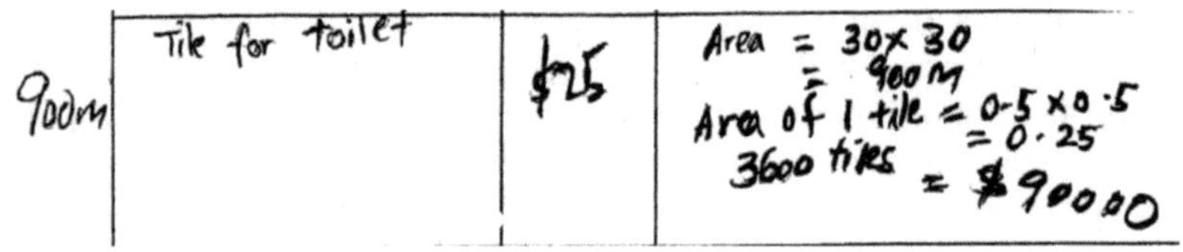

Fig. 16.2 Calculations of floor area

16.3.3 Connections Between School-Based Mathematics and the Real-World

Real-world considerations featured during group discussions among almost all groups but these considerations were not necessarily realised in mathematical calculations. For example, Group 4 as reported in Sect. 16.3.1 did not work out mathematically the desired capacity of the water tank and the required volume to sustain usage in the eco-friendly house. In addition, real-world context was not considered during scale drawings. This was observed from Group 1 above where an inappropriate scale was used for the size of the school hall in their scale drawings. In another instance, real world constraints were not considered during scale drawing when a dominant member in Group 3 did not realise that the scale drawings of different views (i.e., side, back and front elevations) of the eco-friendly house her group was designing had to match so that a coherent scale drawing of the house could be made and then used to make their scale model to match the proposed real house. She encouraged group members assigned different elevations to draw to use their own scales rather than one common scale.

16.3.4 Connections Between ISL and BEC Scores and Qualitative Findings

The ISL and BEC scores of the students whose work or conversations were highlighted in the qualitative results as problematic were comparable with those of their peers who did not have such difficulties and who were able to actively make connections in their work during the PW task. Thus, complementary methods and measures are necessary for the purpose of evaluating the attainment of interconnectedness by PW, particularly in relation to inter-subject connections if we were to understand the quality of interconnected use of mathematical knowledge and skills.

16.4 Discussion

This chapter reported on the extent to which a PW task involving mathematics, science, and geography successfully attained the objective of interconnectedness. There were varying perceptions of interconnectedness displayed by student-groups. This was evident from both the quantitative and qualitative results. There was also a range of interconnectedness perceived by students even within each group as seen in the excerpt from Group 1 above. Whilst there was evidence to show the impact of PW on students' perceptions of inter-subject connections, the connections between mathematical topics, and connections between school mathematics and the real-world, these were mediated by some issues which arose from qualitative findings.

Firstly, when queried, students were able to state inter-subject connections from the task yet there was limited actual integrated use of these particularly during mathematical calculations and decision-making (e.g., water tank example from Group 4). This brings into question whether students are really aware of how to use mathematics in integration with other school subjects. Proponents of integrated learning (e.g., Beane 1995) have argued for the use of authentic real-world projects to streamline mathematical learning because schooling must be relevant to students' lives in order to engage them. The Singapore curriculum is structured according to topics for each subject and there is little room to manoeuvre learning across subjects during lessons unless an overhaul in school-based planning, staff distribution, and teacher education can be carried out. Schools can only embark on PW to highlight inter-subject connections on an annual basis but the impact may not be as significant as envisioned by curriculum planners. This may be because teaching in regular lessons is predominantly subject-focused with limited in-depth links across subjects where teachers show how to infuse mathematical calculations and arguments for decision-making in a real-world problem from the perspective of another subject.

Secondly, echoing findings of van den Heuvel-Panhuizen (1997) in short contextualised tasks, some student-groups proposed their mathematical solution to the PW task without making sense of the solution within the real-world constraints of the problem. There were selected but limited connections between school-based mathematics and the actual coordinated use of mathematics during a real-world context (e.g., the scale example from Group 1). This may be due to many teachers in Singapore mathematics classrooms not emphasising discussion of mathematical problems nor their solution in view of how these make sense in the world. Application problems in textbooks and resources attempt to show connections between the mathematics to be learnt and the context where it can be used; but teaching has mainly focused on calculations and getting the correct answer without much real-world interpretation.

Thirdly, there were obvious challenges faced by students during scale drawings from the PW task used in the research. This was especially evident when student-groups could not draw connections between the chosen scale and the computation of the actual area or length. This could be because of a lack of spatial visualisation promoted in relation to the teaching of scale at secondary levels.

16.5 Conclusion

The findings have implications for teachers and teacher education in Singapore. For teachers, there can be more deliberate attempts to connect real-world use of school-based mathematics during regular classes and more efforts in helping students translate their identification of inter-subject connections into relevant mathematisation and accompanying mathematical calculations. During a mathematical PW, students' attention should be drawn to making "interconnectedness" more explicit through teacher questioning. For mathematics educators, the incorporation of real-world interdisciplinary tasks such as applications and modelling need emphasis in teacher preparation courses, equipping teachers with the confidence and repertoire to discuss mathematics within real-world contexts. Finally, the findings provide an added dimension from the Singapore perspective to the debate on the mathematical educational value of interdisciplinary real-world projects.

References

Beane, J. A. (1995). Curriculum integration and the disciplines of knowledge. *Phi Delta Kappan, 76*(8), 616–622.

Boix Mansilla, V., Miller, W. C., & Gardner, H. (2000). On disciplinary lenses and interdisciplinary work. In S. Wineburg & P. Grossman (Eds.), *Interdisciplinary curriculum: Challenges to implementation* (pp. 17–38). New York: Teachers College Press.

Chan, J. K. (2001). *A curriculum for the knowledge age: The Singapore approach.* Paper presented at the eighth annual Curriculum Corporation conference, Sydney.

Chang, A. S. C. (2004). *Promoting thinking skills through project work.* Paper presented at the annual general meeting of Educational Research Association, Singapore.

Creswell, J. W. (2007). *Qualitative inquiry and research design: Choosing among five approaches.* Thousand Oaks: Sage.

Curriculum Planning and Development Division [CPDD]. (1999). *Project work: Guidelines.* Ministry of Education: Singapore.

Curriculum Planning and Development Division [CPDD]. (2012). *Mathematics syllabus: Primary.* Ministry of Education: Singapore.

Fielding-Wells, J., & Makar, K. (2008). *Using mathematical inquiry to engage student learning within the overall curriculum.* Paper presented at *Topic Study Group: Mathematical knowledge for teaching at 11th International Congress on Mathematical Education*, Mexico. Retrieved from. tsg.icme11.org/document/get/747.

Jacobs, H. H. (1989). The interdisciplinary concept model: A step-by-step approach for developing integrated units of study. In H. H. Jacobs (Ed.), *Interdisciplinary curriculum: Design and implementation* (pp. 1–11). Alexandria: Association for Supervisors and Curriculum Development.

Jacobs, H. H. (1991). The integrated curriculum. *Instructor, 101*(2), 22–23.

Lyle, J. (2003). Stimulated recall: A report on its use in naturalistic research. *British Educational Research Journal, 29*(6), 861–878.

Ministry of Education [MOE]. (2001). *Press release: Project work to be included for university admission in 2005*. Retrieved December 13, 2002, from http://www1.moe.edu.sg/press/2001/pr20062001.htm

Ng, P. T. (2008). Educational reform in Singapore: From quantity to quality. *Educational Research for Policy and Practice, 7*(1), 5–15.

Ng, K. E. D. (2009). *Thinking, small group interactions, and interdisciplinary project work.* Doctoral dissertation, The University of Melbourne.

Ng, K. E. D., & Stillman, G. (2007). Interdisciplinary learning: Development of mathematical confidence, value, and the interconnectedness of mathematics scales. In J. Watson & K. Beswick (Eds.), *Mathematics, essential research, essential practice* (Vol. 2, pp. 533–542). Adelaide: MERGA.

Ng, K. E. D., Stillman, G. A., & Stacey, K. (2007). Interdisciplinary learning and perceptions of interconnectedness of mathematics. In J. H. Woo, H. C. Lew, K. S. Park & D. Y. Seo (Eds.), *Proceedings of the 31st conference of the international group for the Psychology of Mathematics Education* (Vol. 2, pp. 185–192). Seoul: PME.

Organisation for Economic Co-operation and Development [OECD]. (2013). *Draft PISA 2015 mathematics framework.* (10 April 2013). http://www.oecd.org/pisa/pisaproducts/Draft%20PISA%202015%20Mathematics%20Framework%20.pdf

Strauss, A., & Corbin, J. (1998). *Basics of qualitative research: Techniques and procedures for developing grounded theory*. Thousand Oaks: Sage.

van den Heuvel-Panhuizen, M. (1997). *Realistic problem solving and assessment.* Paper presented at the Mathematics Teaching conference, University of Edinburgh.

Venville, G., Rennie, L., & Wallace, J. (2004). Decision making and sources of knowledge: How students tackle integrated tasks in science, technology and mathematics. *Research in Science Education, 34*(2), 115–135.

Wineburg, S., & Grossman, P. (Eds.). (2000). *Interdisciplinary curriculum: Challenges to implementation*. New York: Teachers College Press.

Yin, R. K. (2009). *Case study research: Design and methods* (4th ed.). Thousand Oaks: Sage.

Chapter 17
Mathematical Modelling Tasks and the Mathematical Thinking of Students

Bárbara Palharini and Lourdes Maria Werle de Almeida

Abstract This chapter describes a study that aims to illustrate how the interaction between mathematical modelling tasks and students' mathematical thinking takes place. We based the study upon David Tall's theory about mathematical thinking regarding mathematical modelling as a pedagogical way to teach mathematics. The research took place in the last year of a university course, with education students undertaking a mathematics degree. A qualitative approach was used with an interpretative analysis to infer points from data collected through audio recordings, video, quizzes, and written data. We have concluded that mathematical modelling can provide, and require the development of cognitive processes that promote interactions between "elementary" and "advanced" mathematical thinking through the cognitive processes of representation, abstraction and generalization.

17.1 Introduction

One of the main purposes of learning and teaching mathematics relates to the construction of mathematical knowledge by students. This construction is present in numerous studies in the field of mathematical education focused on the development of students' mathematical thinking. Many students demonstrate understanding difficulties when it comes to mathematics. According to Sfard (1991), there must be something different in mathematical thinking:

B. Palharini (✉)
Department of Mathematics, Northern Paraná State University, Av. Portugal, 340 Cornélio Procópio, PR 86300-000, Brazil
e-mail: barbara.palharini@uenp.edu.br; babipalharini@hotmail.com

L.M.W. de Almeida
Department of Mathematics, University of Londrina, Antonio Pisicchio, 155, apto 303, Londrina, PR 86050-482, Brazil
e-mail: lourdes.maria@sercomtel.com.br

G.A. Stillman et al. (eds.), *Mathematical Modelling in Education Research and Practice*, International Perspectives on the Teaching and Learning of Mathematical Modelling, DOI 10.1007/978-3-319-18272-8_17

> Indeed, since in its inaccessibility mathematics seems to surpass all the other scientific disciplines, there must be something really special and unique in the kind of thinking involved in constructing a mathematical universe. (p. 2)

A study about mathematical thinking is presented in this chapter using David Tall's theory, which traces an association between *elementary mathematical thinking* and *advanced mathematical thinking*. According to Tall (2002), the importance of advanced mathematical thinking studies lies in investigating how mathematical development takes place among people, as well as how people's minds work. This sort of investigation can lead the researcher to developing ways to improve mathematical lessons to help students develop advanced mathematical thinking. In this chapter, we have considered mainly the work of Dreyfus (2002) and Tall (1995, 2002) to address the cognitive theory of mathematical thinking and cognitive processes related to such a process.

We describe a research study reported in Palharini (2010) that took place in the field of mathematical modelling, under the perspective of mathematical education. The research was developed in the final year of graduation at the university with 19 students who were undertaking a mathematics degree to teach mathematics in middle and secondary years of school, from which we analyzed the work of three students. In this context, we considered the insertion of modelling tasks into the discussion regarding their potential for the development of mathematical thinking.

The aim of this research is to answer the question: Does student engagement in modelling tasks promote the development of advanced mathematical thinking? As mathematical thinking is something that happens in people's minds, during mathematical tasks, we view mathematical modelling tasks, following Almeida and Dias (2004) and Almeida and Ferruzzi (2009), under a cognitive perspective described by Kaiser and Sriraman (2006) and used by Borromeo Ferri (2006).

17.2 Theoretical Assumptions

17.2.1 Transitions from Elementary Mathematical Thinking into Advanced Mathematical Thinking

Seeking to understand mathematical development in individuals, and how cognitive development of mathematical thinking occurs, we studied the work of Tall and Dreyfus regarding the cognitive theory of mathematical thinking, which these authors proposed. To speak of mathematical development of an individual, it is necessary to understand how people deal with their daily tasks, in particular those involving mathematical processes. To execute such tasks, we have to think in a mathematical way. Mathematics education deals with the theory regarding mathematical thinking in terms of "elementary mathematical thinking" and "advanced mathematical thinking". According to Tall (1995), it is possible to visualize the cognitive growth from elementary to advanced mathematical thinking:

> The cognitive growth from elementary to advanced mathematical thinking in the individual may therefore be hypothesised to start from "perception of" and "action on" objects in the external world, building through two parallel developments—one visuospatial to verbal-deductive, the other successive process-to-concept encapsulations using manipulable symbols—leading to a use of all of this to inspire creative thinking based on formally defined objects and systematic proof. (p. 3)

Both advanced and elementary mathematical thinking involve cognitive processes that consist of a large array of interacting component processes (Dreyfus 2002). According to Tall (2002), it is necessary to study mental processes that are referred to, in this chapter, as *cognitive processes,* associated with thought. *Representation* and *abstraction* are the most important processes of mathematical thinking.

In addition, representation includes three sub-processes: *visualization*, *switching representations* or *translating*, and *modelling*. Abstraction includes *generalization* and *synthesis*. All of these processes take place in people's minds as cognitive actions performed without perception of such performance. However, in what situations may educators perceive these processes? Are there moments in people's lives in which such processes appear and, if so, can they be refined? Is it possible to create learning environments or scenarios to explore the emergence of such processes in people's mathematical tasks? These are some of the questions permeating the research development, presented in this chapter.

In this context, we used a *theory of three mathematical worlds* formulated by Tall (2004b): "Distinct but interrelated worlds of mathematical thinking each with its own sequence of development of sophistication [. . .] that in total spans the range of growth from the mathematics of new-born babies to the mathematics of research mathematicians" (p. 1). The ideas of the three mathematical worlds are discussed by Tall (2004a, b, 2006) and Gray and Tall (1994).

According to Tall (2004a) the name of the first mathematical world is the "conceptual-embodied world," or "embodied world," and it "grows out of our perceptions of the world, as well as it consists of our thinking of things that we perceive and sense, not only in the physical world, but in our own mental world of meaning" (p. 2). In the said world, things happen by reflection and by the use of increasingly sophisticated language. Here, people can focus on aspects of sensory experience. It is common for the emergence of representation processes to take place, such as visualization, since it is necessary to see objects, manipulate them and think of encapsulating their most important features, as a graphical representation or diagrams, such as a triangle.

The second world is the "proceptual-symbolic world," or simply, "proceptual world." According to Tall (2004a), "it is the world of symbols that we use for calculation and manipulation in arithmetic, algebra, calculus and so on . . . These begin with actions (such as pointing and counting) that are encapsulated as concepts by using symbols" (p. 2). In this world, the use of a symbolic representation is necessary (one that cannot exist without mental representations), as well as processes of abstraction, generalization or synthesis, used to deal with symbols, switching representations or translating, or modelling.

The third world, named by Tall (2004a) as the "formal-axiomatic world," or "formal world," is "based on properties, expressed in terms of formal definitions that are used as axioms to specify mathematical structures (such as group, field, vector space, topological space and so on)" (p. 3). Here, mathematical thinking processes in individuals relate to each other. Mathematical proof is used, and there are increasing mathematical languages in a sophisticated way, using the logic and rigor of mathematics.

During students' engagement with mathematical tasks they move through the Three Worlds of Mathematics. In this movement a few interactions take place between elementary and advanced mathematical thinking. To analyze these interactions it is possible to consider the emergence of cognitive processes of mathematical thinking as studied by Tall (2002) and Dreyfus (2002). Considering the importance for teachers to be conscious of these processes to comprehend some of the difficulties their students face, we will present mathematical modelling as a pedagogical way to teach mathematics from a cognitive perspective.

17.2.2 *Mathematical Modelling*

According to Bean (2003, p. 2), "all forms of 'modelling' require interpretive and creative thinking". When dealing with thought and cognitive processes, a cognitive perspective of mathematical modelling is required. As to modelling perspectives, Kaiser and Sriraman (2006) described, among others, a cognitive perspective, a kind of meta-perspective, in which the research objectives and its psychological goals are, respectively:

> a) analysis of cognitive processes taking place during modelling processes and understanding of these cognitive processes [. . .] b) promotion of mathematical thinking processes by using models as mental images or even physical pictures or by emphasising modelling as mental process such as abstraction or generalization. (p. 304)

Following Kaiser and Sriraman (2006), our research considers the analysis of the emergence of cognitive processes during modelling tasks. When a researcher views mathematical modelling from a cognitive perspective, it is necessary to see students' work in class. In this context, we understand mathematical modelling as a pedagogical way to teach mathematics that suggests an approach through mathematics of a problematic situation, not essentially mathematically-related, according to Almeida and Dias (2004).

To illustrate elements regarding the way that students' mathematical thinking involved in modelling tasks takes place, we used modelling tasks where, from a real situation, students need to translate between reality and mathematics, make assumptions, choose variables, deduce a mathematical model, interpret it and validate it according to the real situation. To indicate tasks for the classroom where teachers can provide students with mathematical thinking processes, we analyze modelling tasks.

17.3 Methodology

This chapter describes a research, reported in Palharini (2010), that was developed in the final year of graduation at the university with 19 students that were undertaking a mathematics degree to teach mathematics in middle and secondary years of school. We analyzed the work of three students who will be referred to as student A, B and C. Data were collected during a school year while students were engaged in modelling tasks. Data collection included audio and video recording of the classes, students' written and verbal response to quizzes and written data about students' solution for modelling tasks. By considering all the procedures in the research we used a qualitative approach following Bogdan and Biklen (1994), and an interpretative analysis regarding students' mathematical thinking processes while they move through the Three Worlds of Mathematics, considering David Tall's theory.

For the development of modelling tasks in class, during the school year, we used a gradual introduction, as described as moments by Almeida and Dias (2004). These moments were used as an approach to introduce modelling tasks and to teach mathematical modelling to students. Firstly, the teacher brings to the classroom mathematical modelling tasks already developed. Containing a real situation and some initial data related with the situation, in this task there is a problem to solve with necessary information to do it. Students, with the support of a teacher, re-create the said task. In the second moment, the teacher, with the students, defines a problem to study from a set of information about something real (here students are responsible for the formulation of hypotheses, development of mathematical model and other modelling stages). Finally, the students themselves create a modelling task, including choosing a theme from the real world, the definition of a problem, the definition of variables and hypotheses, the deduction of the mathematical model, its interpretation and validation in order to answer the initial problem. We analyzed four tasks, and each student participated in the development of two of those (Table 17.1).

The data collected supports two types of analysis undertaken:

1. Specific analysis occurred where we analysed responses of all three students to the *Diazepam in the Body Task*. Responses from the three students were then analysed for the second task each engaged with (as indicated in Table 17.1).
2. Expanded analysis, involved considering the involvement of each of the three students in the two tasks, in order to consider elements relating to what mathematical thinking was involved in mathematical modelling tasks.

Firstly we analyzed qualitatively data collected from all tasks (type 1). In the expanded analysis we conducted an interpretative analysis following the theoretical background.

Table 17.1 Student involvement in the various modelling tasks

Modelling task	Students involved		
	A	B	C
Diazepam in the body	✓	✓	✓
World records	✓	✗	✗
Ethanol production vs the dynamics of vehicle production	✗	✓	✗
Quantitative analysis of nicotine and cadmium accumulated in a smoker's body	✗	✗	✓

17.4 Results and Discussion

To illustrate elements that show us the emergence of advanced mathematical thinking we try to observe and report mathematical thinking processes that appeared during the modelling task development. We observe such processes when students transit through the three mathematical worlds mentioned by Tall (2004b). In all tasks, we can see students creating and using tables followed by use of graphical representations to understand the real problem. To do this it is necessary to create a mental representation of the problem, for example, for the real problem stated in the *Diazepam in the Body Task.*

Diazepam in the Body Task
Diazepam is a medication indicated for patients with anxiety attacks, somatic and psychological anxiety related to the treatment of psychiatric disorders in the relief of muscle spasm due to trauma, such as injury and inflammation. It is a prescription drug (black label) and its improper use or use for long stretches of time can be harmful.

Information package leaflet:

Elimination: the Diazepam plasma concentration/time curve is biphasic: an initial rapid and intense distribution phase, with a half-life that can reach 3 hours and a prolonged terminal elimination phase (half-life of 20–50 hours). The terminal half-life elimination (t1/2b) of active metabolic nordiazepam is about 100 hours. Diazepam and its metabolic effects are primarily excreted in the urine, predominantly in conjugated form. The clearance of Diazepam is 20–30 ml/min.

Thus the *half-life* (Diazepam elimination) is given as follows: In the early stage it can reach 3 hours. In the final stage it can last 20–50 hours.

Treatment should be administered at a starting dose not exceeding 10 mg and the duration should be as short as possible not exceeding 2–3 months including the period of gradual withdrawal of the drug.

From the context, students studied the situation to estimate the concentration of Diazepam in the body over a period of time. During modelling tasks to initiate mathematical development a simplification is necessary, an abstraction of a non-mathematical situation to a mathematical situation. It starts with insights of patterns and regularities that denote a data mathematical behaviour. Such insight is called by Dreyfus (2002) an abstraction process. In this sense, we infer that students develop advanced mathematical thinking, since abstraction use denotes a possible occurrence of advanced mathematical thinking.

In Fig 17.1 we present how students used graphical representation or table representation to obtain an algebraic model that describes the problem of diazepam concentration in the body with numbered arrows 1, 2 and 3 representing the translations that students made to answer the question. In Fig. 17.1 it is possible to note the use of objects, such as graphs and tables that allow the students to see the behaviour of the data. This allows us to infer that students used the cognitive process characterised by Dreyfus (2002) as visualization and that students' thinking transits through the conceptual-embodied world, since such objects are considered by Tall (2004b) as embodied objects. Furthermore, the use of symbols (translations 1 and 3), make us believe that student thinking transits through the conceptual-embodied and proceptual-symbolic world. In this transit the use of three kinds of representation to deal with mathematical concepts is observed, as differential equations and exponential functions, which were possible from a table given to the students (Fig. 17.1). The transitions between representations are from the table to an algebraic representation, using data provided by the table (see translation 1), from an algebraic to a graphical representation, using the model developed (see translation 2), and last from a graphical to an algebraic representation, where students apply the model to estimate when a Diazepam dose of 10 mg is eliminated from the body. We can consider these transitions being possible by the use of representation processes, as characterised by Dreyfus (2002).

The same process took place in the tasks *World Records*, *Ethanol Production vs The Dynamics of Vehicle Production* and *Quantitative Analysis of Nicotine and Cadmium Accumulated in a Smoker's Body* that was developed by the students from choosing the subject to answer the initial question related to this subject, according to the suggestion of the third moment of modelling referred to in Sect. 17.3.

In the formulation of the mathematical model students use mathematical symbols and mathematical operations, theorems and definitions, such as the concept of ordinary differential equations, concept of limits, Cauchy's theorem, the concepts of monotone and convergent sequences, exponential function, and others. These actions operated by students are related to their transit through the proceptual-symbolic and axiomatic-formal world (Tall 2004b), and with the emergence of representation processes. The student written data from the *World Records Task* reveals how the student moves in the axiomatic-formal world, and probably the use of advanced mathematical thinking:

> The data form a decreasing sequence, that is, (R_n) forms a monotonically decreasing sequence. The human body has its limits, so it is impossible for one person to complete the 100 m swimming in one second; we do not know the value of the limit, but it exists and so (R_n) is limited. Given these observations, we believe that the data during the year leads to a sequence (R_n) monotonically decreasing and also limited. Hence, by a theorem of analysis (all limited monotone sequences are convergent), we can say that this is a convergent

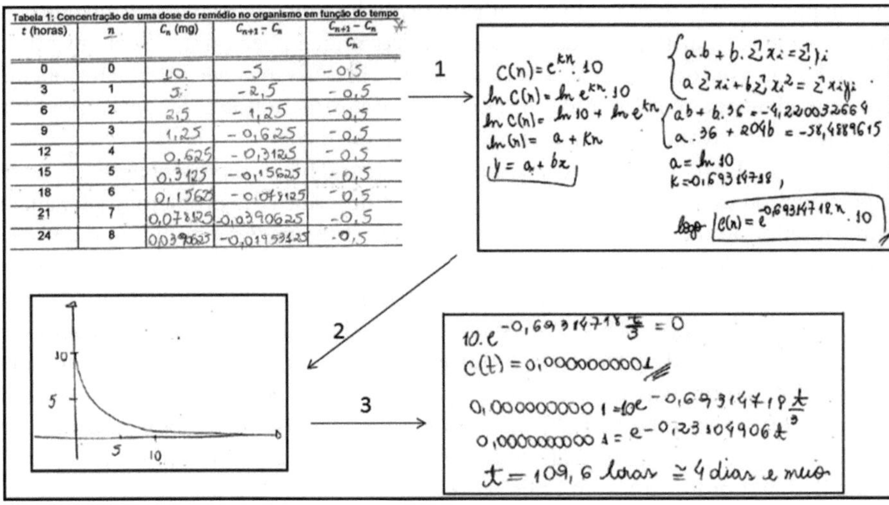

Fig. 17.1 Representations used by student B during the *Diazepam in the Body Task*

> sequence. As the sequence is convergent, then there exists the limit of the sequence of records. That is, there is a real number L such that lim $R_n = L$ (Written data of the student A in the *World Records Task*).

By processing symbols and using objects related to the formal mathematics there is an interaction between the student's transit in the symbolic and axiomatic worlds. This interaction implies an interaction of thoughts, an interaction between elementary and advanced mathematical thinking. In this context, students use procedures consistent with elementary mathematics and subsequently advanced mathematics, based on abstract entities that should be used by deductions and formal definitions.

When students are constructing a mathematical model their focus is in obtaining a symbolic representation which initially requires a mental representation. There is the use of various representations, switching representations (as seen in Fig. 17.1), and the generalization process is also required. This is what we can see in Fig. 17.2, where students make a generalization to estimate the nicotine concentration in the body of a smoker.

What Fig. 17.2 shows is how student C, from the amount of nicotine in one cigar, developed a mathematical model, $C(t) = 0,8 \cdot \frac{1-e^{-0,346573632t}}{1-e^{-0,346573632}}$, to eliminate the amount of nicotine of various cigars of the smoker's body in *t* hours. Whilst student C uses symbols related to the symbolic world, by working with mathematical concepts and deduction based on formal properties, he/she transits into the symbolic and axiomatic worlds. In this context, when students generalize from a simple case to a more general case, they use the process of generalization, related to the abstraction processes, in interaction with the representation processes. According to Dreyfus (2002, p. 34), "if a student develops the ability of making abstractions, consciously, of the mathematical situations, he/she reaches a level of advanced mathematical

t	Nicotine concentration in the Body
0	$C(0) = 0{,}8e^{0}$
1	$C(0) + C(1) = 0{,}8 + 0{,}8e^{-0{,}346573632.1} = 0{,}8(1 + e^{-0{,}346573632})$
2	$C(0) + C(1) + C(2) = 0{,}8(1 + e^{-0{,}346573632}) + 0{,}8e^{-0{,}346573632.2} =$ $= 0{,}8(1 + e^{-0{,}346573632} + e^{-0{,}346573632.2})$
3	$C(0) + C(1) + C(2) + C(3) =$ $= 0{,}8(1 + e^{-0{,}346573632} + e^{-0{,}346573632.2}) + 0{,}8e^{-0{,}346573632.3} =$ $= 0{,}8(1 + e^{-0{,}346573632} + e^{-0{,}346573632.2} + e^{-0{,}346573632.3})$
...	...
m	$C(0) + C(1) + C(2) + \cdots + C(m) =$ $0{,}8\left(\underbrace{1 + e^{-0{,}346573632.1} + e^{-0{,}346573632.2} + e^{-0{,}346573632.3} + \cdots + e^{-0{,}346573632.m}}_{sum\ of\ terms\ of\ a\ geometric\ progression}\right)$

Fig. 17.2 Generalization by student C in the *Quantitative Analysis of Nicotine Accumulated in a Smoker's Body Task*

thinking". In the same way, interactions between abstraction and representation processes may represent the occurrence of advanced mathematical thinking.

17.5 Final Remarks

Considering modelling tasks where, from a real situation, students need to translate between reality and mathematics, make assumptions, choose variables, deduce a mathematical model, interpret it and validate it according to the real situation, from our research it is possible to conclude that to solve problems students need to use different objects of the three worlds. During students' engagement with modelling tasks we could see that they move through the Three Worlds of Mathematics and in these movements interactions between elementary and advanced mathematical thinking took place. Modelling tasks allow students to complete a cycle of development, from one world to another, non-linearly, using cognitive processes related to mathematical thinking. Advanced mathematical thinking occurs when multiple processes interact. So, guided by data analysis, we can assert that students' engagement in modelling tasks promote the development of advanced mathematical thinking. We can understand the transitions between the mathematical worlds as a refinement of thoughts. In such refinement, students can develop mathematical thinking from "elementary" to "advanced" and from "advanced" to "elementary" thinking. According to our study it is possible to see that in situations with modelling tasks educators can perceive and study mathematical thinking processes and, by knowing these, it is possible to create learning environments or scenarios to explore the emergence of such processes.

References

Almeida, L. M. W., & Dias, M. R. (2004). A study on the use of mathematical modelling as a strategy for teaching and learning. *BOLEMA – Boletim de Educação Matemática, 22*, 19–35.

Almeida, L. M. W., & Ferruzzi, E. (2009). A socioepistemological approach to mathematical modelling. *Alexandria Revista de Educação em Ciência e Tecnologia, 2*(2), 117–134.

Bean, D. (2003). Modelling in the perspective of thought. *Annals of the III national conference on modelling and mathematics education*, 1–11. Piracicaba, SP.

Bogdan, R. C., & Biklen, S. K. (1994). *Qualitative research in education*. Porto: Porto Editora.

Borromeo Ferri, R. (2006). Theoretical and empirical differentiations of phases in the modeling process. *Zentralblatt für Didaktik der Mathematik, 38*(2), 86–95.

Dreyfus, T. (2002). Advanced mathematical thinking processes. In D. Tall (Ed.), *Advanced mathematical thinking* (pp. 25–41). Dordrecht: Kluwer.

Gray, E. M., & Tall, D. O. (1994). Duality, ambiguity, and flexibility: A "proceptual" view of simple arithmetic. *Journal for Research in Mathematics Education, 25*(2), 116–140.

Kaiser, G., & Sriraman, B. (2006). A global survey of international perspectives on modelling in mathematics education. *ZDM – The International Journal on Mathematics Education, 38*(3), 302–310.

Palharini, B. N. (2010). *Mathematical modelling and mathematical thinking: A study in the light of the three worlds of mathematics*. Master's thesis, Universidade Estadual de Londrina, Londrina.

Sfard, A. (1991). On the dual nature of mathematical conceptions: reflections on processes and objects as different sides of the same coin. *Educational Studies in Mathematics, 22*(1), 1–36.

Tall, D. (1995). Cognitive growth in elementary and advanced mathematical thinking. In D. Carraher & L. Miera (Eds.), *Proceedings of 19th international conference for the Psychology of Mathematics Education* (Vol. 1, pp. 61–75). Recife: PME.

Tall, D. (2002). *Advanced mathematical thinking*. Cambridge: Kluwer.

Tall, D. (2004a). Introducing three worlds of mathematics. *For the Learning of Mathematics, 23* (3), 29–33.

Tall, D. (2004b). Thinking through three worlds of mathematics. In M. J. Høines & A. B. Fuglestad (Eds.), *Proceedings of the 28th conference of the international group for the Psychology of Mathematics Education* (Vol. 4, pp. 281–288). Bergen: PME.

Tall, D. (2006). A theory of mathematical growth through embodiment, symbolism and proof. *Annales de Didactique et de Sciences Cognitives, 11*, 195–215.

Chapter 18
Measurement of Area and Volume in an Authentic Context: An Alternative Learning Experience Through Mathematical Modelling

Santiago Manuel Rivera Quiroz, Sandra Milena Londoño Orrego, and Carlos Mario Jaramillo López

Abstract This chapter shows the research results of the analysis of how Colombian Year ten students built mathematical models, through the measurement of area and volume, in the context of the flooding of their school. This situation allowed the creation of mathematical models associated with the quadratic and cubic functions, which were obtained from the environmental situation that allowed the students their application. The study was a qualitative case study focussing on two pairs of students. It took into account the everyday environment, communication, prior experiences and the interaction of each member with the group. The aim of the study was to explore and analyse the different ways students build models and apply mathematical elements arising in an authentic context.

18.1 Introduction

This study was developed in the context of floods that occurred in an educational institution in Caucasia (a municipality of the Department[1] of Antioquia, Colombia) because of the overflowing of a river. Moreover, it focused on the analysis of the

[1] Territorial divisions in Colombia are called Departments.

S.M.R. Quiroz (✉)
Institución Educativa Divino Niño (Caucasia), Universidad de Antioquia, A.A 1226 Medellín, Colombia
e-mail: santiagorq@hotmail.com

S.M.L. Orrego
Institución Educativa el Dos, Universidad de Antioquia, A.A 1226 Medellín, Colombia
e-mail: samyjdam@gmail.com

C.M.J. López
Instituto de matemáticas, Universidad de Antioquia, A.A 1226 Medellín, Colombia
e-mail: carlos.jaramillo1@udea.edu.co

G.A. Stillman et al. (eds.), *Mathematical Modelling in Education Research and Practice*, International Perspectives on the Teaching and Learning of Mathematical Modelling, DOI 10.1007/978-3-319-18272-8_18

way a group of Year ten students built mathematical models through the measurement of the area and volume involved in the flood. Therefore, this study addressed these mathematical elements that are strongly associated with the situation in context, making easier the apprehension of concepts due to the use of mathematical modelling. Therefore, from this, we aim for the student to manage data in order to encourage thought about the application of mathematics in their everyday environment.

18.2 The Problem

Currently, floods have increased in Colombia because of diverse biological and environmental factors. In particular, some villages and small towns in Bajo Cauca (a region of Antioquia) have been experiencing this problem and schools are greatly affected by this situation. This is the environment where the students live and that could encourage them to learn mathematics, geometry in particular. Wagner (cited by Planas 2002) claims that the individual path of each student is also a product of social practices and cultural meanings from which they learn to interact with their environment. As a consequence, by observing the learning of mathematics in the situational context of the students, their possibilities to learn the concepts in a more meaningful way are greater. Thus, we consider that from a natural phenomenon, such as floods, that directly affects a population, mathematical modelling situations that promote the notions of area and volume from a metric-geometric thinking can be encouraged. Our objective in this context is to analyse the way students build mathematical models through the measurement of the area and volume arising from an authentic context.

18.3 Mathematical Modelling as a Learning Strategy in the Classroom

Mathematical modelling permits students to integrate mathematical aspects with situations in their context or other disciplines, resulting in the application of mathematical concepts to everyday practices. According to Biembengut and Hein (2007), mathematical modelling ends up being the motivation for creativity in the formulation and resolution of problems, capacity to use computers and to work in groups, orientation to research, and capacity for presentation of research reports. In addition, these authors think that for learning from mathematical modelling to occur, several activities or steps consisting of limiting and stating the problem are necessary, for there are many variables in an authentic context that cannot be totally addressed. By addressing the context from a modelling approach, it is necessary to take mathematical elements that facilitate the understanding of a selected phenomenon. To this end, this study is aimed at establishing the relation between theory and practice given that the students in this study had the opportunity to conduct some fieldwork, to make measurements, to represent different registers and, based on theoretical elements and mathematical concepts, to build models that allowed them to understand the situation they were addressing.

18.4 Research Methods

This research studied the relations between the student, the mathematical modelling work, the context, and the participative environment; all of these taking into account the everyday environment, the communication and the experiences of each participant, as well as those of the group as a whole. These are typical aspects of a study based on qualitative research, undertaken by means of a case study following Stake (1999) and Hays (2004).

18.4.1 Context

The research was developed in an educational institution of the municipality of Caucasia, in the Department of Antioquia, Colombia, which is particularly affected by floods because of the overflowing of the Cauca River. The school is 80 m away from the bank of the river. The families of the students of this institution have a very low income and most of them live close to the river and therefore are directly affected by the phenomenon.

18.4.2 Participants

The study was undertaken with 35 Year ten students with whom we developed some discussions, where their experiences were mutually shared. These conversations created an environment of trust, an essential aspect of the approach of our research. Four of these students, Dani, Angélica, Pereira and Anyi, were selected to create two work teams. These teams were analysed in the case study. Criteria for selection were concerned with the willingness they showed for teamwork, responsibility and interest shown in the activities. These students volunteered to actively participate in the research. Three of them were directly affected by the phenomenon (floods of their households). The investigative design was developed as fieldwork in the following six stages: exploration at the time of the flood phenomenon, delimitation of a particular situation, the construction of a mathematical model for area, the construction of a mathematical model for volume, validation of the mathematical models, and sharing presentations of the work. During these stages the students were involved in different activities such as measuring, drawing and representing of the areas affected by the floods.

18.4.3 Sources of Data Collection

The information was obtained through three sources: (a) the active observation of the context where people live, through audio, videos and written notes—this observation works as a conclusive element for the definition of the research problem; (b) interviews where data were collected through an unstructured conversation that allowed for the interviewee to give, openly and spontaneously, information that could be important for the research; (c) written notes by the students, that allowed the researchers to determine the different processes that the students followed for the construction of models that related to the measurement of the area and volume. Since the objective was to analyse the way in which Year ten students build mathematical models resulting from an authentic context, the analysis of data was made concurrently, that is to say, continuously, also making comparisons between the data collected via written texts, videos and interviews. The data gathered by means of the three sources were validated through triangulation from the interview data.

18.5 Towards the Building of Mathematical Models

In this section we explain the steps that were taken into account for the building of the mathematical models, such as the context and its social impact, the delimitation of the situations in such a context, and the building of models regarding the flooded surface area and its volume.

18.5.1 The Authentic Context: Flood Phenomenon and Social Impact

The interaction of the students with the context of the floods is essential as it was because of their experiences and knowledge of the environment that it was possible for them to approach the learning of school mathematics. This context that affects the *Institución Educativa Divino Nino* (Divino Niño High School) and the community in general, became a key interaction environment in which the students thought about social and economical issues regarding the phenomenon through a series of questions that were posed by themselves as they walked around and observed the different places affected by the flood such as the streets, sewers and different buildings. Some of these questions were: Why do these flood phenomena occur in the country? Why do you think some areas of our municipality flood? What social and environmental problems do you think these floods imply for the municipality? These questions were discussed in the classroom with the teacher. Based on these thoughts about the phenomenon, the students were asked: *Taking into account the flood in the municipality, what are the mathematical notions that can be determined?* The students identified the surface area and water volume of the flooded areas as important mathematical elements; therefore, Angelica's team made the diagram shown in Fig. 18.1. This diagram, which resembles a pie graph in some respects, is a representation of the flooded areas, and a product of the reflections of the students due to their knowledge of the context. It is interesting to note how they perceive the existence of a correlation of flooded areas, not only in the mathematical aspect regarding proportion, but also how those proportions show a great deal of flooded area. Figure 18.1 is also evidence of the way they can delimit the concept.

18.5.2 Delimitation of the Situations in the Context of the Floods

Hein and Biembengut (2006), Villa-Ochoa (2007) and Blum and Borromeo Ferri (2009) consider that it is essential to delimit the problem in a process of mathematical modelling. It is important to highlight that for this particular context several variables are presented, which cannot be totally addressed. Therefore, after addressing several questions, posed along with the students, in which some dependence situations between variables associated with the context and according to the purpose of the research were shown, the students chose the following question: *Could a mathematical expression determine the flooded area from the water level?* This encouraged them to create a map of the school building to identify the different flooded areas at different heights (10 cm, 20 cm and 30 cm) of the water using as a reference a wall next to a sewer inside the building where the water that flooded the different parts of the school flowed. Figure 18.2 shows the map of the institution and a flooded area when the water level was 20 cm with respect to the point of reference.

Thus, in the process of building a model, the students draw a map of the school and use colours to represent the surface area of the flooded area. The representation of a non-mathematical model (the school plan), becomes an important tool within the process, since it is a means that interacts with mathematics for building mathematical models and gives a solution to the problem posed. Students used

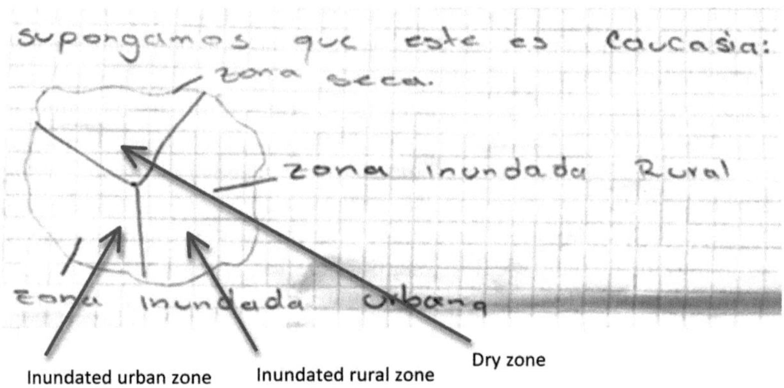

Fig. 18.1 Representation of Angelica's team

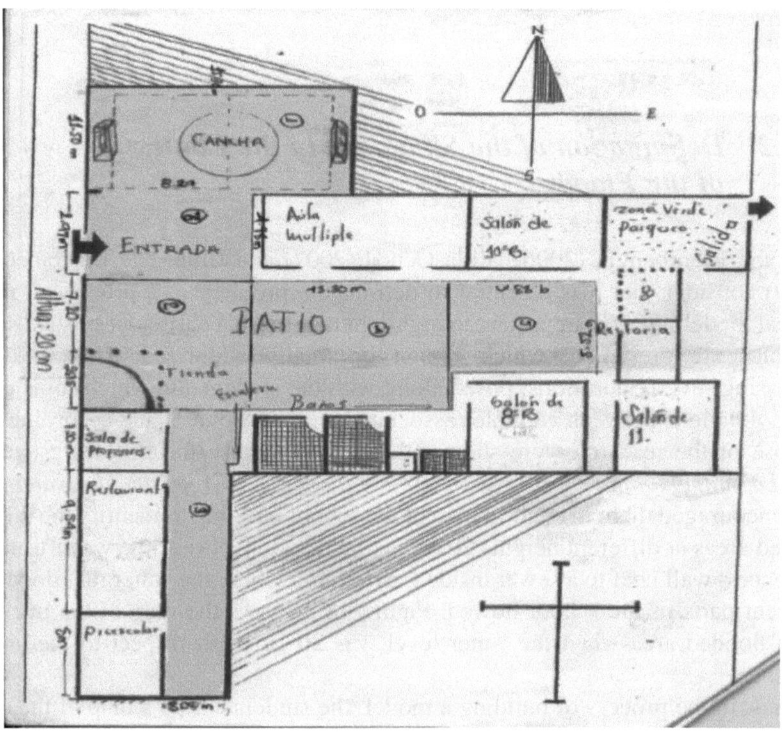

Fig. 18.2 Map by Team 2 (Dani) of the flooded area with a 20 cm height of water level

the process shown by Del Olmo et al. (1993), since they made the transformations and associations of the irregular forms to find the surface area and the volume of the flooded area, making several measurements of the water level according to different heights of water at the reference point.

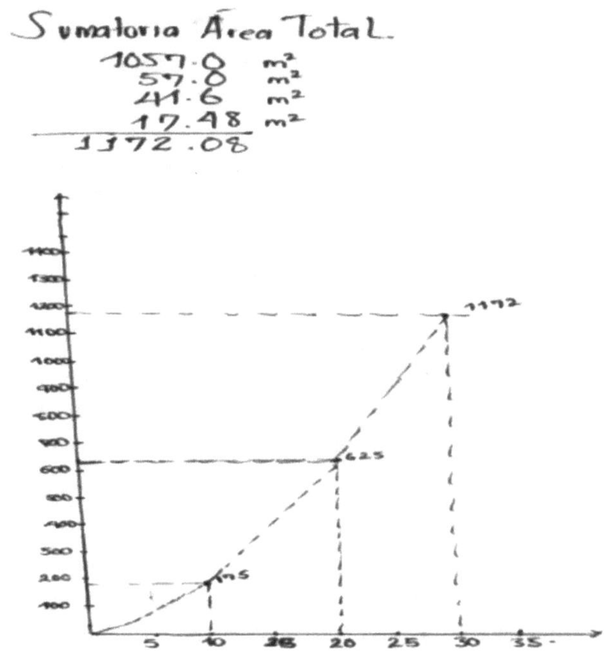

Fig. 18.3 Graph of flooded surface area by Anyi's team

18.5.3 A Mathematical Model: Surface Area Versus Water Level

In order for the students to establish the relation between surface area and water level, they make a representation on a Cartesian coordinate system, and so they obtained a curved line from the different measurements of water level and the respective measurements of the flooded areas as shown by the work from Anyi's team in Fig. 18.3. We can see the attempt of the students to generalize mathematically the results of the graph as they observed a dependence situation between the two magnitudes, so it is suggested for them to attempt the following activity:

> *Propose a mathematical expression that makes it possible to relate the two variables (water level and area) so the flooded area can be found from any water level in the point of reference.*

Firstly, both focus groups had difficulties, since in class they had worked with only linear functions and (2 × 2) systems of linear equations previously. It was suggested they search for different graphs of functions in books and on the internet. The work teams consulted books and the internet finding some quadratic and cubic functions that they related with those they drew on the Cartesian coordinate system. This encouraged them to study the quadratic function, its definition and properties. Under the guidance of the teacher-researcher, the class addressed the mathematical problem that related the side of a square with its area which made the students construct the equation $y = x^2$, which, by replacing the variable x by any

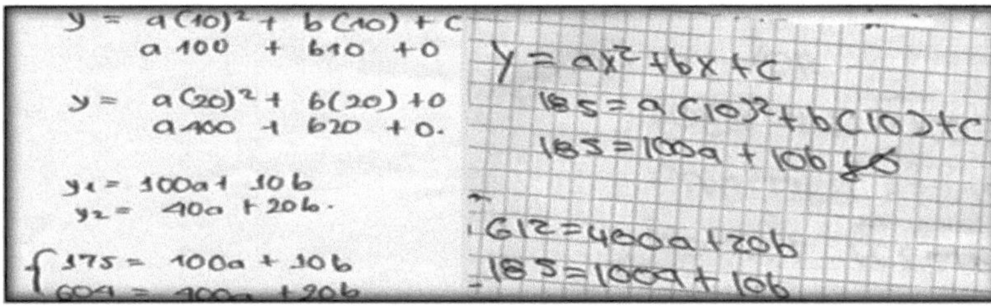

Fig. 18.4 Building the model (Teams 1 and 2)

measurement for the side of a square, it was possible to find its area. Thus, the students observed that the quadratic function approached better to the graph they had made. Therefore, they made an analogy between the surface area flooded with the area of the square and the side of the square with the height of the water level. They considered that the function drawn in relation to the flooded surface area versus the height of the water level matched also a quadratic function that has the general equation: $y = ax^2 + bx + c$.

In order to progress in the process of model building and taking into account the drawn graph, the teacher-researcher posed some questions to the students including: What sort of function is it? What are the dependent and independent variables involved in it? Are they known? The students held a discussion on which were the dependent and independent variables and, how these variables were already known in the graphs they had drawn. Therefore, the students made substitutions for the heights, using the values of 10 cm and 20 cm, which they took as values for x variable and the respective areas of these levels as y. In this way, they established a 2 × 2 system of linear equations, using the variables a and b Then, the students used the method of elimination through equating and found the values for a and b, as shown in Fig. 18.4. By replacing the values they found for a and b in the quadratic function general formula, the work teams were able to establish their mathematical models. Dani and Anyi's model was $y = 1.2\,x^2 + 6.4\,x$ whilst Angélica and Pereira's model was $y = 1.27\,x^2 + 4.8\,x$. The mathematical model for area built by Dani-Anyi was drawn in GeoGebra, allowing the students to visualize the graph and establish the different areas compared to the different water level values. As the models were built using water level heights of 10 and 20 cm, both models were verified by predicting the area for a height of 30 cm from the models and the area for this measured and calculated in the field. The results approximated those found in fieldwork.

18.5.4 A Mathematical Model: Water Volume Versus Height of Water Level

The question that encouraged the creation of a mathematical model where we could predict the volume of water that flooded the school from its level was: *What's the*

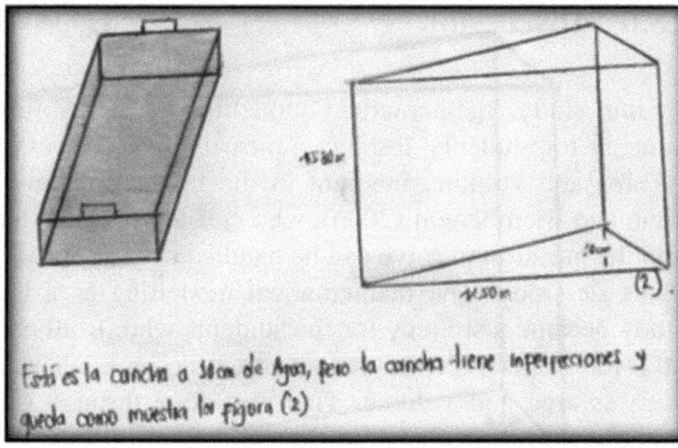

Fig. 18.5 Representation of the flooded soccer field by Dani

relationship between the water level measured at the point of reference and the water volume that is displaced? Similarly to previous activities regarding the surface area, the students began by finding the volume of the water for a height of 10, 20 and 30 cm. They calculated the volume using the formula V = length × height × width. They discussed the possibility of dividing the result by two, given that from the students' experience and their context knowledge they realized that the water level at the "edge" of the flooded area is almost zero. Afterwards, through transformations and associations (see Fig. 18.5) they obtained the corresponding values for volume depending on the height of the water level. To build the model, the students drew a graphic with the volume found versus the height of water level, obtaining a curve similar to the previous one (area versus height). The students knew that they had used an additional magnitude to calculate the volume, that is, the water level height. Also, they discovered in books that cubic functions are curved and share some similarities with the quadratic function. After some discussion about these questions raised by their mathematisations, they concluded that the cubic function was the better match to the graph they drew, because after thinking about the quadratic function $y = x^2$, they concluded that the degree of that function depends on the variables involved. Therefore, to find the volume, they analysed three variables involved (length, width and height) that were related with the degree of the x variable in the cubic function, where the volume is assumed as a one-dimensional magnitude. Likewise, the regular patterns found between the number of variables and the degree of the function allowed the students to take the cubic function as base to build the model. This was found after creating 3 × 3 systems of linear equations using the method of equalization. As such an approach was unknown to the students, it was necessary to handle it in the regular classes at the same time. Figure 18.5 shows the geometrical figures used for calculation of a volume section when the water level was 10 cm at the point of reference. Volumes were verified for levels of 10 cm, 20 cm and 30 cm that resulted in quite approximated data.

18.6 Discussion

In this study, mathematical modelling as a learning methodology becomes a strategy for students, through a measurement process, to understand the concepts of area and volume, inherent in the floods phenomenon. The point of view of Hein and Biembengut (2006), who consider that the teaching of curricular content and the question to solve can be handled concurrently, was shown to be possible in this case study. The mathematical modelling as a learning methodology in this study became a strategy for the students who, from a measuring process appropriated for themselves the mathematical notions inherent in the flood phenomenon such as area and volume. This was done through direct and indirect measures, transformations of irregular figures related to their fieldwork observations of the flooded high school, relations depending on the water level height, which allowed the development, together with the students, of a way of modelling that involved the analysis of the variation and the geometrical elements implied in the context.

Mathematical models for surface area and volume were constructed for two water level heights. They coincided with those found in the fieldwork. Then, they tested for a third height producing precise results. The mathematical model built by one team was drawn in Geogebra, allowing the students to visualize the graph and establish the different areas in relation with different heights. In this sense, Moreno-Armella and Hegedus (2011) point out technology as another example of representative description, especially when it is used with interactive geometry. Therefore, the use of technology to represent results allowed students to interact with the non-determined records in paper and establish relationships and display results in a faster manner, making it easier for them to interpret and simplify those results, since they could observe, for instance, that for a height of 1 m (100 cm), the surface area was 12,800 m^2. So, the development of the model allowed foretelling results.

Some further factors regarding the construction of models and aspects related to them arose from analysis of the data.

The Authentic Context Students were encouraged to participate in the exploration of a natural phenomenon in their context, so it was easier for them to understand situations that happen around them, through mathematical modelling. This sort of situation creates a connection with experiences, everyday life and previous knowledge of the students about the phenomenon that is the object of study.

The Strategies The students, as inquirers, explored, delimited, simplified the context and proposed strategies to build a mathematical modelling, through the study of mathematical concepts such as quadratic and cubic functions that were later compared with the graphs they made. In this aspect, the work allowed the students to understand not only mathematical notions, but also identify aspects of the work that could help them to take action to reduce the effects of the phenomenon as was noted in the audio record.

The Technology An important aspect of the work with the students was the use of technological tools, which helped us to visualize and obtain a general idea of the impact of the phenomenon on the community. By observing the surface area that was flooded for heights over a metre, for example, they estimated, for instance, to what height the surface area of the water could flood the whole municipality.

Mathematical Conceptualization Due to the activities, the students understood the notion of area as the extent of the flooded area, and volume as the space occupied by a certain quantity of water. Furthermore, throughout the development of the measurement activities, the students perceived the area and the volume as magnitudes that can be compared, evaluated, approximated, added, and subtracted, as one-dimensional, two-dimensional and three-dimensional magnitudes.

18.7 Conclusion

The study took into account the individual and group particularities of the students in order to encourage discussion and thought about models built collectively. Likewise, we are aware that the building of models reflects a partial reality of the phenomenon, as according to Skovsmose (1999), "some aspects of reality are not available to mathematics and remain within reach of the descriptions of natural language and vice versa" (p. 185). As well as highlighting the ways of creating knowledge, the way the students related with each other and their environment was observed, with the purpose of facilitating the students not only to find an alternative way to associate mathematics with their context, but also to have a critical attitude towards mathematics that arises all around them and to consider solution alternatives that help to reduce the social and environmental impact of the floods. To this end, the students proposed solutions such as the improving of pipes, building of dykes and docks, and programs for garbage collection for minimizing the impact of a flood. This demonstrates that this activity related to the building of models goes beyond the mathematical context. Likewise, we can claim that mathematical modelling in the context of mathematical education allows students to see themselves as citizens that can read, think, reflect and propose solutions in their own context.

References

Biembengut, M., & Hein, N. (2007). *Modelagem Matemática no Ensino*. Sao Paulo: Contexto.

Blum, W., & Borromeo, R. (2009). Mathematical modelling: Can it be taught and learnt? *Journal of Mathematical Modelling and Application, 1*(1), 45–58.

Del Olmo, M., Moreno, M., & Gil, F. (1993). *Superficie y volumen: ¿algo más que el trabajo con fórmulas?* Madrid: Sintesis, S.A.

Hays, P. (2004). Case study research. In K. DeMarrais & S. D. Lapan (Eds.), *Foundations for research: Methods of inquiry in education and the social* (pp. 217–234). Mahwah: Erlbaum.
Hein, N., & Biembengut, M. (2006). *Modelaje matemático como método de investigación en clases de matemática* (V Festival Internacional de Matemática. De Costa a Costa, pp. 1–25). Costa Rica: Educación Matemática.
Moreno-Armella, L., & Hegedus, S. J. (2011). The emergence of mathematical structures. *Educational Studies in Mathematics, 77*(2), 369–388.
Planas, N. (2002). Nociones Sociales Recontextualizadas en Educación Matemática: el Caso de la Competencia Comunicativa. *VI Simposio de la SEIM*, 175–186.
Skovsmose, O. (1999). *Hacia una filosofía de la educación matemática critica.* (P. Valero, Trad.) Colombia: Universidad de los Andes.
Stake, R. (1999). *Investigación con estudio de casos* (2nd ed.). Madrid: Morata.
Villa-Ochoa, J. A. (2007). La modelación como proceso en el aula de matemáticas. Un marco de referencia y un ejemplo. *Tecno Lógicas, 19*, 51–81.

Chapter 19
Mathematical Modelling and Culture: An Empirical Study

Jhony Alexander Villa-Ochoa and Mario J. Berrío

Abstract This chapter presents partial results of a qualitative case study involving five students from a rural educational institution. The students, motivated by making sense of mathematics beyond the classroom, chose topics for the mathematical modelling of coffee farming. They detected some variables found in the cultural practice of coffee growing and then limited the modelling problem so some of their beliefs began to emerge. Results show that some students' beliefs about culture or contextual knowledge did not remain static throughout the study.

19.1 Introduction

In Colombia, the *Ministerio de Educación Nacional* (Ministry of National Education-MEN) has stated that mathematics teaching in primary and secondary school must consider global and local issues in line with education for all; likewise it must consider the diversity, inter-culturalism and the education of citizens with enough responsibility to enjoy their democratic rights and fulfill their duties (Ministerio de Educación Nacional-Colombia 2006). In this sense, it is reasonable to suggest that school mathematics should recognize the importance of situations, problems and phenomena that emerge from social and cultural contexts thus considering these elements as supporting the constitution of mathematical school knowledge. Thus, mathematical modelling, when involved in the study of such contexts through mathematics, seems to be assumed as an activity that is coherent with the goals and ideals suggested for Colombian mathematical curricula.

To us mathematical modelling is a process of studying a phenomenon or situation through mathematics. In this sense, "Modeling can be understood as a pedagogical approach that emphasizes students' choice of a problem to be investigated in the classroom" (Borba and Villarreal 2005, p. 29). Thereby, mathematical

J.A. Villa-Ochoa (✉) • M.J. Berrío
Facultad de Educación, Universidad de Antioquia; RECOMEM,
Bloque 9-415, Medellín 050010, Colombia
e-mail: jhony.villa@udea.edu.co; marioberrio7@hotmail.com

G.A. Stillman et al. (eds.), *Mathematical Modelling in Education Research and Practice*, International Perspectives on the Teaching and Learning of Mathematical Modelling, DOI 10.1007/978-3-319-18272-8_19

modelling not only defines its scope and some philosophical discussions (mainly those related to the transition from the real world, extra-mathematical considerations, and the nature of knowledge of the nature of production practices), but also gives such phenomena and those situations a constitutive role in the modelling activity. According to Borba and Villarreal (2005), students play an active role in modelling, "instead of being just the recipients of tasks designed by others" (p. 29). These considerations make *project work* a favourable setting for mathematical modelling. In this setting, the topics and problems students deal with emerge as interactions between them, the context and their teacher. This establishes mathematical modelling as an unstructured activity different from contextualised tasks which generally are given in text books. Hence in our study, modelling is not restricted to the task of resolving contextualized tasks, even when these tasks include authentic contexts provided by a teacher or any other means different from students' everyday life and their culture.

This chapter reports part of a much larger study that identified some elements that take place in the (re)construction that students make of mathematical models when they are immersed in a specific cultural context. It analyses some aspects of student performance in mathematical modelling in the culture of coffee farmers. The focus research question was: What interactions occur between knowledge of coffee farming and school mathematics through mathematical modelling?

19.2 Roles of Mathematical Modelling in Culture

In international research, some relationships between mathematical modelling, society and culture have been recognized. In particular, Christiansen (1999) refers to Niss's work (1990) to highlight the importance of individuals being able to reflect critically on models and their applications, as well as, to recognize that mathematics plays an important role in shaping the limits of our activities. Likewise, Christiansen emphasizes the fact that mathematics is implicit in culture and society. In this sense, both Blomhøj (2004) and Blum and Borromeo Ferri (2009) have stated the importance of modelling as a process that involves mathematics learning founded in students' reality; also, through this process, other views of mathematics can be generated; but beyond that, other views can be developed about the contexts themselves. In particular, mathematical modelling makes connections between students' daily life experiences and mathematics, which, besides motivating students, place mathematics in culture as a means for describing and understanding daily life activities (Blomhøj 2004).

The role of mathematical modelling has gone beyond an emphasis on cognitive and conceptual processes of mathematics; this is done in order to lead to aspects linking mathematics with critical societal issues and culture. Both Barbosa (2006) and Araújo (2009) have globally contributed to exhibiting the main aspects of a socio-critical perspective on modelling. Araújo (2009) and Rosa et al. (2012) have emphasized the role of mathematical modelling, not only to explain situations that emerge from reality, but also as a way to enable students to have a critical position facing social demands, as well as modifying, and transforming the world.

19.3 The Study

Since this study was focused on the elements that students take into account in re-constructing mathematical models based on their own cultural situations, it was neccssary to address the research from a qualitative perspective using case studies. According to Stake (2007), this kind of study allows investigation of the particularity and complexity of a single case to understand its activity within important circumstances. This study, in its first stage, involved a group of 29 students from a rural school located in a town whose economy is based on coffee growing. These students took a field trip around areas in the vicinity of their school with the purpose of establishing a dependence relationship between variables. Students gathered in teams. Each team chose its *work project* topic. This chapter analyses the oral and written productions of one team chosen because they could see mathematics involved in the context of coffee growing, which is a part of their everyday life and of their families' lives. The topic that emerged in these discussions was related to the number of trees that could be planted on a plot of land according to topographical features (flat or sloping) typical of this mountainous region.

Other stages the students conducted in this research were related to:

- a review of the mathematical models used by the *Federación Colombiana de Caficultores* (Colombian Federation of Coffee Growers),
- a (re)construction of a new mathematical model (e.g., models with trigonometric functions),
- an approval of a proposed mathematical model,
- a mathematical model tracking what emerges from a triangular planting method, and
- contrasting this research information with an expert.

Questionnaires were used and there were student-researcher (teacher) discussions. The researcher was able to ask questions allowing him to probe in detail students' considerations. Data were collected through logbooks (drawn up by a researcher), and documents (drawn up by students). Student discussions were videotaped. Subsequently, we digitized the documents, and watched the videos to determine which episodes could be analysed. While the data were collected, the information was compared in parallel, so a general idea of the content could emerge (Creswell 2008). Similarly, an emerging categorization process was developed, so that each piece of evidence coming from each source of information was assembled with the other pieces and a triangulation process was carried out.

19.4 Some Findings

This chapter is based on the results obtained in the early stages of the research. In these stages, it was possible to observe that mathematical modelling, not only has implications for learning mathematics, but also it involves relationships with students' culture.

Students, after being involved in their chosen contexts and in discussions regarding contextual mathematics, took on a topic related to the number of trees that could be planted on a slope compared to the number of trees that could be planted on flat land. Once the situation was defined, the students were divided into two teams. Ana and Felipe formed Team 1; both of them were from Year 11, and team 2 included Maria (Year 10), and Karla and Jaime (Year 11). The names used in this chapter are pseudonyms. In the first part of the process, the students were committed to the analysis of a rectangular sloping field; the dimensions of the plot can be seen in Fig. 19.1a. Encouraged by their teacher, students first divided the figure into squares with 100 m sides in order to have one hectare (10,000 m^2, agricultural unit of measurement). Figure 19.1b, c shows this process.

In Fig. 19.1b, c one can see that both teams sketched the hectare on a slope with the dimensions of the land. Thus, they showed that they considered this measure as *a square that suits any land regardless of its slope*. Our knowledge of the region allows us to consider that this is an idea that seems to be widely present among people in the agricultural sector. This idea of land area is related to terrain dimensions, but not to its slope. The idea seems to emerge from the contact that people have with their context.

To obtain more evidence about the characteristics that the students assigned to the notion of area –which we will call "Euclidean surface" – the teacher invited them to think about the following situation: "If you were to buy a rectangular piece of land and you had two options, A or B, which one would you buy and why? You can consider a plot of land to grow coffee and suppose that every square (i.e., a hectare) costs the same". See Fig. 19.2.

Facing this situation, students gathered, discussed, and then gave their opinions. For Team 1, Ana reported her team's work and stated:

> Plot A is 13,200 m^2 and plot B is only one hectare. [We would buy land A] because land A is wider, and despite having a slope, it can be used [for] the "tresbolillos" method [a triangular planting method] so there would be [many] more trees in a plot of land.

For Team 2, Maria reported her team's work and stated:

> We would buy land A, because we calculated the area. The area would be 13,200 m^2 [Land A], while land B would be about 10,000 [square] metres, then more trees would fit into this [Land A].

Both team 1 and 2 agreed that the land that should be bought was land A, because it was 13,200 m^2 (sloping), while land B had only 10,000 m^2 (horizontal). On the other hand, team 2 mentioned the direct relationship between the area and the number of trees: "the larger the area, the greater the number of trees that fit". It is

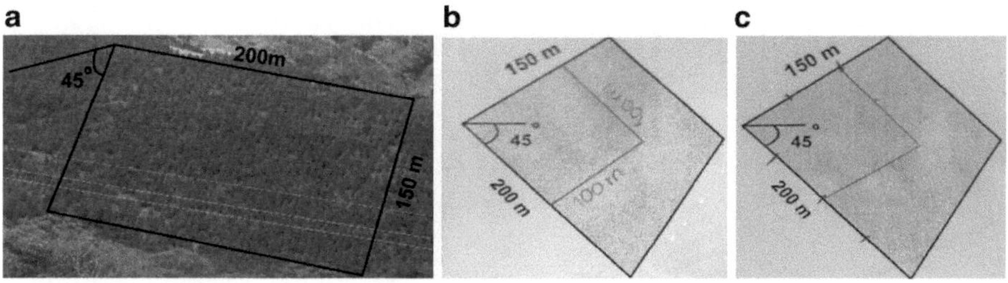

Fig. 19.1 Sloping field (a) and sketching process of Teams 1 (b) and 2 (c)

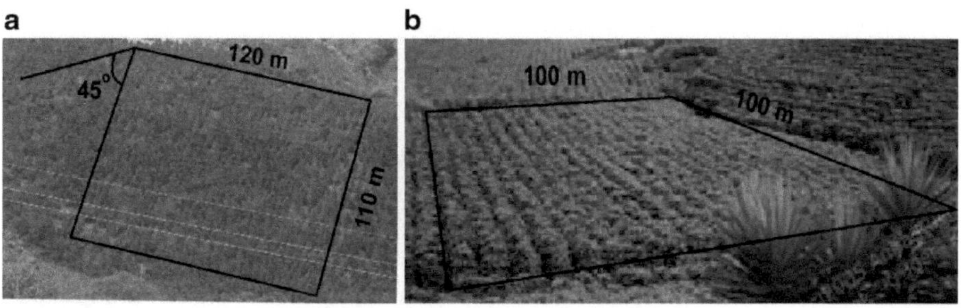

Fig. 19.2 Slope (a) and flat ground (b)

clear that the students noticed other variables that could be taken into account when making a decision regarding which land to buy in terms of advantages and disadvantages according to its topography, yet they only focused on the connection between the area and its relation to the amount of trees that could be planted.

According to students' answers, there is an apparent disconnection between mathematical considerations and cultural experiences. Accordingly, the teacher asked students to research how planting is done on both types of land. Students found in coffee farming books that the horizontal length from one tree to the other is the same in both plots of land (See Fig. 19.3a). The teacher suggested they make some drawings of the process. In one of the proposals, one of the teams had marked one metre lengths in both the hypotenuse and the side representing the flat land (see this mark in Fig. 19.3b). In this situation the teacher asked them to do it the way they would normally do it in their lands in their daily life. Immediately, students changed their layout and placed the trees according to tree projections (Fig. 19.3b). Later they stated:

Felipe: When the land is on a slope, we draw the squares with a one-metre distance, so as not to draw it one metre at ground level. [...] then we make a kind of reflection [projection]. One metre in the air, and then it is calculated one metre below, and it is repeated with the rest. When there is zero slope, it is marked every other metre. (Team 1)

Maria: The distance between each tree is measured; then, there is a metre and it is reflected [projected]. (Team 2)

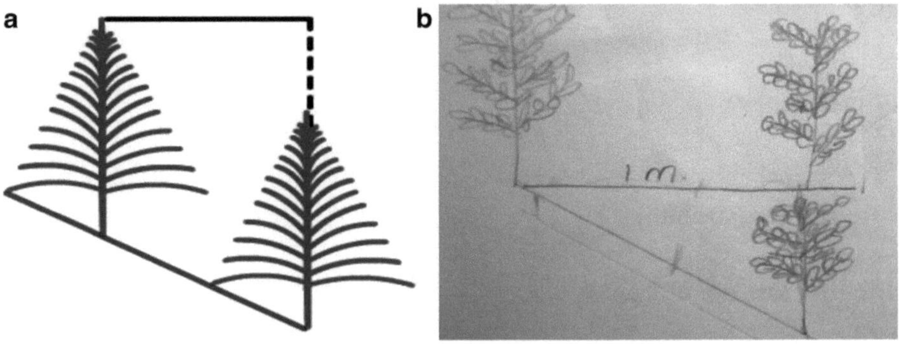

Fig. 19.3 Planting

Next, the teacher suggested the students consider an 8-m distance to plant trees and to assume a diameter of one metre at the top of the tree. With this task, the two teams worked without any trouble and handed in a figure such as Fig. 19.4. Later, the teacher asked them to use the same length to plant trees on a sloping ground; in this same situation team 1 used a different argument, which is shown in Fig. 19.5.

The analysis of Team 1 students is not what the teacher asked for, but it allows us to notice that students also managed to differentiate the "Euclidean surface" from the area suitable for planting. In Fig. 19.5, it seems that the students used the two distances (flat and sloping) with the same size; thus, projecting the trees from the flat land to the slope. That is, students sketched the trees in the sloping land considering the same angles as in the flat land. So the teacher asked them; how many trees fit on the sloping land? Felipe replied:

> . . . well, there are two ways. If we plant considering the projection as we have always done, we plant 6 trees [taking into account the sloping land is 8 metres], because they are the projection, [. . .], it is almost possible to plant seven trees, but actually the seventh one is not possible as it is necessary [to have] an additional piece of land.

In his conclusion Felipe said: "if you want the same number of trees in a sloping plot of land as in a flat one it is necessary to have a longer area".

These findings allow us to see that students were able to overcome the idea they had about a "Euclidean surface" as an equivalent to "agricultural area" concluding that beyond physical distances, all surfaces having minimum horizontal projections are equivalent in agricultural terms (e.g., Fig. 19.6a, surfaces on natural distance, geometric distance and reduced distance have the same tree capacity despite having different physical length). When students were able to understand that a slope is a determining feature in the number of trees that would fit in a plot of land, they were able to produce some mathematical models for flat plots of land (Fig. 19.6b) and for plots of land having a sloping land projection (Fig. 19.6c).

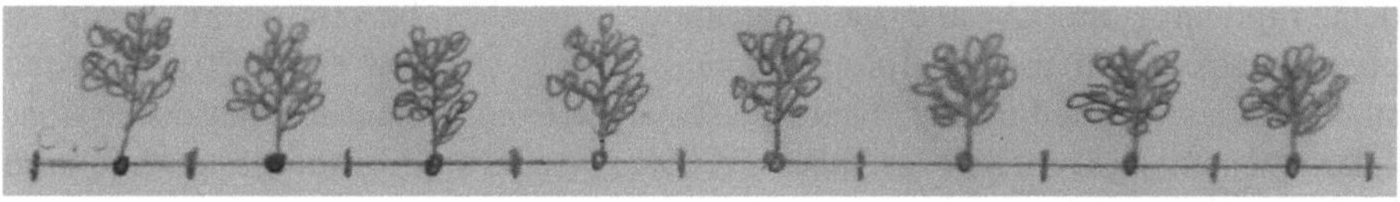

Conclusion: 8 trees fit in 8 metres leaving a one-metre distance from tree to tree

Fig. 19.4 Horizontal planting (Team 2)

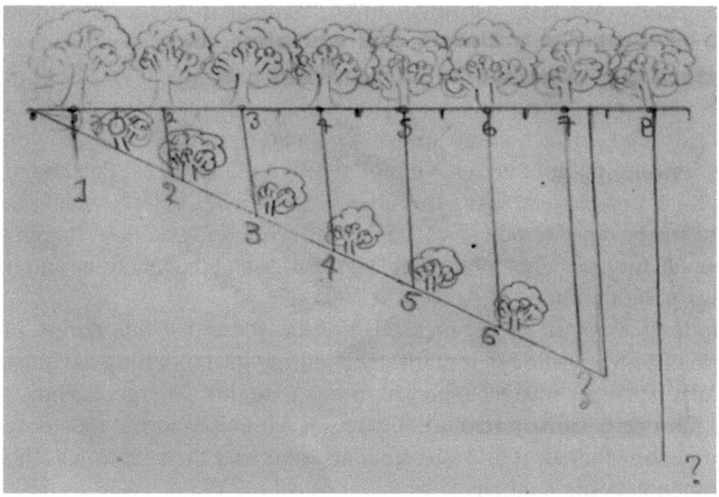

Conclusion: 5 trees can be planted in a plot if they are planted considering their reflections [projections].

Fig. 19.5 Planting on a sloping terrain (Team 1)

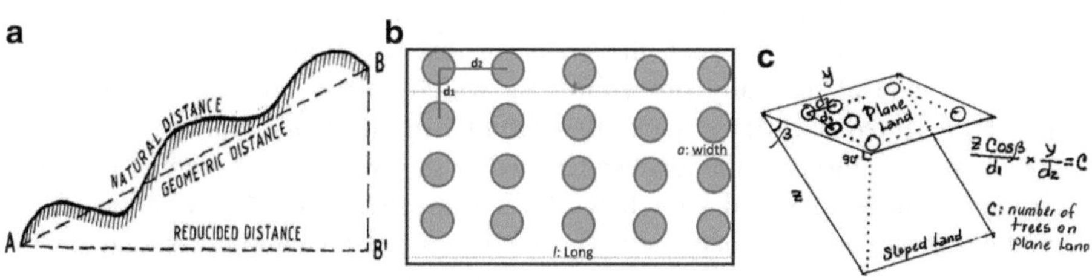

Fig. 19.6 (**a**) Three types of distances (Berrío 2012) (**b**) Model of number of trees $C = al/(d_1d_2)$ (**c**) Model of number of trees on both lands

19.5 Discussion

International research has shown that (figurative) contexts can have very individual, unpredictable effects (Busse and Kaiser 2003) and that in an authentic context, students not only participated and were empowered with such aspects as gathering data, producing models and their meanings, but also, become more aware of phenomena related to the aforementioned context (Muñoz et al. 2014). In the student cases reported in this chapter, tasks arose from student-teacher discussions and agreements; in other words, tasks did not appear as a priori *situations* structured by the teacher.

The data presented above show that at first students believed that a "Euclidean surface" and "arable land" were equal. This fact may be interpreted as an apparent disconnection between the world students live in and contextual situations described in classroom tasks (Stillman 2000). To do an in-depth study to understand this apparent disconnection, students were asked to describe why they believed so. Students argued that "the larger the area [geographical or 'Euclidean'] the greater its tree capacity" in this case, students did not take into account the slope in the plot of land. In fact, some students pointed out that coffee-growers believe that to have a sloping field is to have more [geographical] area; therefore, it can be sold for a higher price. According to students' statements, beliefs such as the above-mentioned can be considered typical of the coffee growing culture; thereby, aspects such as explanation systems, philosophies, theories and daily actions and behaviour must identify with culture (D'Ambrosio 2005).

The findings reported in the previous section show that this notion of equivalence between surfaces did not remain stable during the modelling development. On the contrary, such an idea changed as other thoughts emerged. Thus, students' reflective attitude became a revitalizing action without ignoring the complexity of the phenomenon studied; it allowed students to extend their ideas and thoughts on the phenomenon being studied.

A fact worth mentioning has to do with the situation shown in Fig. 19.4, where, in one of the groups, the students tried to use the same measure between the bases of the trees in both lands. This action seemed to be inconsistent with the daily procedures carried out in their culture, which measures the distance using only horizontal figures. This action was a consequence of the previously explored idea about the equivalences between the areas; in this situation, this idea seemed to overshadow one of the daily procedures on which coffee tree planting is based. Thus, it presented a certain "lack of meaning" of the mathematical activity at school. According to this, the teacher-researcher motivated the students to support their classroom actions and knowledge built from their everyday activities. This way, the task is reoriented, and students can recognise the meanings associated with context, in other words, it becomes a tool to understand the world.

Accepting the ideas students have on certain cultural aspects and not remaining neutral in a mathematical modelling process requires a recognition of the role of context as an establishing element of scientific knowledge at school (i.e., in mathematics, other disciplines and the context itself). This does not only have a utilitarian purpose, that is, so situations in which it is used as a context can promote mathematical content or motivate students to mathematics; nor, once knowledge is obtained, is context set apart solely for focusing on the mathematical knowledge that could have been derived from the situation. In other words, the main role of context in mathematical modelling must be to open the possibility of providing a cultural system of reference for mathematical activity, which gives value to pedagogical modelling which is not only limited to a strategy or a teaching method, or the view of mathematics *set* in a context (mathematics-in-context). However, it does not exclude these elements.

The previous ideas challenge researchers and teachers on the development of skills that allow them to focus not only on the aspects that lead to mathematical model production, but also to recognise other aspects, as shown in this chapter, arising from students' interests. Likewise, they demand a certain sensitivity to recognise the opportunities that a context offers both to include student development skills and to create knowledge based on context. To a certain extent, it is to develop an idea about reality itself (Villa-Ochoa and López 2011).

19.6 Conclusions

This chapter is based on a mathematical modelling thread in which students became involved, supported by a teacher-researcher, in the understanding of a typical situation of their own culture. Thus, the results of this study are consistent with those provided by Rosa et al. (2012) who stated that in mathematical modelling the knowledge was context based as it derived from experiences and it was strengthened by the cultural meanings in which the people were immersed. However, beyond the mathematical knowledge students made use of mathematical modelling (trigonometric models, rational functions, variations and measurements, etc.) in this study, this chapter shows that students' cultural or context knowledge does not remain static throughout a modelling activity. In particular, we show that there are situations in which mathematical modelling allows some considerations, ideas, beliefs, and explanations that are part of students' culture, and we want these situations to be reconsidered and rethought at least at an individual level.

According to D'Ambrosio (2009), over time we would expect individual knowledge to be discussed and analyzed from the perspective of its compatibility until achieving socially shared knowledge. In that sense, this research states that knowledge shared by the group is then socially organized, thus becoming a body of knowledge which is a response to its members' needs and will. In the situation mentioned in this chapter, it could be stated that mathematical modelling activated other dynamics of the individual knowledge of some members of a culture.

The research, from which this chapter derives, found evidence that converges with other studies that highlight the educational and social role in modelling (Barbosa 2006; Araújo 2009). In addition, this study highlights the intention not only to bring a context or situation from culture with motivational purposes, or to introduce or produce a concept or to produce utilitarian ideas for mathematics by showing it is everywhere, but also it has many applications. Hence, without it, scientific knowledge would not have reached its current level of development. This is not just about learning a specific content in context or developing skills to identify "*forms*" from a context comparable to mathematical "*forms* ". In contrast, it is to accept the role of non-mathematical knowledge that emerges in a modelling process. This research encourages researchers to produce other studies showing implications to promote situations in which mathematics and context are linked without being subordinated to each other.

References

Araújo, J. D. L. (2009). Uma abordagem Sócio-Crítica da modelagem matemática: A perspectiva da educação matemática crítica. *ALEXANDRIA Revista de Educação em Ciência e Tecnologia, 2*(2), 55–68.

Barbosa, J. C. (2006). Mathematical modelling in classroom: A socio-critical and discursive perspective. *ZDM, 38*(3), 293–301.

Berrío, M. (2012). *Elementos que intervienen en la construcción que hacen los estudiantes frente a los modelos matemáticos. El caso del cultivo de café.* Unpublished master's degree dissertation. Universidad Nacional de Colombia.

Blomhøj, M. (2004). Mathematical modelling - a theory for practice. In B. Clarke (Ed.), *International perspective on learning and teaching mathematics* (pp. 145–159). Göteborg: National Center for Mathematics Education.

Blum, W., & Borromeo Ferri, R. (2009). Mathematical modelling: Can it be taught and learnt? *Journal of Mathematical Modelling and Application, 1*(1), 45–58.

Borba, M., & Villarreal, M. (2005). *Humans-with-media and the reorganization of mathematical thinking.* New York: Springer.

Busse, A., & Kaiser, G. (2003). Context in application and modelling—an empirical approach. In Q.-X. Ye, W. Blum, S. K. Houston, & Q.-Y. Jiang (Eds.), *Mathematical modelling in education and culture: ICTMA 10* (pp. 3–15). Chichester: Horwood.

Christiansen, I. (1999). Reflexiones críticas sobre modelos matemáticos en la clase: Sueño o realidad? *Revista EMA, 5*(1), 29–50.

Creswell, J. W. (2008). *Educational research. Planning, conducting, and evaluating quantitative and qualitative research.* Upper Saddle River: Pearson, Prentice Hall.

D' Ambrosio, U. (2005). Sociedade, cultura, matemática e seu ensino. *Educação e Pesquisa, 31* (1), 99–120.

D' Ambrosio, U. (2009). Mathematical modeling: Cognitive, pedagogical, historical and political dimensions. *Journal of Mathematical Modelling and Application, 1*(1), 89–98.

Ministerio de Educación Nacional. (2006). *Estándares Básicos de Competencia.* Bogotá-Colombia: Cooperativa Editorial Magisterio.

Muñoz, L., Londoño, S., Jaramillo, C., & Villa-Ochoa, J. A. (2014). Contextos auténticos y la producción de modelos matemáticos escolares. *Revista Virtual Universidad Católica del Norte, 42*, 48–62.

Rosa, M., Reis, F., & Orey, D. (2012). A Modelagem Matemática Crítica nos Cursos de Formação de Professores de Matemática. *Acta Scientiae, 14*(2), 159–184.

Stake, R. E. (2007). *Investigación con estudio de casos.* Madrid: Morata.

Stillman, G. (2000). Impact of prior knowledge of task context on approaches to applications tasks. *The Journal of Mathematical Behavior, 19*(3), 333–361. doi:10.1016/S0732-3123(00)00049-3.

Villa-Ochoa, J. A., & López, C. M. J. (2011). Sense of reality through mathematical modelling. In G. Kaiser, W. Blum, R. Borromeo Ferri, & G. Stillman (Eds.), *Trends in teaching and learning of mathematical modelling* (pp. 701–711). New York: Springer.

Chapter 20
Mathematical Modelling of a Social Problem in Japan: The Income and Expenditure of an Electric Power Company

Noboru Yoshimura

Abstract The problem of the income and expenditure of an electric power company was taken up as a teaching material for mathematical modelling. This teaching experiment focused on students making assumptions and validating results in mathematical modelling when posing their own solutions to the company's financial deficit problem. As a result, it was confirmed that, in general, Japanese students are not able to make assumptions appropriately from complicated real-life problems and that students' modelling skills in handling uncertain numerical values and conditions were acquired through their discussion about the correct answers. In addition, it demonstrated that it is possible to use mathematical modelling teaching materials which deal with the income and expenditure of an electric power company for instructing junior high school students in Year 9.

20.1 Introduction

Japan's electric power supply depends greatly on nuclear generation, which covers about 30 % of the total electric power supply. Since the nuclear disaster at Fukushima in March 2011, 53 out of the 54 nuclear power plants in Japan have been shut down for security checks. This creates worries about power shortages in the large metropolitan areas every year. The electric power company will restart aging thermal power plants and buy electric power from other power companies to make up for power shortages. The cost of producing electricity goes up due to such things as soaring crude oil and coal prices. On the other hand, while the government is giving a subsidy for solar power generation systems, it needs to move forward institutionally with programs for promoting the installation of solar panels in homes

N. Yoshimura (✉)
Tennoji J.H.S. attached to Osaka Kyoiku University, 4-88 Minamikawahori-cho, Tennoji-ku, Osaka 543-0054, Japan
e-mail: noboru@cc.osaka-kyoiku.ac.jp

G.A. Stillman et al. (eds.), *Mathematical Modelling in Education Research and Practice*, International Perspectives on the Teaching and Learning of Mathematical Modelling, DOI 10.1007/978-3-319-18272-8_20

and it obliges power companies to buy electricity produced by solar power. The government and the power companies aim for a stable power supply. As a result, the power companies have a terrible problem with the deficit.

"Essential characterisations of modelling and applications involve posing and solving problems located in the real world, which. . .includes. . .general contexts of living as they impact on individuals, groups and communities" (Niss et al. 2007, p. 17). This is focused on making assumptions and validating results in mathematical modelling of students' own solutions for eliminating the financial deficit borne by utility companies. The aim of this mathematical modelling activity was to develop students with the skills to lead the way in our society in keeping with one of the goals of modelling, namely "to provide experiences. . .that contribute to education for life after school. . .enhancing the quality of life" (Niss et al. 2007, p. 19).

20.2 Teaching Practices

The teaching experiment involved a task that was implemented in Year 9 with 160 students in four classes from 'T' national junior high school. Three hours were allocated to the task on different days in December 2010. The teacher was the usual teacher of the task. The aims of the implementation were (a) training students in mathematical modelling, and (b) allowing students to present their ideas for the elimination of the deficit. At first the teacher made all students find solutions in-class as is the usual traditional Japanese style lesson. Subsequently, he asked them to discuss their ideas and for each group to give a presentation.

Electric Power Company Task
An extract of an explanatory document and a recent newspaper article about power shortages in metropolitan areas every year showed that the power companies have a terrible problem with deficit. The scenario for the power companies was as follows:

In the electric power company, annual energy production is 150 billion kWh. About 30 % of all electricity generation is covered by solar power and the company buys 1kWh for 42 yen. About 70% of all electricity generation is covered by thermal power and fuel cost is 7.4 yen/kWh. The other annual expenditure (employment cost, equipment cost and so on) is 1,200 billion yen. On the other hand, about 40 % of all annual electricity generation sells for 20 yen/kWh for home use and about 60 % sells for 12 yen/kWh for business use.

To consider: Propose an idea to eliminate the deficit.

20.2.1 The First Period

During the first period the teacher presented students with the following tasks about balance of payments in the present situation and gave some conditions regarding the income of the power company.

Task 1: Get an Idea for How Things Really Are

1. Calculate the balance (the difference between income and expenditure).

   Expenditure

   $150 \text{ billion} \times 0.3 \times 42 = 1{,}890 \text{ billion (yen)}$
   $150 \text{ billion} \times 0.7 \times 7.4 = 777 \text{ billion (yen)}$
   $1{,}890 \text{ billion} + 777 \text{ billion} + 1{,}200 \text{ billion} = 3{,}867 \text{ billion (yen)}$

   Income

   $150 \text{ billion} \times 0.4 \times 20 = 1{,}200 \text{ billion (yen)}$
   $150 \text{ billion} \times 0.6 \times 12 = 1{,}080 \text{ billion (yen)}$
   $1{,}200 \text{ billion} + 1{,}080 \text{ billion} = 2{,}280 \text{ billion (yen)}$

   Balance

   2,280 billion in income – 3,867 billion in expenditure = 1,587 billion yen deficit

2. When sales for home use are x yen/kWh and the balance is y billion yen, determine the functional relationship.

   $y = 150 \times 0.4 \times x + 1{,}080–3{,}867 = 60 \text{ x} - 2{,}787$ Ans. $y = 60\,x - 2{,}787$

3. When sales for business use is x yen/kWh and the balance is y billion yen, determine the functional relationship.

   $y = 1{,}200 + 150 \times 0.6 \times x - 3{,}867 = 90\,x - 2{,}667$ Ans. $y = 90\,x - 2{,}667$

4. Propose an idea that if the company raises the electricity price, sales for home use is x yen/kWh and sales for business use is y yen/kWh then the balance becomes 0 yen.

   $300 \times x + 450 \times y = 19{,}335$

Many students substituted x yen/kWh increased from 20 by degrees which is the electricity price sales for home use and calculated the value of y yen/kWh which is sales for business use. Through this approach they gained an idea of the estimated values of x and y. Thus students substituted different values for x and y discussing the ratio of x to y in order not to give both business and home users grievances. As a result, they came up with their own ideas as shown in Fig 20.1.

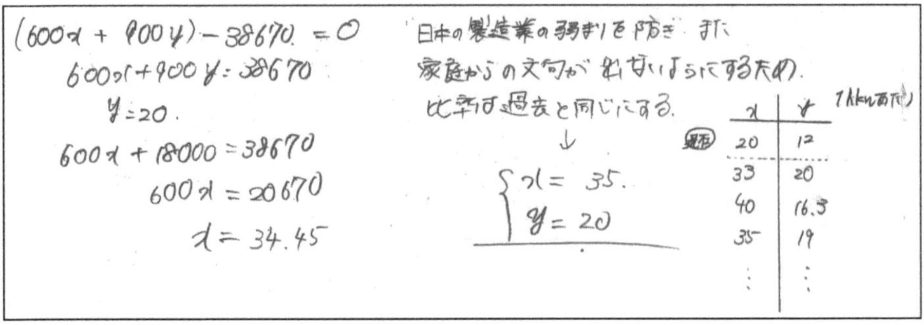

Fig. 20.1 The proposed idea in a student's worksheet

Task 2: If the Conditions of Income Are the Same and the Conditions of Expenditure Change, Eliminate the Deficit

5. The other annual expenditures (employment cost, equipment cost and so on) decrease by half. What is the rate of solar power to eliminate the deficit?

   $150 \times x \div 100 \times 42 + 150 \times (100 - x) \div 100 \times 7.4 + 600 = 2{,}280$
   $x = 10.98$ Ans. About 11 %

6. If the other annual expenditures decrease by a quarter, how much should the company buy 1 kWh of solar power for to eliminate the deficit?

   $150 \times 0.3 \times x + 150 \times 0.7 \times 7.4 + 1{,}200 \times 0.75 = 2{,}280$
   $x = 13.4$ Ans. 13.4 yen/kWh

7. If the other annual expenditures decrease by x % and the rate of solar power is y %, then propose an idea to eliminate the deficit.

   $40 \times x - 173 \times y = 100$

Many students made indeterminate equations. As is the case with Task 1 (4), many students substituted x % increase and decrease by the other annual expenditures and calculated the rate of solar power y %. Again, this gave them an idea of the estimated values of x and y. On the other hand, when solar power covered about 30 % which is y recently, some students got the value of 46 % which is the annual expenditures x % at the beginning. They considered the value of 46 % impossible to attain and realized there is no possible answer with this condition. This is shown by the fact that some students determined an appropriate scope of the values of x and y. Thus they substituted different values for x and y discussing the relevance of these. Each student had his/her own idea of how to eliminate the deficit through a trial and error process. Students' ideas were as shown on their worksheets (e.g., Fig. 20.2).

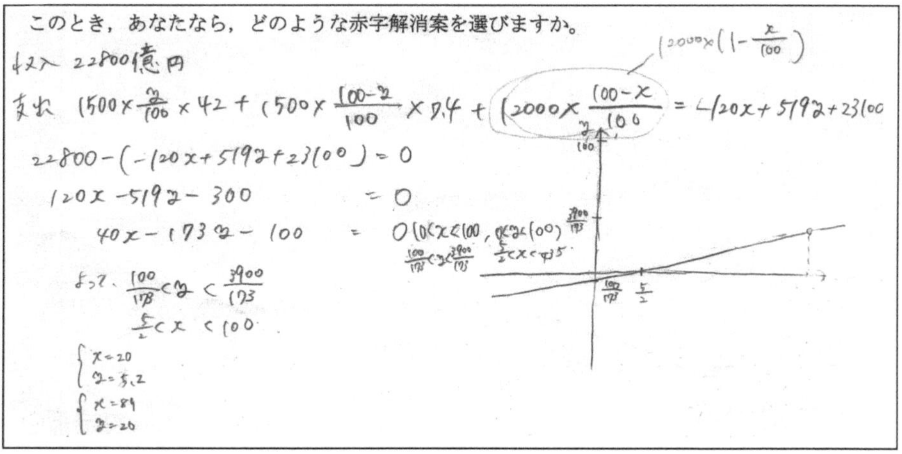

Fig. 20.2 One student's proposal to eliminate the deficit

20.2.2 The Second Period

Task 3: Change the Condition of Both the Income and the Expenditure. Propose an Idea to Eliminate the Deficit Students worked the solutions individually and then each idea was evaluated by each group. When students worked the solutions individually, they began with the income. Students raised the electricity price sales for home use, and business use and for both of them individually. After a careful review of the income values, they worked with the expenditure. They decreased the 42 yen the company buys 1 kWh for step by step and calculated the ratio of the value the other annual expenditures decreased. They obtained a solution for the estimated values. When they were unable to establish an idea, they started these steps all over again. When each idea was evaluated by each group, some of them handled the expenditures not the income at first. They reversed the order and evaluated the idea. This suggests some students acquired the modelling skills in handling uncertain numerical values and conditions through their discussion about the correct answer.

As a result of intensive discussions among students, each group drew up a refined idea to eliminate the deficit. Students' ideas were as shown on their worksheets (e.g., Fig. 20.3).

20.2.3 The Third Period

Each group made a presentation of the solution processes, the relevance of the solution, and reconsidering and making assumptions the hard way.

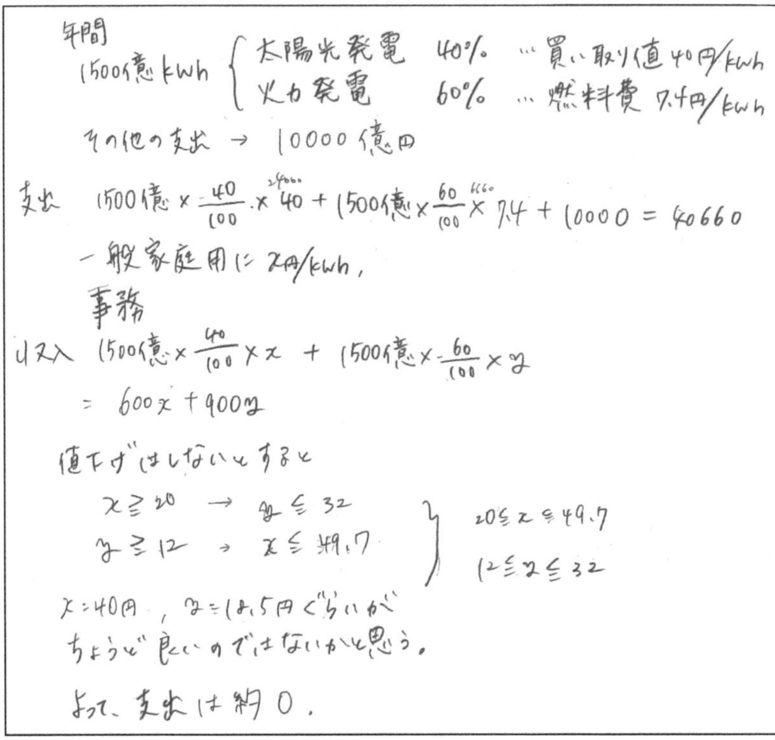

Fig. 20.3 One of the refined ideas from the groups in the worksheets

20.3 Evaluation of Students' Beliefs

In Japan there is a growing trend for students who are not interested in mathematics to be interested only in achieving a good grade in mathematics. Many students also feel that mathematics is not useful in their lives. These are serious issues for mathematical education in Japan. We found the same tendency in a survey of students of 'T' national junior high school students who achieve at a higher level. A survey of 'T' national junior high school students to evaluate if the teaching material for mathematical modelling discussed in this chapter changes some aspects of this situation was conducted.

20.3.1 Utility of Mathematics in the Junior High School

The assessment of students' evaluation of the utility of mathematics in real problems was carried out using a pre-questionnaire before teaching, followed by a post-questionnaire after teaching. Students were asked:

When you think about real problems, do you think that mathematics is useful?

The answer was chosen from the following choices: (A) Very useful; (B) Agreeably useful; (C) Not so useful; (D) not useful at all.

Two tasks for mathematical modelling which were problems about pension tax issues in the 7th grade (Yoshimura and Yanagimoto 2013) and about the choice of electric cars in the 8th grade were given to students in addition to the problem about the electric power company. Three hours were allocated to each task on different days in December 2009 and March 2010. The results for the *Electric Power Company Task* were as shown in Table 20.1 (Year 9) together with results from these cohorts using different contexts for comparison.

The results reveal that the number of students who answered (A) or (B) were 89 % in the pre-questionnaire, 95 % in post-questionnaire for 9th grade students, 87 % in the pre-questionnaire, 94 % in the post-questionnaire for 7th grade students, and 82 % in the pre-questionnaire, 91 % in post-questionnaire for 8th grade students. In Year 9 the students commented that "We used mathematics to solve the problem about the income and the expenditure", "We solved the immediate problem of an electric power company" and so on, as the reasons for changing their answers between the two questionnaires. This confirms more students were able to appreciate the usefulness of mathematics through this learning experience than the previous two contexts.

In the 9th grade, another task about Bluefin tuna was implemented with 157 students in four classes from 'T' national junior high school (see Yanagimoto and Yoshimura 2013). Two hours were allocated to the task on different days in March 2010. The results are shown in Table 20.2 together with results for the current task for comparison.

Table 20.1 Students' evaluation of the usefulness of mathematics (%)

Year			Response (%)			
Level	*N*	Questionnaire	A	B	C	D
7	152	Pre	40	47	11	2
		Post	53	41	6	0
8	155	Pre	36	46	14	2
		Post	47	44	8	1
9	156	Pre	36	53	11	0
		Post	44	51	5	0

Table 20.2 Students' evaluation of the usefulness of mathematics (%)

	Year			Response (%)			
Teaching context	Level	*N*	Questionnaire	A	B	C	D
Bluefin tuna	9	157	Pre	20	46	30	4
			Post	29	55	14	2
Electric power company	9	156	Pre	36	53	11	0
			Post	44	51	5	0

Sixty-six percent of students in the pre-questionnaire chose the answers (A) or (B) before the *Bluefin Tuna Task*, whereas 89 % of students who were involved in the three task sequence chose the same answers before the third task (Pension tax, Electric cars and Electric power company). The post-questionnaire result shows that the teacher guided mathematical modelling helped the students feel that mathematics can be useful.

20.3.2 Students' Interest in This Lesson

Previously, we also used the issue of pension tax in a teaching practice with Year 9 students (Yoshimura and Yanagimoto 2013). This is a serious modern social problem in Japan. The assessment of students' interest in this topic and the electric power and Bluefin tuna topics was carried out using a post-questionnaire after teaching. Students were asked:

Were you interested in learning about the income and expenditure of an electric power company?

The answer was chosen from the following choices: (A) Very much, (B) Rather, (C) Not so much, (D) Never. The results are shown in Table 20.3 together with results from previous cohorts using different contexts for comparison.

This table shows that the students who participated in this study responded in almost the same way as those when Pension tax (Yoshimura and Yanagimoto 2013, p. 249) and when Bluefin tuna (Yanagimoto and Yoshimura 2013, p. 238) was used. Previous work has shown that interest depends on how much students can relate to the learning context (Yoshimura and Yanagimoto 2013, p. 248). The teacher of the Electric power company study was different from the teacher of the Pension tax and Bluefin tuna tasks. Thus, interest levels do not appear to be related to the identity of the teacher or the task context in these examples. In this learning context about 80 % of students chose the answers (A) or (B), and showed as much interest as in the other two studies 82 % and 79 %, respectively.

The reasons given by students undertaking the *Electric Power Company Task* for answering (A) or (B) included: "I was surprised that the electric company had a terrible problem with the deficit and that the energy policy issue was very important", "By thinking about this problem by ourselves. We thought about it more deeply" and "We will face the same problem in future. We felt that mathematics

Table 20.3 Students' interest (%) in different contexts

	Year		Response (%)			
Teaching context	Level	*N*	A	B	C	D
Electric power company	9	156	21	60	13	6
Pension tax	9	157	20	62	15	3
Bluefin tuna	9	157	26	53	14	7

was necessary to solve this social problem." All of these learning contexts include a socially serious problem, topics discussed frequently in Japan. In general, students' interests are influenced by their own involvement and the currency of topics.

20.3.3 *Students' Evaluation of the Pliability of Mathematics*

All groups of students proposed a different idea to eliminate the deficit. An assessment of the students' evaluation of the pliability of mathematics was included in the post- questionnaire for the *Electric Power Company Task* after teaching. We asked:

Is there only one correct answer in mathematics problems?

The answer was chosen from the following choices: (A) Only one answer, (B) Some answers, (C) Many answers, (D) Answers depend on the problems. The results are shown in Table 20.4.

The result reveals that the number of students who answered (C) or (D) was 31 % in the pre-questionnaire, 51 % in the post-questionnaire. In particular, the number of students who answered (D) was 24 % in the pre-questionnaire, 38 % in the post-questionnaire. Reasons given for students' positive evaluation of the pliability (i.e., choice C or D) in the post- questionnaire included: "The answers varied with the different ideas" and "In this study there were a lot of different solutions which the groups proposed." This teaching material allowed students to make assumptions that led to validating results in mathematical modelling. This result confirms more students were able to appreciate the pliability of mathematics through experiencing this aspect of the teaching practice.

20.4 Conclusion

Firstly, it is possible to use mathematical modelling teaching materials which deal with the income and expenditure of an electric power company for instructing junior high school students. It is appropriately placed for Year 9 in the mathematics curriculum in Japan. Students learn about indeterminate equations and linear functions in Year 8, so this content is easy to treat as applied mathematics teaching materials the next year. Students were able to discuss the relevance of the results and the solution processes animatedly. It is thought that group learning was

Table 20.4 Students' evaluation of the pliability of mathematics (%)

		Response (%)			
N	Questionnaire	A	B	C	D
156	Pre	32	27	7	24
	Post	11	38	13	38

effective while using this teaching material of mathematical modelling in handling uncertain numerical values and insufficient conditions.

Secondly, one of the goals of dealing with teaching materials for mathematical modelling is to promote excellence in problem solving and also to make students aware of the usefulness of mathematics. The students' evaluation of mathematics changed for the better. When asked, the percentage of students who answered positively that mathematics was useful for solving real problems increased from 89 % to 95 %. We further claim, based on our continuing teaching experiments, that a continuous focus on mathematical modelling can strengthen belief in the usefulness of mathematics overcoming a perceived disconnection between school mathematics and real-world applications by students. The mathematical modelling of a social problem aroused students' curiosity, made students' open up to social issues more boosting awareness as a member of society. Students who thought this teaching material was interesting accounted for 81 % of the participants. Furthermore, interest levels are not related to the identity of teacher.

This teaching material allows students to make assumptions and to validate results in mathematical modelling. The various assumptions led to a lot of different outcomes so students felt that they discussed mathematical solutions to their satisfaction. When asked, the percentage of students who answered that pliability of mathematics for producing multiple solutions increased from 31 % to 51 % after the task was completed. Despite these encouraging results, these are general surveys. We need to obtain deeper insights into the level of performance of the students on the tasks by further analysis.

With respect to the scheme developed by Maaß (2010) for classifying modelling tasks so other potential users can be guided in terms of task features, objectives and target audience, this teaching material has a general focus in terms of the whole process as a modelling activity, though the teacher led students. There has also been consideration about the contexts which interest students in other literature (e.g., Galbraith et al. 2010). Some of this literature has been related to the modelling of social issues in schooling contexts (e.g., Julie and Mudaly 2007). Indeed, Galbraith et al. (2010) and Caron and Bélair (2007) suggest that social contexts have several benefits as topics for modelling. This teaching material appears to have interested many students because it is a social context and related to their future lives, despite the topic being chosen by teachers.

Our future challenge is to promote the development of mathematical modelling teaching materials from the viewpoint of nurturing pioneers who will be trailblazers in our society. The aim of this mathematical modelling activity is to develop people who can lead the way in our society in keeping with one of the goals of applications and modelling and their systematic embedding in the school curriculum (Niss et al. 2007).

References

Caron, F., & Bélair, J. (2007). Exploring university students' competencies in modelling. In C. Haines, P. Galbraith, W. Blum, & S. Khan (Eds.), *Mathematical modelling ICTMA 12: Education, engineering and economics* (pp. 120–129). Chichester: Horwood.

Galbraith, P., Stillman, G., & Brown, J. (2010). Turning ideas into modelling problems. In R. Lesh, P. Galbraith, C. R. Haines, & A. Hurford (Eds.), *Modelling students' mathematical competencies* (pp. 133–144). New York: Springer.

Julie, C., & Mudaly, V. (2007). Mathematical modelling of social issues in school mathematics in South Africa. In W. Blum, P. L. Galbraith, H.-W. Henn, & M. Niss (Eds.), *Modelling and applications in mathematics education: The 14th ICMI study* (pp. 503–510). New York: Springer.

Maaß, K. (2010). Classification scheme for modelling tasks. *Journal für Mathematik-Didaktik (Journal for Didactics of Mathematics), 31*(2), 285–311.

Niss, M., Blum, W., & Galbraith, P. (2007). Introduction. In W. Blum, P. L. Galbraith, H.-W. Henn, & M. Niss (Eds.), *Modelling and applications in mathematics education: The 14th ICMI study* (pp. 3–32). New York: Springer.

Yanagimoto, A., & Yoshimura, N. (2013). Mathematical modelling of a real-world problem: The decreasing number of bluefin tuna. In G. A. Stillman, G. Kaiser, W. Blum, & J. P. Blum (Eds.), *Mathematical modelling: Connecting to research and practice* (pp. 229–239). Dordrecht: Springer.

Yoshimura, N., & Yanagimoto, A. (2013). Mathematical modelling of a real-world problem: Pension tax issues. In G. A. Stillman, G. Kaiser, W. Blum, & J. P. Brown (Eds.), *Mathematical modelling: Connecting to research and practice* (pp. 241–251). Dordrecht: Springer.

References

[illegible]

Part III
Pedagogical Issues for Teaching and Learning

Chapter 21
The Place of Mathematical Modelling in the System of Mathematics Education: Perspective and Prospect

Henry O. Pollak

Abstract Mathematical modelling is being introduced as a new product into the complex system of mathematics education in many countries. It has to fit with the existing parts and interfaces in this system. As one example, we will consider the effect of mathematical modelling on the transition from secondary to tertiary education. If mathematical modelling is of greater importance to the planners at one of these two levels of education than at the other, stresses may result. As a second example, we propose to examine changes in teacher education necessitated by the introduction of mathematical modelling at the secondary level. Ideally, one might wish to prepare teachers to teach mathematical modelling by concentrating purely on modelling without the distraction of new mathematical ideas, but this cannot always be done. Finally, the effect of mathematical modelling on the relationship between mathematics education and mathematics itself is discussed. Each should gain from cooperation with the other.

21.1 Introduction

When I was at Bell Laboratories, I learned the importance of thinking about the telephone system as a *total system*. Part of the job of Bell Laboratories was to make sure that any new product would work in the presence of everything that was already out there. One simple example is: When the Princess phone, the first pushbutton phone, was being developed, it became clear that a user would produce the digits of a telephone number too rapidly for some of the existing central office switching equipment. This problem could not be ignored.

In my second career as a mathematics educator I see the same problem again, but with one enormous difference: There is no one organization whose responsibility it

H.O. Pollak (✉)
Teacher's College, Columbia University, 524 West 120th Street, New York, NY, USA
e-mail: hpollak@adsight.edu

G.A. Stillman et al. (eds.), *Mathematical Modelling in Education Research and Practice*, International Perspectives on the Teaching and Learning of Mathematical Modelling, DOI 10.1007/978-3-319-18272-8_21

is to see that the whole system works together, and works well together as a system. The total system for mathematics education has many different components that have to work with each other, and many interfaces across which everything has be compatible. We are now introducing a new product to many teachers in many countries: We want to, or we have been told to, teach mathematical modelling. Now think of the parts of the system. Here are a few: There are curriculum planners, textbook publishers, textbook adoption procedures, teachers, the people who run the educational system like principals, superintendents and commissioners of education. There are students, test makers, other disciplines that use mathematics; there is higher education, and there are employers, parents, and politicians. Oh, and we must not forget teacher educators! *And* who would dare to overlook mathematicians?

The effect of mathematical modelling on each of these system components is well worth studying; so is the effect of modelling on the interfaces between pairs of components. So the number of situations to be explored is somewhere between linear and quadratic in the number of components. This chapter will take a look at just three of these potential sources of difficulty. They will be the interface between secondary and tertiary education, teacher education, and the relations between mathematics and mathematics education.

21.2 The Effect of Modelling on the Relationship Between Secondary and Tertiary Education

21.2.1 Modelling in the High Schools: What Will Universities Do?

In the United States, there is a major effort to introduce mathematical modelling into the schools. Will the universities change their placement examinations to take this new ability into account? College placement examinations have the reputation of being among the most immovable features of our mathematical scene. How will the students' ability to model be used at the undergraduate level? Will calculus courses, will elementary science and economics courses, keep this new ability alive, never mind actually use it to good purpose?

21.2.2 Modelling in the Universities: What Will High Schools Do?

The opposite situation could become a problem in China. In recent years, there has been a rapidly increasing interest in mathematical modelling at the tertiary level–see, for example, Xie (2013). On the other hand, the history of the National

University Entrance Examination in mathematics gives the impression that it is a rather immovable feature on China's mathematical scene. Will the national examination change in the direction of inclusion of mathematical modelling, and then will the secondary curriculum reflect such a change? If modelling has become so important at the tertiary level, can the secondary preparation ignore mathematical modelling?

One final comment about the United States: If we are successful in teaching modelling in secondary school; then the problem will be to keep it alive and to use it at the tertiary level. If we fail in teaching modelling at the secondary level, others will *have* to teach it at the tertiary level.

21.3 Teacher Education: Preparing Teachers to Teach Mathematical Modelling

21.3.1 Teaching About the Modelling Process: Can It Involve New Mathematics?

We in the USA have an enormous job preparing in-service teachers to teach mathematical modelling. What are some of the issues: First of all, teachers must see, or better still, they must work through, sample materials that teach mathematical modelling. They will not know what you are talking about without that. What is involved in taking a group of teachers through a modelling experience? It means that they must participate in formulating the problem situation, deciding what to keep and what to ignore in creating an idealized model, do the mathematics in the idealized situation, and then decide if the results make sense in the original situation. A single example of this might take longer than a typical classroom period, which may cause a scheduling problem. Textbooks may not facilitate a need to spend several class periods on one modelling problem. A different kind of difficulty, teacher educators might tell you, is that the mathematics involved in the idealized situation should be completely familiar to the teachers. The argument is that you do not have the time to teach both the modelling process *and* new mathematics as well! You do not want to distract the teachers from the new thing, namely mathematical modelling. And yet, the very fact that this mathematics came from a modelling situation may naturally lead to mathematical questions that you *would not have asked* otherwise! This is a serious difficulty! We will return to this problem when we will look at an example of a modelling situation.

21.3.2 Assessment: How Do You Judge the Success of a Model?

It is natural that teachers' comfort with mathematical modelling will be facilitated by seeing what assessment questions for modelling will look like. Teachers want to know how what they are teaching is going to be tested. But beyond that, assessment has been very much in the news, and has become of significant and even international political interest. But, as we see in every context we examine, mathematical modelling brings a whole new dimension into assessment.

Think about what it means to assess success in modelling. The idealized model must give results that are mathematically correct. Fine. But for the first time in the experience of many teachers, *mathematical* correctness is not enough. The results must also be sensible within the situation which is being modelled: The results must be sensible from *both* points of view. This is a new experience for many teachers. It is a loss of sovereignty for mathematics as it has been taught, that is, for unapplied, pure, mathematics. How will teachers, and other levels of the mathematics education system, react? By the way, how will mathematicians react?

Probably 40 years ago, I was an invited guest at a national summer conference whose purpose was to grade the AP Examinations in Calculus. When I arrived, I found myself in the middle of a debate occasioned by the need to evaluate a particular student's solution of a problem. The problem, as best as I can remember, was to find the volume of a particular solid which was inside a unit 3-dimensional cube. The student had set up the relevant integrals correctly, but had made a computational error at the end and come up with an answer in the millions. (I think he multiplied instead of dividing by some power of 10.) The two sides of the debate: (1) He set everything up correctly, he knew what he was doing, he made a silly numerical error, let's *take off* a point. (2) My God, he must have been sound asleep! How can a solid inside a unit cube have a volume in the millions? It shows no judgment at all. Let's *give* him a point.

My recollection is that side (1) won the argument, by a large margin. But now suppose the problem had been set in a mathematical modelling context. Then it would no longer be an argument just from the traditional mathematics point of view. As I said, in a mathematical modelling situation, pure mathematics loses some of its sovereignty. The quality of a result is judged not only by the correctness of the mathematics done within the idealized mathematical situation, but also by the success of the confrontation with reality at the end. If the result does not make sense in terms of the original situation in the real world, it is not an acceptable solution. Now how would you vote?

I need not remind you of one of the topics that has been much debated in recent years: Is the main purpose of teaching modelling to help motivate the pure mathematics in the idealized model, or is our main purpose to teach mathematical modelling for its own sake? Peter Galbraith has used the wonderfully compact phrases "modelling as vehicle" versus "modelling as content" which he attributes to Cyril Julie (see Galbraith 2007). Is our purpose really one of these two, or is it a mixture of both? Will the various players in the system of mathematics education agree on the relative importance of modelling as vehicle and modelling as content? One of the pieces of the mathematics education system which I said I was *not* going to write about is the planners of the ideal curriculum; but this affects them.

21.3.3 What Do Teachers Believe Modelling Is?

If you are going to work with the mass of teachers who are going to be teaching mathematical modelling, there is an interesting set of questions that you really should consider before you start: What do teachers believe is the meaning of the phrase "mathematical model"? Is a map a mathematical model? Is an architectural blueprint a mathematical model? Is that dodecahedron in the display case in the hall a mathematical model? What do teachers think mathematical modelling means, and what do they think the purpose of teaching mathematical modelling is? Answers may be found in "Teachers' Conceptions of Mathematical Modeling" by Heather Gould (2013). Here are brief statements of some of Gould's results:

21.3.3.1 Mathematical Models

- Almost all teachers believe that mathematical models can be physical manipulatives such as fraction tiles, pattern blocks, 3D solids.
- Teachers agree that mathematical models can be objects such as maps and blueprints.

21.3.3.2 Mathematical Modelling

- Teachers believe that mathematical modelling situations can come from whimsical, unrealistic scenarios.
- The majority of teachers do not believe that making choices and assumptions is always part of the modelling process.
- About a third do not believe that you always have to check the mathematical solution in the context of the modelling situation with which you began.

21.3.3.3 Mathematical Modelling in Education

Teachers are much surer that the curriculum *intends* modelling as vehicle than they are that the curriculum intends modelling as content. Only about one third are actually confident of the latter. As an example of their own beliefs, they strongly believe that modelling helps in understanding scientific phenomena, but only one third see modelling as helping in social sciences and the humanities. Apparently, topics such as fair division, apportionment, and elections have not reached many teachers.

21.4 The Relations Between Mathematics and Mathematics Education

21.4.1 A Modelling Problem

The following modelling problem is readily accessible to students:

> You are a passenger in a car, and you wish to estimate the speed of a vehicle which is going in the same direction as you are, and is passing your car.

Forgive me, this is in the USA, and I state the problem in English units, not in metric units. The scene is a divided, multiple lane highway. First of all, make it very clear that the situation is that you are a passenger in the car, not the driver. We do not want drivers to think about mathematics rather than about their driving. The first part of the modelling situation was the following: "As a second car starts to pass the car in which you are a passenger, suppose you decide you would like to estimate how fast the other car is going". What kind of approach might students take? You are in car A, the passing car is car B. You know the speed $v(\mathrm{A})$ of the car you are in. You see a landmark ahead, a bridge or a road sign for example, you count the time $t(\mathrm{B})$ it takes car B to get there, and compare it to the (presumably longer) time $t(\mathrm{A})$ for your car to get there. You know that distance equals velocity times time for both cars, so you know that $v(\mathrm{A})t(\mathrm{A}) = v(\mathrm{B})t(\mathrm{B})$. So your estimate for $v(\mathrm{B})$ is $v(\mathrm{A})t(\mathrm{A})/t(\mathrm{B})$. Your first act as a modeler is to check that this formula makes basic sense: $t(\mathrm{A}) > t(\mathrm{B})$, so $v(\mathrm{B})$ will be $> v(\mathrm{A})$, so that is okay.

So you have a formula. What assumption did you make to obtain this formula? That both cars keep going, each at a constant speed. How accurate can you expect this computation to be? Here comes the first surprise: If your counting of seconds is not correct, but either too fast or too slow, what will this do to the answer? This will take some time and discussion, because this type of question is probably new to the students. The answer is, it does not matter! Your velocity estimate is not affected. This is a property of homogeneous linear functions, or, if you prefer, of a proportion. Of course if your counting is uneven, that *will* affect your answer. Another aspect of accuracy: Do the syllables in the language in which you count time allow you to estimate half seconds? This will help your accuracy. Is there another method to estimate time? Most people do not carry a stopwatch. Perhaps a student will make the following interesting suggestion: It is raining. Use the windshield wipers to measure time.

Think back to our earlier discussion: Yes, this problem requires only the most familiar of mathematics, but the discussion of accuracy will probably be a new aspect of the pure mathematics in your model. The modelling situation demands that you examine an aspect of homogeneous linear functions, or of proportionality, that perhaps does not arise in the teaching of the pure mathematics! To return to another previous question: Is this modelling as vehicle, or modelling as content?

Another part of the modelling process: What do you mean by the time that either car reaches a landmark? Does it matter whether you try to time the front of the car or the rear of the car when it gets to the bridge? Will that affect your answer more or less than the accuracy of time to at best a half second? Does curvature in the road matter? How about the extra time it takes the passing car to change lanes? Does it matter? Those kinds of questions are part of modelling. What do you have to keep, and what can you afford to ignore?

Now comes the second part of the model: What if you do not decide right away that you want to estimate v(B)? Can you still make an estimate *after* car B has passed car A? Perhaps the students will think about using two landmarks instead of one. What happens to the sensitivity of the model? What, if anything, changes in our previous modelling decisions?

Now a third part of the model: In the US, the longest truck permitted on the highways as of this writing is 53 ft. Suppose your car is being passed, but passed slowly, by such a long truck. You can count how many seconds it takes the whole truck to pass car A. That is another way to make an estimate of speed. Add the rate at which the truck is passing car A to the speed of car A, and you have the speed of the truck. How does this differ from the previous method? Is the method still insensitive to the accuracy of your counting? (Answer, no). And you will have to know how to convert between the feet and seconds in your measurement of the truck and the miles per hour of v(A). And, is the 53 ft the length of the trailer, or is it the length of the trailer plus the cab? What does the 53 painted on the side of the van actually mean?

If both methods are available to you, how would you choose between them? Here is one possibility: If the truck is only a little faster than car A, then the new method will do better; if the truck is much faster, you are better off using a distant landmark. Not many students may have got that far in the discussion. Is there a possible optimal combination of the two methods of estimating? That sounds like a more difficult problem.

A possible assessment problem for the preceding modelling experience: What if you want to estimate the speed of an oncoming car, a car that is going in the *opposite* direction? If you can find a nice model for this, it might be simpler or more general than the one we considered. You do not need a divided super-highway for that, you can do it on an ordinary two-lane road; but maybe this will introduce other complications!

21.4.2 Modeling as Vehicle and Modelling as Content

The mathematics used in the preceding idealized model is very simple and very familiar, but some of the questions about sensitivity of the mathematical results may *not* be very familiar. If you are giving a course on the *teaching* of mathematical modelling, perhaps you cannot always avoid doing some new mathematics?

During the 2011–2012 academic year, a team of students at Teachers College in cooperation with COMAP, the Consortium for Mathematics and its Applications, prepared a handbook consisting of 26 modules (Gould et al. 2012), each of which consisted of a modelling problem ready for classroom use at the high school level. This became available in the summer of 2012. Last fall, Heather Gould and I ran a Saturday workshop on the teaching of mathematical modelling, six half days during which we planned to teach 12 of the modules. We had a group of 31 people, partly current secondary or college teachers of mathematics, and partly mathematics education graduate students at Teachers College. We ended up teaching seven of the modules, but only by combining two of them into one session: It took twice as long for the group to complete a module, roughly four hours rather than the two which we had expected. Prof. Margaret Kidd, who had tried some of the modules working with in-service teachers in California, had warned us that this is exactly what would happen! Even though most members of the group were participating in the workshop voluntarily and were not taking it for university credit, almost all stayed with us for all six sessions and one of the sessions was the Saturday after Hurricane Sandy! A second TC-COMAP handbook (Teachers College 2013) was prepared in the 2012–2013 academic year, namely a handbook of assessments for each of the 26 modules in the first handbook.

21.4.3 Primary and Middle School

When it comes to elementary arithmetic, there are some aspects of the elementary operations which students may not learn in elementary school. If you add two numbers A and b, where A is large and b is small, the precision of b, and probably b itself, do not matter at all. If you subtract B from A, where A and B are almost equal, the answer may be meaningless. Dividing by 0 is forbidden, but dividing by almost 0, while not forbidden, is probably stupid. These instincts should be acquired in elementary school. A very good way to acquire them is from modelling experiences.

Twenty five years ago, in a path-breaking experimental project called "The Regional Math Network", Kay Merseth from Harvard, with support from the National Science Foundation, produced four volumes (Merseth 1987), centered on sports, ice cream, the Quincy Market in Boston, and outer space, which used a huge variety of applied and modelling situations, to teach 7th and 8th grade mathematics. But what was so exceptional is that it took full responsibility for the mathematics traditionally taught at that level. It didn't just motivate, or just apply, it guaranteed to do it all. It was way ahead of its time.

21.4.4 Secondary School

COMAP's "Mathematics, Modeling Our World" series (e.g., Garfiunkel et al. 1998) was created in a similar spirit. As suggested in Sect. 21.3 above, mathematical modelling should sometimes endeavor to take responsibility for a particular mathematical topic: Create the necessity for it, develop it, and fit it into the system. You do not want the body of the mathematics education system to reject something that feels like a foreign invader, you want it to absorb a new part of itself.

21.4.5 The Secondary: Tertiary Interface

One more example: You might cure a long-standing difficulty, at least in the United States. In the traditional curriculum, we teach complex numbers in grade 11, as part of the work with quadratic equations, but then in fact complex numbers never get mentioned again until second year calculus, when students are ready to learn about e and then e^$i\omega$. But you could do a discrete version of an oscillating spring using just DeMoivre's Theorem in high school rather than complex exponentials, and get a nice "modelling as vehicle" example.

21.5 Modelling as a Source of New Insights in Mathematics Itself

A final example will shed additional light on the effect of mathematical modelling on the relation between mathematics and mathematical education. One of the joys of creating a mathematical model for a real-world situation is that you can never be sure what mathematics you are going to get into. The following is taken from Gould and Pollak (2013).

The modelling situation in the module is to estimate the temperature at a location for which a direct reading is not available. We idealize the problem by assuming that the nature of the terrain permits one to assume that linear interpolation from nearby temperature readings is a reasonable way to proceed. In other words, if you had a reading of 71° at one location, and a reading of 75° at another, we assume that at the place halfway between them, a guess of 73° would be acceptable. If you have readings of T_1 at x_1 and T_2 at x_2, our guess for a reading at $\alpha x_1 + (1 - \alpha)x_2$ would be $\alpha T_1 + (1 - \alpha)T_2$, at least if $0 < \alpha < 1$. You are entitled to increasing doubts as α gets further and further away from the unit interval.

How do we know that we have this correct, and not backwards? If we set $\alpha = 1$, then we are at x_1. The temperature is T_1, as it should be. If we set $\alpha = 0$, we are at x_2 and the temperature is T_2, again as it should be. If $\alpha = ½$, we are halfway between x_1 and x_2 and the temperature is halfway between T_1 and T_2.

The problem becomes more interesting if you make it two-dimensional and endeavor to estimate the temperature in the interior of a triangle at whose vertices the temperature is known. Let us assume that for $i = 1$, 2, and 3, we know the temperature T_i at three non-collinear points P_i. We assume for convenience that $T_1 < T_2 < T_3$, and that we wish to know the temperature at a point P inside the triangle determined by the three P_i. One way to proceed would be to draw the line from P_2 to P and extend it until it meets the line segment between P_1 and P_3 at point we will call X. You can find the temperature T(X) at X by measuring by interpolating between T_1 and T_3 in the same proportion as P_1X and XP_3. Now that you know T(X), find T(P) by interpolating between T(X) and T_2 in the same proportion as XP and PP_2.

That is an answer to the modelling problem, but the mathematical fun, now that we have come this far, is just beginning. You could just as well have drawn a line from P_1 through P until it meets P_2P_3 at some point Y. Then you can compute T (Y) by interpolating between T_2 and T_3.

How do you know you will get the same answer? It's all linear interpolation, so you *must* get the same answer, but *how do you prove it?*

Now, forgive me, we are no longer modelling. Let us start to do pure mathematics. We can turn this into a vector space kind of problem by defining three basis functions: For each i, $i = 1$, 2, and 3, we define F_i as the linear function *such* that F_i is 1 at P_i and is 0 at the other two points. Then the original temperature function can be defined as

$$T_1F_1 + T_2F_2 + T_3F_3.$$

But we are not going to work with this, we will just work with one of the F_i, and see what that tells us. So let's acknowledge the fact that we are doing pure mathematics by changing notation.

Here is our new picture:

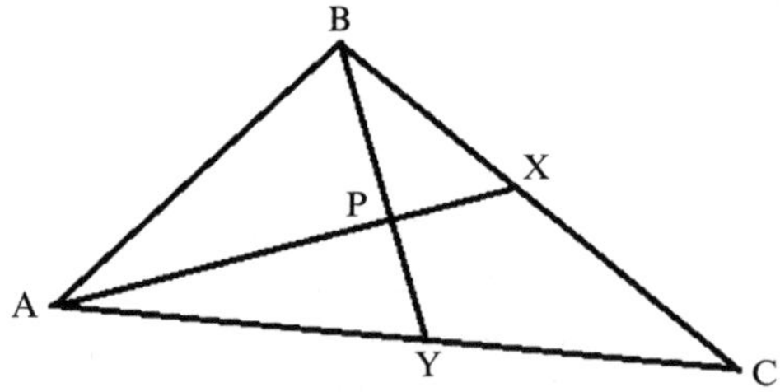

We are looking at the linear function f_A and at its value at P. Remember, this linear function is 1 at A and 0 on the line segment BC. One way to say its value at P is

$$f_A(P) = \frac{PX}{AX}.$$

Another way to say it is

$$f_A(P) = f_A(Y)\frac{BP}{BY}.$$

Thus we have two expressions for f_A(P). Since they are equal,

$$\frac{PX}{AX} = \frac{YC}{AC}\frac{BP}{BY}$$
$$\frac{AC}{YC}\frac{BY}{BP}\frac{PX}{AX} = 1.$$

But this is exactly Menelaus' Theorem in geometry! Menelaus of Alexandria, in about 100 AD, wrote a three-volume work on spherical geometry. This is the planar version of one of his theorems, which was not in Euclid. You think of extensions of the sides of triangle APY such that B, X, and C are collinear, and then the last identity holds. It is usually proved by using many similar triangles.

This has been a small example of a mathematical modelling situation leading to an unexpected approach to a result in geometry. Mathematical modelling can benefit pure mathematics itself! Not that this thought is new – Information Theory is a huge example of this phenomenon.

21.6 Coda

Where do we go from here? I believe that modelling problems should be problems from the real world, and I believe we can teach such problems. There are good problems all around us, many people, including myself, put a lot of effort into finding such problems. But I believe that teaching a true modelling problem takes time. I have seen examples of mathematics education systems where you cannot find a whole period, never mind a week, for students to discuss a modelling situation, formulate an idealized model, do relevant mathematics, and then examine the success of what we have accomplished. We must avoid mathematical modelling acquiring the reputation that it is just a fancier terminology for the same old word problems. We must somehow find the time it takes to go through the complete modelling cycle. Probably not every time, but can we have 3 or 4 h every few months during which to do full-scale modelling? This may be a battle within the mathematics education system, but in my opinion it must be undertaken.

A number of people have written books entitled something like "The Joy of Mathematics". I should like to see a book entitled "The Joy of Mathematical Modelling" with fifty to a hundred examples, taken mostly from everyday human experience. The joy I have had in my life of doing and teaching mathematical modelling should be transmitted. Who will join me?

References

Galbraith, P. (2007). Dreaming a 'possible dream': More windmills to conquer. In C. Haines, P. Galbraith, W. Blum, & S. Khan (Eds.), *Mathematical modelling (ICTMA 12): Education, engineering and economics* (pp. 44–62). Chichester: Horwood.

Garfunkel, S., Godbold, L., & Pollak, H. (Eds.). (1998). *Mathematics: Modelling our world*. Bedford/Cincinnati: COMAP/South-Western Educational Publishing (Annotated Teacher's ed.).

Gould, H. (2013). *Teachers' conceptions of mathematical modeling*. Doctoral dissertation, Columbia University.

Gould, H., & Pollak, H. O. (2013). Estimating temperatures and where it gets you. *Consortium: The Newsletter of the Consortium for Mathematics and Its Applications, 104*, 38–39.

Gould, H., Murray, D. R., & Sanfratello, A. (Eds.). (2012). *Teachers College, Columbia University Mathematical modeling handbook*. Bedford: COMAP.

Merseth, K. (1987). *The regional math network*. Cambridge: Harvard Graduate School of Education.

Teachers College. (2013). *Teachers college Columbia university mathematical modeling handbook II: The assessments*. Bedford: COMAP.

Xie, J. (2013). An introduction to CUMCM: China/contemporary undergraduate mathematical contest in modeling. In A. Damlamian, J. F. Rodrigues, & R. Sträßer (Eds.), *Educational interfaces between mathematics and industry* (pp. 435–443). Cham: Springer.

Chapter 22
Moving Within a Mathematical Modelling Map

Rita Borromeo Ferri

Abstract Comments and thoughts raised by Henry Pollak's plenary are reflected on at the meta-level and on the basis of a "mathematical modelling map" which the reactor created as a simple instrument opening a field between applied and pure mathematics as well as between theoretical aspects and empirical results concerning teaching and learning of mathematical modelling. This map also shows Henry Pollak's and the reactor's position within this map and their "moves" in order to give an insight into two "generations" of mathematical modellers.

22.1 Introducing the Mathematical Modelling Map

The mathematical modelling map (see Fig. 22.1) consists of four parts, with the focus on teaching and learning of mathematical modelling (MM). Two parts of this map represent the applied and the traditional, respectively, pure mathematics, which can be seen on the one hand as contradictions and on the other hand as a symbiosis, because one needs the other, especially when talking about mathematical modelling. Henry Pollak's famous modelling cycle (Pollak 1979) includes these two parts as well, and for him as an applied mathematician in his first career it is necessary to ask, how these parts come together concerning teaching and learning of mathematical modelling. Firstly, Henry Pollak shall be embedded in the parts of pure and applied mathematics.

The two other parts of the map illustrate what is, firstly, meant by having a theoretical background ("theory") about modelling in an educational environment and, secondly, when doing empirical research in this field. Under "theory" the

R. Borromeo Ferri (✉)
Institute of Mathematics, University of Kassel, Heinrich-Plett-Straße 40, 34132 Kassel, Germany
e-mail: borromeo@mathematik.uni-kassel.de

G.A. Stillman et al. (eds.), *Mathematical Modelling in Education Research and Practice*, International Perspectives on the Teaching and Learning of Mathematical Modelling, DOI 10.1007/978-3-319-18272-8_22

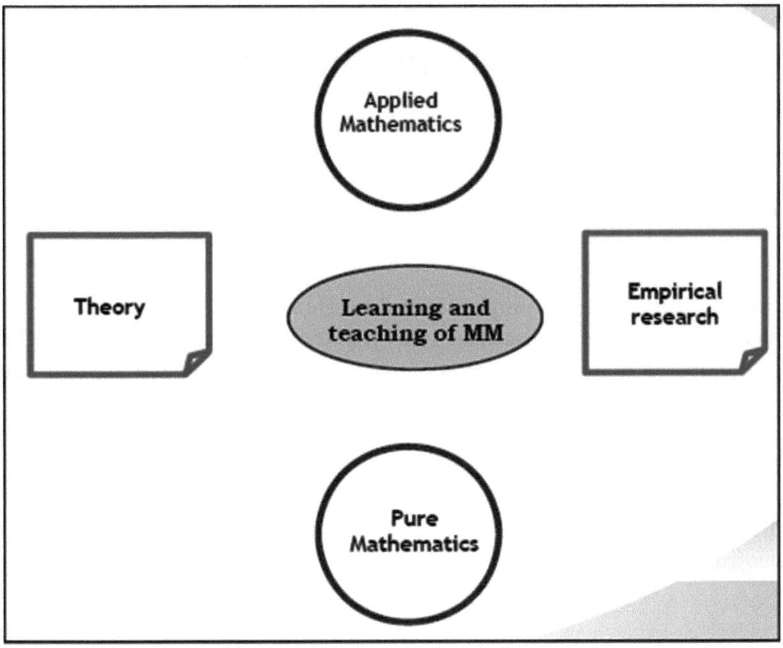

Fig. 22.1 Mathematical modelling map

following aspects are summed up, without pretending to meet any requirement of completeness:

- teleological aspects, including justifications for applications and modelling in school (Blum 2011) and theoretical perspectives on modelling (Kaiser and Sriraman 2006);
- cognitive aspects, including modelling competency and subcompetencies, in particular "blockages" (Stillman et al. 2010), and "modelling routes" (Borromeo Ferri 2007);
- epistemological aspects, including relations between the real world and mathematics; and
- task-development aspect.

In general it is still an open question whether theoretical thoughts were raised first or whether empirical research was "intuitively" done in the late 1970s and 1980s, when looking, for instance, at the early empirical studies of Treilibs, Burkhardt and Low (1980) or Kaiser-Meßmer (1986). "Intuitively" in this context means, for example, the development of modelling tasks and then the investigation of how this works in schools and how students, university students or teachers react or deal with this development or how modelling processes look like, and so on. This procedure was widespread, and so a lot of examples arose out of practice and for practice.

Today, we have made progress concerning how we can design qualitative and quantitative empirical studies and which instruments we use for these to fulfil criteria of good research. Adequate, inventive and very stimulating research questions emerged in this field and over the last 15 years empirical studies increased enormously. Furthermore, there are still a lot of open questions. Many of these research questions have their origins in theoretical and epistemological thoughts of pioneers like Henry Pollak. His modelling cycle was, and still is, a theoretical basis or point of view between applied mathematics and educational thoughts for the current debate and is still cited very often. All young researchers starting in the field of mathematical modelling should know about this cycle for understanding the beginnings and the ideas at this time.

22.2 "Moves" Within the Map

In the following, three key questions, which were raised implicitly in Henry Pollak's plenary are discussed, which causes certain "moves" within the map.

First key question (see Pollak Chap. 21 this volume, Sect. 21.3.2): Is mathematical modelling meant to help to teach mathematics or is the purpose to teach modelling for its own sake? From a theoretical point of view, one would say that mathematical modelling helps to teach traditional, curricular-based mathematics, if we define it like, for example, exercising algorithms. If individuals work within the "world of mathematics" during the modelling cycle they need these competencies and their knowledge, for instance, about formulae or how to solve equations. So mathematical modelling seems for many teachers to be a helpful instrument, which they have to use anyway, because of educational-political requirements.

From empirical studies there is evidence about using modelling problems independent of the current mathematical topic and so not curricular-based for a particular grade – this was done in many studies. But there is still an interesting question, which needs to be proven empirically. This leads to the second key question (see Pollak, Chap. 21 this volume): Are learners worse or better in their modelling processes and in their understanding of mathematics when using modelling tasks topic-independently?

A lot of experiences from modelling weeks in Hamburg (Kaiser and Schwarz 2010), Kassel and other German cities show that especially complex modelling problems help pupils to learn modelling in huge parts for its own sake on the one hand, but also to go deeper into mathematical content on the other hand. It is not an argument which direction is better or not, but it clearly depends on the goals the teacher has with the tasks he or she chose. So what comes first, the mathematics itself or learning mathematics during modelling activities, is a didactical decision anyway. Importantly, in any case, modelling is a "normal" part of mathematics lessons and not an activity only when there is time left from other topics.

Both should be in an adequate balance, but it depends on a lot of factors like different cultures, views on pre- and in-service teacher education, teacher training for mathematical modelling, university curricula and in particular on the school curricula and educational standards in mathematics in different countries. A short example is the current educational developments in Chile. The ministry of education in Chile is now using more or less the mathematics standards from Germany, where mathematical modelling is one compulsory competency and so mathematical modelling is, if it is formulated in Henry Pollak's words, a "new product" for the teachers there. So, the goal in Chile is to spread it all over the country. What I learned in Chile is that mathematical modelling problems there should be used only at the end of a mathematical topic and not when starting, for example, with functions. This is what is written in the Chilean mathematics curriculum. So here one part of the didactical decision is already made by the curriculum developer. These teachers in Chile, like thousand of teachers, for example, in Germany, although the mathematics standards exist now for over 10 years, do not know what mathematical modelling means nor do they have an idea how to deal with modelling in school. This is the reality!

This brings us to the third key question (see Pollak, Chap. 21 this volume): Why should modelling be taught and learnt? Still a lot of researchers in the modelling community work on such teleological and more theoretical questions. The theoretical underpinning of different perspectives and aims of mathematical modelling within cultures or the question which modelling cycle can be used for a specific purpose promotes the discussion. Also, the concept of the "implicit or intuitive model", for example, is a theoretical construct till now and it seems that it can be reconstructed as something like a pre-model between fringe-consciousness and consciousness (Borromeo Ferri and Lesh 2013). For this, more empirical studies are needed for making these theories more fundamental and usable for school purposes. So this constant interplay between theory and empiricism is important for the research field of mathematical modelling.

When talking about this interplay and especially about empirical studies, knowledge about qualitative and quantitative research approaches and about data analysis or interpretation are needed. As a new doctoral student in the field of modelling and also other research fields of course, there is more or less no other way to learn this and to be familiar with empirical methods.

Moving back to the theory and from there to cognitive aspects gives an insight into the modelling processes of students, which was a goal for many empirical studies in the last years: For example to reconstruct cognitive barriers while modelling (Blum 2011), individual modelling routes (Borromeo Ferri 2007) or the question what kinds of blockages appear and how to minimize them (Stillman et al. 2010). The role of meta-cognition within mathematical modelling was, and still is, important (Stillman 2011). The results of these studies were very helpful for teacher education and in particular for training teachers in their diagnostic competencies. Beneath knowledge about diagnoses a lot more dimensions of competencies are needed, like the theoretical, task and instructional dimensions. This model (Borromeo Ferri and Blum 2009), based on experiences from other studies, has no empirical evidence as yet. Moving to the task-development aspect, which is also one part of the model shown before, the importance of this aspect becomes clear:

It includes knowledge and criteria about good modelling problems and ideas, and how to develop or find modelling problems. For pre- and in-service teachers it is not trivial to create adequate modelling problems, in particular complex ones, which are used during the previously mentioned modelling weeks and days.

22.3 Conclusions

Henry Pollak mentioned that mathematical modelling is a "new product" in many countries to fit into existing systems of mathematics education. So mathematical modelling is addressed to the ministry of education, to the curriculum writers, to students, to pre- and in-service teachers, university professors and university students. All these people have to learn or gain ideas now about how to teach modelling. But this is not trivial and cannot be expected as a transfer from other disciplines. For this, all parts of the map are needed, strong theoretical thoughts, well designed and conducted empirical studies and in the background and as a basis the pure and applied mathematics, we need this for the daily work in mathematics lessons or in university seminars as well as for research purposes in the next decades.

References

Blum, W. (2011). Can modelling be taught and learnt? Some answers from empirical research. In G. Kaiser, W. Blum, R. Borromeo Ferri, & G. Stillman (Eds.), *Trends in teaching and learning of mathematical modelling* (pp. 15–30). New York: Springer.

Borromeo Ferri, R. (2007). Modelling from a cognitive perspective. In C. Haines, P. Galbraith, W. Blum, & S. Khan (Eds.), *Mathematical modelling: Education, engineering and economics (ICTMA 12)* (pp. 260–270). Chichester: Horwood Publishing.

Borromeo Ferri, R., & Blum, W. (2009). Mathematical modelling in teacher education – Experiences from a modelling seminar. In European Society for Research in Mathematics Education (Eds.). *Proceedings of CERME 6* (pp. 2046–2055). Lyon: European Society for Research in Mathematics Education.

Borromeo Ferri, R., & Lesh, R. (2013). Should interpretation systems be considered to be models if they only function implicitly? In G. A. Stillman, G. Kaiser, W. Blum, & J. P. Brown (Eds.), *Teaching mathematical modelling: Connecting to research and practice* (pp. 57–66). Dordrecht: Springer.

Kaiser-Meßmer, G. (1986). *Anwendungen im Mathematikunterricht. Vol. 1 – Theoretische Konzeptionen. Vol. 2 – Empirische Untersuchungen*. Bad Salzdetfurth: Franzbecker.

Kaiser, G., & Schwarz, B. (2010). Authentic modelling problems in mathematics education – examples and experiences. *Journal für Mathematik-Didaktik, 31*(1), 51–76.

Kaiser, G., & Sriraman, B. (2006). A global survey of international perspectives on modeling in mathematics education. *ZDM – The International Journal on Mathematics Education, 38*(3), 302–310.

Pollak, H. O. (1979). The interaction between mathematics and other school subjects. In UNESCO (Ed.), *New trends in mathematics teaching IV* (pp. 232–248). Paris: OECD.

Stillman, G. (2011). Applying metacognitive knowledge and strategies in applications and modelling tasks at secondary school. In G. Kaiser, W. Blum, R. Borromeo Ferri, & G. Stillman (Eds.), *Trends in teaching and learning of mathematical modelling: ICTMA14* (pp. 165–180). New York: Springer.

Stillman, G., Brown, J., & Galbraith, P. (2010). Identifying challenges within transition phases of mathematical modeling activities at year 9. In R. Lesh, P. Galbraith, C. R. Haines, & A. Hurford (Eds.), *Modelling students' mathematical modeling competencies: ICTMA13* (pp. 385–398). New York: Springer.

Treilibs, V., Burkhardt, H., & Low, B. (1980). *Formulation processes in mathematical modelling*. Nottingham: Shell Centre for Mathematical Education.

Chapter 23
Negotiating the Use of Mathematics in a Mathematical Modelling Project

Jussara de Loiola Araújo and Ilaine da Silva Campos

Abstract Our goal, in this chapter, is to present the negotiation between a student – Maria Estela – and the teacher, on how to use mathematics in a mathematical modelling project. The study is based on a qualitative analysis of empirical data from a meeting of Maria Estela's group with the teacher to progress the modelling project. It was the first time that Maria Estela had participated in such an activity and she, as well as her group, did not know how to model mathematically the phenomenon they were studying. From our analysis, we were able to understand that the negotiation between the teacher and Maria Estela was crucial to the continuation of the group project. The negotiation, at the time the group was making decisions, revealed different understandings about how to use mathematics and gave rise to a *negotiation space*, leading to an approach different from those considered appropriate, at the beginning, by both the teacher and Maria Estela.

23.1 Introduction

Some authors such as Araújo (2010); Blomhøj (2009); Blum et al. (2007); Kaiser and Sriraman (2006) discuss the growing interest of teachers and researchers in mathematical modelling in mathematics education in several countries including Brazil. In these studies, the existence of a multiplicity of understandings about modelling, and of objectives regarding its implementation in school spaces, stands out. In this chapter specifically, our understanding of modelling is based on Barbosa (2006), who conceives it as a learning milieu in which students are invited to

J.L. Araújo (✉)
Departamento de Matemática, Universidade Federal de Minas Gerais, Av. Presidente Antônio Carlos, 6627, Pampulha, CEP: 31.270-901, Belo Horizonte, Minas Gerais, Brasil
e-mail: jussara.loiola@gmail.com

I.S. Campos
Universidade Federal de Minas Gerais, Rua Panamérica, 44. Pampalona, CEP: 44031-435, Feira de Santana, Bahia, Brasil
e-mail: ila_scampos@yahoo.com.br

G.A. Stillman et al. (eds.), *Mathematical Modelling in Education Research and Practice*, International Perspectives on the Teaching and Learning of Mathematical Modelling, DOI 10.1007/978-3-319-18272-8_23

investigate problem situations from other areas of knowledge or everyday situations using mathematics. Our interest in discussing modelling is in harmony with the socio-critical perspective (Barbosa 2006), which aims to develop students' competence to question the role of mathematical models in society, amongst other goals.

Based on the analyses of Kaiser and Sriraman (2006) and Blomhøj (2009), it is evident that the socio-critical perspective is widespread in Brazil, from where this chapter originates. Supporting this contention is the fact that most of the studies cited by these authors, in this perspective, are from Brazilian authors such as Araújo (2009), Barbosa (2006, 2009) and Caldeira (2009).

Part of the large number of studies on modelling in mathematics education is related to students' practices in these learning milieus. This is also the case with research in which modelling is conceived from the socio-critical perspective. In modelling activities, as we understand them, in which students are invited to investigate problem situations in order to reflect on the role of mathematical models in society, the activities proposed are open-ended, without a predetermined strict orientation. When students develop mathematical modelling projects, or some other open-ended activities, doubts about how to proceed may arise.

Alrø and Skovsmose (2002) claim that learning milieus in which students are invited to investigate are still rare in mathematics classes. Therefore, students may take some time to understand, or sometimes they do not even understand, what is expected of them when activities different from the usual ones are proposed in mathematics classes (Araújo and Barbosa 2005; Julie and Mudaly 2007). Alrø and Skovsmose (2002) state that new patterns of dialogue and interaction are established in activities of this nature, validating research that focuses on these dialogues or interactions.

Sometimes, differences between what is expected and what is accomplished by the students are only perceived after the task is completed (Borba et al. 1999). On the other hand, when the teacher follows the development of these activities while they are being carried out, negotiations can happen between teacher and students (Leiß 2005; Oliveira et al. 2009) aiming at a greater agreement between their interpretations.

Barbosa (2007, p. 239) calls *negotiation spaces* the encounter between the teacher and the students to negotiate meanings related to the modelling task. "Negotiation implies dialogue, which assumes that none of the parties wishes to impose his/her perspective, but rather put it up for discussion." We understand that negotiation can take place to meet the different demands of the activity: to define an agreement between the teacher and students, to meet the needs of the pedagogical practice of the teacher, to guide the students' own performance in the development of the modelling activity, among other reasons.

Our goal, in this chapter, is to present the negotiation between a student – Maria Estela – and her teacher, on how to use mathematics in a mathematical modelling project. It was the first time that Maria Estela participated in an activity such as this and she, as well as her group, did not know how to model mathematically the phenomenon they were studying.

In order to achieve this goal, we start by presenting the context and participants of the study, the methodological aspects, as well as a description of the modelling project developed by Maria Estela and her group. Excerpts from moments of negotiation between the student, Maria Estela, and her teacher are presented in Sect. 23.3. Section 23.4 presents a discussion of data based on the theoretical framework. Finally, Sect. 23.5 brings some final considerations about the study.

23.2 Context, Participants, Methodological Aspects and the Modelling Project

In this chapter, we will examine empirical data from the second meeting of a group of students from the first semester of the Public Management course at the Federal University of Minas Gerais (UFMG), Brazil, with a mathematics teacher, the second author of this chapter. This meeting aimed to follow up on the development of a modelling project.

The project was one of the evaluative activities of the subject Mathematics A, where the teacher is the first author of this chapter[1]. The subject, under the responsibility of the Mathematics Department of UFMG, aims to address functions, derivatives and integrals, with a class load of 60 h in one academic semester. As part of the activities of Mathematics A, the teacher of the subject requested that students develop modelling projects from the socio-critical perspective. In this perspective, the modelling activity is open-ended and it is common that students have the freedom to choose the topics that they will address in their projects. The topics were chosen and the groups were formed according to interest in the topics. Nine groups were formed, each consisting of between five and eight students.

One student, Maria Estela, who is also a journalist, suggested the topic "the controversy over health accounts in Minas Gerais"[2]. This topic is related to expenses with the *Sistema Único de Saúde* – Unified Public Health System – (SUS), which, by law, must ensure free and universal health care in Brazil. Other students joined her, forming a group. Within this topic, the purpose of the modelling project of Maria Estela's group was to understand the accounts regarding the health system in the State of Minas Gerais, from 2004 to 2008.

Maria Estela's interest in this topic was related to her work as a journalist and she had been researching this subject for quite some time. She suspected that the Government of the State of Minas Gerais did not comply with the law on health

[1] The first author of this chapter was the teacher responsible for the subject Mathematics A and will be named "teacher of the subject". The second author followed the development of the subject and acted as advisor to two groups. She worked also as a researcher and was collecting data for her masters research (Campos 2013). In this chapter, the second author is called "teacher".

[2] Brazil is a federative republic formed of 26 states plus the Federal District, and Minas Gerais is one of the states.

Table 23.1 The ratio of health spending to revenue in the state of Minas Gerais

Year	Minimum percentage established by EC-29	Percentage informed by the government of Minas Gerais	Percentage calculated by SIOPS
2004	12	12.16	8.66
2005	12	12.33	6.87
2006	12	13.20	6.04
2007	12	13.30	7.09
2008	12	13.12	8.65

care spending. According to constitutional amendment 29 (EC-29), the Government should allocate 12% of its revenue to health. However, information provided by the Government and other organizations was conflicting with respect to this subject, as can be seen in Table 23.1. This information was obtained by Maria Estela who then presented it to the other group members.

Table 23.1 displays the percentages of revenue spent on health from 2004 to 2008 by the Government of Minas Gerais, based on three different sources. The second column shows what should have been spent (12%) according to EC-29; the third column shows what the Government reports having spent; and, the fourth column shows what the Public Health Budget Information System[3] (SIOPS) shows the Government of Minas Gerais spent.

Table 23.1 summarises well what inspired the group's modelling project – "the controversy over health accounts in Minas Gerais" – since it shows a mismatch of information relating to health spending. Although the Government claims to have invested a percentage greater than that established by EC-29, according to SIOPS, there is a health deficit. This aspect was emphasized by Maria Estela when she presented this information to her group.

This study is qualitative in nature (Bogdan and Biklen 1994) as our objective is to analyse the negotiation between a student – Maria Estela – and the teacher on how to use mathematics in a mathematical modelling project. The main methodological procedure adopted was unstructured participant observation. According to Vianna (2003), unstructured observation takes place in a natural context, which enables the observer to become part of the culture of the subjects and to see the world from their point of view. The data obtained through this method, are presented in the next section.

[3] The SIOPS, organization linked to the Ministry of Health, checks compliance with Amendment 29, which is based on the resolution n°. 322, of 2003, of the National Health Council (CNS), which regulates what can be regarded as expenditure on health in Brazil.

23.3 "It Isn't My Fault If I Already Have All the Data!"

The negotiation between Maria Estela and the teacher on how to use mathematics in the mathematical modelling project took place in the context described previously. At the beginning of the meeting, Maria Estela presented to the group various data, quantitative and qualitative, on the topic of the project, which she had been collecting for some time as part of her work as a journalist. That's when the teacher asked the group:

Teacher: What's your idea? I want to see if I can help you.
Maria Estela: So, the question is if they complied with Amendment 29, as provided for in article 198 of the Constitution, and if the parameters of the CNS were taken into consideration. What is the actual investment in health?

Maria Estela continued explaining the data until the teacher questioned what the group wanted to do:

Teacher: [. . .] Yes, people. And what about the project? What I can't see is . . . I can interpret these data, I can see what you are questioning. And for this project? What do you want to say? You just want to show that?
Teacher: See, Maria Estela, the issue I was discussing with José [another group member] is that if you are only interested in interpreting, then . . . Is this the project proposal?
Maria Estela: No, our proposal is to take these data and develop mathematical relationships between them.
Teacher: Which kind of relationships do you want? [. . .]. These data are ready.

Maria Estela, supported by José, explained how she understood the use of mathematics in the development of the modelling project:

Maria Estela: [. . .] Well, here's the thing: the mathematical modelling will build relationships, create formulas. We have to create formulas for it, right?
Teacher: Would the idea be to calculate the deficit?
Maria Estela: It is exactly a formula for the deficit. And that means what? Is it an exponential function of [logarithmic] napierian growth? [laughter]. Napierian decrease?
José: What I understand [. . .] We understand that it is ready.
Maria Estela: That's it, it's ready. All we have to do now is to create a mathematical relationship.
Teacher: The relationship to express that situation?
Maria Estela: [. . .]. That's it. And that is what? A formula.

The teacher explained how she understood the use of mathematics in the development of the modelling project:

Teacher: You may have a problem. Based on that problem, we get to these relationships. Because only getting the data and plotting it on the graph and comparing it, to me this is not the kind of the work that was proposed, do you understand? It is to investigate a situation. In fact, what you're doing is just a comparison of something you are already sure of, that you've already analysed. In other words, you just changed the language, from table language to graph language.

Maria Estela: Well then! But, it's not my fault if I already have all the data and I didn't have to research it, is it?

The group continued discussing and Maria Estela continued disagreeing with the teacher:

Maria Estela: But, we need to create a formula.

Maria Estela: [...] No, no. This is a work that is ready, a work that is virtually ready. It is an investigation work, a modelling project, that doesn't mean that I haven't researched these data.

The discussion continued until the teacher made a suggestion with which Maria Estela and José agreed:

Teacher: I'll give an idea: we could think about the percentage for this year, what was spent, and about the deficit for this same year. Then, we think about the deficit for the next year and calculate the total deficit that the State has with health.

Maria Estela: Hummm! We made a projection.

José: It was a different approach.

Maria Estela: No she didn't use a different approach. She developed a mathematical relationship.

In this episode, the teacher wanted to know about the information that the group already had in order to guide the students on how to proceed with the data obtained previously. For Maria Estela, all numerical data related to the project had already been collected and, to complete the modelling project, "All [they had] to do now [was] to create a mathematical relationship." In other words, they had to find a formula to relate the data that had already been collected. So, for Maria Estela, modelling is finding or developing (she uses both words) a mathematical formula that the collected data fit.

The teacher, in turn, insisted that "only getting the data and plotting it on the graph and comparing it [...] is not the kind of the work that was proposed". She realized that the only thing the students intended was to create a graph from the tables to confirm the certainty that they had about the topic of the project. However, for the teacher, the proposal was "to investigate a situation" and, to this end, they would have first to elaborate a problem. So, for the teacher to develop a mathematical modelling project is to propose a problem related to the topic chosen by the group and to investigate, to search for information and data to obtain, by means of mathematical models, answers to the problem.

23.4 The Negotiation Between Maria Estela and the Teacher: Data Through a Theoretical Lens

In the episode presented earlier, there was a divergence between Maria Estela and her teacher, but neither of them seemed to want to impose their point of view. On the contrary, each presented arguments to support their views. Therefore, the *negotiation space* was established (Barbosa 2007). According to the author, although the voice of the teacher constitutes an authority in these spaces, he/she does not seek to direct the students' actions. According to Barbosa (2007), "if we are interested in the students having an 'authentic' modelling experience, [...] it [is] necessary to hear what [is] being said by the student and use it as a type of lever to formulate the next comment" (p. 239, emphasis in original).

In harmony with this orientation, the teacher made a proposal that took into account what Maria Estela intended to do, but which also included the way she herself understood the development of a modelling project. In her proposal, students would have to do something more than a simple reinterpretation of the data. They would have to analyse the data, search for some mathematical formula to correlate them (in the way Maria Estela understood modelling), but that would allow future analyses, projections, which would require some research (as the teacher understood modelling).

Since the teacher and Maria Estela disagreed about what should be done, both had to give up some aspect of her point of view. Maria Estela agreed that the work was not ready, that there was still something to be done, and the teacher agreed that the group would not do something that she assumed to be part of the development of a project of mathematical modelling – the search for, or gathering of, information and data.

As well as hearing each other, we believe that giving up of some aspect of his/her point of view is also a characteristic of the *negotiation space*. This style of communication in mathematics classes is similar to what Alrø and Skovsmose (2002) call *inquiry cooperation model* (IC-Model), which is "a particular form of student-teacher interaction when exploring a *landscape of investigation*" (p. 46). In the *landscape of investigation*, students are invited to explore situations in which mathematics is involved and to look for explanations. It is in this sense that, in the episode analysed in this chapter, the teacher emphasized that it was necessary "to investigate a situation" when we develop modelling projects. Such landscapes have, as one of their main objectives, to destabilise what Borba and Skovsmose (1997) call *the ideology of certainty of mathematics*. According to these authors, the ideology of certainty sustains the character of neutrality of this science, imbuing it with the power of the holder of the definitive argument in various debates in society.

In the episode, the teacher claims that "In fact, what [the group was] doing [was] just a comparison of something [they were] already sure of, that [they had] already analysed." We consider, then, that for the teacher, the level of certainty that Maria Estela had about what she had proposed to study had influenced her to use mathematics just to validate her arguments, corroborating the *ideology of certainty of mathematics*.

In this sense, the divergences between Maria Estela and the teacher and the negotiation between them on how to use mathematics in the modelling project seem to originate from different conceptions of mathematics. Although we have not carried out a study focusing on conceptions of mathematics, it seems that the development of modelling projects in *negotiation spaces*, in which students are invited to participate in the *landscape of investigation* of situations with reference to the reality, can reveal different conceptions of mathematics and destabilize the *ideology of certainty of mathematics* that is often supported by such conceptions.

23.5 Final Remarks

In our analysis of the dialogue between students and the teacher, we could verify how much the negotiation between them was determinant on the continuity of the project. The presence of the teacher, at the time the group was deciding on how to continue the modelling project, showed different understandings about how to use mathematics and gave rise to a *negotiation space*, leading to an approach different from those considered appropriate, at the beginning, by both the teacher and Maria Estela. We raised the hypothesis that different modelling conceptions originate in different conceptions of mathematics, mainly from the actions of Maria Estela. In this regard, we emphasize the role of the teacher in the *negotiation spaces* to destabilize certainty with respect to mathematics. Another issue that was left open is whether the negotiation between Maria Estela and the teacher could also be interpreted as a negotiating of power between them, since the (mathematics) teacher held the power that Maria Estela wished to support her arguments against the Government of the State of Minas Gerais, while she herself held the power to have the information used in the project.

Based on these questionings, we want to emphasize the importance of such research which focuses on *negotiation spaces* in mathematical modelling milieus, both from the point of view of the organization of modelling activities and the development of the scientific field. From the point of view of organization of activities, these spaces allow students, even in their first experience with modelling, to have authentic experiences (Barbosa 2007). From the perspective of the scientific field, to analyse negotiations in these spaces can help us understand what happens when different activities of investigative character are incorporated into the daily life of the classroom, allowing discussions regarding conceptions of mathematics and issues of power.

References

Alrø, H., & Skovsmose, O. (2002). *Dialogue and learning in mathematics education: Intention, reflection, critique*. Dordrecht: Kluwer.

Araújo, J. L. (2009). Formatting real data in mathematical modelling projects. In M. Blomhøj & S. Carreira (Eds.), *Mathematical applications and modelling in the teaching and learning of mathematics*. Proceedings from TSG21 at the ICME11 (*IMFUFA-text* no. 461, pp. 229–239). Roskilde: Department of Science, Systems and Models, Roskilde University.

Araújo, J. L. (2010). Brazilian research on modelling in mathematics education. *The International Journal on Mathematics Education, 42*(3–4), 337–348. doi:10.1007/s11858-010-0238-9.

Araújo, J. L., & Barbosa, J. C. (2005). Face a face com a modelagem matemática: Como os alunos interpretam essa atividade? *Bolema – Boletim de Educação Matemática, 23*, 79–95.

Barbosa, J. C. (2006). Mathematical modelling in classroom: A sociocritical and discursive perspective. *ZDM_The International Journal on Mathematics Education, 38*(3), 293–301.

Barbosa, J. C. (2007). Teacher-student interactions in mathematical modelling. In C. Haines, P. Galbraith, W. Blum, & S. Khan (Eds.), *Mathematical modelling: Education, engineering and economics* (pp. 232–240). Chichester: Horwood.

Barbosa, J. C. (2009). Mathematical modelling, the socio-critical perspective and the reflexive discussions. In M. Blomhøj & S. Carreira (Eds.), *Mathematical applications and modelling in the teaching and learning of mathematics*. Proceedings from TSG21 at the ICME11 (*IMFUFA-text* no. 461, pp. 133–143). Roskilde: Department of Science, Systems and Models, Roskilde University.

Blomhøj, M. (2009). Different perspectives in research on the teaching and learning of mathematical modelling – Categorising the TSG21 papers. In M. Blomhøj & S. Carreira (Eds.), *Mathematical applications and modelling in the teaching and learning of mathematics*. Proceedings from TSG21 at the ICME11 (*IMFUFA-text* no. 461, pp. 1–17). Roskilde: Department of Science, Systems and Models, Roskilde University.

Blum, W., Galbraith, P. L., Henn, H.-W., & Niss, M. (2007). Preface. In W. Blum, P. L. Galbraith, H.-W. Henn, & M. Niss (Eds.), *Modelling and applications in mathematics education: The 14th ICMI study* (pp. xi–xiv). New York: Springer.

Bogdan, R., & Biklen, S. (1994). *Investigação qualitativa em educação: Uma introdução à teoria e aos métodos*. Porto: Porto Editora.

Borba, M. C., & Skovsmose, O. (1997). The ideology of certainty in mathematics education. *For the Learning of Mathematics, 17*(3), 17–23.

Borba, M. C., Meneghetti, R. C. G., & Hermini, H. A. (1999). Estabelecendo critérios para avaliação do uso de modelagem em sala de aula: Estudo de um caso em um curso de Ciências

Biológicas. In E. Fainguelernt & F. Gottlieb (Eds.), *Calculadoras gráficas e educação matemática* (pp. 95–113). Rio de Janeiro: Editora Art Bureau.
Caldeira, A. D. (2009). Mathematical modelling and environmental education. In M. Blomhøj & S. Carreira (Eds.), *Mathematical applications and modelling in the teaching and learning of mathematics*. Proceedings from TSG21 at the ICME11 (*IMFUFA-text* no. 461, pp. 145–157). Roskilde: Department of Science, Systems and Models, Roskilde University.
Campos, I. S. (2013). *Alunos em ambientes de modelagem matemática: caracterização do envolvimento a partir da relação com o background e o foreground*. Unpublished master dissertation, Belo Horizonte: Universidade Federal de Minas Gerais.
Julie, C., & Mudaly, V. (2007). Mathematical modelling of social issues in school mathematics in South Africa. In W. Blum, P. Galbraith, H.-W. Henn, & M. Niss (Eds.), *Modelling and applications in mathematics education: The 14th ICMI study* (pp. 503–510). New York: Springer.
Kaiser, G., & Sriraman, B. (2006). A global survey of international perspectives on modelling in mathematics education. *ZDM_The International Journal on Mathematics Education, 38*(3), 302–310.
Leiß, D. (2005). Teacher intervention versus self-regulated learning? *Teaching Mathematics and Its Applications, 24*(2-3), 75–89.
Oliveira, A. M. P., Campos, I. S., & Silva, M. S. (2009). As estratégias dos professores para desenvolver modelagem matemática em sala de aula. *Boletim GEPEM, 55*(3), 175–192.
Vianna, H. M. (2003). *Pesquisa em educação: a observação*. Brasília: Editora Plano.

Chapter 24
Moving Beyond a Single Modelling Activity

Jonas B. Ärlebäck and Helen M. Doerr

Abstract In this theoretical chapter, we draw on a models and modelling perspective on teaching and learning to elaborate on the components of model development sequences using the lens of variation theory. We give empirical examples of how model exploration activities and model application activities can be described and understood from a variation theory perspective. The chapter concludes by presenting tentative principles for the design of such activities within a model development sequence.

24.1 Introduction

Despite the steadily growing body of resources and research on the teaching and learning about and through mathematical modelling, the uptake and impact of mathematical modelling in classroom practice remains limited (Niss et al. 2007). Traditional methods of teaching prevail in most classrooms, and often the adopted textbook dictates the form and content of the implemented curriculum. The approach to modelling in many textbooks can at best be described as what Blum and Niss call "the island approach" (1991, p. 60), where a given mathematical content is first treated purely from a mathematical point of view, and application and modelling tasks coupled to the specific content are addressed after the formal mathematics has been established. In some textbooks applied problems are used as vignettes to introduce mathematical content. Typically however, modelling tasks are like isolated islands in the ocean of pure mathematics.

J.B. Ärlebäck (✉)
Department of Mathematics, Linköping University, 83 Linköping SE-581, Sweden
e-mail: jonas.bergman.arleback@liu.se

H.M. Doerr
Department of Mathematics, Syracuse University, 215 Carnegie Hall, Syracuse, NY 13244–1150, USA
e-mail: hmdoerr@syr.edu

G.A. Stillman et al. (eds.), *Mathematical Modelling in Education Research and Practice*, International Perspectives on the Teaching and Learning of Mathematical Modelling, DOI 10.1007/978-3-319-18272-8_24

In this chapter, we want to move beyond an isolated, single modelling activity and focus on sequences of modelling tasks that facilitate the development of learners' mathematical ideas and modelling competencies. One approach taken in this line of research is that adopted by Barquero and colleagues (Barquero et al. 2007; Barquero et al. 2013). These researchers draw on the Research and Study Course methodology and praxeologies from the Anthropological Theory of Didactics (Chevallard 2004, 2006) as didactical devices to teach mathematics and mathematical modelling at the university level to first year technical engineering students. Another approach that moves beyond a single modelling activity is that of Dominguez and colleagues, who integrate first year university courses in mathematics and physics (Dominguez et al. 2013; Dominguez et al. 2015). In their research, modelling tasks form the basis for simultaneously learning mathematical models and their applications in the physical world. A research study at the upper secondary level by Galbraith and Clatworthy (1990) has demonstrated how a modelling strand conducted in parallel with conventional mathematics coursework over a 2-year period can develop students' abilities to use modelling skills within and outside of mathematics. This study also points to the need for a pedagogy that supports this development. Our approach to this line of research has been to examine and investigate model development sequences proposed by Lesh and colleagues (Lesh and Doerr 2003b; Lesh et al. 2003). While model development sequences have received some attention in the research literature, we want to elaborate on the design and conceptualization of model development sequences using variation theory (Marton and Booth 1997; Marton and Tsui 2004). Our goal in this chapter is to draw on variation theory and our own empirical research to formulate tentative design principles for model exploration and model application activities, within a model development sequence.

24.2 Models, Modelling, and Model Development Sequences

Model development sequences are situated within a contextual perspective on modelling (Kaiser and Sriraman 2006) in which learners are confronted with activities where they need to develop mathematics to make sense of meaningful situations. The goal of such activities is for the learners to develop a conceptual system or model that can be shared with others and used for sense making in structurally similar contexts. More precisely, "models are conceptual systems (consisting of elements, relations, operations and rules governing interactions) that are expressed using external notation systems, and that are used to construct, describe, or explain the behaviours of other system(s)" (Lesh and Doerr 2003a, p. 10).

Much research within this perspective has focused on *model eliciting activities (MEAs)* that confront learners with the need to initially develop a model to make

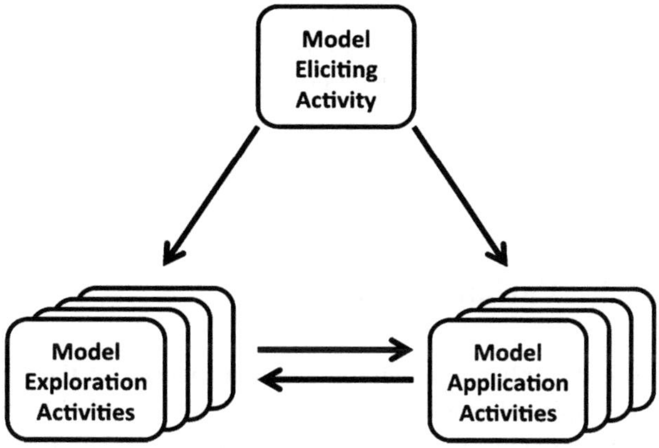

Fig. 24.1 The general structure of a model development sequence

sense of a meaningful situation. There are six well-established principles for designing MEAs that are grounded both theoretically (Lesh and Doerr 2003b; Lesh and Zawojewski 2007) and empirically in multiple settings (Diefes-Dux et al. 2006; Iversen and Larson 2006; Lesh et al. 2000; Yoon et al. 2010). In model development sequences, MEAs are the starting point, designed to elicit learners' initial ideas about a problem situation in such a way that their thinking is made visible both to the learners themselves and to their teachers. In a model development sequence, the MEA is followed by one or more structurally related *model exploration activities (MXA)* and *model application activities (MAA)*, as shown in Fig. 24.1. These subsequent activities are intended to support learners in developing their initial ideas or models. However, design principles for these activities have not yet been theoretically developed and empirically grounded; we aim to contribute to that development and grounding in this chapter.

The focus of a model exploration activity (MXA) is on the underlying mathematical structure of the elicited model. These activities provide learners with opportunities to make the model an explicit object of thought. This is achieved as learners develop representational systems and language to use in thinking about their models. The explorations of structure can involve investigating algebraic representations, tables, interactive graphics, diagrams, animations or simulations. A goal for MXAs is for the learner to experience the different strengths of various representations, the structural similarities and differences between and among representations, and ways of using different representations purposefully and productively. In model application activities (MAAs), learners apply their initially elicited model to new phenomena, situations and contexts. By applying their model in new contexts, learners are often led to make further adaptations to the model, to extend representations, to deepen their understandings or to refine language for describing situations (Lesh et al. 2003; Lesh and Doerr 2003b).

Throughout a unit of instruction designed as a model development sequence, learners develop their models through interacting with other learners and engaging in multiple cycles of conjecturing, interpreting, describing and explaining (Lesh and Doerr 2003a). We have recently been examining how variation theory might help us to better understand how to design effective MXAs and MAAs to support student learning. The methodology to reach this goal has been an iterative dialectic approach, where we have coordinated our empirical results with the tenets and constructs of variation theory.

24.3 Variation Theory

Rooted in phenomenography, which focuses on the nature of experiences and "is the empirical study of the limited numbers of different ways in which various phenomena in, and aspects of, the world around us are experienced, conceptualized, understood, perceived, and apprehended" (Marton 1994, p. 4424), variation theory is an experiential theory (Ko and Marton 2004). A central tenet of variation theory is to consider what knowledge learners are intended to learn in a particular setting, *the object of learning,* when designing the teaching and learning (Runesson 2005). Learning comes about through discerning critical aspects and features of the object of learning, along dimensions of variation in experiences. It is the variation that opens up the possibilities for the learner to become aware of the critical aspects of certain content and facilitates the development of targeted capabilities for solving problems.

All capabilities have both *specific aspects* and *general aspects* (Marton et al. 2004). The specific aspect is concerned with the knowledge of a particular content of a given subject, whereas the general aspect focuses on the capability or values that learners can develop through learning the content. It is through the general aspect that multiple specific aspects are woven together to a more coherent body of knowledge (Lo 2012).

From a variation theoretical point of view, it is imperative to pay attention to what varies and what stays invariant in any learning situation. Variation makes certain critical aspects appear in the foreground, while other aspects remain in, or are pushed into, the background. This opens up the possibilities for the learners to discern the critical aspects in a learning situation. Marton et al. (2004) discuss the following four "patterns of variation" (p. 16), or types of variation: *contrast*, *generalization*, *separation*, and *fusion*. *Variation by contrast* stems from the premise that in order to experience something in the first place, there must be a comparison with something else. Hence, contrast variation makes it possible to discern a new aspect of a particular object of learning. When an aspect is discerned and has manifested itself in the learners' awareness, *generalization* needs to occur for the learner to understand the discerned aspect as fully as possible; this will enable the learner to identify the discerned aspect in different situations, settings and contexts. This type of variation that develops a more general understanding of the object of learning is called generalization. Equally important as obtaining a generalized understanding of an aspect of the object of learning is for the learner to crystallize the discerned critical aspect with respect to another critical aspect. This *variation of separation* means varying the critical aspect while keeping other critical aspects invariant, so that the critical aspect discerned is separated from other critical aspects of the object of learning. When the object of learning is complex and construed by multiple critical aspects, these aspects need to be *fused*, so that the learner can gain a holistic understanding of the object of learning rather than experiences of isolated critical aspects. By varying multiple critical aspects at the same time, the learner experiences the complexity of the object of learning by *fusing* together multiple critical aspects.

In variation theory, designing learning environments is about creating a space that opens up dimensions of variation challenging the taken-for-granted nature of learners' experiences. This *space of learning* provides opportunities and makes it possible for the learner to experience the designed variation in an instructional sequence, and thereby learn. The experienced variation enables learners to discern the distinctions realized in language, and linguistic distinctions enable learners to discern the variation. In this way, language is a key means by which a shared space of learning is constructed by teachers and learners (Marton and Tsui 2004).

24.4 Model Development Sequences in the Light of Variation Theory

We examine model development sequences in terms of variation theory by identifying the objects of learning within a given sequence. In so doing, we will identify different patterns of variation that occur in MXAs and MAAs within a model development sequence and we will draw on those patterns in formulating tentative design principles for MXAs and MAAs. The examples given below come from a model development sequence centred on the concept of average rate of change (see Ärlebäck et al. 2013 for further details). This particular sequence begins with a MEA that elicits students' initial models of rates of change by using the learners' bodily motion and motion detectors. This is followed by model exploration activities using a computer simulation of motion along a straight path. The sequence ends with two model application activities: one investigating how the light intensity varies with the distance from a point source, and one examining how the voltage drop changes across a discharging capacitor in a simple resistor-capacitor circuit. The goal with the MEA is for the learners to *think up their models* of rates of change and to collectively participate in a shared space of learning where the objects of learning can be subsequently explored and applied. Our tentative principles for designing model exploration and application activities will be discussed in the following section.

24.4.1 Principles for Designing Model Exploration Activities

The purpose of an MXA is for the learner to explore the mathematical structure of the elicited model, that is, *to think about the model.* The fundamental dialectic relationship between the learners' experiences and language development stressed in variation theory align with the models and modeling perspective's very definition of models, where a central feature is for the learners to be able to express their conceptual systems (models) using external notation systems. Language in general and mathematical representational systems in particular are the key external notational systems that enable learners to express and share their evolving mathematical thinking in a shared space of learning, and render it possible for new critical features to be experienced and discerned through different patterns of variation. Hence, in designing MXAs within a model development sequence, special attention should be paid to the evolving space of learning initially spanned by the MEA, engaging and to furthering the learners' precise and careful use of language and representations. This suggests the following principles for designing MXAs. A model exploration activity should engage the learners in

1. contrasting strengths and weaknesses of various representations;
2. using language precisely and carefully;
3. using representations purposefully and productively.

The intent of these three design principles is to guide the development of modelling activities to engage learners in *thinking about their models.* As the first principle indicates, contrast variation is key in comparing strengths and weaknesses of various representations. The variation of separation is key in gaining an understanding of when and why a particular representation can be more productively used than another.

24.4.1.1 An Example of Contrast Variation

We illustrate the pattern of contrast variation in the model exploration activity in the model development sequences centred around average rate of change. We designed an activity focusing on the difference between the concepts of speed and velocity. Learners' difficulties in understanding these notions are well established in the research literature (see Ärlebäck et al. 2013 for an overview). Contrast variation makes it possible to discern a new aspect of a particular object of learning by comparing it with some other aspect. The learners had worked with and established relationships between piecewise linear position graphs and their corresponding velocity graphs for scenarios with non-negative velocities. When velocity is non-negative, speed and velocity appear as identical quantities. In this MXA, negative velocities were introduced to contrast different aspects of the corresponding speed and velocity graphs for the position graph of an object moving with a negative velocity.

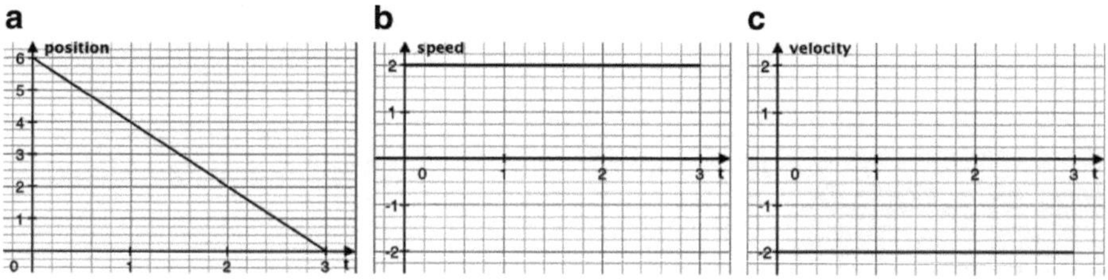

Fig. 24.2 Related position (**a**), speed (**b**), and velocity graph (**c**)

Given the position graph in Fig. 24.2a, the learners are asked to construct both the velocity graph and the speed graph, thereby beginning to sort out the structural differences and similarities between speed (Fig. 24.2b) and velocity (Fig. 24.2c). The relationship between speed and velocity can then be further explored by using situations described by more complex piecewise linear position graphs.

24.4.1.2 An Example of Variation of Fusion

Multiple aspects of complex objects of learning need to be experienced simultaneously at the same time in order for the learner to fuse them together into a coherent whole. In the model development sequence focused on average rate of change, an intended object of learning is for the learners to coordinate and understand the relationship between a moving character's animated motion, its position graph and its velocity graph (see Fig. 24.3). Dynamically linking the character's walking with its graphical representations, the learners explore different walking scenarios, which thereby opens up the possibility for the learners to fuse these three aspects of this particular object of learning (i.e., the average rate of change).

24.4.2 Principles for Designing Model Application Activities

The purpose of a MAA is for students *to think with their model* in a range of situations. By applying their models in new contexts, learners are often led to make further adaptations to their model, to extend representations, to deepen their understandings or to refine language for describing the situation. Introducing new contexts expands the space of learning, integrates and connects new models (alongside with the structural aspects, use of language, and representation that comes with the connected model). Hence, central for the design of MAAs is to systematically vary the context in which the learners are to apply their models. This suggests the following principles for designing a MAA. A model application activity should engage the learners in

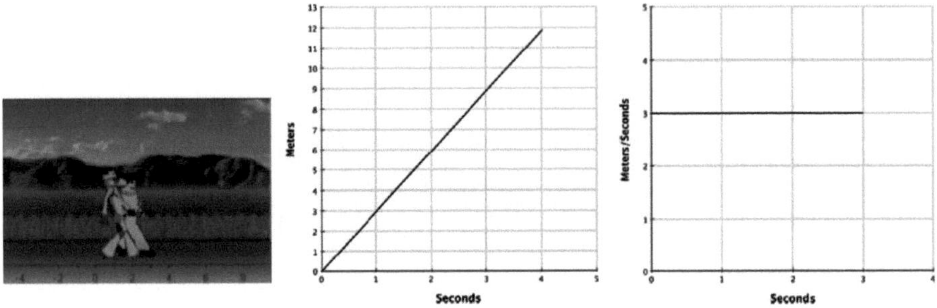

Fig. 24.3 The coordination of simulated motion, position and velocity graphs (Kaput & Roschelle 1996)

1. using their model in a variety of distinct contexts;
2. extending their models or connecting their models to other models;
3. using language to interpret, describe and/or predict behaviour of a real phenomenon.

These three design principles are intended to guide the development of model application activities that support the learner in *thinking with their model* so as to develop a generalized model (or conceptual system) that can be productively used in a range of contexts. We give brief examples suggesting how these principles are informed by variation theory.

24.4.2.1 An Example of Variation of Separation

By varying a discerned critical aspect while keeping other critical aspects invariant, the discerned critical aspect is separated and kept distinct from other critical aspects of the object of learning, so as to crystallize the learner's understanding of that aspect. In the model application activity with the discharging capacitor, the learners developed exponential models of decay to describe and predict how the voltage across the capacitor changes as the capacitor discharges. By varying only the resistance or the capacitance in the circuit, the learners investigate the effect that different values of resistance and the capacitance had on the rate at which the capacitor discharged. Since other aspects of the circuit remained invariant, the learners had the opportunity to interpret the effects of the change in resistance on the graphical representations and in the exponential algebraic expressions modelling the voltage drop across the discharging capacitor.

24.4.2.2 An Example of Variation of Generalization

To develop a generalized understanding, the learner needs to identify a discerned aspect (in this case, average rate of change) in different situations, settings and contexts. In the model development sequence, the aspect of generalisation is manifested in the learners being exposed to the three contexts of (a) enacted and animated motion along a straight path; (b) how light intensity from a light source varies with the distance from the source; and (c) how the voltage across the discharging capacitor varies with time. In all these contexts, the learners used their models of average rate of change to make sense of the changing phenomena in each of these situations, by applying, extending, connecting, and further developing their models of average rate of change.

24.5 Discussion and Conclusions

Our discussion of variation theory and model development sequences suggests that the point of departure for any instructional unit should be grounded in the learners' experiences. Model eliciting activities provide the learners with opportunities to begin with their experiences and express their thinking using the language and representations they have at hand, thus making public the elements of a shared space of learning. By systematically using variation of contrast and separation in model exploration activities, this shared space of learning can then be modified and developed as new critical features of the model structure are discerned and explored. Explicitly attending to different patterns of variation opened up the possibilities for the learners to extend the space of learning and deepening their mathematical learning. The collectively created shared spaces of learning then support the learners in generalizing and fusing multiple aspects of more complex objects of learning.

The tentative principles for designing model exploration activities and model application activities we have presented are both theoretically grounded and supported by previous and on-going empirical research. Taken together, these tentative design principles potentially provide the foundation of a framework for a pedagogy of modelling that is both theoretically and empirically driven. One implication of this work is, that for learners to develop useful mathematical knowledge and modelling competencies, we need to move beyond sporadically implemented islands of isolated good modelling tasks (Blum and Niss 1991), and to focus research on the design and implementation of sequences of modelling activities. However, implementing model development sequences in schools provides challenges for the teacher, especially if different types of variation are to be an integral part of what drives the learning. Hence, studies focusing on how to characterize productive teaching practices drawing on both model development sequences (and units of instruction that move beyond single modelling activities) and variation theory are needed.

Model development sequences can be seen as a general framework for designing and implementing learning activities that allows extensions by drawing on and connecting to other theoretical frameworks. In this context, the variation theoretical framework can be applied in a self-similar way to the model development sequence at different levels. Foremost, variation theory can inform and shed light on the sequence of modelling tasks taken as a whole where the different types of modelling tasks (the MEA, MXAs, and MAAs) are regarded as the constitutive parts of a model development sequence. The individual tasks in the sequence all address special aspects of the object of learning. The development and ability to productively apply and modify a given model in different situations and settings, is the general aspect of the object of learning, whereby learners develop the capability to have a dynamic and flexible understanding of the model in question. Another possibility variation theory offers, is to take the development of the learners' modelling competences as the general aspect of learning and the specific aspect to be the development and the learners' ability to apply a particular model in different situations and settings. The latter application of variation theory, as well as the synergy between applying variation theory at the two levels of model development sequences, are challenging endeavours, but nevertheless seems a promising line for future research.

References

Ärlebäck, J. B., Doerr, H. M., & O'Neil, A. H. (2013). A modelling perspective on interpreting rates of change in context. *Mathematical Thinking and Learning, 15*(4), 314–336.

Barquero, B., Bosch, M., & Gascón, J. (2007). Using research and study courses for teaching mathematical modelling at university level. In D. Pitta-Pantazi & G. Philippou (Eds.), *Proceedings of the fifth congress of the European Society for Research in Mathematics Education* (pp. 2050–2059). Larnaca: University of Cyprus.

Barquero, B., Serrano, L., & Serrano, V. (2013). Creating necessary conditions for mathematical modelling at university level. In B. Ubuz, C. Haser & M. A. Mariotti (Eds.), *Proceedings of the eighth congress of the European Society for Research in Mathematics Education* (pp. 950–959). Antalya: Middle East Technical University.

Blum, W., & Niss, M. (1991). Applied mathematical problem solving, modelling, applications, and links to other subjects: State, trends and issues in mathematics instruction. *Educational Studies in Mathematics, 22*(1), 37–68.

Chevallard, Y. (2004, May). Vers une didactique de la co disciplinarité. Notes sur une nouvelle épistémologie scolaire. Paper presented at Journées De Didactique Comparée, Lyon. Retrieved from yves.chevallard.free.fr/spip/spip/article.ph3?id_article = 45

Chevallard, Y. (2006). Steps towards a new epistemology in mathematics education. In M. Bosch (Ed.), *Proceedings of the IVth congress of the European Society for Research in Mathematics Education* (pp. 22–30). Barcelona: Universitat Ramon Llull Editions.

Diefes-Dux, H. A., Hjalmarson, M. A., Zawojewski, J. S., & Bowman, K. J. (2006). Quantifying aluminum crystal size part 1: The model-eliciting activity. *Journal of STEM Education: Innovations and Research, 7*(1&2), 51–63.

Dominguez, A., Zavala, G., & Alanis, J. A. (2013). *Integrated physics and math course for engineering students: A first experience*. Proceedings of the American Society for Engineering Education Annual Conference and Exposition, Atlanta.

Dominguez, A., de la Garza, J., & Zavala, G. (2015). Models and modelling in an integrated physics and mathematics course. In G. A. Stillman, W. Blum, & M. S. Biembengut (Eds.),

Mathematical modelling in education research and practice: Cultural, social and cognitive influences (pp. 513–522). Cham: Springer.

Galbraith, P. L., & Clatworthy, N. J. (1990). Beyond standard models – Meeting the challenge of modelling. *Educational Studies in Mathematics, 21*(2), 137–163.

Iversen, S. M., & Larson, C. J. (2006). Simple thinking using complex math vs. complex thinking using simple math – A study using model eliciting activities to compare students' abilities in standardized tests to their modelling abilities. *ZDM, 38*(3), 281–292.

Kaiser, G., & Sriraman, B. (2006). A global survey of international perspectives on modelling in mathematics education. *ZDM, 38*(3), 302–310.

Kaput, J., & Roschelle, J. (1996). *SimCalc: MathWorlds* [Computer software].

Ko, P. Y., & Marton, F. (2004). Variation and the secret of the virtuoso. In F. Marton & A. B. M. Tsui (Eds.), *Classroom discourse and the space of learning* (pp. 43–62). Mahwah: Erlbaum.

Lesh, R. A., & Doerr, H. M. (2003a). Foundations of a models and modeling perspective on mathematics teaching, learning, and problem solving. In R. A. Lesh & H. M. Doerr (Eds.), *Beyond constructivism: Models and modeling perspectives on mathematics problem solving, learning, and teaching* (pp. 3–33). Mahwah: Erlbaum.

Lesh, R. A., & Doerr, H. M. (Eds.). (2003b). *Beyond constructivism: Models and modeling perspectives on mathematics problem solving, learning, and teaching*. Mahwah: Erlbaum.

Lesh, R. A., & Zawojewski, J. S. (2007). Problem solving and modeling. In F. Lester (Ed.), *Second handbook of research on mathematics teaching and learning* (pp. 763–804). Greenwich: Information Age Publishing.

Lesh, R. A., Hoover, M., Hole, B., Kelly, A. E., & Post, T. (2000). Principles for developing thought-revealing activities for students and teachers. In A. E. Kelly & R. A. Lesh (Eds.), *Handbook of research design in mathematics and science education* (pp. 591–645). Mahwah: Erlbaum.

Lesh, R. A., Cramer, K., Doerr, H. M., Post, T., & Zawojewski, J. S. (2003). Model development sequences. In R. A. Lesh & H. M. Doerr (Eds.), *Beyond constructivism: Models and modeling perspectives on mathematics problem solving, learning, and teaching* (pp. 35–58). Mahwah: Erlbaum.

Lo, M. L. (2012). *Variation theory and the improvement of teaching and learning* (Gothenburg studies in educational sciences 323). Göteborgs: Acta Universitatis.

Marton, F. (1994). Phenomenography. In T. Husén & T. N. Postlethwaite (Eds.), *The international encyclopedia of education* (2nd ed., pp. 4424–4429). Oxford: Pergamon.

Marton, F., & Booth, S. (1997). *Learning and awareness*. Mahwah: Erlbaum.

Marton, F., & Tsui, A. (2004). *Classroom discourse and the space of learning*. Mahwah: Erlbaum.

Marton, F., Runesson, U., & Tsui, A. B. M. (2004). The space of learning. In F. Marton & A. B. M. Tsui (Eds.), *Classroom discourse and the space of learning* (pp. 3–40). Mahwah: Erlbaum.

Niss, M., Blum, W., & Galbraith, P. (2007). Introduction. In W. Blum, P. Galbraith, H. W. Henn, & M. Niss (Eds.), *Modelling and applications in mathematics education. The 14th ICMI study* (pp. 3–32). New York: Springer.

Runesson, U. (2005). Beyond discourse and interaction. Variation: A critical aspect for teaching and learning mathematics. *Cambridge Journal of Education, 35*(1), 69–87. doi:10.1080/0305764042000332506.

Yoon, C., Dreyfus, T., & Thomas, M. O. J. (2010). How high is the tramping track? Mathematising and applying in a calculus model-eliciting activity. *Mathematics Education Research Journal, 22*(2), 141–157.

Chapter 25
The Possibility of Interdisciplinary Integration Through Mathematical Modelling of Optical Phenomena

Jennifer Valleriano Barboza, Luana Tais Bassani, Luciano Lewandoski Alvarenga, and Lucilaine Goin Abitante

Abstract Currently, several methodologies emerge with the intent of contributing to meaningful learning, among which mathematical modelling stands out, making it possible to relate mathematics to everyday problem situations and other disciplines. An extension program was developed using this methodology to approach optical phenomena in an interdisciplinary teaching process for mathematics and physics. The target audience was teachers and future teachers of these disciplines and the purpose was to promote practices that contribute to their training, encouraging them to organise activities with their high school students, integrating these disciplines. Therefore, four experiments were conducted involving physical and mathematical concepts, allowing the creation of generic models and practices that promote meaningful learning.

J.V. Barboza (✉)
Instituto Federal Catarinense – Campus Concórdia, Rua Dr Maruri, Apto 303. Barrio Centro, Concórdia, SC 89700-000, Brazil
e-mail: jenni_sud@hotmail.com

L.T. Bassani
Instituto Federal Catarinense – Campus Concórdia, Rua Ledoino Berté, 09. Barrio Salete, Concórdia, SC 89700-000, Brazil
e-mail: luanataisbassani@gmail.com

L.L. Alvarenga
Instituto Federal Catarinense – Campus Concórdia, Rodovia SC 283, Km 08, Bairro Fragosos, Concórdia, SC 89700-000, Brazil
e-mail: Luciano.alvarenga@ifc-concordia.edu.br

L.G. Abitante
Instituto Federal Farroupilha – Campus São Borja, Rua Eligio Cavedon, 56. Vila Cinquentenário, Tuparendi, RS 98940-000, Brazil
e-mail: lucilaine@sr.iffarroupilha.edu.br

G.A. Stillman et al. (eds.), *Mathematical Modelling in Education Research and Practice*, International Perspectives on the Teaching and Learning of Mathematical Modelling, DOI 10.1007/978-3-319-18272-8_25

25.1 Theoretical Framework and Justification

Currently, one can see that many elementary education students demonstrate an aversion to the discipline of mathematics, that happens because of how the subject is introduced in schools, considering it is, at times, disconnected from reality or it is seen by students as something final and hardly understandable (Bassanezi 2002). A possibility to reverse this scenario in the educational environment would be developing a way to educate through commitment to connecting the reality surrounding individuals and their society (Bassanezi 2002). After all, the way of teaching and intellectualising students is related to their time of existence and surrounding society (Kuenzer 2007).

Based on this, it is necessary that teachers look for alternatives, with the objective of including mathematical instruments, which should be connected to the other areas of human knowledge (Bassanezi 2002). Besides, it is known that the learning process, when contextualized with other knowledge areas, makes it possible for students to mobilize competences and solve real-world practical problems, and, also, it stimulates their creativity and curiosity, a fact that may be assured considering that:

> Given the worldwide impending shortage of youngsters who are interested in mathematics and science it is necessary to discuss possible changes of mathematics education in school and tertiary education towards the inclusion of real world examples and the competencies to use mathematics to solve real world problems. (Kaiser and Stillman 2011, p. v)

On the other hand, several methodologies arise with the objective of contributing to meaningful learning, among which is the approach of making it possible to relate mathematics to everyday problem situations and to other disciplines (Scheffer et al. 2006), by promoting the learning of abstract concepts in many areas of knowledge, since, according to Bassanezi (2002), it is a process that connects theory to practice. By varying the pedagogical practice, the teacher can contribute to the teaching-learning process, if not the students might become distracted and lose their concentration and will to participate and learn (Braguim 2006).

Thus, it is possible to note the importance that teachers be in continual professional development, in order to follow pedagogical changes that may be introduced into the classroom. For this reason, it is possible to talk about initial, continuous and specialised graduation of teachers and still, emphasise continuous qualification, due to its importance, because the professional is part of on-going knowledge construction and professional development (Anfope cited in Nunes 2000).

Furthermore, it is also verified how important it is for the teacher to develop interdisciplinary activities. After all, as mathematics and physics are related to several human activities and natural phenomena, it is also convenient that they are related in the classroom, consequently altering the situation currently observed in several schools in Brazil, where the curriculum separates the teaching into disciplines (Couto 2011). When it comes to interdisciplinarity, the individuality of the subjects is maintained, but they are brought together by demonstrating the multiplicity of factors which influence reality and the necessary language in certain knowledge domains needed to be used (Brasil MEC 2002), which is supported by Jackson and Davis (2000, p. 49) who note that "for schools that understand the power of the big ideas for deepening the curriculum within disciplines, using that power to show connections across disciplines is a logical step". However, these are practices that require the dedication of those involved, since these practices are not built overnight, and the professional needs are of a reflexive consciousness (Fazenda 1994).

In accordance with Fazenda (1996), interdisciplinary work is compromised by society in general, and by considering the intentionality in which the participants propose to teach, and to interfere in their daily lives and, thus, produce knowledge. Knowledge is produced, after all, by the union of subjects and knowledge areas, and the learning process shapes a quality educational system, thus approximating the learned subjects to students' daily lives, ignoring the education fragmentation. Therefore, the curriculum should be integrated and interdisciplinary, so that, it is possible to reduce isolation and fragmentation, since it is necessary to recognize the constant knowledge construction, in an interdisciplinary and contextualized way, resulting from individual and collective actions from the participants (Veiga 2008).

In this context, one can also justify the use of teaching mathematical modelling, as it allows the expression of everyday problems in a mathematical language and finding their solution by interpreting while relating them to the real world (Bassanezi 2002; Biembengut and Hein 2011). Through mathematical modelling, the teacher becomes a mediator in the construction of knowledge and is able to facilitate the student in searching for solutions, modelling them mathematically, and creating mathematical models which represent the proposed situation (Scheffer et al. 2006). For Wessels (2014, p. 2),

> various research studies reported on in the literature point out the feasibility and success of a modelling approach in mathematics wherein rich, complex, open tasks are used to construct meaningful mathematical knowledge to prepare learners for everyday life, tertiary studies and their future careers.

In this context a project for extending the professional learning of teachers was developed using this methodology with an optical phenomena approach to teaching mathematics and physics. The project involved conducting a literature search on geometric optics and mathematical modelling, fabricating roadmaps of activities, developing and building experiments that enabled mathematical modelling by observing optical phenomena, as well as organising and developing workshops that led to the ongoing training of mathematics and physics teachers. This chapter reports how the workshop dynamics happened and what was learnt by the teachers and the participants.

25.2 Methodology

The extension program began in October 2012 and was developed through mathematical modelling when approaching optical phenomena in the teaching of mathematics and physics. The target audience was the teachers from these disciplines, from public and private schools from the city of Concórdia, Brazil, and surroundings.

To accomplish this, the workshops were organized around four experiments involving constructed concepts developed in physics related to the reflection of light, specifically angular association, translation and rotation of flat mirrors, as well as mathematical concepts and processes such as: recognition of variables; directly and inversely proportional magnitudes; angles; one angle bisector; tabulation and interpretation of data; graphical representation; understanding of mathematical functions and relationships, with the goal of creating generic models; and being able to view an application of limit of a function.

Some of these concepts were approached as an introduction and others were discussed by the participants while handling the experiments and interacting in groups. These discussions treated, at various times, hypotheses, findings and validation, which helped the participants know the phenomena and model them mathematically. At the end, participants created mathematical models and then judged their validity.

25.3 Workshop Details and Observations

From the achievements of the course, the materials used for the proposal are described in this section, as well as in the subsequent section where the analysis and considerations obtained are described.

25.3.1 Participants

The workshop was delivered on four different occasions. Initially, the target audience was the future teachers and teachers who were already working at public schools (Licentiate students in physics – Campus Concórdia). It was also implemented during the Academic week at the Rio do Sul Campus, with Licentiate Students in Mathematics and Physics, and it was proposed in the form of a short course at the International Congress of Mathematics Education (VI CIEM), which was conducted in October 2013. When thinking about applying this practice for basic school students, the researchers intended to offer the courses for the teachers, in a way that every teacher involved, after familiarising themselves with the proposed activities, would develop it with their students.

25.3.2 *Methods and Materials*

The proposed activities were implemented in groups of 5–10 people, in a way that each group received one of the experiments. Data collected for the evaluation reports involving the workshop, included discussions raised by the groups and photographs of the participants' scripts developed to obtain the proposed model.

25.3.2.1 Flat Mirror Movement

The phenomenon of flat mirror movement can be considered the easiest among the four phenomena considered in the workshop. In order to build it (see Fig. 25.1), it was necessary to use a mirror (20 cm × 20 cm × 4 mm), a wooden board marked with lines of measurement covering approximately 90 cm and three objects. Initially the participant is asked to manipulate the experiment –object/mirror, in order to perceive the existing relations between the distances from the mirror to the object and the virtual images obtained with and without displacement, which can be modified.

25.3.2.2 Combination of Flat Mirrors and Parallel Flat Mirrors

In order to approach the next phenomenon, an experiment was initially proposed involving the mirror combination, with two manipulated angle mirrors and after another one, related to the combination of two flat parallel mirrors (0° angles to each other).

The first experiment (Fig. 25.2) required two mirrors (one 40 cm × 40 cm × 4 mm and another 40 cm × 39 cm × 4 mm), a hinge, silicone gel, a circular wooden board with a diameter of 40 cm with an angle representation of 0° to 360° and a cut

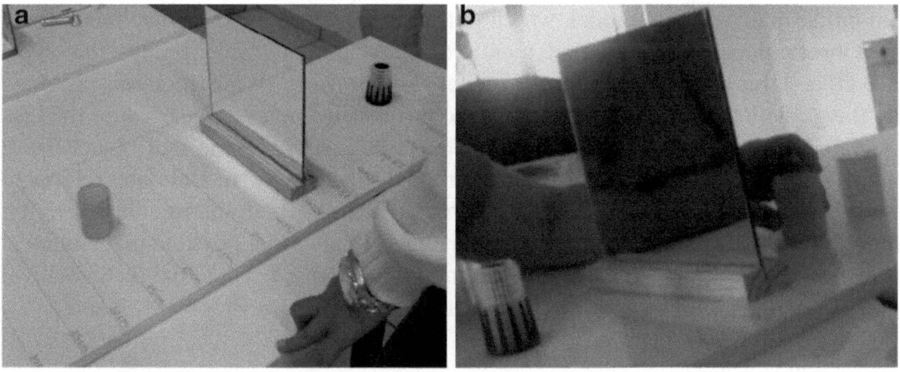

Fig. 25.1 Mirror movement (Abitante et al. 2013)

Fig. 25.2 Combination of flat manageable mirrors (Abitante et al. 2013)

of 20 cm forming a radius of a circle (to balance the mirror). With this experiment, participants manipulated it in order to change the angle and make the first observations and notes, making it possible to understand which quantities were varying (i.e., number of images and angle). When the mathematical model was obtained, the participants manipulated a similar experiment, which allows an interesting application (see Sect. 25.4 for details).

During the initial activities the participants conducted the experiment by altering the angle and making their first observations and notes, allowing them to notice what the varying magnitudes were (i.e., images and number of angles). The next task was to build a chart. From the following activities, it is intended that the student observes that the smaller the angle, the bigger will be the number of images. Thus, it is possible to verify the nature of the proportion involved (i.e., inverse) and the operation in which the variables are related (i.e., division). Through subsequent activities, it can be noticed that there are angles at which there are no images and the relation between the number of images (denoted as N) and the angle (α), comes to an initial mathematical model, which will probably need to be complemented.

Observations from the implementations reveal there were groups that have developed this project and built the model without using all the questions provided in the guideline booklets, being able to confirm their impressions, because the task solver is asked to observe that one of the visualizations is the real object itself, implying it will be subtracted from the model, making the model come closer to satisfying tests to be carried out in a subsequent task. In an intermediate activity, through the creation of a chart, the behaviour is related to a possible mathematical function that is to be expressed by the model discovered. Thus, the mathematical law is obtained which expresses the situation, namely $\mathrm{N} = \frac{360^\circ}{\alpha} - 1$.

Fig. 25.3 Combination of flat parallel mirrors (Abitante et al. 2013)

A final task refers to an infinitely small angle so as to tend to zero, allowing the introduction and application of the limit of a function concept starting with the second experiment (see Fig. 25.3) – the materials involved were similar (according to Fig. 25.2). Through this experiment the student observes that when the angle among the mirrors tends to zero, infinitely many images are formed, a phenomenon, which is justified by the model. After all, when zero is substituted for α, division is undefined. Thus, it is apparent that there is no division of a number by zero; instead there is the limit, or, when the angle tends to zero the number of images tends to be infinite, or in other words, when the independent variable is in the denominator and tends to zero, the limit of the function tends to infinity.

25.3.2.3 Mirror Rotation

For an experiment dealing with mirror rotation (see Fig. 25.4), a mirror (20 cm $\times$ 20 cm $\times$ 4 mm) shaped as a quadrant of the circle and another square-shaped, a laser source and mathematical tools such as protractor and ruler were used. In this experiment, it is possible to manipulate the incident ray and the reflected ray, as well as the mirror's rotation angle, which allows the participant to establish mathematical relations that address diverse mathematical concepts.

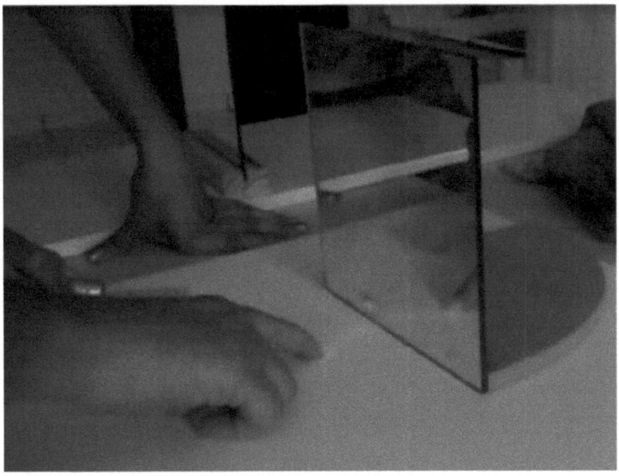

Fig. 25.4 Laser focusing on the mirror (Abitante et al. 2013)

25.4 Results Analysis

In this section, aspects such as participation in the process of manipulation of the experiment, considerations discussed in attempts to develop a model and conclusions drawn in each of the experiments are analysed, observed in different groups with whom the workshop has been used.

In the *Flat Mirror experiment*, participants were required to manipulate it. From this point, the observations were discussed; from this it is possible to see that the object and its image have a symmetric distance to the mirror, and that when changing the location in which the mirror is inserted, the distance between the two virtual images is twice the displacement from the mirror. These conjectures allowed the establishing of relations to represent them by using mathematical language. After analysing what was observed, the groups tried to use a new distance to validate their model. This guidance was aimed at ensuring that, in the mathematical model, the group had not operated with the sum of the number by itself, instead of doubling the distance, thus, ultimately testing the elaborated model. In this way, it was observed that the participant was able to obtain a mathematical law to adequately describe the model and predict other results, to manipulate object and /or mirror. Another observation was that the groups who started with this experiment were able to understand and develop the mathematical model of the subsequent experiments with greater ease than groups who started the other two experiments without this knowledge a priori

As for the *Combination of Flat Mirrors and Parallel Flat Mirrors experiment* it is possible to report that to obtain different data with the initial trials, data on the number of images and angle were organized in a table. For Bassanezi (2012) the modelling can start with an obtained data chart thus differing from Biembengut and Hein (2011)'s assertion, which indicates that a picture or constructing a graphical representation can already be considered as a model.

Next, the participants observed that the smaller the angle between the mirrors, the larger the number of images. Thus, it was verified that there was a relation considering that while the measure of the angle rises, the number of images lowers and there are certain angles where there is no image and no finite result for the operation in which the variables are related (division). When analysing the table, it was possible to relate the number of images and the angle obtaining an approximation of the mathematical model, something similar to $N = \frac{360^\circ}{\alpha}$, which needed a complementation so that a final validation was possible, since this mathematical law did not satisfy the data from the table. We noted that there were groups who developed this experiment and built the model making use of other perceptions attributed by the manipulable material, since one of the views is the actual object itself, which implies being subtracted and thus satisfies the table data when substituted into the model. Thus, the mathematical law that expresses the situation: $N = \frac{360^\circ}{\alpha} - 1$ was elaborated. In addition, there were occasions when there was time for the group to plot the graph of the data table, which also contributed to the analysis and validation of the developed model.

Finally, the participants were guided to observe what would happen if they used an angle as small as possible, with, the smallest angle obtained experimentally being approximately 20°, depending on the hinge between the mirrors. After the discussions, the second part of the experiment (Fig. 25.3) was introduced, which refers to a considerably small angle in order to tend to zero, allowing the introduction and application of the concept of limit of a function. With this, the participant visualized and concluded that, when the angle between the mirrors tends to zero, infinite images are formed, which is also justified by the model as mentioned before. This was considered one of the most important outcomes by the teachers and future teachers who participated.

For the *Mirror Rotation experiment* previous concepts were discussed in the group, such as what the incident ray, reflected, normal and bisector line were. Then, the participants were oriented to direct the laser onto the mirror, from the different angle of 90°; after all, with this angle the incident ray forms a straight line which coincides with the line segment formed by the reflected ray. This way, the groups recognized and traced the incident and reflected rays, as well as the normal line perpendicular to the surface. With this, the angle of incidence of the laser is maintained and movement is performed with the mirror so as to create a rotation angle, from which the rays are traced back and Normal. Subsequently, it was verified that, when prolonging the rays, some hypotheses about the existing relation between angles may be implied. When measuring them, it was concluded that the obtained angle equals the double of the rotation angle.

Moreover, it should be noted that participants appreciated the workshop, and expected good results when developing it with their students in the future. However, they also discussed how much it would take them to build the experiments, and considered if it was feasible using less material, thus lowering the cost. Participants were invited to use the teaching laboratory at the institution that funded the development of the project in order to make the materials available for teachers to use with their class.

25.5 Final Considerations

The construction of the optical experiments was intended to bring knowledge areas, such as mathematics and physics together and provide in-service and future teachers opportunities for activities that join these two areas. By handling the experiments, the participants were able to define the problems, elaborate the hypothesis, deduce and validate mathematical models in accordance with the objectives of mathematical modelling, outlined by Silva and Almeida (2005). Through these activities it was noticed that the mathematical models have great importance in the teaching of physics, because participants understood the theory through practice. Furthermore, it was possible to notice that in order to elaborate physical experiments, it was necessary to understand the underlying mathematical concepts. With respect to this, Campos (2000) states that these sciences complement each other, so that the mathematics enables a consistent physics, which is not used only to formulate models.

In developing the workshops it was noted that both the insertion of mathematical modelling as an interdisciplinary process and the use of manipulative experiments contributed to the learning of mathematical and physical concepts involved and stimulated the interest of the participants. It was found that the interdisciplinary proposal approached both disciplines while it allowed the students to expand their knowledge and broaden their vision on everyday situations. As much as some proposals – as the interdisciplinary ones or the ones involving mathematical modelling – are able to challenge teaching practice, since they take time, dedication and the collaboration of all the people involved, it is up to the teacher to accept the challenge and become involved, to facilitate the student's understanding while facing problem situations and increasing the possibilities of understanding physical and mathematical concepts through mathematical models.

Through the first experiment, it was observed that participants often used the manipulation of the mirror and the object to validate their results, find and establish relationships. Moreover, it was possible to realize that the model is easy to understand, making it possible for the teacher to perform this activity with their students. The second experiment improved the participants' mathematical knowledge, because it allowed them to better comprehend mathematical concepts that are complex, such as the limit of a function. For this activity, the discussion generated made it possible for participants, after group discussion, to establish relations that, without the experiment, would probably not have been perceived. With the third experiment, it was possible to perceive that physics concepts and basic mathematics may be understood by students, such as definitions of angles, reflection, and refraction, amongst others.

In general, it was observed that the proposed interdisciplinary activities provided participants with the application of physical and mathematical concepts, which helped them analyse aspects, without which, the experiment would be difficult to understand, besides stimulating the participant group to discuss possible scenarios and solutions in order to obtain the appropriate mathematical model for each situation. Finally, it would be possible to complement the experiments by using the laser in an aquarium with water, since it is possible in this context for a teacher to approach the trigonometric relationships through the phenomenon of refraction of light.

Acknowledgement Thanks to teachers and students involved during project development including Alexssandra Pasuch for her contributions and IFC-Concórdia for financial aid.

References

Abitante, L. G., Barboza, J. V., Bassani, L. T., Lewandoski, L. A., & Pasuch, A. (2013). *Modelagem Matemática de fenômenos ópticos*: *relato de experiência interdisciplinar entre matemática e física*. Anais do VIII Conferência Nacional sobre Modelagem na Educação Matemática – VIII CNMEM, Santa Maria.

Bassanezi, R. C. (2002). *Ensino aprendizagem com modelagem matemática: Uma nova estratégia*. São Paulo: Contexto.

Bassanezi, R. C. (2012). *Temas e modelos*. Campinas: UFABC.

Biembengut, M. S., & Hein, N. (2011). *Modelagem Matemática no ensino*. São Paulo: Contexto.

Braguim, R. A. (2006). *Abordagens metodológicas no ensino da Matemática: perímetros e áreas*. Universidade Cruzeiro do Sul. http://www.educadores.diaadia.pr.gov.br/arquivos/File/2010/artigos_teses/MATEMATICA/Dissertacao_Braguim.pdf. Accessed 2 June 2013.

Brasil, M. E. C. (2002). *Parâmetros curriculares nacionais: Ensino médio*. Brasília: MEC/SEF.

Campos, C. R. (2000). *O ensino da Matemática e da Física numa perspectiva integracionista*. PUC. http://www.pucsp.br/pos/edmat/ma/dissertacao/celso_campos.pdf. Accessed 15 Sept 2013.

Couto, R. M. S. (2011). Fragmentação do conhecimento ou interdisciplinaridade: ainda um dilema contemporâneo? *Revistafaac*. http://www2.faac.unesp.br/revistafaac/index.php/revista/article/view/34. Accessed 2 June 2013.

Fazenda, I. (1994). *Interdisciplinaridade: história, teoria e pesquisa*. Campinas: Papirus.

Fazenda, I. (1996). *Práticas interdisciplinares na escola* (3rd ed.). São Paulo: Cortez.

Jackson, A. W., & Davis, G. A. (2000). *Turning points 2000: Educating adolescents in the 21st century*. New York: Teachers College Press.

Kaiser, G., & Stillman, G. Series preface. (2011). In Kaiser, G., Blum W., Borromeo Ferri, R., & Stillman, G. *Trends in teaching and learning mathematical modeling* (pp. v–vi). New York: Springer.

Kuenzer, A. Z. (2007). *Ensino médio profissional: As políticas do estado neoliberal* (4th ed.). São Paulo: Cortez.

Nunes, C. S. C. (2000). *Os sentidos da formação contínua. O mundo do trabalho e a formação dos professores no Brasil*. Unicamp. http://www.bibliotecadigital.unicamp.br/document/?code=vtls000215996. Accessed 15 Sept 2013.

Sheffer, N. F., Nava, A. L., Aimi, S., Dallazen, A. B., Andretta, F. C., & Corrêa, R. M. (2006). *Matemática e tecnologias: Modelagem matemática*. Erechim: Edifapes.

Silva, A. G. O., & Almeida, L. M. W. (2005). *Modelagem matemática no contexto da matemática e cidadania*. http://www.sbempaulista.org.br/epem/anais/Comunicacoes_ Orais/. Accessed 2 June 2013.

Veiga, I. P. A. (2008). Projeto político-pedagógico: novas trilhas para a escola. In I. P. A. Veiga & M. F. Fonseca (Eds.), *As dimensões do projeto político-pedagógico* (6th ed.). Campinas: Papirus.

Wessels, H. (2014). Levels of mathematical creativity in model-eliciting activities. *Journal of Mathematical Modelling and Application, 1*(9), 22–40.

Chapter 26
Activation of Student Prior Knowledge to Build Linear Models in the Context of Modelling Pre-paid Electricity Consumption

José Luis Bossio Vélez, Sandra Milena Londoño Orrego, and Carlos Mario Jaramillo López

Abstract This chapter presents partial results from a research project developed through a case study of ninth grade students attending a country secondary school in Colombia. The purpose of the study was to analyze mathematical modelling activity in the classroom addressing the students' social and family situations. From the products and interpretations of some students in the creation of particular linear models in a school context, we could observe the reality the students live, brought to the classroom through the activities of the modelling process and their arguments built around problem solving. The study shows that it is not necessary for the teacher to suggest or make up real-world problems for modelling situations for students to work on in class as the students pose appropriate situations to model in the classroom from their our prior everyday experiences.

26.1 Introduction

This case study is an analysis of the mathematical modelling process applied by secondary school students to situations from their sociocultural contexts addressed in their classrooms. The ninth grade students activated prior knowledge of their

J.L.B. Vélez (✉)
Mathematics, Institución Educativa el Dos, Universidad de Antioquia, A.A 1226, Medellín, Antioquia, Colombia
e-mail: copiando2010@gmail.com

S.M.L. Orrego
Institución Educativa el Dos, Universidad de Antioquia, A.A 1226 Medellín, Colombia
e-mail: samyjdam@gmail.com

C.M.J. López
Instituto de matemáticas, Universidad de Antioquia, A.A 1226 Medellín, Colombia
e-mail: carlos.jaramillo1@udea.edu.co

G.A. Stillman et al. (eds.), *Mathematical Modelling in Education Research and Practice*, International Perspectives on the Teaching and Learning of Mathematical Modelling, DOI 10.1007/978-3-319-18272-8_26

particular real-world contexts to build their own linear models. Likewise, we consider the discussion about ways to facilitate students to describe school mathematics with examples from their life experience, through a mathematical modelling process from an educational perspective.

26.2 Problem

The Ministry of National Education (MEN), the institution responsible for the regulation of education in Colombia, points out that:

> The learning of mathematics begins with informal mathematics from students in real-world contexts and from school every day. It is necessary to mix the learning process for the construction of contexts and situations that allow moving towards formal mathematics. (MEN 2006, p. 78)

We understand this as an effort to recognize the reality of students, which involves semantic and conceptual relationships that arise from their particular cultural and historical context, so that new social purposes may be added to school mathematics education. On the other hand, it is also a starting point for students to recognize definitions, properties of mathematical objects, axioms, theorems and algorithmic procedures from formal mathematics.

In the *Lineamientos Curriculares para la Educación Matemática* (Curricular Guidelines for Mathematics Education, MEN 1998), the teaching of school mathematics is presented as follows:

- *Basic knowledge*: numerical thinking and numerical systems, spatial thinking and geometrical systems, metric thinking and measurement systems; probalistic thinking and data systems; and "variational thinking and algebraic and analytical systems"
- *Processes*: reasoning, problem posing and problem solving, modelling, communication; development, verification and execution of procedures
- *Contexts*: from mathematics itself, in everyday life and other sciences.

Accordingly, we think it is important to consider mathematical modelling aimed at supporting variational thinking, and algebraic and analytical systems, in relation to situations in the context of the everyday life of the student. Taking the example of mathematical modelling in Colombia, it is still in a preliminary stage, even though it is one of the processes for the mathematical classroom advocated by MEN since 1998 (Villa-Ochoa et al. 2009).

In our study, students developed a mathematical modelling process in the classroom to generate linear models from an everyday experience. The participants in the study lived in a rural area of the region of Urabá. Therefore, the question that guided our research was:

How does a group of ninth grade students generate linear models in a process of mathematical modelling from a situation in context?

In order to answer and analyse this question, from the particular everyday prior experiences of the student, we observed aspects such as the use of context, building of linear models and their use. This enabled us to describe how social and cultural factors engage students to use school mathematics.

26.3 Theoretical Considerations

In our study, theoretical referents from Niss et al. (2007) were considered regarding connections between the real world and mathematics through a mathematical modelling process, from the educational perspective related with the teaching and learning of school mathematics; although there are other perspectives for mathematical modelling as outlined in a classification scheme developed by Blomhøj (2009). An educational perspective will allow us to describe and analyse the elements involved in the process of development of mathematical models in a certain situation in the context of the student's everyday life, as well as the utilization of mathematical concepts, the utilization of the models (i.e., process) and the context of the situation (i.e., everyday life). It will be possible then for us to analyze and describe the elements involved in that process applied to a particular experience of their everyday context. Likewise, a space in the classroom to discuss the situation concerned with the students consistent with a process of mathematical modelling is established from this perspective.

Regarding the approach to modelling as a process, in teaching and learning mathematics, we have taken the mathematical model as a structure appropriate to a context-specific situation based on considerations by Niss et al. (2007): "Sometimes a model will be an idiosyncratic, ad hoc building, but often it will be a variation of a standard type (e.g. inverse proportionality, linear, exponential or logistic growth, the harmonic oscillator, a Poisson probability process, etc.)" (p. 10). Accordingly, a mathematical model made by the student, according to the variables of the situation may not necessarily be related with a mathematical concept (i.e., it is idiosyncratic), but with the use of a standard model for the student to model a phenomenon. In this sense, a context-specific situation described in context is referred to by Niss et al. (2007) as being of the real world meaning "everything that is to do with nature, society or culture, including everyday life, as well as school and university subjects or scientific and scholarly disciplines different from mathematics" (p. 8).

We recognize that by addressing social and cultural issues in the modelling process, the reality perceived by the teacher may limit the different mathematical relations built and linked to the school context of the students. In this case, we have taken on board considerations of Villa-Ochoa and Jaramillo (2011) who refer to the need for the teacher to develop a *sense of reality* as a tool to make easier the interaction between socio-cultural contexts and school mathematics through mathematical modelling when confronting students with the reality of identifying and manipulating data, and the simplification and abstraction of quantities and variables in order to build models for use.

26.4 Methodological Design

The nature of this research was a qualitative case study. In this sense, Stake (1999) describes such research as the study of the particularity and complexity of a single case to get to understand the activity in important circumstances. This method enabled us to consolidate the description of a mathematical modelling process to create linear models from a situation in a context for three (3) students, whom we have given the pseudonyms: San, Ezel and Rita. At the time of the research, they were ninth grade students, between the ages of 14 and 15 years old, They were chosen because of the interest they showed in solving a problem regarding their families' economy. These students had no prior experience in mathematical modelling processes. For the purpose of this chapter, however, we will focus on the description of only the process developed by Rita in the context of pre-paid energy, as San and Ezel addressed a situation in the context of banana crops, and there is no room to describe this other modelling. Therefore, we chose the situation in the context of pre-paid energy as an exemplar from practice in the Colombian context of a paradigmatic case (Freudenthal 1981) to argue the case for mathematical modelling examples arising naturally in the classroom as students activate prior knowledge from their everyday experiences.

The analysis of the data was performed according to considerations by Stake (1999), where he refers to the necessity for coding, categorization, and triangulation of data from interviews, direct observations and written documents. We isolated the categories oriented to issues related to the use of context, construction of linear models and their use, in order to build the themes of the final report.

Data were collected on eight occasions, lasting 2 h each, distributed over 3 months and supplemented with extracurricular activities, guided by the students themselves. The following elements were used as instruments to collect data: the written utterances to record the built models, recordings on tape as a means to keep a record of the responses to questions asked in class, and comments written by the teacher, who in turn, was a member of the research team.

26.5 Findings

26.5.1 Context of Pre-paid Energy: A Context in the Student's Life

The families of the Urabá region face the difficult situation of suspension of electric service for nonpayment. Therefore, the power company in the region (Empresas Públicas de Medellín – EPM) installed in their homes a prepaid electricity meter, a device that allows purchasing of a number of kilowatts hours of electricity in advance that lasts until the credit is expended. This is conducted through cards ranging from COP $3,000 pesos upwards.

From the situation described above, we can see what is known as a *situation in context* according to Niss et al. (2007) when it comes to the *real world*, in which a student is involved in her own family context. She described the pre-paid energy situation as follows:

Rita: Well, I think the most important thing came up from a question my aunt asked me. The thing is, she sells ice, but she realized that she doesn't make enough money out of the sales to pay the electricity bills. She has a modem at home to charge credit. So she asked me 'How come I ran out of credit so soon?' The sales of ice weren't enough to pay the credits.

We can see that the problem stems from excessive power consumption, while trying to find additional income to improve the economy of the family through the sale of ice buckets. Therefore, Rita's aunt poses the problem from her needs and her business, but in this case, the student takes it as the starting point of the modelling process. *How come I ran out of credit so soon?* In this question, we find issues that are involved in matters, unknown to the student, regarding the subject of electrical energy that will serve them as a guide to describe, explain, understand or build parts of the world (Blum et al. 2002).

With this in mind, we can visualize the conditions that led to the posing of the problem for the modelling process, so the fact the student knows of this credit system (that the credit reload grants through money) is important for this process. It allowed her to acquire the meanings and to identify the abstraction of quantities involved in the situation. This can be noted in the following dialogue:

Teacher: How does the reload of power credits work?
Rita: If you pay $5,000 for a reload of power that contains 30 kWh, this unloads as the power is consumed. When it reaches zero, the power is gone.

Up to this point, we understand the need for Rita's family to understand the reason for the instability of the household economy when they question the excessive power consumption. Subsequently, we observe how Rita, being the one responsible for managing the purchase and use of the PIN (Personal Identification Number) that provides the power credit, comes to identify the variation in the context of power consumption of the electrical appliances and consider using letters to represent variables and build models.

26.5.2 Building of Linear Models Adjusted to the Context of Pre-paid Energy

Immediately after Rita understands the relation between the value of the recharge and the consumption in kilowatts hour, based on her experience, she opportunistically decides to use a digital device (the modem) to observe the average power consumption in kilowatt hours for each appliance. Thus, she experiences variations

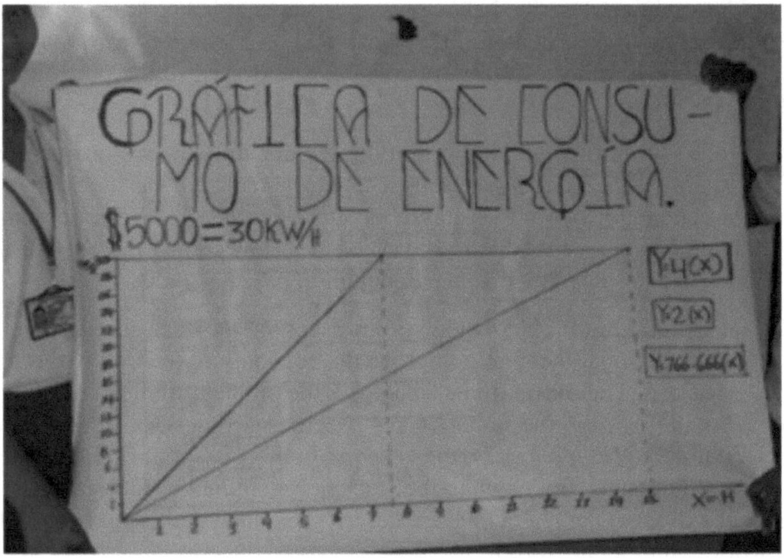

Fig. 26.1 Graphical representation of power consumption of the freezer and fridge

in power consumption. She made this measurement by plugging only one device in at a time in order to observe the average amount of consumption per hour of the freezer and the refrigerator, and then compare them by representing them as two straight lines in the Cartesian plane (see Fig. 26.1), and using the letters *x* and *y* to represent the variables. This is evident in the following:

Teacher: Why should you use the letters *x* and *y*?

Rita: Because we can relate cause and effect, then *x* and *y* are the common points that lead to an approximation. So I used *y* and *x*, and a line going up, and what happens? On one side, I put consumption per hour, and on the other side, I got the amount spent. For example, the hours are represented by *x* and the consumption in kilowatt hours [are represented] by *y*. Well, so, if the freezer is spending 4 kWh each hour, in 2 h that'd be 8 kWh and so on.

Rita just described *cause and effect* as a way to relate the different changes in their everyday context. We can understand this from Sierpinska's (1992) perspective, who postulates *cause and function* as a human effort for trying to explain the changes in the world. At the same time, use of letters in functional relations (Trigueros et al. 2000) to generate a correspondence between the context of the situation and mathematics is observed. "x = H" "kwh = Y" are written next to the axes. The representation of the straight lines can be seen in the graph shown in Fig. 26.1.

By representing and describing the use of the freezer in relation to the use of the refrigerator with straight lines, Rita was able to interpret the slopes of the lines in the plane in relation to the variation observed in the prepaid electricity meter (i.e., the Modem), and conclude that the higher the slope the greater the power consumption. This action shows how a mathematical concept can be linked to everyday issues in a school student's life context, in this case the slope of a straight line in the plane. The student describes it like this:

Rita: The consumption agrees with the inclination of the line, so the line of the freezer slopes closer to the y axis than the line of the fridge, this shows us that the freezer consumes more power than the fridge.

By identifying variations in power consumption from the graph, Rita was able to build linear algebraic expressions to describe the power consumption of both the refrigerator and the freezer, and another expression to calculate the consumption cost of each appliance. This is related to what Niss et al. (2007) describe as the use of standard models to explain a situation in the real world. This is obvious in the following descriptions.

Rita: Well, in this graphic (Fig. 26.1), I am showing the freezer and the fridge consumption. In order to calculate the consumption of the freezer, I had to make this formula: y equals to 4 kWh ($y = 4x$). The formula for the fridge is y equals 2 kWh ($y = 2x$). Here we can see that the freezer consumes 4 kWh; in 2 h it consumes 8 kWh; in 3 h, 12 kWh... while the refrigerator consumes 2 kWh each hour, 4 kWh each 2 h.... and a [power] reload of \$5,000, containing 30 kWh equals \$166.66 pesos each ($y = 166{,}66x$).

From this, we can say that Rita uses linear functions to model the power consumption in both the refrigerator and the freezer. According to (Posada et al. 2006) "it is important to determine from the statement if the rate of change is implicit or explicit because, in both cases, if the rate of change is constant, then it can be associated with a linear function" (pp. 140–141). In this sense, linear models are generated when the student associates the constant rate of change with the average hourly power consumption of both the freezer and refrigerator.

26.5.3 Use of Models to Make the Household Economy More Stable

The use of linear models by the student served the purpose of calculating the power consumption in kilowatt/hours through models for the freezer ($y = 4x$) and the refrigerator ($y = 2x$). With these results, calculations were made with the model $y = 166{,}66x$ to find the cost of electricity. These calculations were validated from the linear graphical representations, approaching the mathematical concept called an equilibrium point (see Fig. 26.1, where the two rising lines intersect the line $y = 5{,}000$ parallel to the x axis). Rita describes the results of the use of the linear models as follows:

Rita: For a reload of \$5,000, which contains 30 kW, the freezer would use them up in only 7 h and a half, while the fridge would take 15 h.

According to the above, we can understand how the student uses *variational thinking* and *algebraic and analytic systems* when integrating the situation in context, graphical representations and linear models, with the purpose of building the arguments needed to solve the problem. This observation is supported from the following commentary:

Rita: As we could see, the freezer spent more [power]. As we had the fridge and as the sales are of approximately five trays a day, we could put them in the fridge, as its upper side contains a freezer itself. Regarding the freezer, well, we wouldn't use it anymore, besides, it was very old anyways; I told her to sell it and so she would use only one electrical appliance. I mean, there would be only the fridge left because before there were two.

We can observe how Rita solved the problem, she inferred from her graphical representation which appliance generated a high power consumption (the freezer) and depleted the credit more quickly. At the same time, her aunt managed to keep selling ice trays (as a way to generate additional income for the household) with only the fridge to keep the trays frozen and stopped using the freezer.

26.6 Discussion

In social and cultural contexts of rural communities, there are situations in which students can develop a process of mathematical modelling in the classroom, due to the richness of the meanings, the relations built by the life experience of the students, and their particular economic and social conditions. This way, an educational perspective of mathematical modeling makes it possible, in the classroom, to lead to a process of school mathematics teaching and learning in order to transform the social and family life of students in these rural contexts. Thus, classroom activities are able to meet extra social objectives in addition to the mathematical formation purposes of the curriculum as intended by the Ministry of National Education (MEN 2006). In particular, by starting with the students' real-world contexts giving opportunities for prior experiential knowledge to be activated in school, it is possible to link the learning process *with going forward to formal mathematics* (MEN 2006, p. 78) through use of mathematical modelling opportunities in class.

Some of the concerns of the students' families regarding their household economic difficulties come through questions of intellectual character, which are taken as challenges by the students, motivating them to find out possible solutions. These solutions become significant to initiate a process of mathematical modelling with students in the classroom, so it is not necessary for the teacher to pose a problem for a modelling process, far from the social context of students.

Some technological devices that relate to a situation in context are used by students with a certain level of knowledge as tools to observe changes and make measurements that would not be possible otherwise. The mathematics teacher does not always understand the use of these objects. In this sense, the teacher is the one who should make an effort to recognize the use of these devices in order to support strategically a possible modelling process suggested by students.

26.7 Conclusions

The mathematical modelling process addressed in this study, under the guidance of the teacher, was developed in the classroom from an educational perspective with students from a rural context. It is noteworthy that the problem was posed by the students in order to benefit the household economy issues of their everyday lives. From this perspective, we see that it is not necessary for teachers to suggest or imagine real-world problems to pose to students, but rather, the same students are able to engage in the development of a modelling process based on their life experiences, that is, it is necessary for the teacher to recognize the sense of reality (Villa-Ochoa and Jaramillo 2011) of the school context situations when students approach their social and cultural environment.

The modelling process made by the students involved technological tools of their environment that consisted in capturing the variation in the consumption of two electrical appliances and then graphing them, by building linear models, using the respective variables and their relations. From these models and their graphical representations, they were able to find the approximate cost of the power consumption of both the refrigerator and freezer, deducting which of these appliances exhausted the power sooner. This allowed them to construct necessary arguments to resolve the problem: the need to stabilize the household economy. Therefore, this strategy of linear models building from a situation of the students' own social and cultural context creates a solution to foster societal life, by enabling the student to relate their everyday life experience and the use of school mathematics in the classroom.

References

Blomhøj, M. (2009). Different perspectives in research on the teaching and learning of mathematical modelling. In M. Blomhøj, & S. Carreira (Eds.), *Mathematical applications and modelling in the teaching and learning of mathematics* (IMFUFA Technical Report No. 461, pp. 1–18). Roskilde: Department of Science, Systems and Models, Roskilde University Centre.

Blum, W., et al. (2002). ICMI Study 14: Applications and modelling in mathematics education – discussion document. *Educational Studies in Mathematics, 51*(1/2), 149–171.

Freudenthal, H. (1981). Major problems of mathematics education. *Educational Studies of Mathematics, 12*(2), 133–150.

MEN. (1998). *Lineamientos Curriculares: Matemáticas*. Bogota: Ministerio de Educacion.

MEN. (2006). *Estandares Básicos de Competencia*. Bogota: Ministerio de Educación Nacional.

Niss, M., Blum, W., & Galbraith, P. (2007). Introduction. In W. Blum, P. Galbraith, H.-W. Henn, & M. Niss (Eds.), *Modelling and applications in mathematics education. The 14th ICMI Study* (pp. 3–32). New York: Springer.

Posada, F. A., et al. (2006). In Primera (Ed.), *Pensamiento Variacional y Razonamiento Algebraico*. Medellín: Artes y Letras Ltda.

Sierpinska, A. (1992). In E. Dubinsky & G. Harel (Eds.), *The concept of function: Aspects of epistemology and pedagogy*. Washington, DC: Mathematical Association of America.

Stake, R. (1999). *Investigación con estudio de casos*. Madrid: Morata.

Trigueros, M., Ursini, S., & Lozano, D. (2000). La conceptuaclización de la variable de la Enseñanza en los medios de comunicación. *Educación Matemática, 12*, 27–48.

Villa-Ochoa, J. A., & Jaramillo, C. M. (2011). Sense of reality through mathematical modelling. In G. Kaiser, W. Blum, R. Borromeo Ferri, & G. Stillman (Eds.), *Trends in teaching and learning mathematical modelling* (pp. 701–711). New York: Springer.

Villa-Ochoa, J. A., Bustamante, Q., Berrio, A. D., Osorio, C., & Ocampo, B. (2009). Sentido de Realidad y Modelación Matemáticas: el caso de Alberto. *ALEXANDRIA Revista de educação em Ciencia e Tecnologia, 2*(2), 159–180.

Chapter 27
Mathematical Modellers' Opinions on Mathematical Modelling in Upper Secondary Education

Peter Frejd

Abstract This chapter examines and discusses how professional mathematical modellers have learned about modelling as well as their opinions on the teaching of mathematical modelling in upper secondary education. An interview study showed that they developed most of their knowledge about mathematical modelling during their PhD studies and through their occupation by working with 'real modelling'. According to the interviewees mathematical modelling should be a part of mathematics education in upper secondary school, and in particular it should be more emphasized as a part of general education to develop students' critical awareness about how models are used in society. They also gave suggestions for approaches to teach modelling and examples of modelling problems to work with from their own workplace.

27.1 Introduction

Mathematical modelling is considered as a bridge between the mathematics learned and taught in schools and the mathematics used at the workplace (Sträßer et al. 2012). This view is also found in school mathematics curricula, in the section on the aim of the subject mathematics, as for example in Sweden where the subject syllabus for upper secondary school emphasizes, the use of mathematics in relation to workplace situations and to use investigating activities in an environment close to practice (Skolverket 2012). One such investigative activity is mathematical modelling, described as one of seven main teaching goals, so as to develop students' ability to "interpret a realistic situation and design a mathematical model, as well as use and assess a model's properties and limitations" (p. 2). These descriptions

P. Frejd (✉)
Department of Mathematics, Linköping University, SE-581, 81 Linköping, Sweden
e-mail: peter.frejd@liu.se

G.A. Stillman et al. (eds.), *Mathematical Modelling in Education Research and Practice*, International Perspectives on the Teaching and Learning of Mathematical Modelling, DOI 10.1007/978-3-319-18272-8_27

suggest the use of realistic modelling activities in mathematics classrooms with a relation to workplaces, at least if the modelling problem is chosen adequately.

However, even if the problems are chosen adequately there seems to be an accepted view among educational researchers in mathematics education that workplace mathematics is not identical to school mathematics. Workplace mathematics is more complex and strongly situation dependent. It also includes specific technologies, social, political and cultural dimensions that are not found in educational settings (e.g., Noss and Hoyles 1996; Wedege 2010). Mathematical modelling applied by different actors at the workplace also seems to be workplace specific and quite different from school situations (Mouwitz 2013). Frejd (2013) found that some professional modellers work in groups where the division of labour is specific and predefined (numerical analyst, meteorologist etc.) in contrast to school settings in Sweden where students spend much time on individual work with exercises from textbooks (Jablonka and Johansson 2010) with the goal to learn 'everything' about the modelling process. In addition, there are other aspects of mathematical modelling that appear in a limited way in Swedish mathematics classrooms, but are large parts of workplace practice, such as programming and that the consumer's (the company's) purpose of developing the model must be taken into consideration (Frejd 2013).

Assuming there is a gap between school mathematics and mathematics used in the workplace regarding mathematical modelling, it is of interest to chart and analyse how people who use mathematics in the workplace view mathematical modelling, especially professional mathematical modellers in different occupations. There is also a need to observe different types of modelling activities found in non-educational settings in search of potential links between the two practices. To this end, this chapter presents empirical research aiming to describe how professional mathematical modellers have learned mathematical modelling and their opinions on how it should be included in upper secondary school. The analysis has been guided by the following research questions:

- How do professional mathematical modellers describe their own learning of mathematical modelling?
- What opinions do professional mathematical modellers express in terms of goals of mathematics education in upper secondary school, goals of modelling in upper secondary school, suitable examples for use in secondary school, and mathematical modelling as a part of a general education?

Exploring and seeking answers to these research questions may contribute to development of new insights into pedagogy and curricula, links between school and the workplace, and how mathematical meanings are created in and out of school contexts as well as to informing curriculum developers and others about the role of modelling in the workplace.

27.2 Workplace Mathematics, Modelling and Modellers' Opinions

One goal of research on workplace mathematics is to explore how and what mathematics is used in specific professions (Noss and Hoyles 1996). Other goals concern the identification of similarities and discrepancies between what mathematics is needed in the workplace and what mathematics is taught in school (Triantafillou and Potari 2010), the analysis of communication between employers and visitors (Williams and Wake 2007), and the search for strategies that will improve a general curriculum that better prepares students for future work (Wake 2012).

One central part of workplace mathematics concerns the use of mathematical models and modelling. At the Educational Interfaces between Mathematics and Industry-ICMI study conference several research papers related to engineering and modelling were presented (see Damlamian et al. 2013). Other examples of research literature including modelling and the use of mathematical models in the workplace, focus on employers that make their working decisions based on mathematical models in technological artefacts, with input and output of numerical values, like bankers (Noss and Hoyles 1996), telecom technicians (Triantafillou and Potari 2010) and operators in a chemical plant (William and Wake 2007). A common finding in these studies is that the underlying mathematical structure of the models used is not considered. Despite the facts the mathematical models sometimes are hidden in technology and the linguistic conventions of representing mathematical models used in the workplace (formula, graph, table) (Triantafillou and Potari 2010; Williams and Wake 2007) differ from those in mathematics education, it is argued that mathematical models together with metaphors and gestures facilitate communication of mathematics between workers and clients (Williams and Wake 2007). One of the "principles for strategic curriculum design" that support workplace mathematics (Wake 2012, p. 1686) emphasises communication about development and validation of mathematical models. Other principles suggested by Wake (2012) are: to take mathematics in practice into account; facilitate activities that pay attention to technology; and, to let students critique mathematics used by others.

There are researchers who discuss implications for education based on results from their own empirical research (observations and interviews) of the working practice of professional modellers (e.g., Drakes 2012; Gainsburg 2003). However, not much research is explicitly focused on modellers' opinions on mathematical modelling in secondary mathematics education. Previous research has shown that modellers in the Netherlands are sceptical of the use of 'messy' modelling problems in secondary education (Spandaw 2011). They argued that modelling is too complex and time consuming for students and that modelling projects are too complicated for teachers to supervise. Instead the aim of secondary education should be to teach basic skills in mathematics (i.e., algebra and analysis).

27.3 Methodology

Research within the field of workplace mathematics should consider two closely linked approaches, a *general approach* and a *subjective approach* (Wedege 2010). A *general approach* focuses on (general) demands from the labour market and the society for "formal" (school) mathematical competencies needed in a workplace, whereas a *subjective approach* focuses on the workers' abilities and their (subjective) needs in their specific workplace. Salling Olesen (2008) addressed both these approaches in a heuristic theoretical model, which is suggested as a helpful research tool for investigating the dynamics of workplace learning.

The model in Fig. 27.1 illustrates a relation between the three components *the societal work process* (division of labour, type of tasks and work organisation), *the knowledge available* (discipline, craft, methods and skills used in a workplace), and *the subjective working experiences* (individual/collective life history and their subjectivities like values, norms, emotions, etc. that appear to be profession specific). In the centre of Fig. 27.1 the words learning, experiences, practices, identifications and defences illustrate "that learning in the workplace occurs in a specific interplay of experiences and practices, identification and defensive responses" (Salling Olesen 2008, p. 118). "A teacher in mathematics may say *modelling activities are time consuming* (learned by experience), *but we don't have the time we need the classroom time to prepare for the final tests* (learned through practice) and *in mathematics we use the notation y*" (learned through identification), *but in*

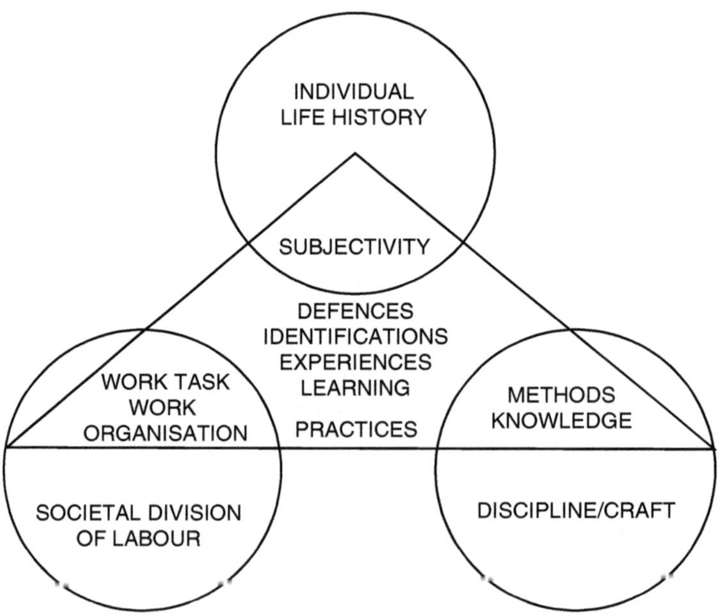

Fig. 27.1 A model to analyse workplace learning (Salling Olesen 2008, p. 119)

physics we use ÿ (learned through a defensive response). The suggested theoretical model is helpful for this chapter, since "the model pays particular attention to the cultural nature of the knowledge and skills with which a worker approaches a work task, whether they come from a scientific discipline, a craft, or just as the established knowledge in the field" (p. 118). The three components (*the societal work process, the knowledge available* and *the subjective working experiences*) may indicate the origin of the given opinions on how modelling should be taught and learned.

In this research project semi-structured interview questions were developed to pay attention to Salling Olesen's (2008) model (for further details see Frejd 2013). The interview questions addressing the research questions are:

1. Is mathematical modelling something that was a part of your education in school or something you learned in your vocation? How?
2. What are the goals of mathematics education in upper secondary school?
3. Is mathematical modelling something that should be brought up in mathematics education at upper secondary school? (if yes) How?
4. Do you have examples from your own practice that may be suitable to use in secondary education?
5. Modelling as a part of a general education (personal finance, global warming, etc.), should that be more emphasized in school and should that be a part of mathematics education?[1]

Nine mathematical modellers in different areas of expertise were invited and accepted to participate in this study. The following abbreviations are used in this chapter when referring to the modellers: climate modeller [CLI], military defence modeller [MD], modeller in physics [PHY], finance modeller [FIN], insurance modeller [INS], construction engineering modeller [CE], traffic simulations modeller [TS], biology modeller [BIO], and scheduling modeller [SCH]. Participants were a convenience sample as two persons were previously known to the authors, four others were either recommended by colleagues or interview participants to be invited and three were found on a web search. The interviews that were conducted and audio taped lasted from 40 min up to 90 min and were later transcribed, summarized and analysed based on the five questions together with the three components of Salling Olesen (2008).

All of the participants have at least a PhD degree in either mathematics, financial economics, physics, biology, or technology and all participants claimed during the interviews that mathematical modelling plays a major role in their profession. The participants' workplace problems were initially derived from reality with the aim to describe, predict, explain or create physical or social reality. However, they did not present coherent descriptions of what modelling means and how they work with modelling.

[1] The last part of question 5 was added after the three first interviews.

27.4 Results and Analysis

The following sections (Sects. 27.4.1, 27.4.2, 27.4.3, 27.4.4, and 27.4.5) address the research questions in relation to the results and analysis of interview questions 1–5.

27.4.1 Source of Learning of Mathematical Modelling

Four of the participants indicated that they studied university courses only in pure mathematics and no explicit courses in mathematical modelling. Three of the participants had some formal teaching in modelling in either a course in applied mathematics, a set of seminars dealing with problems from industry, or in courses that included some aspects of modelling like simulations with Matlab. The other two participants indicated that they had experienced a more extensive teaching in modelling, referring to courses in optimization and courses on how to solve differential equations related to realistic tasks. All participants were insistent they had learnt mathematical modelling mainly during their PhD studies and in their occupation, by working with 'real modelling' problems either through schooling by supervisors or by working individually or in a group. Some participants also claimed that they had learned about modelling by reading scientific papers and books. Opinions expressed were that modelling problems in education and in the workplace and in research are quite different, where problems in education often are restricted and limited, whilst the problems in workplaces are more complex. The following excerpts from interviews illustrate that workplace learning of modelling (*the societal work process*) is the fundamental part in the experts' opinions on learning modelling due to the nature of their working problems:

> When you get to the real problems [in research and in the workplace], not the idealised and clarified problems that can be solved, then you need modelling. (PHY)

> When you enter the workplace you "encounter problems that you never have encountered before, you will not simulate three processes rather 5,000". (INS)

> In teaching modelling the problems are clarified, 'they must be solvable within a reasonable time' and you have the data, but in reality they are more complex. (TS)

The participants' opinions on the extent to which modelling was a part of their own education may originate from Olesen's (2008) three components *the societal work process, the knowledge available* and *the subjective working experiences*. For the climate modeller, for instance, *subjective working experiences* are mainly based on working in teams where his particular task is to solve well-defined problems (e.g., solve differential equations), *societal work process,* and *the knowledge available* or the skills needed are methods for solving PDEs, also emphasised as a skill he required during his university studies.

27.4.2 Goals of Mathematics Education in Upper Secondary School

The modellers' opinions of the goals of mathematics education in secondary education seem to be mainly *subjective*. Examples of opinions are: to gain knowledge to handle everyday situations (INS); to train mathematical skills (CE); to get a tool or a language to apply in different situations (BIO); to understand the role of mathematics in society and as a part of our cultural heritage (FIN); to learn logical thinking (SCH); and knowledge for further studies (TS). In addition, the goal of mathematics education to develop democratic citizens is addressed:

> we [citizens] will be deceived by politicians with different agendas if we don't know mathematics,. . ., everyone needs this [mathematics] or else our democracy will erode (PHY)

The opinions expressed may be found in school documents like syllabuses (e.g., Skolverket 2012). There was also a variety in what participants stressed as important to learn in terms of concepts versus processes. For example, one participant emphasized understanding:

> The possibility of understanding is more important than memorizing... understand why and how it [mathematics] is used is more important than to understand how to do it. (INS)

Another participant focused more on the skills, stating that

> The goal is to train a skill to be able to [make] unhindered use [of] things like standard functions,. . .you should know the laws of logarithms and the trigonometric identities, it is not something that you need to look up every time, if you forget you should be able to derive them. (CE)

27.4.3 Goals of Modelling in Upper Secondary School

According to six of the participants, modelling should be a part of secondary education, the other three being uncertain. A repeated argument for including modelling in upper secondary school is that modelling motivates students to learn and apply mathematics in different situations. A doubt expressed may relate to their working tasks, *societal work process*, which is that the students are not "mature [enough] and do not have practical experience" (MD) since "if it is going to be realistic it gets very complex" (INS). The participants gave a range of suggestions for how to include modelling in mathematics education that seem to be grounded in *subjective working experiences*. The most extreme suggestion was a paradigm shift in mathematics education, "instead of teaching mathematics I will teach mathematics as a tool to solve problems" (BIO). Other less extreme approaches, such as to explicitly express to the students that they actually use models, which now are implicit (SCH) or that the teacher may adapt the models based on students knowledge (CE). A suggestion of using interdisciplinary weekly assignments was put forward (FIN), which may include both general knowledge of how mathematics is used in society and more deepened practice on how to model. One idea that may be implemented in a classroom setting is for the students to work in small groups as described in the following excerpt (INS):

> [in these groups] the students can meet a practical problem, like a new pharmaceutical product has been developed and the question is can we use it? How can we design an experiment to evaluate that? [Let students] ask people, so they will meet the reality… I don't think that the problem always needs to be solved mathematically, they don't need to know all the mathematical models, but they need to reason about them. Then the teacher, with help of the students, presents a way to solve the problem with mathematics… and hopefully the students will realise that mathematics actually supports them.

The role of the teacher was described as an important factor for adequate teaching. The teacher "needs to be passionate about it [modelling], because it is nothing you do like this [snapping his fingers] you need to think carefully" (FIN). One way to make teachers more interested and gain inspiration is to let them listen to presentations given by modellers and researchers on how they work with modelling (TS). Another way is to change the problems in the textbooks (TS).

27.4.4 Examples from Practice Suitable for Use in Secondary School

Most of the participants' opinions (seven out of nine) were that it is relatively easy to adapt some modelling problems from their own *societal work process* to be used in secondary school. One suggestion for upper secondary students is to investigate the models used for score cards for loans, for which the following situation might be adaptable:

> If you go into a bank and apply for a loan, then they will collect information about you. For example your age, your income, where you live, you savings, your loans etc. There is loads of information that they put into a model, which the bank has developed, and the output is a p/d number. The p/d number is probability of default, which is the likelihood that you will be unable to meet its debt obligations and if that probability is not too high, compared to a set number, then you will not get the loan. But if you get the loan the p/d number [...] will effect the interest you have to pay. (FIN)

Other examples are: to explore the predator–prey relation with Excel (BIO); to analyse how temperature affects the magnetism of a piece of iron (PHY); work with 'realistic' linear optimisation problems and networks problems (SCH); estimate safety distances when cars are following each other on the freeway (TS); trying to identify factors that should be part of the development of pension funds and how a pension fund might be organised (INS).

27.4.5 *Mathematical Modelling as a Part of a General Education*

All participants indicated that modelling should be emphasized more in school as a part of general education. It is important to understand results, and be able to form an opinion and critically examine statistics to become a democratic citizen (CLI). This is expressed in the following interview excerpt where it is claimed that

> [Modelling] is the key response to the question, why do we need mathematics? If someone wants to understand and be a part of the society, affect society and take part in decisions made in everyday life, in family, at work, in society, in the world then you must know these things (PHY)

In a society decision-making often depends on economic considerations, which creates a need for awareness of mathematics and modelling (TS). The subjective opinions above of the climate modeller, physics modeller and traffic simulation modeller are examples of arguments for modelling being more emphasized in school.

Opinions about whether modelling as a general education should be a part of mathematics education or not seemed to be *subjective* and differed between the participants. Three participants expressed that modelling should be taught mainly in mathematics education.

> I definitely think that it [modelling] should be central part of mathematics education (INS).

The other participants suggested that modelling may be used also in relation to other subjects (FIN) like in physics, chemistry, biology and social science (CE) as well as in home economics (TS).

27.5 Discussion and Implications

To summarize, the modellers interviewed have mainly learned mathematical modelling during their doctoral studies and through their occupation, by working with 'real modelling' problems. The modellers have used different learning strategies either through guidance by supervisors or working in a group but a few were self-taught through personal reading. One (*societal*) reason for this not being through their education is the nature of their working problems – they are not the idealised and clarified problems like the ones they found in their education. The modellers' opinions on how modelling should be taught and learned differed. However, most of the modellers' (*subjective*) opinions are that mathematical modelling should be a part of mathematics education in upper secondary school. They gave examples of how to implement modelling in mathematics education and gave suggestions as to how their own workplace problems may be adaptable to be used in secondary education. All modellers agreed that mathematical modelling should be more emphasized as a part of general education to develop students' critical views on how models are used in society.

The results found in this study contradict to some extent those by Spandaw (2011), where a majority of 12 scientists interviewed in the Netherlands were sceptical about using modelling problems in secondary mathematics education. A few of the Swedish modellers also expressed a concern that the modelling might be too complex for upper secondary education, but the overall opinion was that modelling may be very useful and motivational for students. The modellers in the Netherlands argued that the main goal for mathematics education in secondary school is to learn basic skills in algebra and analysis (Spandaw 2011), an opinion raised also in this study. However, other more general goals mentioned were related to modelling, such as becoming a democratic citizen and opinions that the emphasis on mathematics education should not only be on procedures but also on why and how mathematics is used in society. The need to consider these and other social aspects of modelling in future mathematics curriculum design is discussed by Jablonka (2010).

As the modellers in this study had limited experience of the present secondary mathematics education, their opinions must be taken as *opinions* and *subjective*. Nevertheless, they gave suggestions from their workplace for approaches to teach modelling as well as proposals of suitable modelling problems to work with in school that may be investigated in further research studies, indicating how modelling can function as a link between school mathematics and workplace mathematics (cf. Sträßer et al. 2012). What was not discussed during the interviews, however, was the difference in objectives for using modelling. In education, modelling is a mathematical classroom activity either as an aim in itself (to develop modelling competencies) or as an aim to develop a broader mathematical ability (modelling as a didactical tool to learn mathematics) (see e.g., Blum and Niss 1991), whereas in the workplace mathematical modelling is "the gateway into the use of mathematics" (Sträßer et al. 2012, p. 7872). The role of the teacher as mediator between workplace mathematics and school mathematics objectives is therefore crucial for all efficient collaborations between school and the workplace (Wake 2013), which was also highlighted by the modellers' opinions in this study on mathematical modelling in upper secondary education.

References

Blum, W., & Niss, M. (1991). Applied mathematical problem solving, modelling, applications, and links to other subjects. State, trends and issues in mathematics instruction. *Educational Studies in Mathematics, 22*(1), 37–68.

Damlamian, A., Rodrigues, J. F., & Sträßer, R. (Eds.). (2013). *Educational interfaces between mathematics and industry. Report on an ICMI-ICIAM-study*. Cham: Springer.

Drakes, C. I. (2012). *Mathematical modelling: From novice to expert*. Doctoral dissertation, Simon Fraser University.

Frejd, P. (2013). *Mathematical modelling discussed by mathematical modellers*. Paper presented at CERME8, Manavgat-Side, Turkey, 6–10 Feb 2013. (To appear in the proceedings.)

Gainsburg, J. (2003). *The mathematical behavior of structural engineers*. Doctoral dissertation, Stanford University.

Jablonka, E. (2010). Reflections on mathematical modelling. In H. Alrø, O. Ravn, & P. Valero (Eds.), *Critical mathematics education: Past, present and future. Festschrift for Ole Skovsmose* (pp. 89–100). Rotterdam: Sense Publishers.

Jablonka, E., & Johanssson, M. (2010). Using texts and tasks: Swedish studies on mathematics textbooks. In B. Sriraman et al. (Eds.), *The first sourcebook of Nordic research in mathematics education* (pp. 269–282). Charlotte: Information Age Publishing.

Mouwitz, L. (2013). Rationality of practice and mathematical modelling: Connections, conflicts, and codifications. In G. A. Stillman, G. Kaiser, W. Blum, & J. P. Brown (Eds.), *Teaching mathematical modelling: Connecting to research and practice* (pp. 563–572). Dordrecht: Springer.

Noss, R., & Hoyles, C. (1996). The visibility of meanings: Modelling the mathematics of banking. *International Journal of Computers for Mathematical Learning, 1*(17), 3–31.

Salling Olesen, H. (2008). Workplace learning. In P. Jarvis (Ed.), *The Routledge international handbook of lifelong learning* (pp. 114–128). London: Routledge Mental Health.

Skolverket. (2012). *The subject syllabus for mathematics*. Retrieved April 1, 2013, from http://www.skolverket.se/polopoly_fs/1.174554!/Menu/article/attachment/Mathematics.pdf

Spandaw, J. (2011). Practical knowledge of research mathematicians, scientists and engineers about the teaching of modelling. In G. Kaiser, W. Blum, R. Borromeo Ferri, & G. Stillman (Eds.), *Trends in teaching and learning of mathematical modelling* (pp. 679–688). New York: Springer.

Sträßer, R., Damlamian, A., & Rodrigues, J. F. (2012). Educational interfaces between mathematics and industry (ICMI-ICIAM-Study 20). *Pre proceedings of ICME12, The 12th international congress on mathematics education*, Seoul, July 8–15, 2012, pp. 7863–7874.

Triantafillou, C., & Potari, D. (2010). Mathematical practices in a technological workplace: The role of tools. *Educational Studies in Mathematics, 74*, 275–294.

Wake, G. (2012). Seeking principles of design for general mathematics curricula informed by research of use of mathematics in workplace contexts. *Pre proceeding of ICME12, The 12th International congress on mathematics education*, Seoul, 8–15 July 2012, pp. 1679–1688.

Wake, G. (2013). Drawing on understanding of workplace practice to inform design of general mathematics curricula. In A. Damlamian, J. F. Rodrigues, & R. Sträßer (Eds.), *Educational interfaces between mathematics and industry. Report on an ICMI-ICIAM-study* (pp. 309–316). Cham: Springer.

Wedege, T. (2010). Researching workers' mathematics at work. In A. Araújo, A. Fernandes, A. Azevedo, & J. F. Rodrigues (Eds.), *Educational interfaces between mathematics and industry. Proceedings of EIMI 2010 Lisbon conference* (pp. 565–574). Portugal: Centro International de Matemática & Bedford: COMAP.

Williams, J., & Wake, G. (2007). Metaphors and models in translation between college and workplace mathematics. *Educational Studies in Mathematics, 64*(3), 345–371.

Chapter 28
Modelling, Education, and the Epistemic Fallacy

Peter Galbraith

Abstract National curriculum statements continue to espouse the ability to solve problems arising in everyday life, society, and the workplace as a major goal of mathematics education. But support for such high sounding rhetoric is typically weakened (or absent) when the specifics of curricula are elaborated. The epistemic fallacy relates to the conflating of ontology and epistemology – confusing the nature of an underlying reality with knowledge of it. Here the underlying reality concerns mathematical modelling as real world problem solving. It is argued that manifestations of the fallacy occur in critiques of modelling theory, in debates about the authenticity of models and approaches, and in considering whether issues concerning practice are most fundamentally a curriculum or a pedagogical matter.

28.1 Introduction

It is more than 40 years since Pollak (1969) challenged the mathematics education community to more seriously engage with genuine applications of mathematics, by drawing attention to the artificial nature of text book examples that claimed to provide illustrations of its practical use. In the period since, the number of papers and research reports addressing the theory and/or practice of mathematical modelling with some form of connection to education has continued to grow astronomically.

At the ICME 12 Congress in Korea in 2012 mathematical modelling featured as the substantive content of a Plenary Lecture, two Regular Lectures, a Topic Study Group, a Special Interest Group, and the Affiliated Study Group meetings of ICTMA[1]. Small wonder then that the literature contains a plethora of views

[1] International Community for the Teaching of Mathematical Modelling and Applications.

P. Galbraith (✉)
School of Education, University of Queensland, Brisbane, QLD 4072, Australia
e-mail: p.galbraith@uq.edu.au

G.A. Stillman et al. (eds.), *Mathematical Modelling in Education Research and Practice*, International Perspectives on the Teaching and Learning of Mathematical Modelling, DOI 10.1007/978-3-319-18272-8_28

concerning the theory and practice of mathematical modelling as it appears within educational settings. Yet the outcomes continue to be mixed.

> While applications and modelling also play a more important role in most countries' classrooms than in the past, there still exists a substantial gap between the ideals of educational debate and innovative curricula on the one hand, and everyday teaching practice on the other hand. In particular genuine mathematical modelling activities are still rather rare in mathematics lessons. (Blum et al. 2002, p. 150)

> The problem of making modelling. . .a reality in every classroom is far from solved, even when there is support at policy level. The challenge is always underestimated both by governments and professional leadership. (Burkhardt 2006, p. 187)

Real world problem solving expertise as an espoused educational goal continues to be reinforced internationally – as in the following:

- Australian Curriculum Assessment and Reporting Authority (ACARA 2010):

> mathematics aims to ensure that students are confident, creative users and communicators of mathematics, able to investigate, represent and interpret situations in their personal and work lives and as active citizens. (p. 1)

- Common Core State Standards Initiative – USA (CCSSI 2012):

> Mathematically proficient students can apply the mathematics they know to solve problems arising in everyday life, society, and the workplace. (p. 1)

However such abilities can only develop if mathematical experiences are drawn genuinely from these same areas of personal, vocational, and civic contexts. When the Australian curriculum statement (ACARA 2013, p. 13) elaborates subsequent curricular content it includes, for example, the following for a pre-University Mathematics curriculum:

> - identify contexts suitable for modelling by exponential functions and use them to solve practical problems.
> - use trigonometric functions and their derivatives to solve practical problems.
> - use Bernoulli random variables and associated probabilities to model data and solve practical problems. (p. 13).

Purposes expressed in this way pay no more than lip service to goals of promoting student ability to apply their mathematical knowledge. They can be met trivially through a token interpretation of what *practical problems* mean and their presence over many years indicates that they are not sufficient to promote capability of mathematical application in the sense described. That requires additional abilities – including to identify a feasible problem from a real context in the first place, and to decide which mathematics (from among that available to an individual) is appropriate to address it. These involve different attributes, from those that look at examples of applications within a topic area that has already been identified.

In this chapter I want to address issues deemed relevant to the disappointing long-term outcomes (Blum et al., 2002; Burkhardt 2006) referred to above. To this purpose the structure of the chapter is as follows. Different approaches to modelling are identified in terms of their ontological roots and the epistemological consequences that flow from them. Sample critiques of modelling, and considerations of authenticity are then framed in terms of properties intrinsic to a modelling genre, rather than in terms of user defined assumptions about the nature of educational settings, and a modelling problem is used to instantiate cognate principles. Finally, Valsiner's Zone Theory (Valsiner 1997) is invoked to address pedagogical implications that arise when the integrity of a modelling approach is challenged by competing classroom priorities.

28.2 Epistemic Fallacy

Accepting that *ontology* is concerned with definitions of fundamental categories of reality we distinguish between *formal* and *domain* ontology. The former implies something general about reality; the latter is concerned with different areas of reality. *Epistemology* defines how we know and reason about a reality in question, so that each 'domain ontology' will have a specific epistemology associated with it. For example 'maps' used by a biologist studying colonies of bees will have a different meaning from 'maps' used by a geographer in studying human settlement. The epistemic fallacy (Bhaskar 1975; Bryant 2011) concerns the conflating of ontology and epistemology – in confusing the nature of an underlying reality with knowledge of it. (The 'fallacy' occurs when statements about 'being' are analysed in terms of 'knowledge of being' – so that ontological questions are avoided through being transposed to epistemological ones.) In the present context this involves the replacement of the consideration of what 'mathematical modelling' is, by how the term is interpreted both for itself, and with respect to its place in educational actions and settings.

In terms of our interests the underlying reality (ontology) is that in our world exist things that cause difficulty (problems), and some of these can be addressed productively through the application of 'mathematics'. This is the fundamental 'reality' of modelling as real world problem solving. There is also a belief/assumption that individuals can be helped to become better at this activity, through the agency of 'education'. Elaborating through an epistemological lens, Niss et al. (2007, p. iv) describe a mathematical model as consisting of an extra-mathematical domain, *D*, of interest, some mathematical domain *M*, and a mapping from the extra-mathematical to the mathematical domain (see Fig. 28.1).

Objects, relations, phenomena, assumptions, questions, etcetera in *D* are identified and selected as relevant for the purpose and situation and are then mapped – translated – into objects, relations, phenomena, assumptions, questions, etcetera pertaining to *M*. Within *M*, mathematical deliberations, manipulations and inferences are made, the outcomes of which are then translated back to *D* and interpreted as conclusions concerning that domain. This so-called *modelling cycle* may be iterated several times, on the basis of validation and evaluation of the model in relation to the domain, until the resulting conclusions concerning *D* are deemed satisfactory in relation to the purpose of the model. The ubiquity of a cyclic process

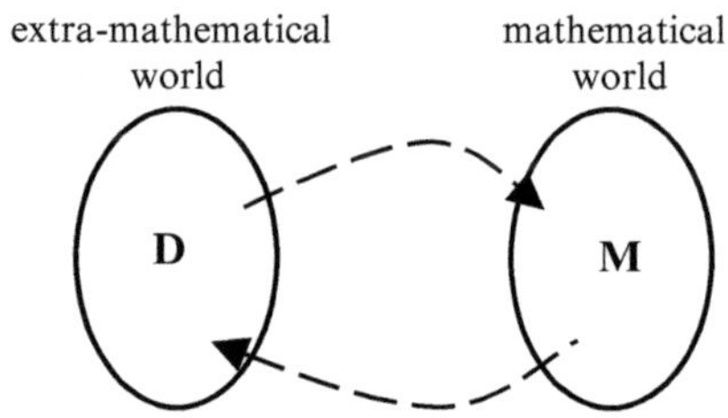

Fig. 28.1 Mathematics and the rest of the world

is noted in many sources (e.g., Borromeo Ferri 2006), with no fewer than 17 variations included in Perrenet and Zwanefeld (2012). We note the important aside that the solution path through a particular problem taken by an individual is usually anything but smoothly cyclic – a host of to-ing and fro-ing occurs between various stages of the overall process. This is mentioned because the literature suggests that distinctions between the cyclic modelling process and individual solution pathways are sometimes confused.

28.3 Models of Modelling in Education

Mathematical modelling activity does not occur in a vacuum, but in some setting involving those engaged in it. Always learning will be involved, whether by an individual from her/his personal activity, informally as when collaborating team members learn from each other, or novice modellers are mentored by experienced colleagues – or in formal educational contexts involving classrooms, students and teachers. Within the field of formal education we note that various epistemological lenses have been used to argue for a multitude of nuances of modelling. These may legitimately be called different perspectives, but there are issues in describing them as distinct genres. Fundamentally, it is argued that all variations of mathematical modelling that appear in educational settings can be incorporated within one of two basic genres as distinguished by Julie and Mudaly (2007) – *modelling as content* (empowering students to become independent users of their mathematics) or *modelling as vehicle* (modelling used to serve other curricular needs).

> Central to the debate is whether mathematical modelling should be used as a *vehicle* for the development of mathematics or treated as *content in and of itself*. A common notion associated with mathematical modelling as a vehicle is that mathematics should be represented in some context. The purpose for embedding mathematics in context is not the construction of mathematical models per se but rather the use of contexts and mathematical models as a mechanism for the learning of mathematical concepts, procedures… Mathematical modelling as content entails the construction of mathematical models of natural and social phenomena without the prescription that certain mathematical concepts, procedures or the like should be the outcome of the model-building process. (p. 504)

The vehicle perspective is portrayed unequivocally in Zbiek and Conner (2006).

> The curricular context of schooling in our country (USA) does not readily admit the opportunity to make mathematical modeling an explicit topic in the K-12 mathematics curriculum. . . .Engagement in classroom modeling activities is essential in mathematics instruction only if modeling provides our students with significant opportunities to develop deeper and stronger understanding of curricular mathematics. (pp. 89–90)

The distinction is thus elaborated as two poles of a duality (Niss et al. 2007, pp. v–vi).

Curricular statements lauding the goal of students able to use mathematics to address problems in their "personal and work lives and as active citizens" implicitly endorse the treatment of mathematical modelling as *content*. On the other hand the listing of applications and modelling related features as one of a plethora of objectives buried in curriculum detail ensures that they become just another competing priority in the classroom struggle for survival – a *vehicle* that can be used productively, but also compromised, or discarded according to circumstances.

The introduction of an instructional setting that imposes its own contextual priorities fundamentally changes the foundations upon which modelling can be anchored. The respective dominance of modelling integrity or other educational imperatives stand as definitive influences and potential antagonists with respect to what mathematical modelling in education can expect to achieve.

28.4 Modelling Critique

An absence of clear ontological positioning, with consequential lack of epistemological clarity, is a likely major reason for the mixed bag of criticisms that attend the role of applications and modelling in education. Criticisms in any field may be error-detecting, constructive, ignorant, or mischievous, and ours is no exception. Error-detecting and constructive criticisms are important to accept and act upon as they stand to correct mistakes in theory and/or practice and to enhance the field. In contrast criticisms resulting from ignorance, or with mischievous intent, need to be dealt with, so that misleading messages do not remain unaddressed.

The modelling field is muddied by the presence of a multitude of epistemological positions (some conflicting), that impact with confusing effect on parties not well versed in applications and modelling as such. For example the terms 'authentic' and 'authenticity' loom large in both modelling literature and curriculum documents. Moreover individuals tend to have a private sense of what authentic means to them. An example of the cross-purposes that have emerged in this area has been provided by Sfard (2008), who argued that the minute an 'out-of-school problem' is treated in school it is no longer an out-of-school problem and in this sense the search for 'authentic problems' to be modelled is necessarily in vain – as they lose their authenticity. This is a particularly useful example for it lies at the heart of what mathematical modelling in education is about.

The above important assertion needs to be addressed on two levels. Firstly counterexamples exist to demonstrate it is misplaced as a generalisation. The most potent evidence for authenticity is when students, having been taught modelling in school, independently apply the skills learned to problems of choice in their personal world. Burkhardt (2006) gives an example of a junior secondary student who successfully won a case made to his parents for increased pocket money by presenting an argument based on a specific model. Another example involving a 12 year old girl and a pony is also found in Burkhardt (1981). The present author has previously shared examples in which a student employed the modelling process to redesign his hydroponic cultures, and a primary school class successfully presented a case to their local council for a new crossing on the basis of statistical data collected, interpreted, and synthesised.

Secondly, and more profoundly, the passage demonstrates how the imposition of 'modelling as vehicle' values distorts perceptions of what can be achieved through 'modelling as content' approaches. Ontological differences are starkly revealed through the imposition of conceptions of what mathematics classrooms must be like, and what mathematics teaching is allowed to be. This privileging of perceived conservative classroom conceptions violates the ontological foundations of mathematical modelling as the content of real world problem solving.

28.5 Authenticity

The significance of *authenticity* is at the heart of constructions of mathematical modelling as real world problem solving. As argued in Galbraith (2013) the term has been used too globally, without sufficient regard to its scope and implications. Authenticity was described in terms of four dimensions, which are important whether the setting involves the activity of an individual on a private project, a workplace example, or a classroom modelling project. These are:

Content authenticity: Does the problem satisfy realistic criteria – involve genuine real world connections? Do the students possess mathematical knowledge sufficient to support a viable solution attempt?

Process authenticity: Does a valid modelling process underpin the approach?

Situation authenticity: Brings conditions necessary for a valid modelling exercise into direct contact with the workplace, classroom, laboratory, private workspace, or other environment within which the modelling enterprise is conducted. The essential characteristic is that 'the requirements of the modelling task drive the problem solving activity'. This means that pedagogical (and other) educational choices will be decided by the needs of the problem solving process – not vice versa.

Product authenticity: Given the time available: Is the solution mathematically defensible? Does it suitably address the real world question asked?

The following example, it is argued, fulfils basic requirements for an authentic problem, and also provides a focus for discussing subsequent issues relevant to the *education-modelling* interface.

28.6 A Modelling Example

Population Prediction

Australian population to top 23 million tonight (Updated 23 April 2013, 10:55 AEST)

Australia's population will reach an estimated 23 million people some time tonight, and demographers say it's on track to hit 40 million by the middle of the century.

The Australian Bureau of Statistics says the projection is based on last year's population estimate and takes into account factors such as the country's birth rate, death rate and international migration. ABS figures show that around 180,000 people move to Australia each year.

It estimates that with a birth every 1 min and 44 s, a new migrant arriving every 2 min and 19 s, and a death every 3 min and 32 s, the 23 million mark will be reached just after 10:00 pm (AEST).

That means our population increases by one person every minute and 23 s.

Source: http://www.radioaustralia.net.au/international/2013-04-23/australian-population-to-top-23-million-tonight/1120164

Investigate the claim that "demographers say it (the population) is on track to hit 40 million by the middle of the century." What are some societal implications?

From the given data: Births/year = 303,439 and Deaths/year = 148,858

Birth rate: $b = 303{,}439/23{,}000{,}000 = 0.0132$ per year

Death rate: $d = 148{,}858/23{,}000{,}000 = 0.00647$ per year

Immigrants (net): $\mathrm{I} = 24 \times 60 \times 60/139 \times 365.25 \times 365.25 = 227\,032$ per year

Assume, as implied by the news report that the given values of birth, death, and immigration rates apply into the future.

Let P_0 = initial population (in 2013); Let P_n = population in year n

Let $r = (b - d)$ = natural growth rate; Let I = average net annual immigration intake

Solution 1. By Spread Sheet (no specialist mathematical knowledge needed)

$P_0 = 23000000;\ b = 0.0132;\ d = 0.00647;\ r = 0.00673;\ I = 227\,032$

$P_1 = P_0 + rP_0 + I = P_0(1 + \mathrm{r}) + I = P_0R + I(\text{where } R = 1 + r)$

$P_2 = P_1R + I$ and copy down

$P_{40} = 40\,459\,252$

Solution 2. By Geometric Series (proceeding from above)

$$P_n = P_0R^n + I\left(R^{n-1} + \ldots R^2 + R + 1\right)$$
$$P_n = P_0R^n + I(R^n - 1)/(R - 1);\ P_{40} = 40\,459\,252$$

Solution 3. By Calculus (continuous approximation)

$$dP/dt = Pr + I$$
$$P(t) = P_0e^{rt} + I(e^{rt} - 1)/r \text{ where } P_0;\ P_{40} = 40\,526\,191$$

A prediction of around 40 million in 40 years seems reasonable!

Evaluate If a population has an average life expectancy of L then on average a fraction $1/L$ of the population dies each year. Using the figures provided above then an estimate of average lifetime = 1/0.00647 = 154.6 (years)!

So while the given figure might apply over a 24 h period (as here), or even over short periods of time, it is no basis for robust population predictions.

Refinement To estimate the death rate we need to research robust statistical data like the 81.5 years life expectancy found on the website: http://www.australiandoctor.com.au/news/latest-news/australia-on-top-in-life-expectancy

For this life expectancy the long term average death rate = 1/81.5 = 0.0122699 year^{-1}, and over time the actual value must stabilise to a value consistent with this. We also note that the net immigration rate is quoted at about 180,000 per year rather than the higher figure derived from the overnight data. Using the revised death rate and immigration figures: P_{40} = 31,200,000 (approx.), a substantially different prediction.

Discussion Points There are implications from population projections for the future provision of jobs, housing, health, education and the list goes on. Are the demographers wrong? Are the data robust? Has the media been creative with facts? These are all issues raised for discussion by the modelling outcomes.

Following from the above how might matters next evolve in a classroom situation, given that the refinement problem has been identified? *If real world problem solving is the driving force* then authentic subsequent actions would be determined by the needs of the refinement – so pursue the modelling activity; examine implications as above; pose further questions... *But if modelling is viewed as a vehicle* then the problem is vulnerable: We can't afford more time on this problem. We need to introduce the next course topic...etcetera.

28.7 Valsiner's Zone Theory

In considering pedagogical issues such as the above that occur regularly at the *education – real world* interface we need frameworks to guide the development of theoretically consistent approaches to teaching and learning. This must involve the development of relevant person – environment relationships as they impact on the context of modelling. We see value in the approach of Valsiner (1997) who defined the Zone of Free Movement (ZFM) and the Zone of Promoted Action (ZPA) to complement the Vygotskian Zone of Proximal Development (ZPD) as frameworks for theorising and structuring teaching and learning.

- The ZPA is oriented to defining and promoting the acquisition of valued new skills, and hence can encompass either narrow or broad visions of what is intended.
- The ZFM structures an individual's access to different areas in the environment, the availability of different objects within such areas, and the ways the individual is permitted or enabled to act on accessible objects in accessible areas.

Both the ZPA and ZFM are culturally determined, and can be loosely associated respectively with ends and means. While the ZPA identifies goals and purposes to be attained, the ZFM determines which actions on the part of learners are possible or permitted in the educational setting.

Within a learning context the ZFM for students is fashioned by environmental constraints such as the existence of technological resources, or time pressures exerted by curriculum and assessment requirements; by the particular learning experiences provided or materials or devices available; and the 'rules' by which a classroom or other situational context is run. Thus the ZFM has inhibiting or enabling properties, with respect to the ZPA in facilitating or inhibiting the achievement of new skills and abilities. It directs attention to activities, objects, or areas in the environment necessary to the purpose for which individuals' learning is being promoted.

What is provided within a ZFM has a definitive impact on how successful the intended learning, represented by the ZPA can be. *When the goal is to nurture the development of real world modelling skills*, then a teacher committed to the 'content' approach will ensure that the ZFM provides what is required and more in terms of support for the associated ZPA (Fig. 28.2a). Within a 'vehicle' approach the incursion of other curricular priorities stands to restrict the ZFM to a level below that needed to support a ZPA geared to supporting modelling as content (Fig. 28.2b).

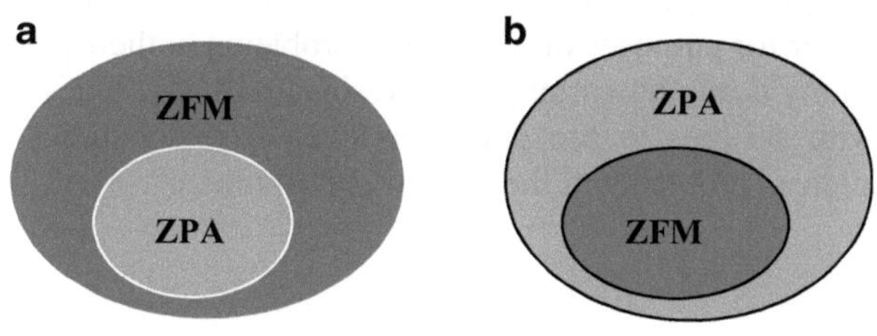

Fig. 28.2 (**a**) Modelling as content; (**b**) modelling as vehicle

28.8 A Curricular Imperative

Empowering students to use their mathematical knowledge to solve real problems, involves much more than trying to find opportunities to infiltrate application related activity (often contrived) into a crowded schedule. Significantly that does not assure proficiency when a problem requires relevant mathematics to be first identified, and then applied – essential if students are to apply their knowledge proficiently to problems located in personal, work, or civic contexts, or other discipline areas. As has been noted, the extent to which these capabilities have been provided to students through modelling programs has fallen well short of what has been hoped for.

Traditionally mathematics teaching goals have been overwhelmingly aligned with the presentation of curricular mathematics, and an essential tension lies behind reluctance to undertake modelling that seems so different from what has become accepted school practice. To change this, mandated objectives for students to formulate and solve problems located in their world must be as explicit as goals associated with learning curricular mathematics. This would legitimise coexistence of complementary strands in mathematics programs, so application and modelling goals are afforded an integrity independent of that associated with learning of conventional material. It does not mean strands need be equally time weighted, but it recognises fundamental distinctions (in the ontologies) between the respective purposes of learning *new* curricular mathematics, and learning to apply *existing* mathematical knowledge to solve real or life-like problems fulfilling situation authenticity.

Supporting students to become proficient users of mathematics goes far beyond 'feel good' statements about the importance of being able to apply mathematical knowledge, for the belief that modelling can be somehow integrated as one competency strand among many curricular imperatives has implications for what can be pursued as modelling competencies. Educational traditions have led to the privileging of certain conceptions concerning school mathematics, and what mathematics teaching and classrooms are allowed to be – traditions that impact severely when modelling initiatives are required to fit the stereotype. By contrast, what modelling as real world problem solving can do is to challenge some of those norms, assumptions, and stereotypes: mathematical, situational, and pedagogical.

John Rennie when editor of *Scientific American* in September 2000 made the telling observation: *"When confronted with a crisis, you will not rise to the occasion; you will sink to the level of your training."* Future citizens cannot call on knowledge, abilities or experiences they have not been supported to develop and foster for purposes of addressing problems in their world. Furthermore, in terms of equity is anything more empowering and socially desirable than providing students with abilities to use whatever level of mathematics they possess, to genuinely address problems in their living environment?

References

ACARA. (2010). Australian Curriculum, Assessment and Reporting Authority. *Mathematics: Draft consultation version 1.1.0 Australian Curriculum*. Formerly available from www.australiancurriculum.edu.au/Documents/Mathematicscurriculum.pdf

ACARA. (2013). Australian Curriculum, Assessment and Reporting Authority. *D12/3780: Draft Senior Secondary Curriculum – Mathematical Methods Curriculum May 2012*. http://www.acara.edu.au/default.asp?id=site_search&query=mathematical+methods

Bhaskar, R. (1975). *A realist theory of science*. Brighton: Harvester.

Blum, W., et al. (2002). ICMI study 14: Applications and modelling in mathematics education – Discussion document. *Educational Studies in Mathematics, 51*(1–2), 149–171.

Borromeo Ferri, R. (2006). Theoretical and empirical differentiations of phases in the modelling process. *ZDM, 38*(2), 86–95.

Bryant, L. R. (2011). *The democracy of objects*. Ann Arbor: Open Humanities Press.

Burkhardt, H. (1981). *The real world and mathematics*. Glasgow: Blackie-Birkhauser.

Burkhardt, H. (2006). Modelling in mathematics classrooms: Reflections on past developments and the future. *ZDM, 38*(2), 178–195.

CCSSI. (2012). *Mathematics: Standards for mathematical practice – Model with Mathematics*. Retrieved March 16, 2013 from http://www.corestandards.org/Math/Practice/MP4

Galbraith, P. (2013). From conference to community: An ICTMA journey – The Ken Houston inaugural lecture. In G. A. Stillman, G. Kaiser, W. Blum, & J. P. Brown (Eds.), *Teaching mathematical modelling: Connecting to research and practice* (pp. 27–45). Dordrecht: Springer.

Julie, C., & Mudaly, V. (2007). Mathematical modelling of social issues in school mathematics in South Africa. In W. Blum, P. Galbraith, H.-W. Henn, & M. Niss (Eds.), *Modelling and applications in mathematics education: The 14th ICMI study* (pp. 503–510). New York: Springer.

Niss, M., Blum, W., & Galbraith, P. (2007). Introduction. In W. Blum, P. Galbraith, M. Niss, & H.-W. Henn (Eds.), *Modelling and applications in mathematics education: The 14th ICMI study* (pp. 3–32). New York: Springer.

Perrenet, J., & Zwanefeld, B. (2012). The many faces of the mathematical modelling cycle. *Journal of Mathematical Modelling and Application, 1*(6), 3–21.

Pollak, H. O. (1969). How can we teach applications of mathematics? *Educational Studies in Mathematics, 2*(2/3), 393–404.

Sfard, A. (2008). *Thinking as communicating: Human development, the growth of discourses, and mathematizing*. Cambridge, MA: Cambridge University Press.

Valsiner, J. (1997). *Culture and the development of children's action: A theory of human development* (2nd ed.). New York: Wiley.

Zbiek, R., & Connor, A. (2006). Beyond motivation: Exploring mathematical modeling as a context for deepening students' understandings of curricular mathematics. *Educational Studies in Mathematics, 63*(1), 89–112.

Chapter 29
Reconsidering the Roles and Characteristics of Models in Mathematics Education

Toshikazu Ikeda and Max Stephens

Abstract A model is generally assumed to be built by translating a real world problem into a mathematical representation. We attempt to re-construct this interpretation and point to at least two distinct meanings: (Role 1) models as hypothetical working spaces, and (Role 2) models as physical/mental entities for comparing and contrasting. This leads us to draw attention to four different perspectives of modelling: (a) where modelling is interpreted as interactive translations among plural worlds not between two fixed worlds; (b) where models have the potential to incorporate scenarios beyond the initial problem situation; (c) where the mathematical world is used as a source of mental entities for comparing and contrasting; (d) where modelling competency means knowing how to balance between these different roles. Perspectives (a) and (d) are concerned with Role 1 and (b), (c) and (d) are concerned with Role 2.

29.1 Research Background

Up to now, a number of trends have been identified globally with regard to modelling approaches (Blum and Niss 1991; Kaiser and Sriraman 2006). Various approaches based on different cultures and objectives exist and these include a significant diversity in the interpretation of models and modelling (Gravemeijer 2007; Lesh and Yoon 2007). It has also been pointed out that the act itself of defining the process of putting real-world problems into quantifiable terms to create a mathematical model may not be appropriate for the problem being considered

T. Ikeda (✉)
College of Education and Human Sciences, Yokohama National University, 79-2, Tokiwadai, Hodogaya-ku, Yokohama 240-8501, Japan
e-mail: toshi@ynu.ac.jp

M. Stephens
Mathematics Education, Melbourne Graduate School of education, The University of Melbourne, Parkville, VIC 3010, Australia
e-mail: m.stephens@unimelb.edu.au

G.A. Stillman et al. (eds.), *Mathematical Modelling in Education Research and Practice*, International Perspectives on the Teaching and Learning of Mathematical Modelling, DOI 10.1007/978-3-319-18272-8_29

(Skovsmose 2005; Skovsmose and Nielsen 1996). Thus, the question of how to think about models and modelling is an old yet new issue. In instruction focused on mathematical modelling, a model has been regarded in the limited sense that models arise when a problem in the real world is translated into, and considered as, a mathematical model (Blum et al. 2007), but in the present chapter we temporarily re-construct this framework to reconsider the meaning and role of models in mathematics education in a broader sense, after which we once again refer to their influences about modelling. Therefore, the aim of this chapter is to elicit new perspectives about modelling by re-constructing the interpretation of model in mathematics education.

29.2 Models in Mathematics Education and Their Role

At the beginning, we define the meaning of a model in mathematics education by referring to the ideas of Pinker (1981) and Fischbein (1987) as follows:

When he/she judges that *M* has the following three points, *M* is a model of *O* for him/her,

(1) *M* has a similar structure with *O* in regarding a particular aim.
(2) *M* is relatively autonomous to *O*.
(3) *M* can be applied in order to get meaningful results about *O*.

Based on the above definition, it is not necessary that a model should be built up in a mathematical world by translating a real world problem in mathematics education. When a problem is derived in a real world, a model is built up in a mathematical world. But, when a problem is derived in a mathematical world, a model will be built up in a real world. Models can be built up in both a real world and a mathematical world. In these two instances, it is assumed that a model is used as a hypothetical working space. Fischbein (1987) noted "The main advantage of an autonomous model is that the subject may rely on the model alone in order to solve various problems posed by the original" (p. 124).

In contrast we can see an alternative role of model in science. Hesse (1966) ascertained: "Now the important thing about this kind of model-thinking in science is that there will generally be some properties of the model about which we do not yet know whether they are positive or negative analogies; these are the interesting properties, because, as I shall argue, they allow us to make new prediction" (p. 8). Here, a model is not used as a hypothetical working space, but as a device to create new predictions or problems. It can be said that a model is used here as a source of physical/mental entities for comparing and contrasting.

Therefore, we propose there are two ways for learners to set up models, as explained in the following subsections, and then we proceed to a general description of their roles in the context of the learner. In mathematics education, models can be broadly classified as having two roles. The first role of a model is as a hypothetical working space; and the other role of a model is as a physical/mental entity for comparing and contrasting. These two roles for models serve important functions in mathematics learning and for cultivating mathematical thinking. That is to say, the use of models when thinking serves to aid us in our own thought. These two roles can be described as follows.

29.2.1 Role 1: Model as a Hypothetical Working Space

When it is difficult to consider the problem in the context in which it arises, it can be replaced by a problem in another context that is easier to think about, and one can attempt to produce an answer by processing the problem in this other context. By "easier to think about" here, we mean that there are some factors in the second context that facilitate human thought. In this case, the model refers to the thing in the other context, or "other world", that can be substituted for the problem at hand and considered in its place.

There are at least three reasons for this. The first reason is that a model enables us to consider abstractly with visual and manipulative tools. When it is difficult to consider the problem in the context of the world in which it arises, operable models can be created or selected, utilizing, for example, blocks, or number lines, or graphs, allowing the problem to be considered visually and/or manipulatively. For example, when students consider addition or subtraction in a real world, they can do so in an operational way through the use of blocks or counters. When considering the multiplication and division of rational numbers, we can do so in an operational way by using a number line model to perceive the numbers visually, or think about it using a graphical representation so that the mechanisms of change can be understood. Thus, thinking about the numbers in a visual and operational way, in this situation, a model is built from a concrete world into an abstract world.

Second, the idea of a model as a hypothetical working space can be also effectively utilized when constructing new mathematical knowledge. When formulating new mathematical knowledge, an imaginable concrete model with a structure the same as the original mathematical problem is set up, and mathematics knowledge can be developed by analysing the concrete model. When we think of multiplication and division of decimals, fractions, or positive and negative numbers, we construct a system within the scope of mathematics by taking a problem from a real situation as a model, and solving that problem (Ikeda and Stephens 2011). One such case, for example, is the multiplication and division of fractions, where a painting problem in a real world can be used as a model. Another case is where the addition and subtraction of positive and negative numbers can be considered on the basis of models such as a person walking east/west in a real world. This model is built from an abstract world into a concrete world.

Third, a mathematical model enables us to consider systematically implications arising within the resulting mathematical system. Typically, in mathematical modelling, problems in the real world can be thought about in a mathematical world. By translating a real world problem into a mathematical world, it becomes possible to utilize the systems of mathematics to consider the problem through formal mathematical processing. For example, if we can make up a quadratic function model about a real world situation, then we can get the maximum or minimum value by transforming the formula systematically. However, the role of a model in modelling is not limited to this standpoint, as will be explained.

To summarize, the idea of a model as a hypothetical working space provides the following three advantages at least, enabling us to consider the situation being modelled:

(R1-1) abstractly with visual and manipulative tools,
(R1-2) concretely with images in a familiar world,
(R1-3) systematically with a developed mathematical system.

29.2.2 Role 2: Model as a Physical/Mental Entity for Comparing and Contrasting

The idea of Hesse (1966), who advocated that thinking develops dynamically, precisely because there are models, threw new light on the role of models in science. The model here is treated contradistinctively, as physical/mental entities for comparing and contrasting, and it is used as a place from which to draw out analogies. This kind of role for a model is clearly different from that mentioned earlier, where the model serves as a hypothetical working space.

When it is unclear how a problem at hand should be thought about within the world in which it arises, facts are discovered in another familiar world that is similar to the one in which the problem arose. Then, one can attempt to explore the problem at hand, taking the structures of, and relationships between, elements in the other world as physical/mental entities for comparing and contrasting. For example, when we build up the system of taxes according to a person's level of income, we pick up a proportional function in our mathematical world as a model and contrast it with a real world problem. In this example, the mathematical world is considered more familiar providing the problem solver with a thorough knowledge of the relationships between the elements involved in the problem. The mathematical representation allows other analogies to come into play because of the addition of further areas that are likely to parallel the problem. In this case, the model refers to the structures and relationships between the elements in a mathematical world that serve as entities for comparing and contrasting and for drawing out analogies.

There are three main reasons that can be identified in this second role. The first involves the creation of a mathematical model as a hypothetical working space, which is then used for the purpose of contradistinction. In other words, we see that in mathematical modelling, the two roles of models, as a hypothetical working space and as a physical/mental entity for comparing and contrasting, are both simultaneously fulfilled. Once a mathematical model is created that represents the relationship between quantities in the real world, the potential to question the consistency between the real and mathematical worlds arises. This in turn has implications on the perspective from which the mathematical model can be examined and revised. Further, we can also see the case when making a concrete model based on the context of real phenomena related to the system in question.

Secondly, once a mathematical model has been created, one can consider the scope to which the new model is both applicable and effective to expand and clarify the real world situations being modelled. It is apparent that, in generalising the mathematical model, the role of a model is not as a hypothetical working space, but more as a context to draw out analogies through comparison.

The third situation is when one attempts to construct a new system within mathematics. This can be done by using as a model a system which has already been constructed within a mathematical framework. It can be a case of constructing a new system, for example, by thinking of positive and negative numbers in terms of a model of the addition, subtraction, multiplication, and division of natural numbers, or by thinking of the addition and subtraction of vectors as a model for the addition and subtraction of positive and negative numbers.

In summary, when one selects and contemplates a model from the other world, in mathematics education, there are at least three principal cases that may be considered as enabling us:

(R2-1) to contrast an original problem situation with a developed model,
(R2-2) to expand and clarify real world situations satisfying a developed model,
(R2-3) to consider a new mathematical system by selecting some already developed mathematical system as our exemplary model.

29.3 Two Roles of Models in Teaching of Modelling

It is possible to see how new modelling perspectives can be developed, if we reflect on the teaching of modelling from these two roles. We describe four new perspectives about modelling.

29.3.1 Modelling Is Interpreted as Interactive Translations Among Plural Worlds, Not Simply Between Two Fixed Worlds

When instructional approaches to modelling are perceived on an international level, two traditional trends are identified: namely, a pragmatic trend and a scientific/humanistic trend (Kaiser 1991). We note that Kaiser and Sriraman (2006) have identified additional trends. In this paper, different approaches to modelling are described, and attention is given to the differences in how models and worlds are regarded. In the case of the pragmatic trend, people mathematise in order to enable formal processing using systems of mathematics, and it is mathematisation for the creation of mathematical models that is emphasised. A dualism between the “real world” and the “mathematical world” underlies this approach. On the other hand, in the case of the scientific/humanistic tendency, the emphasis is placed not on a mathematical model, but on the model as a medium for promoting mathematisation. As Freudenthal (1991) noted,

> According to my terminology, a model is just the – often dispensable – intermediary by which a complex reality or theory is idealized or simplified in order to become accessible to more formal mathematical treatment. . . . I lay so much stress on the role of the model as an intermediary because people are all too often unaware of its indispensability. Much too often mathematical formulas are applied like recipes in a complex reality that lacks any intermediate model to justify their use (p. 34).

Both cases have in common the idea that the model functions as a space for thinking that enables problems to be contemplated in a separate space. However, the intermediate model for Freudenthal seems to be a model built between a real world and purely mathematical world in order to overcome the difficulty of mathematisation. This distinguishes Freudenthal’s usage from the idea of a pragmatic trend in which mathematical modelling is interpreted in terms of two fixed worlds (dualism). We have the term “real model” in the pragmatic trends (Kerr and Maki 1979), and this term usually refers to a model that is expressed in words as an actually arisen problem. However, as opposed to a model for facilitating mathematisation, a model expressed in words may be more appropriately interpreted as a model that acts as a communicable device to understand the original problem. Being able to represent and describe what is being modelled is an effective means for overcoming the difficulty of mathematisation. Therefore, the idea of an intermediate model leads us into a new perspective where mathematical modelling is interpreted as an interactive translation, not between fixed two worlds (between a real world and mathematical world), but among plural worlds. If we can represent an original action (or operation) with a new action (or operation) explained with elicited properties, we can interpret that two corresponding actions (original and new) each existing in a different world.

Consider, for example, the problem of determining the height of a balloon (Ikeda et al. 2012). The problem for students is: “We can see the rising balloon. Let’s devise a method to measure both the height of the balloon and the horizontal distance from here to the point just below the balloon!”. Creating an effective 3D model allows students to manipulate concretely using various mediums such as drawing paper, which assist them to visually and operationally contemplate the relationships between variables. Further, considering the problem geometrically by drawing and measuring enables students to develop this kind of model and to manipulate it symbolically. In this example, modelling is executed through interactive translations among at least four worlds. The first is the real world; the second is a representational world which they can manipulate; the third is a world which allows them to consider geometrically by drawing and measuring; and the fourth is a world which allows them to consider symbolically and algebraically.

In this way, when we examine the relationships between quantities within a given space, contemplation using a visually operable model is useful for investigating what the important variables are, and for exploring what kinds of relationships exist between the variables. Further, contemplation with a geometrical reduction model by drawing and measurement is useful for developing the trigonometric relations among the variables. By regarding the modelling process as a series of interactive translations among plural worlds, we can ensure that students make progress over time, not necessarily in one defined teaching episode.

In summary, we propose a new perspective in which modelling is interpreted as interactive translations among plural worlds, where the role of intermediate models in plural worlds is intended to assist students to make progress in mathematical modelling. If we cannot determine the role of a model in a certain world, we have to delete that "world" since it is of no benefit to students. We need to study the possibilities and limitations of this perspective.

29.3.2 Models Having the Potential to Incorporate Scenarios Even Beyond the Initial Problem Situation

A model by its nature serves to emphasize the fact that not all things are equivalent. When we focus on the role of a model as a physical/mental entity for comparing and contrasting, this point comes into much sharper relief. When examining a mathematical model that is intended to describe a phenomenon in a real-world setting, there are two points to consider. The first point is that the mathematical model may not – some would say cannot – include all the important variables required in the initial problem scenario. The second point is that the mathematical model may include points that lie beyond the original scope of the problem. In the teaching of modelling, the former has probably received more emphasis, but we suggest that the latter should be given equal emphasis. This point is one of the virtues of thinking about a model as a physical/mental entity for comparing and contrasting. Accordingly, after creating a mathematical model of a phenomenon present in a real situation, an activity to clarify the extent of the model's applicable range should be encouraged. To illustrate this point, let us look at the mathematical model of quantifying the hours of night (Ikeda 2013).

Although the mathematical formula has been created and described in the Northern hemisphere as shown Fig. 29.1, if we consider a more general scenario, we can ask what the situation would look like in the Southern hemisphere. Then, if we scroll along the graph of the formula created, we notice that the function is also expressed in terms of negative values as shown in Fig. 29.2. The negative values for the latitudes can be considered to represent and quantify the hours of night time in the Southern hemisphere, thus extending the applicability of the original formula.

Perceiving models in this way suggests that we can explore formats of modelling instruction that can be generalized, without having to settle for one narrow problem solution and interpretation.

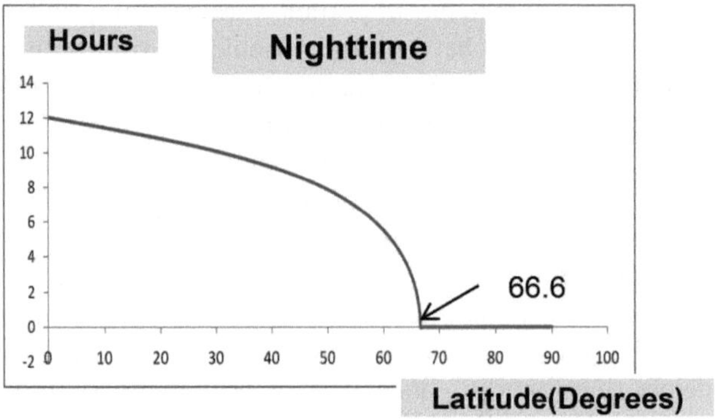

Fig. 29.1 Night time in the northern hemisphere

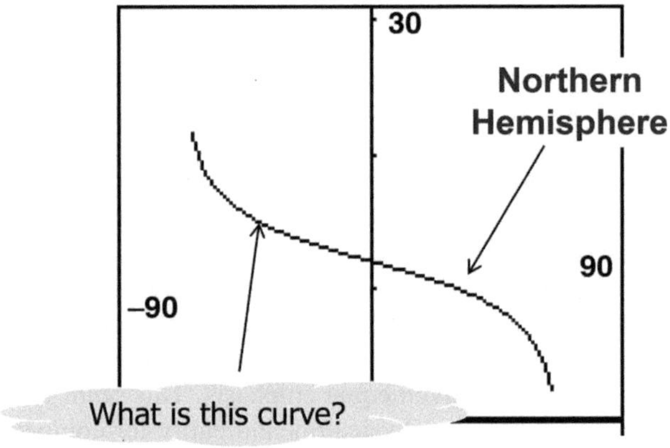

Fig. 29.2 Night time in both hemispheres

29.3.3 *Where a Mathematical World Provides Entities for Comparing and Contrasting Different Design Standards*

In mathematical modelling, sometimes the objective is to use mathematics in the real world to establish fair standards or for design (Niss 2008). Some examples referred to by Niss (2008) are taxation systems, examination-grading systems, IQ tests, life insurance, and so on. In these kinds of problem situations, we must bear in mind that the world of mathematics is being used as a model to test out different design principles. According to critical mathematics education, the goal here is not to produce solutions within a conventional mathematical modelling process. We can interpret this as the action of taking knowledge from a mathematical world that models a situation present in the real world, and then of applying and fine-tuning

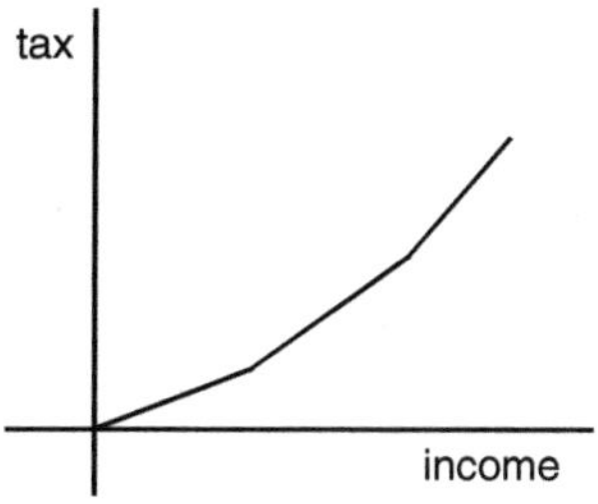

Fig. 29.3 A revised model

this model according to certain design principles or standards. For example, consider the tax problem "How should we impose taxes according to persons' incomes?" To deal with this real world problem, we could select a single proportional function, $y = ax$, from a mathematical world as our model. By comparing and contrasting the single $y = ax$ model with a real world situation, the following critical question arises: "Is it desirable to impose the same rate of tax on both low income persons and on high income persons?" By refining the model to take account of this question, an alternative model providing for progressive rates of income tax according to bands of income can be developed by combining three linear proportional functions as shown in Fig. 29.3. This new model allows us to see that this three-stage model is not necessarily the end of the story. It also allows us to evaluate the merits of using a series of increasing linear functions as opposed to a continuously changing rate. What we need to highlight here, therefore, is that the world of mathematics is not being used as a hypothetical working space, but rather as a model, in the sense of mental entity, for comparing and contrasting different design principles.

This kind of situation forces us to distinguish the role of a model as a hypothetical working space in conventional mathematical modelling from the role of a model as a mental entity for comparing and contrasting using different design principles. This distinction also has implications for mathematical modelling instruction, as well as how to evaluate different schematic depictions of modelling.

29.4 Modelling Competency Interpreted as How to Balance Between Two Roles of Models

In the teaching of modelling, we argue that the two roles of a model – as a hypothetical working space and as a mental entity for comparing and contrasting – are both required. Consequently, our focus needs to shift to the question of how both roles should be meta-cognized and contemplated. Several earlier studies, such as by Moscardini (1989) and Ikeda and Stephens (1998), have highlighted the need to start out using an essentially simple model, and then gradually and realistically, to examine and to revise one's initial model by taking the real-world scenario into account. This can be understood as a suggestion regarding the balance between the two roles of models. Specifically, this can be interpreted as suggesting that the more complex the problem, the greater should be an emphasis on using a model in its role as mental entity for comparing and contrasting, rather than simply as hypothetical working space. Of course, it is necessary to investigate how to make students aware of this point, and as a consequence to investigate the optimal forms of instruction to foster such awareness. This is a challenge for further research and study.

References

Blum, W., & Niss, M. (1991). Applied mathematical problem solving, modelling, applications, and links to other subjects: State, trends and issues in mathematics instruction. *Educational Studies in Mathematics, 22*(1), 37–68.

Blum, W., Galbraith, P. L., Henn, H.-W., & Niss, M. (Eds.). (2007). *Modelling and applications in mathematics education, the 14th ICMI-study*. New York: Springer.

Fischbein, E. (1987). *Intuition in science and mathematics*. Dordrecht: D. Reidel Publishing Company.

Freudenthal, H. (1991). *Revisiting mathematics education*. Dordrecht: Kluwer.

Gravemeijer, K. (2007). Emergent modelling as a precursor to mathematical modelling. In W. Blum, P. L. Galbraith, H.-W. Henn, & M. Niss (Eds.), *Modelling and applications in mathematics education: The 14th ICMI study* (pp. 137–144). New York: Springer.

Hesse, M. B. (1966). *Models and analogies in science*. Notre Dame: University of Notre Dame Press.

Ikeda, T. (2013). Pedagogical reflections on the role of modelling in mathematics instruction. In G. A. Stillman, G. Kaiser, W. Blum, & J. P. Brown (Eds.), *Teaching mathematical modelling: Connecting to research and practice* (pp. 255–275). Dordrecht: Springer.

Ikeda, T., & Stephens, M. (1998). Some characteristics of students' approaches to mathematical modelling in the curriculum based on pure mathematics. *Journal of Science Education in Japan, 22*(3), 142–154.

Ikeda, T., & Stephens, M. (2011). Making connections between modelling and constructing mathematics knowledge: An historical perspective. In G. Kaiser, W. Blum, R. Borromeo Ferri, & G. Stillman (Eds.), *Trends in teaching and learning of mathematical modelling* (pp. 669–678). New York: Springer.

Ikeda, T., Stephens, M., & Wada, Y. (2012). A teaching experiment in modelling through scale reduction methods: A bridge to later trigonometric methods. *Journal of Mathematics and System Science, 2*(6–7), 359–367.

Kaiser, G. (1991). Application-orientated mathematics teaching: A survey of the theoretical debate. In M. Niss, W. Blum, & I. Huntley (Eds.), *Teaching of mathematical modelling and applications* (pp. 83–92). Chichester: Ellis Horwood.

Kaiser, G., & Sriraman, B. (2006). A global survey of international perspectives on modelling in mathematics education. *ZDM, 38*(3), 302–310.

Kerr, D., & Maki, D. (1979). Mathematical models to provide applications in the classroom. In S. Sharron & R. E. Reys (Eds.), *Applications in school mathematics, 1979 yearbook* (pp. 1–7). Reston: The National Council of Teachers of Mathematics.

Lesh, R., & Yoon, C. (2007). What is distinctive in (our views about) models & modelling perspectives on mathematics problem solving, learning, and teaching? In W. Blum, P. L. Galbraith, H.-W. Henn, & M. Niss (Eds.), *Modelling and applications in mathematics education: The 14th ICMI study* (pp. 161–170). New York: Springer.

Moscardini, A. (1989). The identification and teaching of mathematical modelling skills. In W. Blum, M. Niss, & I. Huntley (Eds.), *Modelling, applications and applied problem solving* (pp. 36–42). Chichester: Ellis Horwood.

Niss, M. (2008). Perspectives on the balance between applications and modelling and 'pure' mathematics in the teaching and learning of mathematics. In M. Menghini, F. Furinghetti, L. Giacardi, & F. Arzarello (Eds.), *The first century of the international commission on mathematical instruction (1908–2008) reflecting and shaping the world of mathematics education* (pp. 69–84). Rome: Encyclopedia Italiana.

Pinker, A. (1981). The concept 'model' and its potential role in mathematics education. *International Journal of Mathematical Education in Science and Technology, 12*(6), 693–707.

Skovsmose, O. (2005). *Travelling through education-uncertainty, mathematics, responsibility*. Rotterdam: Sense.

Skovsmose, O., & Nielsen, L. (1996). Critical mathematics education. In A. J. Bishop, M. A. Clements, C. Kietel, J. Kilpatrick, & C. Laborde (Eds.), *First international handbook of mathematics education* (pp. 1257–1288). Dordrecht: Kluwer.

Chapter 30
Developing Statistical Numeracy: The Model Must Make Sense

Janeen Lamb and Jana Visnovska

Abstract Teaching statistical numeracy in middle school classrooms requires high quality instruction that promotes opportunities to use mathematics in modelling problem situations. In this chapter we report on a professional development session that involved nine teachers from six rural and remote high schools in Queensland, Australia. Results indicate that some teachers focused on the mathematics they would teach, limiting numeracy opportunities, while others focussed on making sense of the problem by modelling, thereby promoting statistical numeracy. This research suggests that ongoing learning opportunities where such differences become the point of professional discussions are needed to support teachers' understanding and appreciation of the role of modelling in promoting statistical numeracy.

30.1 Introduction

The Australian Curriculum has seven General Capabilities embedded across each subject area. Numeracy is one of these capabilities. The Rationale for the Australian Curriculum states that the national curriculum

> develops the numeracy capabilities that all students need in their personal, work and civic life, and provides the fundamentals on which mathematical specialties and professional applications of mathematics are built... These capabilities enable students to respond to familiar and unfamiliar situations by employing mathematical strategies to make informed decisions and solve problems efficiently (ACARA 2013 para 1).

J. Lamb (✉)
Faculty of Education and Arts, National School of Education, Australian Catholic University, 1100 Nudgee Road, Banyo, QLD 4014, Australia
e-mail: janeen.lamb@acu.edu.au

J. Visnovska
School of Education, The University of Queensland, Social Sciences Building (24), St Lucia, QLD 4072, Australia
e-mail: j.visnovska@uq.edu.au

G.A. Stillman et al. (eds.), *Mathematical Modelling in Education Research and Practice*, International Perspectives on the Teaching and Learning of Mathematical Modelling, DOI 10.1007/978-3-319-18272-8_30

Numeracy has been discussed in these terms for many years (AAMT 1997) and research reporting on realising these goals recommends learning that is set in experiences at school, at work and in civic life (Greer et al. 2007). However, as Galbraith (2013) points out, being able to do this requires a new approach and therefore new knowledges such as being able to select appropriate problems from these contexts and to decide which mathematics could be used. Moreover, Greer et al. (2007) argue that without a strong emphasis on context students will tend to suspend sense making, concentrating instead on using mathematical procedures without any reflection on the context of the problem. To promote sense making Verschaffel et al. (2010) conceptualise problem-solving as a six-phase modelling process. The phases are: (1) understanding the key elements in the problem situation, (2) constructing a mathematical model of the relevant elements and relations embedded in the situation, (3) working through the mathematical model to derive mathematical result(s), (4) interpreting the outcome of the computational work, (5) evaluating if the interpreted mathematical outcome is appropriate and reasonable, and (6) communicating the obtained solution of the original real-world problem (p. 10). Beyond the notions of context focus and sense making, Verschaffel et al. (2010) also call for a different teaching approach.

One such teaching approach called 'ambitious instructional practices' has become synonymous with high quality instruction (Franke et al. 2007). This approach builds conceptual understandings in mathematics by using problems that are cognitively demanding (Lampert et al. 2010) and where tasks are accessible by all (Boaler and Staples 2008; Jackson and Cobb 2010). Consequently, the teacher needs to manage students as they work on these tasks resulting in the orchestration of productive whole class discussions of students' solutions (Stein et al. 2008). During the discussion phase of the lesson, teachers are to 'press' students for evidence or justification while students make connections between their own solution and the solutions of their peers (Staples 2007). For these changes to pedagogy to be realised, professional development (PD) needs to incorporate access to ambitious instructional practices.

Current research recommends PD that provides 'safe' opportunities for teachers to work on problems as students while the PD provider leads, demonstrating ambitious instructional practices (Lampert et al. 2013). The priority is for teachers to work collaboratively through intended classroom tasks that should be cognitively demanding, non-procedural in nature (Boston and Smith 2009) and have multiple entry points resulting in multiple solutions. While engaging in these activities, teachers develop different solutions to be compared during the discussion phase while at the same time building a strong professional community. This gives teachers opportunities to experience as students how this phase of the lesson can promote depth of understanding of the key mathematical ideas (Stein et al. 2008). It also gives teachers an opportunity to reflect on these big mathematical ideas and intended learning goals (Boston and Smith 2009) and provides opportunities to rehearse ways to 'press' students, to revoice and to link student contributions. The important issue here is that these pedagogical elements are considered teachable

(Lampert et al. 2013). Moreover, ambitious instructional practices are well suited to promote instruction aimed at sense making proposed by Verschaffel et al. (2010).

With this research literature in mind and the need to support teachers in their teaching of statistical numeracy, the PD goal of our study was to enhance instructional practices in the teaching of statistics for teachers of 13 and 14 year old students. Our research questions focused on characteristics of PD activities that spark initial teacher interest, collaboration, and engagement in professional conversations, and help make teachers' current instructional practices visible to the research team. We conjectured that when provided with opportunities, teachers would develop different models to represent data and subsequently compare models, consider the relative strengths and weaknesses of these models, and explore mathematical ideas that the models make available for classroom discussion.

30.2 Research Design

This project is set in a rural and remote region of Queensland, Australia, where newly registered teachers are most often contracted to teach for 3 years. This practice results in high teacher turnover, ensuring a constant flow of new, inexperienced teachers to these regions (Heslop 2003). Making this situation more complex for the teaching of mathematics in years 8, 9 and 10 is the usual practice that this teaching is often delegated to novice teachers without a mathematics background. The participants in this longitudinal project included 1 regional mathematics advisor and 11 teachers from 6 high schools spread over an area of 500 km^2. Three of these participant teachers had more than 3 years teaching experience. Pseudonyms have been used for all participants.

A design research approach (Cobb et al. 2003) was adopted to support and study teachers' learning. This approach involves iterative "micro" cycles of designing PD activities and conducting ongoing analyses of teacher learning that inform the design of subsequent PD sessions. It also involves "macro" cycles where resources developed in prior design studies (Visnovska and Cobb 2013; Visnovska et al. 2012) are tested and modified as they are adapted to new PD contexts. Given our understanding of the PD context we were entering, the themes of statistical numeracy and modelling appeared broadly relevant to the teachers and provided a promising starting point for collaboration.

30.2.1 The Task

We report on the first PD activity, where the *Student Exercise Task* was introduced to the teachers on the first day of the project. This sense making activity involved numerical data and was pitched at middle school student level so that the teachers could use it in their own classrooms. Prior to the commencement of the activity, the teachers engaged in a discussion on statistical numeracy, mentioning a broad range of relevant issues. Importantly, the discussion concluded that statistics without context is meaningless and that we need context in order to make decisions about data. The *Student Exercise Task* was then introduced.

Student Exercise Task
The data shows the number of paces per day for a group of 28 Grade 8 students. Teachers and the school principal are concerned that students are not getting enough exercise. Your job is to organise and represent the data in some way so that it can be presented to parents at a school open day. Find a way to represent the data so that you can make some conclusions about the exercise habits of these students.
12000, 10000, 4000, 5000, 800, 500, 5500, 7000, 9000, 11000, 5500, 4500, 5500, 5000, 12000, 7500, 4500, 4500, 5500, 5500, 11500, 9500, 5000, 9000, 5500, 4000, 6000, 5000.

Our intention with this task was twofold. Firstly, it was to provide us with an opportunity to document teachers' statistical understandings and inform the next cycle of PD design. Secondly, it was to provide the teachers with an opportunity to work collaboratively on a problem where aspects of statistical numeracy and sense-making could be discussed while building professional relationships. To achieve these intentions we asked the teachers to form groups of three and decide if the Grade 8 students were getting enough exercise. They were to present their arguments to the rest of the group as if we were parents. In this chapter we present the work of two teacher groups. The first group had all teachers within the first 3 years of teaching while the second group had two more experienced teachers in it.

30.2.2 Data Collection and Analysis

Data collected from the initial PD session analysed here included our research planning log, field notes of one of the authors and photographs of teachers' work. The overarching approach to analysis is an adaptation of the constant comparative method tailored for analysing data sets that are generated during design experiments (Cobb and Whitenack 1996). In analysing the initial teacher participation, we compared teachers' data analyses with the six phases of problem solving activity (Verschaffel et al. 2010). We then documented the mathematical and pedagogical themes that became a topic of the PD discussion, how these topics were initiated in the group, and how they were linked to ambitious instructional practices (Lampert et al. 2013). Lastly, we noted the diversity of opinions within the group related to the discussed topics. Our primary goal was to develop insights into supporting the work of newly formed PD groups of mathematics teachers and inform subsequent PD design.

30.3 Results and Discussion

The researchers were prepared for an array of different models to represent the *Student Exercise Task* knowing that some are better at providing insights into this particular context. Moreover, the same graph can result in very different classroom activities: The focus can be on the graphing conventions, or, more importantly for developing students' sense-making and numeracy, the activity can focus on how the graph models the problem at hand, provides an insight to it, and helps us propose a response (Greer et al. 2007).

The teachers enthusiastically worked on the task in their groups, developing different models. During this time the researchers decided on the order in which the groups would be invited to present, aiming to advance the PD agenda (cf. Stein et al. 2008).

30.3.1 Teacher Models Give Opportunities for Discussion

During the discussion, teachers were given opportunities to justify their reasoning, critique their own model and the models presented by other groups. The teachers who developed the graph displayed as Fig. 30.1 were asked to present first. They were selected as we believed it would be difficult to present a strong argument using this graph to decide whether the students were getting enough exercise or not. The teachers explained that they had decided to draw a histogram to represent the data. They oriented the PD group to the graph explaining that the *X*-axis displays bins at 1,000 pace intervals while the *Y*-axis represents student frequency. They also pointed out that all the important features of their graph were present including the title, the positioning of zero and the labelling of each axis. This presentation did not include an answer to the question: Are the students getting enough exercise?

One teacher, while trying to understand the graph and what it indicated about the exercise habits of the students, pointed out that the height of bars did not correspond to how many paces the students walked and that the model could mislead others to think that it did. When the teachers in the presenting group were 'pressed' to answer the question that was central to the task, they realised how difficult it was to use their graph to present an argument. The 'press' for further explanation stimulated a discussion of the importance of students knowing how to construct a graph that included all the important elements on the one hand, and how to construct a graph that purposefully assists in the interpretation of data thereby building numeracy skills on the other hand.

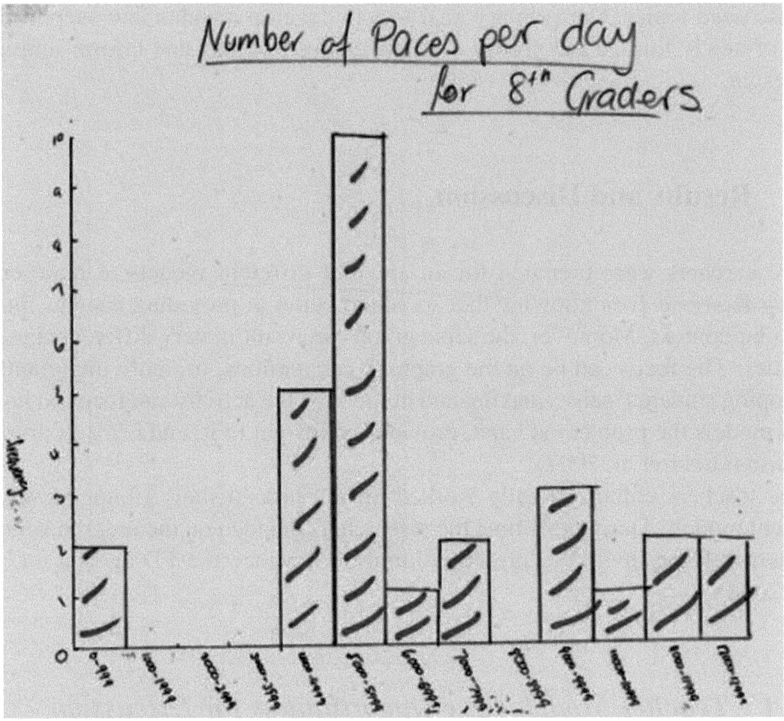

Fig. 30.1 First group's histogram model of the *Student Exercise Task*

By placing particular emphasis on the mathematical conventions of graphing, the first presenting group did not use the developed graph as a model and never progressed beyond the second modelling phase (Verschaffel et al. 2010). Importantly, other teachers expected to progress further and held the first group accountable for addressing the task question. The ensuing discussion helped us all see how easy it is to suspend sense making and to concentrate solely on the mathematical elements of the problem (cf. Greer et al. 2007).

The second group of teachers selected a pie chart to display the data and make their case (Fig. 30.2). The presenting teachers oriented PD group to their model highlighting that they knew that 10,000 paces per day was the recommended benchmark and that the section of the pie graph with 23 represents the number of students who walked less than 10,000 paces per day. The smaller section with 2 represents those students who walked less than 800 paces per day. The shaded segment with 3 represents those students who had met or exceeded the recommendation. To accentuate the size of the last group, the teachers visually cut this section out and placed it to the side of the pie graph. They concluded their presentation saying that with only three students meeting the benchmark per day, the students in grade 8 do not exercise enough.

Fig. 30.2 Second group's pie chart model of the *Student Exercise Task*

A teacher started the discussion pointing out that there were recording errors in this graph. For example, there were five students who met the benchmark, not three, and thus the numbers marked in all sections were inaccurate. Another teacher stated that it was more intuitive to read the 'less than 800 paces' group as a subgroup of the 'less than 10,000 paces' group, making her believe that the legend was inaccurate. Once the graph accuracy issues had been addressed the researchers pressed the group to consider whether this graph modelled the situation well and whether it could support an argument about the grade 8 students' exercise habits. Teachers all agreed that the graph clearly and convincingly illustrated the magnitude of students who did not meet the recommended benchmark.

When asked to compare the two responses captured by the graphs displayed as Figs. 30.1 and 30.2, the teachers quickly concluded that the first response was more closely aligned with teaching of traditional mathematics with a greater focus on the elements of the graph, whilst the second response was more closely aligned with modelling and required sense making so that an argument could be presented. From the perspective of the researchers this was an important discussion. It allowed the teachers to explore the difference in how the two graphs were used and to identify that each use would target different learning goals for students. It also demonstrated that not setting specific solution methods permitted rich discussions that teachers could conceivably have with their own students when using this or similar open-ended tasks. Further, even though the second group made an error in their graph, the contribution was treated as valuable because it developed a clear argument about the students' exercise habits and allowed for model comparisons. Thus, important pedagogical issues of use of graphs in modelling, use of open-ended tasks, and treatment of errors became available for the group discussion in the context of this PD activity.

30.3.2 Effects of Teaching Experience on Modelling

A deeper issue that deserves consideration is that the difference in the amount of teaching experience between the two groups was noticeable in how they approached the task and whether they used graphs as modelling tools. Students in rural and remote schools where there is only one mathematics teacher depend largely on this teacher for opportunities to develop the numeracy skills intended in the Australian Curriculum. The teachers are often novices but because they are in the school for a limited time, investment by schools in their continuing professional learning is not easily justifiable under the notion of developing local capacity. As a result, if a PD is provided it is likely to target immediate teacher needs rather than instructional practices suitable for modelling that require prolonged investment.

The difference between novices and more experienced teachers was further highlighted when the group with more experienced teachers subsequently developed a second model. The teachers were dissatisfied with their pie graph because it did not show how many students were close to the 10,000 paces per day benchmark. Consequently, they constructed a further model, a graph (Fig. 30.3). When presenting their second model, they pointed out that each bin along the *X*-axis represented one student with the number of paces recorded on the *Y*-axis. They also pointed out that the horizontal marker for 10,000 paces showed which students met this benchmark, concluding that the students in grade 8 do not get enough exercise. They made the point that there were clusters of students below the 10,000 pace marker, not just two groups "below" as it appeared from the pie chart.

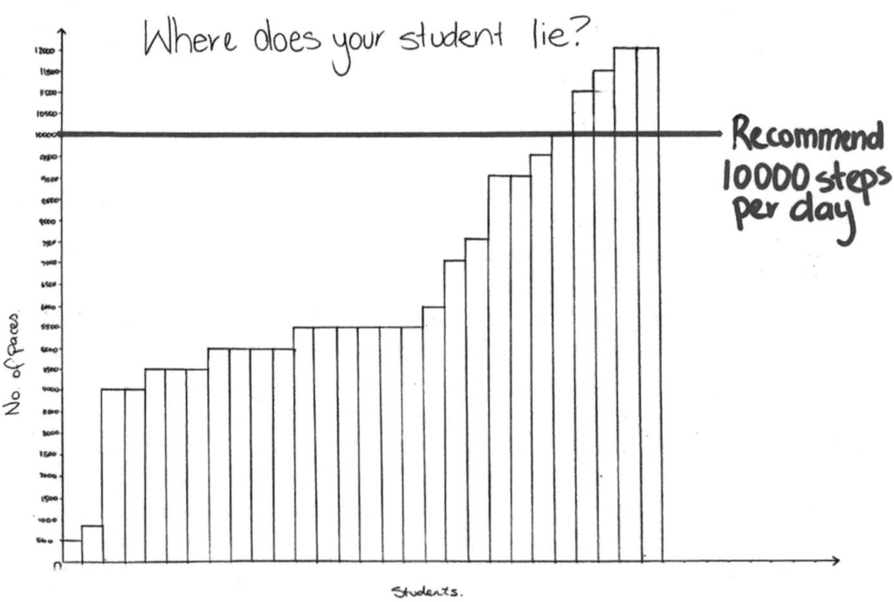

Fig. 30.3 Second group's graph of the *Student Exercise Task*

The researchers 'pressed' the teachers to explain why Fig. 30.3 allowed for a stronger argument regarding the student exercise habits. Teachers noted that ordering student data made it easy to see where students were in relation to one another as well as the benchmark. They commented that the graph allowed for insights that were not easily accessible in Figs. 30.1 or 30.2 and it was a representation that students would likely understand. It was clear that this group used graphs to model and make sense of the problem situation, and had progressed through all modelling phases (Verschaffel et al. 2010).

30.4 Conclusions

The design research approach employed in this study allowed initial interactions with teachers from rural and remote Queensland to provide some insights into the teachers' current instructional practices and ideas related to the teaching of statistical numeracy, and to using graphs to model and make sense of problem situations. First we identified that teachers may well focus on the mathematics of the problem suspending sense making in the same way Greer et al. (2007) reported for students. This suggests that teachers need to be supported to ensure that students in their classrooms have access to problem situations that are set in suitable contexts and demand them to develop the level of numeracy skills espoused in the Australian Curriculum.

Second, the teachers did develop a range of different models and they collaboratively critiqued and discussed these models. In particular, these discussions highlighted the differences between learning to draw the graphs correctly and learning to use graphs as models when interpreting and responding to contextual problem situations. In addition to supporting teachers' insights into ways in which graphs may be used in mathematics classrooms, we conjecture that these discussions laid the groundwork for teachers to have similar whole-class conversations in their classrooms.

We aimed to demonstrate aspects of ambitious instructional practices as we orchestrated the mathematical activities. We believe that both *the selected task* and *the order in which teachers' responses were presented* played important roles in the type and quality of discussion that resulted. On initial reading, the task we decided to use may appear ill structured. It was designed to provide multiple entry points, to be non-procedural, and to allow for teacher explorations of graphs in a modelling context (Boston and Smith 2009; Greer et al. 2007). As anticipated, the task was suitable for these purposes. In addition, we identified and upheld cognitive demands of the task, demonstrated 'press' where further explanations were required, revoiced teachers' contributions, and asked them to make connections between models (cf. Lampert et al. 2013). Our orchestration of the *Student exercise* PD activity became a shared reference point in the subsequent PD sessions, when teachers engaged in planning and rehearsal activities before implementing data analysis and modelling tasks in their own classrooms (Lamb and Visnovska 2013).

Lastly, in addition to learning about teachers' views and practices, the teachers' interactions during the initial PD activity provided insights into supports needed for the group to become a professional learning community with 'safe' opportunities to interact with colleagues (Lampert et al. 2013). Teachers' interactions suggested that most of them were keen to collaborate on and share mathematical ideas, and that challenges to mathematical ideas were accepted and handled respectfully. Pedagogical ideas and values remained somewhat private and largely unchallenged. For instance, there were no challenges to acceptability of graphing conventions as a learning focus in a mathematics lesson, although there seemed to be some difference of opinions about this issue in the group. Such dynamics is not surprising at the beginning of a group collaboration where group members are yet to develop mutual trust and shared goals (Visnovska et al. 2012). This initial phase of community development might be a particularly difficult task for rural remote teachers who are often physically isolated by large distances from colleagues who teach middle school mathematics. Developing PD activities in which teachers start sharing and challenging their pedagogical views was an important goal for the research team.

Implementing the General Capability, numeracy, requires new knowledge (Galbraith 2013) where context needs to be prioritised. The initial PD session highlighted for us that this would not happen by osmosis, particularly for isolated, inexperienced teachers in rural and remote regions. Therefore we suggest that considerable work to support teachers in this endeavour is needed.

Acknowledgements The research reported here was supported by the Queensland Association of Mathematics Teachers and by the Department of Education and Teaching, Central Queensland.

References

ACARA. (2013). *Australian Curriculum Mathematics General Capabilities*. http://www.australiancurriculum.edu.au/Mathematics/General-capabilities. Accessed 29 Aug 2013.

Australian Association of Mathematics Teachers. (1997). *Numeracy = everyone's business. Report of the numeracy education strategy development conference*. Adelaide: AAMT.

Boaler, J., & Staples, M. (2008). Creating mathematical futures through an equitable teaching approach: The case of Railside School. *Teachers College Record, 110*(3), 608–645.

Boston, M. D., & Smith, M. S. (2009). Transforming secondary mathematics teaching: Increasing the cognitive demand of instructional tasks used in teachers' classrooms. *Journal for Research in Mathematics Education, 40*(2), 119–156.

Cobb, P., & Whitenack, J. (1996). A method for conducting longitudinal analyses of classroom videorecordings and transcript. *Educational Studies in Mathematics, 30*, 213–228.

Cobb, P., Confrey, J., diSessa, A. A., Lehrer, R., & Schauble, L. (2003). Design experiments in education research. *Educational Researcher, 32*(1), 9–13.

Franke, M. L., Kazemi, E., & Battey, D. (2007). Mathematics teaching and classroom practice. In F. K. Lester (Ed.), *Second handbook of research on mathematics teaching and learning* (pp. 225–256). Greenwich: Information Age Publishers.

Galbraith, P. (2013). Students and real world applications: Still a challenging mix. In V. Steinle, L. Ball, & C. Bardini (Eds.), *Mathematics education: Yesterday, today and tomorrow*.

Proceedings of the 36th annual conference of the mathematics education research group of Australasia (pp. 314–321). Melbourne: MERGA.

Greer, B., Verschaffel, L., & Mukhopadhyay, S. (2007). Modelling for life: Mathematics and children's experience. In W. Blum, P. Galbraith, M. Niss, & H.-W. Henn (Eds.), *Modelling and applications in mathematics education: The 14th ICMI study* (pp. 137–144). New York: Springer.

Heslop, J. (2003). Living and teaching in aboriginal communities. In Q. Beresford & G. Partington (Eds.), *Reform and resistance in aboriginal education*. Crawley: University of Western Australia.

Jackson, K., & Cobb, P. (2010). *Refining a vision of ambitious mathematics instruction to address issues of equity*. Paper presented at the Annual Meeting of the American Educational Research Association, Denver.

Lamb, J., & Visnovska, J. (2013). On comparing mathematical models and pedagogical learning. In G. A. Stillman, G. Kaiser, W. Blum, & J. P. Brown (Eds.), *Teaching mathematical modelling: Connecting to research and practice* (pp. 457–466). Dordrecht: Springer.

Lampert, M., Beasley, H., Ghousseini, H., Kazemi, E., & Franke, M. L. (2010). Using designed instructional activities to enable novices to manage ambitious mathematics teaching. In M. K. Stein & L. Kucan (Eds.), *Instructional explanations in the disciplines* (pp. 129–141). New York: Springer.

Lampert, M., Franke, M. L., Kazemi, E., Ghousseini, H., Beasley, H., Chan, A., Cunard, A., & Crowe, K. (2013). Keeping it complex: Using rehearsals to support novice teacher learning of ambitious teaching. *Journal of Teacher Education, 64*(3), 226–243.

Staples, M. (2007). Supporting whole-class collaborative inquiry in a secondary mathematics classroom. *Cognition and Instruction, 25*(2), 161–217.

Stein, M. K., Engle, R. A., Smith, M. S., & Hughes, E. K. (2008). Orchestrating productive mathematical discussions: Five practices for helping teachers move beyond show and tell. *Mathematical Thinking and Learning, 10*(4), 313–340.

Verschaffel, L., van Dooren, W., Greer, B., & Mukhopadhyay, S. (2010). Reconceptualising word problems as exercises in mathematical modelling. *Journal of Mathematical Didaktics, 31*, 9–29.

Visnovska, J., & Cobb, P. (2013). Classroom video in teacher professional development program: Community documentational genesis perspective. *Zentralblatt für Didaktik der Mathematik, 45*(7), 1017–1029.

Visnovska, J., Cobb, P., & Dean, C. (2012). Mathematics teachers as instructional designers: What does it take? In G. Gueudet, B. Pepin, & L. Trouche (Eds.), *From text to 'lived' resources: Mathematics curriculum material and teacher development* (pp. 323–341). Dordrecht: Springer.

Chapter 31
Mathematical Modelling and Cognitive Load Theory: Approved or Disapproved?

Jacob Perrenet and Bert Zwaneveld

Abstract The research question for this theoretical study is: What is the use of Cognitive Load Theory (CLT) for mathematical modelling education (MME)? CLT is a notable theory about the consequences of human memory structure for teaching, claiming specific relevancy for mathematics education. A set of scientific CLT research papers at secondary level mathematics teaching is compared with the *Dutch Handbook of Mathematics Didactics* on pedagogical perspective as well as on concrete directives for modelling education. Cognitive scheme theory is the common base, but concrete directives strongly differ. Most CLT directives appear to be of limited use for MME, but some are interesting. A discussion of CLT's rejection of constructivism leads to the importance of structured support in teaching use of the mathematical modelling cycle.

31.1 Introduction

Learning mathematical modelling is a complex task, for students as well as for teachers. In our opinion (Perrenet and Zwaneveld 2012), mathematical modelling in secondary mathematics education should be more than teaching how to solve a real world problem by use of a given mathematical model. The students also should actively construct the model, use it and improve it, as part of the whole modelling cycle. Many representations exist for the series of actions comprising the mathematical modelling cycle. We have described this variety and end up with our own representation as seen in Fig. 31.1. Students have to learn to analyze, conceptualize,

J. Perrenet (✉)
Eindhoven School of Education, Eindhoven University of Technology, Jacob Perrenet, Randwycksingel 11c, Maastricht 6229 EG, The Netherlands
e-mail: j.c.perrenet@tue.nl

B. Zwaneveld
Welten Institute (Research Center for Learning, Teaching, and Technology), Open University of The Netherlands, Bert Zwaneveld, Bieslanderweg 18, Maastricht 6213 AJ, The Netherlands
e-mail: G.Zwaneveld@uu.nl

G.A. Stillman et al. (eds.), *Mathematical Modelling in Education Research and Practice*, International Perspectives on the Teaching and Learning of Mathematical Modelling, DOI 10.1007/978-3-319-18272-8_31

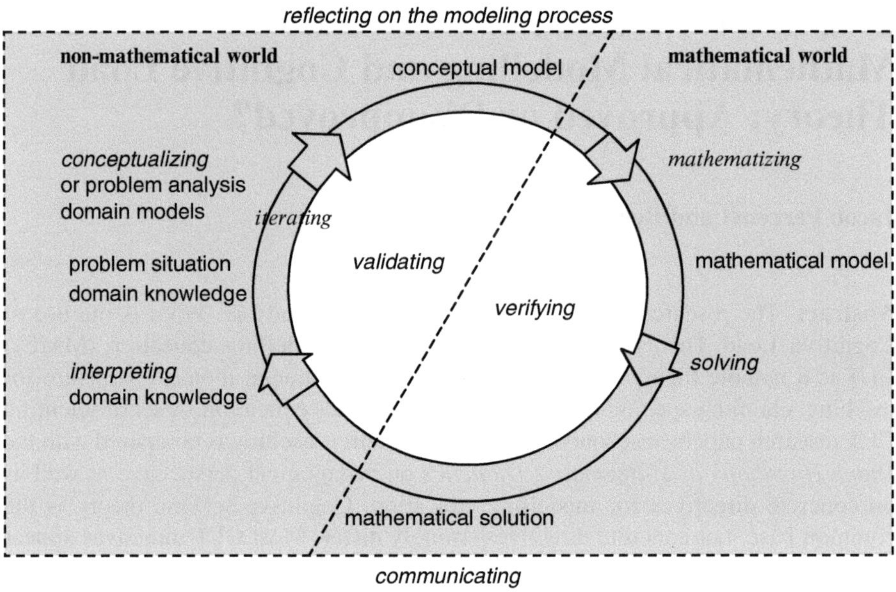

Fig. 31.1 The mathematical modelling cycle

solve problems, interpret, verify, validate, reflect, communicate, use domain models as well as mathematical models, act in the mathematical world as well as in the non-mathematical world, and more. In addition, teachers have to teach, facilitate, coach, test it, and more. This is a complex task for students and teachers; in cognitive psychological terms: it causes, for students as well as for teachers, a *heavy cognitive load*, so maybe *Cognitive Load Theory* could be helpful here.

Cognitive Load Theory (CLT) is an influential psychological theory, based upon the limited capacity of human working memory and the process of scheme construction in long term memory. It has produced a series of evidence based directives for learning task characteristics. It claims specific relevance for mathematics education, because of the high degree of interconnectivity of the elements of the mathematics domain (Sweller 1994). We will give an overview of the theory in Sect. 31.2 of this chapter.

Despite the claim of relevance, CLT does not seem to be known or accepted by the mathematical didactical community. For example, in the new *Dutch Handbook of Mathematics Didactics* (Drijvers et al. 2012) extensive attention is given to the international knowledge base of (cognitive) psychology and although the concept of cognitive load gets some attention, Cognitive Load Theory and its results are not mentioned. Specifically, nowhere is CLT related to learning mathematical modelling, one of the chapters of the *Handbook* (Spandaw and Zwaneveld 2012). In Sect. 31.3 of this chapter we will summarize the content of the *Handbook* in general and the content of the chapter about modelling education specifically.

The absence of CLT in the *Handbook* stimulated us to undertake a literature review to answer the question, *What are the similarities and differences between both perspectives?* In Sect. 31.4 we describe this review and its results (Perrenet and Zwaneveld 2014). In Sect. 31.5 we apply these results to the learning and teaching of mathematical modelling in order to answer the central question of this chapter: *What is the use of CLT for modelling education?*

CLT originally disapproved constructivism and related educational perspectives (Kirschner et al. 2006). In Sect. 31.6 we start from this stance and describe the reaction from researchers from the school of Problem Based Learning (PBL), as we are convinced that this discussion can be applied to mathematical modelling education. The example of the PBL process leads to suggestions of how to decrease the cognitive load of students and teachers in the educational process of using the mathematical modelling cycle. In Sect. 31.7 we summarize our argumentation and reflect upon our conclusions.

31.2 Cognitive Load Theory

CLT distinguishes two main parts of the human memory: long-term memory where information and knowledge are stored, and working memory that processes incoming information. Processing information means storing incoming information and linking it to information already present in the long-term memory or the working memory. CLT builds on experimental results of Miller (1956) that working memory has a limited capacity of only about seven elements. However, the human brain can create bigger wholes out of related elements, called 'chunks' or 'schemes'. This process increases the capacity of working memory, as former loose details become part of schemes. According to Sweller (1994), one of the founding fathers of CLT, the process of learning has two essential aspects, scheme acquisition and the transfer of learned procedures from controlled to automatic processing. Working memory is actively involved in these mechanisms.

Sweller (1994) distinguishes two types of learning tasks: 'low element-interactivity' and 'high element-interactivity'. A low element-interactivity task consists of learning loose elements, for example words in another language. A high element-interactivity task consists of learning more complex material, consisting of several elements which are relevant on their own, but which together connectively contribute to what Sweller calls a problem solving scheme. Specifically, learning mathematics is a high element-interactivity task.

Sweller et al. (1998) distinguish three kinds of cognitive load. *Intrinsic* cognitive load refers to the inherent difficulty of learning a specific subject. This load cannot be reduced by the instruction design. *Extraneous* cognitive load is determined by the instructional material and therefore can be influenced by the instructional design. *Germane* cognitive load is caused by the learning process itself; it is directed at the development of cognitive schemes. The general directive of CLT for instructional design is to restrict the extraneous cognitive load in order to create space in working memory for germane and intrinsic load.

CLT research has shown a number of *effects* of specific types of (mathematics) instructional design leading to various directives. The effects are proven by a shorter learning time and better results at similar problems as well as at transfer problems (Clarke et al. 2005; Paas and Van Merriënboer 1994). The directives are aimed at decreasing extraneous cognitive load. We conclude this section with a short summary of the directives:

- *Use Worked-out Examples*: Show the whole solution process of a problem.
- *Use Completion*: Present students with a part of the solution and ask to complete it.
- *Use Goal Free*: Give a mathematical situation; ask what can be derived.
- *Avoid Split Attention*: Do not spread but combine information over text and illustration.
- *Avoid Redundancy*: Do not repeat information if it is already given before or known.
- *Use More Modalities*: Present complementing information audible as well as visual.

31.3 Mathematics Didactics and Mathematical Modelling Education in the *Dutch Handbook*

The *Handbook* (Drijvers et al. 2012) offers for a target audience of teachers and teacher education students a summary of the existing international knowledge on the learning and teaching of mathematics. Cognitive psychology, including cognitive schemes, is an important part of the knowledge base. Successively treated general themes are the goals of mathematics education, cognitive schemes and the development of mathematical thinking, problem solving and routines, and instruction strategies. We summarize the text in the *Handbook* about modelling (Spandaw and Zwaneveld 2012). It treats the knowledge, skills and attitudes, teachers and students should have for modelling education.

The objective is that students know the (or a) modelling cycle and be able to use that as a heuristic method to solve problems in a non-mathematical domain. The core of a modelling process, with various stages, activities and forms of knowledge is represented by the modelling cycle of Fig. 31.1 in the introduction of this chapter. Moreover, students should be able to justify their strategy of attacking a modelling problem. They have an investigating attitude and insight into the role of mathematics in non-mathematical problems; they see mathematics as more than a set of algorithmic rules and they have an image of subsequent education.

The teacher should, before (s)he really starts a modelling activity with the students, be aware of possible strategies the students could use (depending on experiences with modelling). The instruction goals should be clear: for instance, one or more stages of the modelling cycle or the whole cycle. In connection with these goals: what do the students already know of the situation, is their domain knowledge enough? A possible choice for the teacher is to divide the activity: first the students try to make a conceptualization and after a discussion there is agreement about one common conceptualization. The same procedure could be applied to the other stages of the cycle. Another strategy is to let the students go further through the stages of the cycle with their own conceptualization. In this case the models could be different and also the solutions. This enables discussion about the validity of the different models.

Modelling is such a complex activity that testing is, generally speaking, not possible by a traditional individual testing paper in a session of 50 min or so. Another argument against individual testing is that mostly students do their modelling activities in groups. Modelling as a project with assessment by a (group) report seems more suitable.

Students should be aware that an important part of modelling is to use domain knowledge. This gives restrictions to the situations and questions, students have to model. The more domain knowledge is used, the better the resulting models.

31.4 Cognitive Load Theory and Mathematics Didactics

As mentioned before, CLT is unknown, or more probably, considered irrelevant in mathematics didactics. This triggered us to undertake a literature review (Perrenet and Zwaneveld 2014), to answer the more general research question: *What are the similarities and differences between the perspectives on mathematics education of CLT and Mathematics Didactics?* We analyzed the similarities and differences between the themes from the *Handbook* – goals of mathematics education, cognitive schemes and the development of mathematical thinking, problem solving and routines, instruction strategies – and the themes from CLT research, that is, the previously mentioned instruction strategies to decrease extraneous cognitive load and the underlying perspective of mathematics education. In Sect. 31.5 we apply the results of this earlier study to MME.

Using the search terms ‘cognitive load’ AND (‘mathematics’ OR ‘mathematical’) we consulted the databases PiCarta, ERIC and PsycINFO over the period from the first CLT publications (1988) until 2012. This resulted in about 100 hits. We selected the research articles about mathematics in secondary education, since that is the *Handbook’s* focus. The result was a set of 36 articles. We analyzed these publications using the central mathematics didactical themes from the *Handbook*. For each theme from the *Handbook* we investigated whether it played a role within CLT and if so, whether or not the results and directives of CLT agreed with those from the *Handbook*. The main results (Perrenet and Zwaneveld 2014) are the following: Both CLT and the *Handbook* have the development of coherent cognitive schemes as their point of departure, but they have different goals. CLT mainly targets short-term instruction of a specific subject and automation of procedures. The *Handbook* targets the long-term development of mathematical thinking by the development of rich, meaningful and coherent schemes. This difference culminates in a different view on expert mathematical thinking: automation versus meaningfulness. CLT deduces its recommendations in a straightforward way from theory and experiments; the *Handbook* is more eclectic. While CLT mainly limits itself to the teaching of solving routine problems, the *Handbook* focuses on teaching the use of heuristics in problem solving. The *Handbook* recommends many heuristic methods; CLT recommends only the goal-free strategy (see Sect. 31.2). The instruction strategies recommended by CLT, using worked examples or completion, are considered of limited usability by the *Handbook*; use of the goal-free method as well as avoiding split-attention and avoiding redundancy might be small but interesting extras to mathematics didactics; the advice to use the modality effect is in the *Handbook*.

31.5 Cognitive Load Theory and Mathematical Modelling Education

Based on the results of our literature search we will give a first answer to the central question of this chapter: *What is the use of CLT for MME?* MME is conceived here as a part of mathematics didactics. Clearly CLT gives no overall solution to the problem of too much cognitive load in teaching and learning mathematical modelling. However its directives could improve the situation, that is decrease extraneous cognitive load in favour of intrinsic and germane cognitive load. We will discuss the application of the directives to MME.

Use Worked-out Examples: Show the whole solution. This would mean showing a whole modelling problem solution. This looks useful but is not really new. Showing a solution is a well-known mathematical instruction strategy, potentially combined with the teacher thinking aloud.

Use Completion: Present a part of the solution and ask the student to complete it. In the modelling chapter of the *Handbook* doing part of the cycle is mentioned, but not which part. The directive means presenting a big part from the start of the cycle on and asking for only the last part by the students, for example, only validation or a second whole cycle. Another example would be starting with what Spandaw and Zwaneveld (2012) call 'small' modelling, that is starting with giving the model or with giving a few standard models.

Use Goal Free: Give a mathematical situation; ask what can be derived. This directive would mean presenting a non-mathematical situation and asking for possible models. This probably would be too open an assignment, as without a problem question the set of possible models is often large. However, it could show the relation between a model and the underlying problem question.

Avoid Split Attention: Do not spread information over text and illustration, but combine it at the same location. This might be useful when relevant in modelling texts and illustrations.

Avoid Redundancy: Do not repeat information, if it is given already in the problem description or known by the student. Just like the directive before, it might be useful in a relevant situation. (Note that the directive does not mean 'avoid irrelevant information'. This interpretation would go against the core of modelling, i.e. separating relevant from irrelevant information in the first phase of modelling.[1])

As a first answer we conclude, that CLT offers, despite its limited perspective on mathematics education in general, some directives that could be tried out as valuable suggested activities related to modelling education.

31.6 Cognitive Load Theory and Problem Based Learning: A Relevant Debate

This section gives a second answer to the question of the use of CLT for MME. The answer is not derived from our own research but from a discussion within the field of psychology.

In 2006 an article appeared written by three well-known CLT-researchers with the provocative title *Why Minimal Guidance During Instruction Does Not Work: Analysis of Failure of Constructivist, Discovery, Problem-Based, Experiential, and Inquiry-Based Teaching* (Kirschner et al. 2006). Their argument was as follows: Unguided or minimally guided instructional approaches are very popular and intuitively appealing. However, the structures of human cognitive architecture as well as evidence from empirical studies are ignored. Approaches that place a strong emphasis on guidance of the student learning process are more effective and more efficient. Learners are only ready for less guidance when they possess sufficiently high prior knowledge to provide internal guidance (Kirschner et al. 2006). The crux in their argument of course is how to handle the various types of cognitive load.

In our opinion, MME is related to, or at least uses, aspects of the series of educational perspectives summed up in the title of the article mentioned above. So the second answer to the question of the usefulness of CLT to MME would be that, according to CLT, learning only limited forms of mathematical modelling would be possible. It would disapprove MME as far as it is related to approaches like Constructivist Learning, Discovery Learning, Problem Based Learning, Experiential Learning, and Inquiry Based Learning. It would disapprove what could be called 'big' modelling: exploring new mathematics when needed for the task, although proven to be useful in higher mathematics education (Perrenet and Adan 2002, 2011). But would the conclusion be to use only 'small' modelling (Spandaw and Zwaneveld 2012), that is, providing the model in advance, using only standard models, using only very limited validation, and using only one iteration of the

[1] Verbal discussion with Paul Kirschner at the Onderwijs Research Dagen (Educational Research Days) in Louvain, May, 2013).

modelling cycle? Should secondary school level MME avoid authentic modelling from the real, messy world, because of human cognitive architecture and the risk of cognitive overload?

The argument of Kirschner et al. was countered from an unexpected angle. Former CLT researchers, Paas and Van Gog, now working in the context of Problem Based Learning (PBL), published an answer, titled *PBL is Compatible with Human Cognitive Architecture: Commentary on Kirschner et al. (*2006*)* (Schmidt et al. 2007).

At first, we will explain the characteristics of PBL. In PBL groups of students (originally in higher medical education) start with a problem situation. They list what they know; then what they should learn; they look for this new knowledge individually and come back to the group to report, compare and reflect. The group process is facilitated by a tutor. The steps of this process can be described by the so-called seven-jump: (1) clarifying terms and concepts of the problem description, (2) identifying the issues of the problem, (3) brain storming about possible explanations, (4) setting learning objectives, (5) studying privately, (6) sharing results, (7) evaluating results. See for further details Perrenet et al. (2000).

The counter argument of Schmidt et al. (2007) is as follows: PBL is not unguided nor minimally guided, as PBL has a supporting action plan, the seven-jump, and PBL uses scaffolding for student independence. The small groups are trained in group collaboration skills prior to instruction. The learning task is to explain phenomena described in the problem in terms of its underlying principles or mechanisms; initially the problem at hand is discussed, activating prior knowledge. The tutor facilitates the learning, using a tutor instruction, consisting of relevant information, questions, etcetera, provided by the problem designer. There are resources for student self-directed study: books, articles, and other media. Thus PBL is well-guided, which decreases extraneous cognitive load.

The point for this chapter is not to compare the modelling cycle with the seven-jump but to use an analogical argument: MME is not unguided nor minimally guided, as MME has a supporting action plan, the modelling cycle. In addition, MME uses scaffolding for student independence. Small groups could be trained in group collaboration skills prior to instruction. The learning task is to solve problems from outside mathematics using mathematics; initially the problem at hand could be discussed activating prior knowledge (from the domain and from mathematics). The teacher facilitates learning. Resources for self-directed study could be used by students: books, articles, etcetera on mathematical as well as domain knowledge. In this way the students' cognitive load would be decreased. The teacher could use a teacher instruction consisting of relevant information (from mathematics and from the domain), hinting questions, possible student errors and misconceptions, etcetera, provided by the problem designer; the teacher instruction could also explain the principles of group work and adequate testing. In this way the teacher's cognitive load would be decreased. Thus MME could be well-guided and this guidance would decrease extraneous memory load. We use the term 'could', as we are not sure that all measures described with 'could' are really present in MME practice and Spandaw and Zwaneveld (2012) do not explicitly mention them. Eventually, the second answer to the question of the use of CLT for MME is, that it stresses the importance of giving supporting structure for students and teacher in the modelling process. An important difference with PBL and its seven-jump is, that there is no uniform representation of the modelling cycle (Perrenet and Zwaneveld 2012) and that even within a given modelling cycle students choose individual paths (Blum and Borromeo Ferri 2009).

31.7 Conclusions and Discussion

At one hand the use of CLT for MME is to ask attention for the aspect of cognitive load in complex tasks and the concepts of extraneous, intrinsic and germane cognitive load. Some interesting minor suggestions to decrease extraneous cognitive load are the use of Worked-out Examples, Completion, Goal Free, and avoiding Split Attention and Redundancy. A major suggestion is to support the student as well as the teacher with structure and resources, such as small group training in collaboration skills, discussing the problem, activating prior knowledge, and using resources for self-directed study for the students, and possibilities as a teacher instruction consisting of relevant information, hinting questions, possible student errors, explanation of the principles of group work and adequate testing to support the teacher. On the other hand, CLT should not be followed in its rather limited perspective on mathematics education in general. So, we are more positive about CLT for MME than for mathematics didactics in general.

In our analysis we compared CLT publications with the *Dutch Handbook*. We do not pretend that the perspective of this *Handbook* represents all mathematics education or all modelling education specifically. However, we feel confident that many aspects of modelling education as described are shared with the ICTMA community's vision.

The characteristic of CLT as being narrowly focused and dogmatic has really changed in recent years. Various CLT researchers now have more attention for complex problem solving, reflection and longer learning perspectives, as especially Van Merriënboer showed us.[2] So, the image we sketched of CLT is only true for what we would call the 'classical' CLT.

During the time our opinion of CLT has changed also. Originally, we mostly disapproved CLT's directives for mathematics education, just like our colleagues of the *Handbook* did. Later on, we supported some of CLT's directives as interesting minor suggestions to decrease cognitive load, but still we disapproved the directive to avoid redundancy. We mistakenly interpreted redundancy as referring to the aspects of a problem situation not essential for the problem, and we considered the activity of separating relevant and irrelevant aspects as an important modelling activity. When this was clarified, we still disapproved CLT's banning of all

[2] Several discussions with Jeroen van Merriënboer by e-mail and by bike, Spring, 2012.

constructive-like education. However, the debate about Problem Based Learning broadened our view of CLT. We will end with a citation of Leron and Hazzan (2006, p. 122): "Building bridges between the conceptualizations used in mathematics education and in cognitive psychology may benefit both communities."

Acknowledgements We thank Jeroen van Merriënboer and Paul Kirschner for the discussions on CLT and mathematics education.

References

Blum, W., & Borromeo Ferri, R. (2009). Mathematical modelling: Can it be taught and learnt? *Journal of Mathematical Modelling and Application, 1*(1), 45–58.

Clarke, T., Ayres, P., & Sweller, J. (2005). The impact of sequencing and prior knowledge on learning mathematics through spreadsheet applications. *Educational Technology Research and Development, 53*(3), 15–24.

Drijvers, P., Van Streun, A., & Zwaneveld, B. (Eds.). (2012). *Handboek wiskundedidactiek.* Utrecht: Epsilon [in Dutch].

Kirschner, P. A., Sweller, J., & Clark, R. E. (2006). Why minimal guidance during instruction does not work: An analysis of the failure of constructivist, discovery, problem-based, experiential, and inquiry-based teaching. *Educational Psychologist, 41*(2), 75–86.

Leron, U., & Hazzan, O. (2006). The rationality debate: Application of cognitive psychology to mathematics education. *Educational Studies in Mathematics, 62*(2), 105–162.

Miller, G. A. (1956). The magical number seven, plus or minus two: Some limits on our capacity for processing information. *Psychological Review, 63*(2), 81–97.

Paas, F. G. W. C., & van Merriënboer, J. J. G. (1994). Variability of worked examples and transfer of geometrical problem-solving skills: A cognitive load approach. *Journal of Educational Psychology, 86*(1), 122–133.

Perrenet, J., & Adan, I. (2002). From mathematical modelling to design based learning; a bridge too far? *International Journal of Mathematical Education in Science and Technology, 33*(2), 187–197.

Perrenet, J., & Adan, I. (2011). Fifteen years of experience with modelling courses in the Eindhoven program of applied mathematics. *Journal of Mathematical Modelling and Application, 1*(4), 3–19.

Perrenet, J., & Zwaneveld, B. (2012). The many faces of the mathematical modelling cycle. *Journal of Mathematical Modelling and Application, 1*(6), 3–21.

Perrenet, J., & Zwaneveld, B. (2014). *Cognitive load theory and mathematics didactics: Excluding or complementing? A literature study*. Unpublished manuscript.

Perrenet, J. C., Bouhuijs, P. A. J., & Smits, J. G. M. (2000). The suitability of problem-based learning for engineering education: Theory and practice. *Teaching in Higher Education, 5*(3), 345–358.

Schmidt, H. G., Loyens, S. M. M., van Gog, T., & Paas, F. (2007). Problem-based learning *is* compatible with human cognitive architecture: Commentary on Kirschner, Sweller, and Clark (2006). *Educational Psychologist, 42*(2), 91–97.

Spandaw, J. H., & Zwaneveld, B. (2012). Modelleren. [in Dutch]. In P. Drijvers, A. van Streun, & B. Zwaneveld (Eds.), *Handboek Wiskundedidactiek* (pp. 235–264). Utrecht: Epsilon Uitgaven.

Sweller, J. (1994). Cognitive load theory, learning difficulty, and instructional design. *Learning and Instruction, 4*, 295–312.

Sweller, J., van Merriënboer, J. J. G., & Paas, F. (1998). Cognitive architecture and instructional design. *Educational Psychology Review, 10*(3), 251–296.

Chapter 32
Social-critical Dimension of Mathematical Modelling

Milton Rosa and Daniel Clark Orey

Abstract Among innovative teaching methodologies, it is important to highlight the use of the social-critical dimension of mathematical modelling to solve problem situations that afflict contemporary society. Research related to mathematical modelling and the social-critical dimension of this approach has redefined its objectives, and is developing a sense of its own nature and potential of research methods in order to legitimize its pedagogical action. In this regard, it is necessary to discuss the importance of philosophical and theoretical perspectives found in social-critical dimensions of mathematical modelling and its epistemology as well. The importance of a learning environment that helps students to develop their social-critical efficacy is supported by the use of mathematical modelling.

32.1 Introduction

In order to reflect on the social-critical dimension of mathematical modelling, two questions are necessary:

- What is the role of schools in promoting students' social-critical efficacy?
- How can pedagogical practices currently used in the process of teaching and learning mathematics impact students' social-critical efficacy?

This allows us to determine the main goals for schools that relate to the development of creativity and criticality that will help students apply different tools for them to solve problems faced in their daily lives as well as competencies, abilities, and skills to help them to live in society.

Unfortunately, in most cases, these goals are established in school curricula without the participation of community input. This curricular aspect contributes to an authoritarian education which demotivates and promotes passivity in parents,

---

M. Rosa (✉) • D.C. Orey
Centro de Educação Aberta e a Distância (CEAD), Universidade Federal de Ouro Preto (UFOP), Campus Morro do Cruzeiro-Bauxita, Ouro Preto, MG 35.400-000, Brazil
e-mail: milton@cead.ufop.br; oreydc@gmail.com

G.A. Stillman et al. (eds.), *Mathematical Modelling in Education Research and Practice*, International Perspectives on the Teaching and Learning of Mathematical Modelling, DOI 10.1007/978-3-319-18272-8_32

teachers and students. The focus of education must be to prepare both teachers and students to be active, critical, and reflective participants in society. However, in order to reach this objective, it is necessary that the entire community supports teaching and learning processes that help students to develop their social-critical efficacy. This means that teachers should be encouraged and supported to adopt pedagogical practices that allow their students to critically analyze problems that surround them in order to promote social justice.

32.2 Conceptualizing Social-critical Efficacy

One of the most important characteristics of teaching for a social-critical efficacy is the emphasis on developing the students' critical analysis of the specific role mathematics plays in the power structures of society (Mellin-Olsen 1987; Skovsmose 1990). Another important feature of this kind of teaching is related to the students' own reflective abilities about social elements that underpin life in a globalized world. Thus, critical perspectives in relation to social conditions affect students' own experiences and may help them identify common problems and collectively develop strategies to solve them.

This is a form of transformatory learning based on students' previous experiences, which aims to create conditions that help to challenge predominant and harmful values. By using their own experiences combined with critical reflection on these experiences, students are able to develop their own rational discourse in order to create meanings necessary for the structural transformation of society (D'Ambrosio 1990; Fasheh 1997; Mellin-Olsen 1987).

Rational discourse is a special form of dialogue in which all parties have the same rights and duties to claim and test the validity of their arguments in an environment free of social and political domination. In so doing, rational discourse provides an action plan that allows participants to dialogue, resolve conflicts, and engage collaboratively to solve problems in accordance with a set of specific rules. In this type of discourse, intellectual honesty, elimination of prejudices, and critical analysis of the facts are important aspects that allow dialogue to happen rationally (Rosa and Orey 2007). This rational transformation context encourages a critical analysis of social phenomena, which can be based on the implications from studies regarding the relation among mathematics, mathematical models, and society (Barbosa 2003). In this kind of educational environment, discourse, conscious work, intuition, creativity, and criticality present with important elements that help students to develop their own social-critical abilities.

32.3 Teaching for Social-critical Efficacy

Education towards students' social-critical abilities places them at the centre of the teaching and learning process. In this regard, classrooms become learning environments in which students develop and create critical abilities by applying transformatory pedagogical approaches. However, in order for this kind of education to be implemented in classrooms, it is necessary to discard some traditional approaches (Jennings 1994). Teaching then becomes a social-cultural activity that should induct students into the creation of knowledge instead of its transmission. This means that pedagogical transformation approaches are the antithesis of pedagogical approaches that seek to transform students into containers filled with information in a *banking mode* of education (Freire 2000).

Currently, the debate between these two teaching approaches continues, but the discussions are centered in relation to content to be taught and limited in relation to the time required to teach this content. In this regard, there is a need to elaborate a mathematics curriculum that promotes critical analysis, active participation, and social transformation (Rosa and Orey 2007). There is a need for curriculum changes that seek to prepare teachers and students to become critical and responsible citizens. This mission aims to use mathematics to find practical solutions to problems faced by society, which must be in accordance with the values and beliefs practised by communities. This means that it is impossible to teach mathematics or other curricular subjects in ways that are neutral and sensitive to the true reality experienced by students (Fasheh 1997).

Thus, an important objective for schools in a democratic society is to provide the necessary tools and information through relevant activities so that students have necessary tools to discuss and critically analyze curricular content by enabling them to solve daily problems and phenomena. From our point of view, mathematical modelling is a teaching methodology focused on students' social-critical efficacy because it engages them in relevant and contextualized activities, which allow them to be involved in the construction of mathematical knowledge.

32.4 Theoretical Basis for the Social-critical Dimension of Mathematical Modelling

The theoretical basis for the social-critical dimension of mathematical modelling has its foundations in *Sociocultural Theory* and the *Critical Theory of Knowledge*.

32.4.1 Sociocultural Theory

Learning occurs through socialization because knowledge is better constructed when students work in groups and act cooperatively in order to support and encourage each other. This approach allows students to reflect on complex problems embedded in real situations that help to construct knowledge by connecting it to other knowledge areas in an interdisciplinary way. According to Sociocultural Theory, students' engagement with a sociocultural environment helps them to be involved in meaningful and complex activities. It is through social interaction (Vygotsky 1986) among teachers and students from distinct cultural groups that learning is initiated and established.

Thus, in the mathematical modelling process, the social environment also influences cognition in ways that are related to cultural context. In this regard, collaborative work between groups of teachers and students makes learning more effective as it generates levels of mathematical thinking through the use of socially and culturally relevant activities Thus context allows the use of a *dialectical constructivism* because the source of knowledge is based on social interactions between students and environments in which cognition is the result of cultural artifacts in these interactions (Rosa and Orey 2007).

32.4.2 Critical Theory of Knowledge

Studies of Habermas' *Critical Theory of Knowledge* reinforce the importance of social context for the teaching and learning processes because this theory promotes the development of the critical consciousness of students so that they are able to analyse how social context shapes their lives (Barbosa 2006; Rosa et al. 2012). This analysis occurs through intellectual strategies such as interpersonal communication, dialogue, discourse, critical questionings, and proposition of problems taken from reality.

The effects of social structure influence distinct knowledge areas purchased by individuals in the social environment, and are partly determined by interests that stimulate and motivate these same individuals. Thus, in this theory it is recognized that there are three knowledge domains (Habermas 1971):

(a) *Technical Knowledge (prediction)* is defined by the way individuals control and manipulate the environment. It is gained through empirical investigations and governed by technical rules. In the mathematical modelling process, students apply this instrumental action when they observe the attributes of specific phenomena, verify if a specific outcome can be produced and reproduced, and know how to use rules to select different and efficient variables to manipulate and elaborate mathematical models (Brown 1984).
(b) *Practical Knowledge (interpretation and understanding)* identifies individuals' social interactions through communication. In the mathematical

modelling process, students communicate by using hermeneutics (written, verbal, and non-verbal communication) to verify if social actions and norms are modified by communication. It is in this kind of knowledge that meaning and interpretation of communicative patterns interact to construct and elaborate the community understanding that serves to outline the legal agreement for the social performance.

(c) *Emancipatory Knowledge (criticism and liberation)* is defined by the acquisition of insights that seek to emancipate individuals from institutional forces that limit and control their lives. It is necessary to determine social conditions that cause misunderstandings in the communication process, tactics that may be used to release particular oppressive and repressive forces, and risks that are involved in these tactics. The objective of this kind of knowledge is to emancipate individuals from diverse modes of social domination. In the mathematical modelling process, insights gained through critical self-awareness of the elaboration of mathematical models are emancipatory in the sense that students may be able to recognize the correct reasons to solve problems faced by their communities. During this process, knowledge is gained by self-emancipation through reflection leading to a transformed consciousness.

However, learning begins to be generated in technical knowledge in conjunction with social existence through interactive and dialogical activities. In the mathematical modelling process, this approach helps students to take ownership of emancipatory knowledge. In this perspective, knowledge is translated in interdisciplinary and dialogical ways so they can be used as instruments for social transformation.

32.5 Determining an Epistemology of the Social-critical Dimension of Mathematical Modelling

Currently, there is no real consensus on specific epistemologies for social-critical dimensions of mathematical modelling, which we describe as a process that involves the elaboration, critical analysis, and validation of a model that represents a system taken from reality. In so doing, students need to work in a motivating learning environment so that they are able to develop and exercise their creativity and criticality through reflexive analysis along with the generation and production of knowledge.

One important objective of the educational system is to provide necessary information that enables students to develop their mathematical thinking in order to critically discuss and analyse mathematical curricular content. In this context, modelling is a teaching methodology that focuses on the development of social-critical efficiency of the students because it helps them to reflect on the role of mathematics in society through the elaboration of mathematical models. It is important to emphasize that models "are not neutral descriptions about an independent reality, but that the modelling process has devices that are usually concealed to the general public" (Barbosa 2006, p. 294).

This involves a critical analysis of social phenomena through the elaboration of mathematical models. This process develops a sense of conscious work, intuition, creativity and emotion that helps students to develop their own social-critical efficacy. In the academic context, this critical engagement is important because it relates mathematical curricular activities with existing social problems in the school and community (Skovsmose 1990). In the social-critical dimension of mathematical modelling, students are expected to understand, reflect, comprehend, analyse, and take action to solve problems taken from their own reality. Starting from familiar problem-situations and contexts, students learn to make hypotheses, test them, correct them, generalize, analyse, and make decisions about the phenomena under study (Rosa and Orey 2007).

The social-critical dimension of mathematical modelling searches to explain different ways of working with reality. Thus, reflecting about reality becomes a transformational action that seeks to reduce its complexity by allowing students to explain it, understand it, manage it, and find solutions to the problems that arise therein. This learning environment allows students to work with real problems by using mathematics as a language for understanding, simplifying, and solving these situations in an interdisciplinary fashion (Bassanezi 2002). From this perspective, there are at least three distinct mathematical modelling pedagogical practices that may be used in the school curriculum (Barbosa 2003).

(a) *Case 1*: Teachers choose a problem, a situation, or a phenomenon and then describe it to the students. According to the curriculum content to be developed, teachers provide students with necessary mathematical tools that are suitable for the elaboration of the mathematical models in order to solve the proposed problem. In our opinion, this is the first step to integrate mathematical modelling into teaching and learning processes. However, for the development of social-critical efficacy, there is a need for active involvement of students in the process of teaching and learning (Rosa and Orey 2007).
(b) *Case 2*: Teachers suggest and elaborate the initial problem. Students need to investigate the problem by collecting data, formulating hypotheses, and making necessary modifications in order to develop the model. Students themselves are responsible for conducting the activities proposed in order to develop the modelling process. One of the most important stages of the modelling process refers to the elaboration of a set of assumptions, aiming to simplify and solve the mathematical model to be developed. In order to work with activities based on the social-critical dimensions of modelling, it is necessary that students relate these activities to problems found in their own communities (Rosa et al. 2012).

For example, teachers may propose students investigate the following problem:

River Pollutants Task
A company discharges its effluent into a river located near their facilities. These waters contain dissolved chemicals, substances that can affect the environment in which the river flows. How can we determine the concentration of pollutants in that river? How can you make sure that pollutant concentration in the river is below the standard limit allowed by law?

Students need to investigate the problem by collecting data and are responsible for conducting the activities proposed in order to develop the modelling process. One of the most important stages of the modelling process refers to the elaboration of the set of assumptions, which aims to simplify and solve the mathematical model to be elaborated as well as the development of a critical reflection on the data that will be collected.

In order to critically reflect on these assumptions, it is important to discuss if the average velocity of the river water is constant, or what happens in the eventuality when there is no seasonal change in the water level, or if the rate of flow is constant, the rate of pollutant concentration in the river is constant, the pollutant and the water are completely miscible regardless of the seasonal change in temperature, there is no precipitation during the period of data collection, the pollutant and water mix completely, the pollutant does not solidify in the sediments of the river, the solid particles are deposited in the sediments of the river, the pollutant is volatile because it can be reduced to gas or vapour at ambient temperatures, the pollutant is chemically reactive, and the shape of the river bed is uneven (Rosa et al. 2012).

These assumptions are related to Halpern's (1996) *critical thinking* that involves a wide range of thinking skills leading toward desirable outcomes and Dewey's (1933) *reflective thinking* that focuses on the process of making judgments related to what has occurred. This approach allows students to solve word problems by setting up equations in which they translate a real situation into mathematical terms, involve observation of patterns, testing of conjectures, and estimation of results by elaborating distinct mathematical models.

It is also necessary to determine what the key questions are that may affect the final concentration of pollutants in the river as well as on the rate of flow of pollutants on its waters. This activity helps students to reflect on the mathematical aspects involved in this problem, enabling them to understand this phenomenon so they can critically solve this situation by turning it into the wellbeing of the members of their community.

(c) *Case 3*: Teachers facilitate modelling processes by allowing students to choose their own themes that have particular value and interest. Subsequently, students are encouraged to develop a project in which they are responsible for all stages

> of the process, that is, from formulation of the problem to validation of its solution. The teachers' role is mediator of the modelling process because they step back and only intervene when necessary. This approach enables students' social-critical engagement in proposed activities in the classrooms.

On the other hand, even though there may be some disagreement regarding the use of a specific mathematical modelling pedagogical practice, it is possible to conduct activities, experiments, investigations, simulations, and research projects that interest and stimulate students at all educational levels. Thus, the choice of the approach to be used by teachers depends on the content involved, the maturity level of the students and the teachers' confidence with the modelling process itself. However, we emphasize that the critical analysis of the results obtained in any approach must be highly encouraged and developed.

During the development of the modelling process, the problems chosen and suggested by teachers or those selected by students must be used to facilitate their critically reflecting on all aspects involved in the situation to be modelled. These aspects are related to interdisciplinary connections, the use of technology, and the discussion of environmental, economic, political, and social issues. Thus, the use of content in the social-critical context becomes the stimuli for students' critical reflection towards the analysis of problems faced by the community. Its aim is to create conditions that help students to challenge dominant worldviews and values; and then give them the tools to transform them to critically reflect on them in order to develop a rational discourse (Freire 2000).

Reflective aspects are related to an *open* approach to mathematics curriculum because its pedagogical practices offer activities that apply multiple perspectives, and require constant critical reflection on the solutions to these activities. However, the open nature of modelling activities may make it difficult for students to initially understand and develop a model that satisfactorily represents the problem under study (Barbosa 2003). In this regard, the social-critical dimension of mathematical modelling may be considered as an extension of the Critical Theory of Knowledge. In this context, the emancipatory approach directs the educational objectives by addressing social and political issues in the pedagogical practices used in educational systems (Horkheimer 1982; Skovsmose 1990).

According to the Brazilian National Curriculum for Mathematics (Brasil 1998), students need to develop their own autonomous ability to solve problems, make decisions, work collaboratively, and communicate effectively. This approach is based on abilities, which help students face challenges posed by society by turning them into flexible, adaptive, reflexive, critical, and creative citizens. This aspect emphasizes the societal role of mathematics by highlighting the necessity to analyse the relevance of critical thinking about the nature of mathematical models and the function of modelling in solving everyday challenges.

32.6 The Process of the Social-critical Dimension of Mathematical Modelling

Mathematical modelling provides opportunities for students to discuss the role of mathematics and the nature of their models as they study systems taken from reality (Kaiser and Sriraman 2006). From this viewpoint, mathematical modelling may be understood as a language to study, understand, and comprehend community problems (Bassanezi 2002). For example, mathematical modelling is used to analyse, simplify, and describe daily phenomena in order to predict results or modify the phenomena. In this process, the purpose of mathematical modelling becomes the ability to develop critical skills that enable teachers and students to analyse and interpret data, to formulate and test hypotheses, and to develop and verify the effectiveness of mathematical models. In so doing, reflections become a transforming action, seeking to reduce the degree of complexity through the choice of a system that can represent it (Rosa and Orey 2007). By developing strategies that encourage students to explain, understand, manage, analyse, and reflect on all parts of this system, the process optimizes pedagogical conditions for teaching and learning so that students understand a particular phenomenon in order to act effectively and transform it according to community needs.

The application of the social-critical dimension of modelling allows mathematics to be seen as a dynamic and humanised subject. This process fosters abstraction, the creation of new mathematical tools, and the formulation of new concepts and theories. Thus, an effective way to introduce students to mathematical modelling in order to lead them towards understanding of its social-critical dimension is to expose them to a wide variety of problems or themes. To this end, questioning is used to explain or make predictions about the studied phenomena through the elaboration of models that represent these situations (Rosa and Orey 2007). Figure 32.1 shows the social-critical mathematical modelling cycle.

However, the elaboration of mathematical models does not mean they will develop a set of variables that are qualitative representations or quantitative analysis of the system because models are understood as approximations of reality. Since mathematical modelling is a process that checks whether parameters are critically selected for solutions to models in accordance with their interrelationship with selected variables from holistic contexts of reality, it is not possible to explain, know, understand, manage, and cope with reality outside of a holistic context (D'Ambrosio 1990).

From a social-critical context, mathematical modelling is impossible to work without the theories and techniques that facilitate solutions for models that are not simply memorized and then forgotten. Traditional teaching often prevents time for students to learn creativity, conceptual elaboration, and the development of logical and critical thinking. According to this perspective, social-critical dimensions of mathematical modelling facilitate the competencies, skills, and abilities necessary for teachers and students to play a transformative role in society (Rosa and Orey 2007).

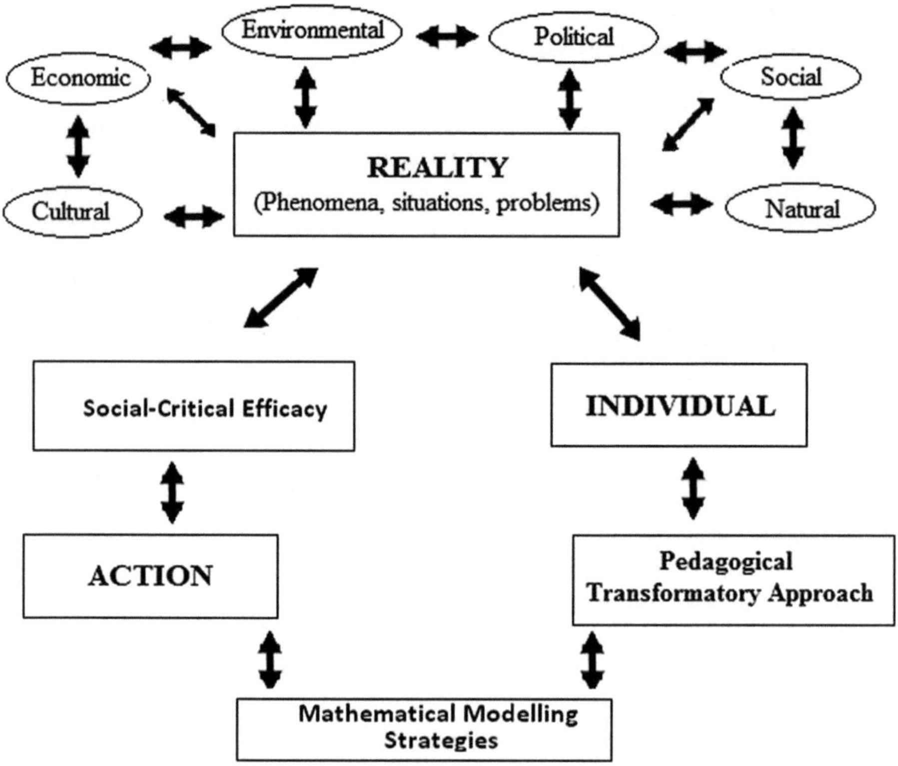

Fig. 32.1 Social-critical mathematical modelling cycle

32.7 Final Considerations

Fundamental characteristics of teaching towards social-critical efficacy emphasize a critical analysis of societal power structures through modelling. As well, modellers are encouraged to reflect on social elements that underpin our increasingly globalised world. Thus, a students' critical perspective in relation to social conditions affect their own experiences and help them to identify common problems and collectively develop strategies (D'Ambrosio 1990). This unique and transformatory form of learning creates conditions that help teachers and students work together to challenge worldviews and values dominant in society. Through these experiences, students are guided towards developing rational discourse by creating meanings that are necessary for the structural transformation of society (Freire 2000). This transformation involves critical analysis of social phenomena through the elaboration of data-based mathematical models.

In this context, mathematical modelling becomes a teaching method that focuses on the development of social-critical efficacy that engages students in a contextualized teaching-learning process that allows them to be involved in the construction of solutions of social significance (Rosa and Orey 2007). This social-critical dimension of mathematical modelling is based on comprehending and understanding reality, in which students reflect, analyze and take action on this reality. When we borrow systems from reality, students begin to study the symbolic, systematic, analytical and critical contexts to their work. Starting from real problem situations, students learn to make, test and correct hypotheses, make transfers, generalize, analyze, complete and make decisions about the object under study. Thus, using social-critical mathematical modelling can explain ways to work with reality through a transformational action that reduces its complexity and allows students to explain, understand, manage, and find solutions for their own interests and problems.

References

Barbosa, J. C. (2003). What is mathematical modelling? In S. J. Lamon, W. A. Parker, & S. K. Houston (Eds.), *Mathematical modelling: A way of life. ICTMA11* (pp. 227–234). Chichester: Horwood.

Barbosa, J. C. (2006). Mathematical modelling in classroom: A socio-critical and discursive perspective. *ZDM, 38*(3), 293–301.

Bassanezzi, R. C. (2002). *Ensino-aprendizagem com modelagem matemática* [Teaching and learning with mathematical modelling]. São Paulo: Contexto.

Brasil. (1998). *Parâmetros curriculares nacionais (PCNs): matemática* [National curricular parameters: mathematics]. Brasília: MECSEF.

Brown, M. M. (1984). *Needed: a critical science perspective in home economics*. Paper presented at the Meeting of the American Home Economics Association, Anaheim.

D'Ambrosio, U. (1990). *Etnomatemática* [Ethnomathematics]. São Paulo: Editora Ática.

Dewey, J. (1933). *How we think*. Lexington: D. C. Heath and Company.

Fasheh, M. (1997). Is math in classroom neutral or dead? A view from Palestine. *For the Learning of Mathematics, 17*(20), 24–27.

Freire, P. (2000). *Pedagogy of the oppressed*. New York: Continuum.

Habermas, J. (1971). *Knowledge and human interests*. Boston: Beacon.

Halpern, D. F. (1996). *Thought and knowledge: An introduction to critical thinking*. Mahwah: Erlbaum.

Horkheimer, M. (1982). *Critical theory*. New York: Seabury Press.

Jennings, T. (1994). Social justice in the elementary classroom. *Social Studies and the Young Learner, 7*(1), 4–6.

Kaiser, G., & Sriraman, B. (2006). A global survey of international perspectives on modelling in mathematics education. *ZDM, 38*(3), 302–310.

Mellin-Olsen, S. (1987). *The politics of mathematics education*. Dordrecht: Reidel.

Rosa, M., & Orey, D. C. (2007). A dimensão crítica da modelagem matemática: ensinando para a eficiência sóciocrítica [Critical dimension of mathematical modelling: Teaching for social-critical efficacy]. *Horizontes, 25*(2), 197–206.

Rosa, M., Reis, F. S., & Orey, D. C. (2012). A modelagem matemática crítica nos cursos de formação de professores de matemática [Critical mathematical modelling in the mathematics teacher education program]. *Acta Scientiae, 14*(2), 159–184.

Skovsmose, O. (1990). Mathematical education and democracy. *Educational Studies in Mathematics, 21*, 109–128.

Vygotsky, L. (1986). *Thought and language*. Cambridge, MA: The MIT Press.

Chapter 33
Pedagogical Actions of Reflective Mathematical Modelling

Morgana Scheller, Paula Andrea Grawieski Civiero, and Fátima Peres Zago de Oliveira

Abstract In this chapter, pedagogical actions of reflective mathematical modelling are presented and discussed. These actions took place at Instituto Federal Catarinense – campus Rio do Sul, Santa Catarina, Brazil, with high school students undertaking technical courses in Agroecology and Agriculture. For development of reflective mathematical modelling, two environments were created, one in Scientific Initiation (i.e., Foundation Science) classes and the other one in mathematics classes. The latter occurred by means of reflective didactic transposition. Data analysis identified two categories of knowledge: mathematical and reflective. Results from a qualitative research study indicate that reflective mathematical modelling, promotes a critical interpretation of reality, aids in elaboration of mathematical concepts, and gives opportunity to develop human and social aspects.

M. Scheller (✉)
Instituto Federal Catarinense – Campus Rio do Sul, Pontifícia Universidade Católica do Rio Grande do Sul – PUCRS, Rua São José, 384, Centro, Atalanta, SC 88410-000, Brazil
e-mail: morganascheller@yahoo.com.br

P.A.G. Civiero
Instituto Federal Catarinense – Campus Rio do Sul, Universidade Federal de Santa Catarina – UFSC, Rua México, 255, casa 3, bairro Sumaré, Rio do Sul, SC 89165 643, Brazil
e-mail: paulaciviero@ifc-riodosul.edu.br

F.P.Z. de Oliveira
Instituto Federal Catarinense – Campus Rio do Sul, Universidae Federal de Santa Catarina – UFSC, Rua Eleonora Budag, 71, casa, bairro Sumaré, Rio do Sul, SC 89165 580, Brazil
e-mail: fatima@ifc-riodosul.edu.br

G.A. Stillman et al. (eds.), *Mathematical Modelling in Education Research and Practice*, International Perspectives on the Teaching and Learning of Mathematical Modelling, DOI 10.1007/978-3-319-18272-8_33

33.1 Introduction

We live in a technological society, in which social implications of science and technology are directly related to construction of mathematical models. Thus, mathematics is built up with social[1] and technological change. Within this context, school might be a space in which, apart from conditions for creating knowledge, environments for discussion and reflection on reality are made possible, in order to comprehend implications of mathematics on social constructs. Therefore, as educators, we commit to insert pedagogical actions that contribute to the formation of a reflective, curious student, who knows his/her history and reality, and is able to doubt, question and make decisions. Among these actions, we consider mathematical modelling within the context of discerning mathematical education an instigating proposal for student and teacher education.

From this perspective, two pedagogical actions (i.e., lesson sequences) are presented and discussed within a case study, both carried out with high school students. The first lesson sequence was conducted within a Scientific Initiation subject (SI) (i.e., a Foundation Science unit) and the second one within mathematics. This second lesson sequence searched elements of the Scientific Initiation lessons to approximate the content taught in mathematics. In both, mathematical modelling was present, either as a method of research, or as a teaching method with research, as advocated by Biembengut (2009). These lesson sequences were articulated with critical mathematics education, in order to promote reflections and attitude changes by students through a discussion of reality explored with mathematics.

33.2 Mathematical Modelling and Didactical Transposition: Thinking Process

The movement of mathematical modelling in Brazilian education has already existed for three decades of research, having become more intense within the last years by means of studies from several research groups in Brazil. These studies have generated significant literature, such as research work and teaching materials (sequels or work units), part of which concerns themes developed in or for high school (Biembengut 2009). Since the beginning of this movement, modelling has been understood as a set of necessary procedures to build a model where the process can be used in any field of knowledge. According to Niss et al. (2007), to build and use a model is to solve a real world problem, a problem that describes, explains or designs parts of the world, with reality as the starting point for modelling. For Bassanezi (2006), mathematical modelling is a process involving theory and

[1] Fordism, Toyotism, among others.

practice, leading the researcher to interact and understand the reality present within the research, and which may result in action affecting it in order to transform. Mathematical modelling allows for integration of questions of reality with mathematical language, and, thus, "formulat[ing], solv[ing] and elaborate[ing] expressions valid not only for a particular solution, but also serv[ing], in the future, as support for other applications and theories" (Biembengut 2009, p. 13). For Niss et al. (2007), Biembengut (2004, 2009, 2012), Bassanezi (2006), Maki and Thompson (1973) and Oke and Bajpai (1982), mathematical modelling starts from a problem and aims at the creation of a model to solve it, and this occurs through several interactions, with coming and going movements.

Within the context of education, Biembengut (2004) defines mathematical modelling as a research method used especially in the Sciences. As it follows the steps of scientific research, the use of modelling has been defended in education. The purpose is to encourage and involve students in doing research and learning mathematics at the same time, which can be used in any phase of school education. For Biembengut (2009), modelling can be used both as a research method and as a mathematics teaching method. The latter is defined by her as mathematical modelling in education.

In mathematical modelling, the teacher, in his/her pedagogical action, develops program content by redesigning mathematical models applied to a certain field of knowledge, and, simultaneously, through orientation of students towards research (Biembengut 2004). Teacher and students share tasks aimed at learning mathematical concepts, and, at the same time, develop a research exercise. As stated by Skovsmose (2005), "thus, they experience what actions based on Mathematics can mean and notice the importance of thinking" (p. 96).

In preparing a pedagogical action (i.e., lesson sequence) aimed at modelling mathematical content is adapted. These adaptations may be related to *Didactical Transposition* (DT), which, according to Chevallard (1991), is understood as a process in which "a content referred to as knowing how to teach suffers from then on, a set of adaptive changes that will enable him/her to play a role among teaching objects" (p. 45). DT is undoubtedly seductive, a seduction which is not free from ambiguities and ambivalences. In a strict sense, DT refers to the step from known knowledge to taught knowledge, allowing for an articulation of epistemological analysis with didactic analysis.

When thinking about the importance of the practice of mathematical model creation by means of DT, we conclude that it is necessary to acknowledge reality as a collective coexistence space. Therefore, it is not enough for students to understand model construction. It is necessary to know its assumptions bound to the comprehension of the world in which it is inserted and to be able to make decisions based on elements that constitute reality. According to Skovsmose (2001), "a model is not a model of 'reality' itself, it is a model of a conceptual system, created by a specific interpretation, based on a theoretical framework more or less elaborate and based on certain specific interests" (p. 42). That is the reason for the need to include reflective knowledge in the development of a mathematical modelling by means of *Reflective Didactic Transposition* (RDT). This knowledge refers to the competence of thinking about the use of mathematics and of evaluating it.

Nowadays, both mathematical modelling and DT should be carried out with a critical view. According to Civiero and Sant'Ana (2013), DT may occur in a reflective manner, that is, it occurs during the teaching learning process, in order to instigate thinking by introducing needs established by students. Thus, knowledge undergoes adaptation connected to the desire of students and designed by means of a critical conscience of knowledge produced in developing the proposed activity. This approach, together with methodological and theoretical contributions of discerning mathematical education is referred to as "Reflective Didactic Transposition, as it is not enough to transpose knowledge, it is necessary to incite thought about mathematical matters bound to reality" (p. 49).

DT without thinking may contribute to dissemination of science and mathematics as socially neutral. Scientific knowledge needs an environment of reconstruction of learning in which knowledge is processed in the interaction between classroom practice and the teaching subject. This aims at providing the student with the power to doubt, propose, consider, and respond with justification. With this vision, we suggest an approach that develops modelling integrated to critical aspects of reality, that is, a *reflective mathematical modelling*.

33.3 Context of the Study

The study was conducted in Instituto Federal Catarinense – campus Rio do Sul, a public school that offers high school integrated in technical and college courses. Data were collected in two steps: first within the SI subject, and, later, in a mathematics subject. The classes in which both lesson sequences were developed were different. Both subjects were taught by the authors of this chapter in a high school course concurrent to Agroecology and Agriculture Technician.

In High School at this institution, SI is understood as a subject in which all students go through experiences by means of a basic research project,[2] which is created and developed with orientation. Aligned with Bazin (1983), we defend that SI cannot reproduce selective, elitist and limited conceptions.

In SI project development, promotion of subject integration is sought. When necessary, mathematics stands out. In this context, modelling was used as a research method in Science that gives opportunity to study and comprehend existing phenomena/problems in other subjects, as well as in everyday life. SI classes had taken place for 2 h a week during the first three terms of the course. Students, in groups of up to two, developed projects on a certain theme of their interest with guidance of teachers.

[2] We understand that basic research occurs when the student is incited to develop an inquiring attitude, from a dialogical, problematizing and reconstructive perspective of knowledge, and which allows for an understanding of the relations between science, technology, and society, leading the student to show an active attitude, interested in change. School research cannot be reduced to just anything, as for instance: book reports; data collection or news from newspapers, magazines, books; or superficial reading and writing about a theme, among other superficial activities called research.

Mathematics classes had taken place for 3 h a week, during the first three terms of the course. In this space, two mathematical models obtained in the project resulting from the first pedagogical action (i.e., lesson sequence) were used and submitted to the didactic transposition process (Chevallard 1991), enhanced by Civiero and Sant'Ana (2013) as Reflective Didactic Transposition.

33.4 Methodological Procedures

This qualitative study (Bogdan and Biklen 2003) is an attempt at comprehending details of meanings emerging from pedagogical actions and characteristics of situations presented by researchers, instead of producting quantitative measures of characteristics or behaviours. The purpose was to investigate the problem in its natural environment (i.e., direct data source), as it manifests in activities, procedures and interactions between teacher and student.

A group of two male students (age 16) was selected for observation during 30 classes in the Scientific Initiation class within the period March to July 2007. In the second pedagogical action (lesson sequence) – 14 mathematics classes – two mathematical models arising from the project developed by students of the first pedagogical action were used, involving 30 students (ages 15 and 16), within the period September to November 2008. Unstructured observations were recorded in a logbook. Materials written by students during the lessons were used as well.

For data analysis, procedures of Content Analysis (Bardin 1979) were used, aimed at explaining and systematizing message content and the meaning of this content by means of logical and justified deduction, having as a reference their origin and message context or effects of these messages. This analysis process was performed in three steps: firstly, a reading of the records of observations performed and material produced by students took place, resulting in indicative elaborations for result construction; secondly, codification, sorting and categorisation of information contained within the records were undertaken; and, thirdly, inference and construction were performed, aimed at making data more reliable and significant according to literature.

33.5 Presentation of Pedagogical Actions and Findings

33.5.1 First Pedagogical Action

The first pedagogical action (i.e., lesson sequence) took place in SI classes, in which students chose and investigated a problem: Which is the best waste for composting used in lettuce production? For this, they followed the steps of scientific research, a process similar to modelling defended by Biembengut (2004).

As the following student transcript shows, the theme was chosen because of its proximity to the course, because it was economically and socially feasible and because there was a possibility of extension to the grower.

Student A: I wish to investigate something relating to something in the course and which we can later show to others outside school as well, like organic producers. To be used by producers.

It consisted of a research about composting from several kinds of organic waste, which sought to analyse if the rotting process of five kinds of waste, disposed in distinct windrows, occurred in a similar way or not for further use in lettuce production.

Based on their curiosities, this group of students felt the need for using mathematics to analyse the behaviour of the height of compost windrows. They elaborated mathematical models recorded in tables, graphs and algebraic models, permeated with natural and mathematical language. This led them to the perception that the height of the four windrows of waste and fibre material, in the rotting process, presented a constant variation (linear behaviour). On the other hand, the height of the windrow with only fibre material showed quadratic behaviour. In this phase, mathematical content was used to obtain answers for the query mentioned above and was necessary for thinking about the theme studied. This process is evident in student exchanges like the following:

Student A: Oh, this coefficient *b* refers to height of windrows in the beginning of the process, and, with time, compost is formed and this height diminishes, and, thus, there is a decrease of the height.

Student B: Initial and final heights of the two windrows were the same, but, during the process, they displayed different behaviour.

Different mathematical concepts were involved in obtaining a relationship for curve formation, such as variation rate, determinant, trigonometry, amongst others.

33.5.2 Second Pedagogical Action

The second pedagogical action (i.e., lesson sequence) took place by means of RDT of knowledge produced in the first pedagogical action to knowledge to be taught in the classroom in order to instigate thinking. In this pedagogical action, mathematical modelling was used as a teaching method (Biembengut 2009) for the development of the concepts of polynomial functions of first and second degree.

Initially, mathematics teachers of the group suggested four[3] SI projects developed from different themes to be treated in mathematics classes. The students chose

[3] Suggested themes were: Influence of ancestry on performance of broilers; Conservation of onion bulbs; Composting from different kinds of organic waste; Response of corn in direct and conventional plantation system.

the theme: *Composting from different kinds of organic waste*, accepting the invitation to uncover mathematical models, which resulted from the research in the first action (i.e., lesson sequence). Following the research, with the tabular data (evolution of height as a function of time), students were prompted to reconstruct reality involved in the study, thus uncovering mathematical concepts existing in it. This resulted in the mathematical models, with y being the height of a windrow after x weeks, described by:

$$y_a = -1,2x + 30 \text{ and } y_b = 0,102x^2 - 2,777x + 29,843$$

The modelled real situation allowed for the explored mathematical concepts to acquire meaning, as evidenced in the statements below:

Student G: The windrow is lowering always by the same measurement.
Teacher K: What does this mean for what you are investigating?
Student C: Look at the table, each week it lowered by 1.2 cm, it's the same amount that appears in the function. It's negative because the windrow is decreasing in size.
Teacher K: Are you talking about angular coefficient? And what does the linear coefficient represent in this formation law?
Student D: It's because of these $-1.2x$ that the graph is a straight [line].
Teacher K: Are you talking about variation of height over time?
Student E: It came to a situation – it stayed 12, 12, 12 always. I think it stopped decreasing. Then, it was time to remove the compost, it should be ready.

At every developed step, new mathematical relations arose. To unveil formation laws of functions obtained by Excel software, it was necessary to acknowledge the curve adjustment process, as well as its relations and symbols. Therefore, students used different methods, intensifying thinking and decision-making.

33.6 Discussion

Reflective mathematical modelling action provided, before data, initiation of discussions and representations. Data analysis identified two categories of knowledge: mathematical knowledge and reflective knowledge. Concerning mathematical knowledge arising from actions of reflective modelling, we point out that students go through abstraction of mathematical knowledge in this process, acquiring its meanings. These meanings are specific for mathematics as well as for the model being elaborated in the reality being investigated as is evident in the second statement of Student A.

Concerning reflective knowledge existing in the first pedagogic action, it made an environmental work that is easy to perform and economically feasible possible for students, becoming one more alternative for vegetable growers. Another highlight refers to reality experienced during the project that provided students with an agroecological management, transcending modelling. Students, after solving parts of the problem, identified the behaviour of waste that would provide compost for lettuce production. In this moment, more than just identifying the best compost and using it, they carried along all knowledge involved, mathematical or not, providing for an enhanced vision of elements that constitute composting, beyond the height of the windrow.

Despite the fact that students of the second action sequence had not experienced the whole project, in accepting the invitation of the teachers and making the choice of the theme to be treated, the reflective processes were evident. Discussion about model creation provided a meaning for mathematical elements constituting each one of the formation laws of functions, according to the explanations of students C, D, and G. Reflective mathematical modelling led them to notice other elements of the rotting process, presenting reflective knowledge that led to decision-making, as shown by the statement of student E. In the *model meaning* step, as described by Biembengut (2009), analysis of these results deepened human and social discussion of development of this project, in a manner that emphasizes the need for reflective mathematical modelling. In the reflective mathematical modelling process, there was need for acquisition of mathematical knowledge in order to be able to think about it and about the reality it is inserted into. In acquiring mathematical knowledge, students took possession of knowledge for thought, as is implicit in the statement of Student B.

The reflective mathematical modelling that occurred in the second action sequence described depended on the education conception orienting practice of the teacher. Confirming the claims of Civiero and Sant'Ana (2013), performance of RDT occurred when teachers developed classes in an environment of questions, "inciting" students to seek knowledge which would sustain, or not, assumptions defined for problem solution. In this manner, the teacher does not bring ready-made models to students, but conducts a reflective process for acquisition of such, in which the student learns how to research and enhance mathematical concepts interwoven with reality.

The role of the teacher in both actions (i.e., lesson sequences) was different. In the first one, the teacher shared the task and guided, without being sure of what would occur during the mathematical modelling process and the result it would bring. In the second one, the teacher "incited" students to discovery, but already had an idea of several possible paths to be followed by students. The teacher intended to explore certain mathematical concepts previously stipulated, following a script, which did not stop other content arising. However, both ensure thinking in the mathematical modelling process. Themes approached by mathematics can lead to thinking and critique in several moments of action. Thus, in debating questions pertinent to reality, students showed that, aside from competence to create mathematical models, they were equally prepared to think about their discoveries.

33.7 Conclusion

This chapter sought to present and discuss two pedagogical mathematical modelling actions (i.e., lesson sequences) developed with high school students. Data analysis provided evidence that reflective mathematical modelling actions allow for the presence of reflective knowledge and mathematical knowledge during the process. Mathematical modelling only makes sense if it is integrated into a process of thinking about reality and of knowledge articulated with social issues, providing the student with a search for alternatives in a solution, in researching, elaborating, making questions, rebuilding knowledge and arguing with autonomy to exercise citizenship based on science. With this focus, the reflective mathematical modelling process was made effective. Mathematical modelling, as a research method or as a mathematics teaching method (Biembengut 2009), when developed from a reflective perspective, allows for extrapolation of boundaries in the classroom. It also allows for students, aside from acquiring mathematical concepts, to be the subject of an action that demands autonomy and decision-making based on their thoughts about the problems they are researching.

Within this context, mathematical modelling developed in the first action (lesson sequence) by students composed the initial content for the second lesson sequence, being adapted for the mathematics class by means of RDT (Civiero and Sant'ana 2013), maintaining the investigation process. In this manner, the process of understanding a problem from the students' reality had as a result real meaning in mathematical concept development.

As a continuation of this study, we suggest actions that make an enhancement of the scientific glance possible for this reflective and mathematical knowledge made possible by reflective mathematical modelling. Another question to be treated further is the observation of how the student's naive curiosity is overcome, in the development of reflective mathematical modelling and/or RDT, as a start of reflective mathematical modelling, for epistemological curiosity (Freire 1996).

References

Bardin, L. (1979). *Análise de conteúdo*. Lisboa: Edições 70.

Bassanezzi, R. C. (2006). *Ensino-aprendizagem com modelagem matemática* (3rd ed.). São Paulo: Contexto.

Bazin, M. J. (1983). O que é Iniciação Científica. *Revista do Ensino de Física, 5*(1), 81–88.

Biembengut, M. S. (2004). *Modelagem matemática & Implicações no Ensino e na Aprendizagem de Matemática* (2nd ed.). Blumenau: Edifurb.

Biembengut, M. S. (2009). 30 Anos de Modelagem Matemática na Educação Brasileira: das propostas primeiras às propostas atuais. *Alexandria Revista de Educação em Ciências e Tecnologia, Florianópolis, 2*(2), 7–32.

Biembengut, M. S. (2012). Concepções e Tendências de Modelagem Matemática na Educação Brasileira. *Cuadernos de investigación y Formación Matemática, Costa Rica, 7*(10), 195–204.

Bogdan, R. C., & Biklen, S. K. (2003). *Qualitative research for education: An introduction to theories and methods* (4th ed.). New York: Pearson Education.

Chevallard, Y. (1991). *La Transposition didactique: du savoir savant au savoir enseigné*. Paris: Ed. La Fenseé Sauvage.

Civiero, P. A. G., & Sant'ana, M. F. (2013). Roteiros de aprendizagem a partir da transposição didática reflexiva. *Bolema, 27*(46), 681–696.

Freire, P. (1996). *Pedagogia da autonomia: saberes necessários à prática educativa*. São Paulo: Paz e Terra (Coleção Leitura).

Maki, D. P., & Thompson, M. (1973). *Mathematical models and applications*. Englewood Cliffs: Prentice Hall.

Niss, M., Blum, W., & Galbraith, P. (2007). Introduction. In W. Blum, P. L. Galbraith, H.-W. Henn, & M. Niss (Eds.), *Modelling and applications in mathematics education – The 14th ICMI study* (pp. 3–32). New York: Springer.

Oke, K. H., & Bajpai, A. C. (1982). Teaching the formulation stage of mathematical modelling to students in the mathematical and physical sciences. *International Journal of Mathematical Education, Science and Technology, 12*(6), 797–814.

Skovsmose, O. (2001). *Educação matemática crítica: A questão da democracia*. Campinas: Papirus.

Skovsmose, O. (2005). *Travelling through education: Uncertainty, mathematics, responsibility*. Rotterdam: Sense Publishers.

Chapter 34
Context Categories in Mathematical Modelling in Fundamentals of Calculus Teaching

Mara Kessler Ustra and Sandro Rogério Vargas Ustra

Abstract In this chapter we present results of a qualitative evaluation conducted in the Fundamentals of Calculus classes in four service courses, Administration, Biology, Accounting and Physics, from a federal university in Brazil. The objectives were to analyze the main difficulties the students had regarding mathematical modelling, especially with functions and the relationship of these difficulties with the different fields of study. The analysis of the results identified some of the main teaching difficulties were related to contextualization in mathematical modelling, as well as the teaching of mathematical concepts for different courses, whether these were related to the Exact Sciences or not.

34.1 Introduction

Despite recent technological and pedagogical advances most students from any level of education fear mathematics (Furner and Berman 2005). Even though mathematics has been one of the disciplines that have an important role in the curriculum since Elementary School, it has a high failure rate, quite evident in the results of national and international assessments (e.g., OECD 2014). Fundamentals of Calculus teaching, whether or not related to an area of the Exact Sciences, also presents many strengths and difficulties, which subsequently affect undergraduate students' failure rates at university. Thus, mathematical modelling has been really important in making mathematics teaching more interesting and motivating.

M.K. Ustra (✉) • S.R.V. Ustra
Campus Ituiutaba, Universidade do Estado de Minas Gerais – UEMG, Belo Horizonte, MG, Brazil

Faculdade de Ciências Integradas do Pontal – FACIP, Universidade Federal de Uberlândia, Rua 20, 1600 – Bairro Tupã, Ituiutaba, MG 38304-402, Brazil
e-mail: maraustra@pontal.ufu.br; mara_kessler_ustra@hotmail.com; srvustra@pontal.ufu.br; srvustra@hotmail.com

G.A. Stillman et al. (eds.), *Mathematical Modelling in Education Research and Practice*, International Perspectives on the Teaching and Learning of Mathematical Modelling, DOI 10.1007/978-3-319-18272-8_34

According to Brito and Almeida (2005), there is a consensus in the literature that the mathematical model is a pedagogical tool, which helps link school mathematics with an activity that is not essentially a mathematical problem.

In this chapter we present results from an evaluation study developed with four groups, which take the subject of Fundamentals of Calculus in the courses of Management, Biological Sciences, Accounting and Physics at a Federal University located in Triângulo Mineiro, Minas Gerais, Brazil. This study aimed to analyze the main difficulties these students had regarding mathematical modelling in functions, as well as their relationships with their fields of study, whether these are inside or outside the Exact Sciences. Thus, we developed various mathematical modelling activities considering the concept of function and analysed the students' performance and the main difficulties they faced. The evaluation conducted is part of a broader project, in which we intend to investigate contexts, possibilities and prospects for mathematical modelling in university courses in order to encourage students to effectively learn mathematical content in its various areas of application or vocational training.

34.2 Theoretical Framework

The use of mathematical models in various areas has become commonplace (Neunzert 2013). Typical examples are their use in technical processes mainly in engineering and social applications (Ferruzzi and Almeida 2013; Laudares and Lachini 2005), where they are important resources, since they provide conditions to determine the optimal operation of a particular process (e.g., planning, optimization, improvement, failure diagnosis, control, and so on). Several studies in literature suggest different modelling techniques to be adopted for specific or global processes (Lima and Saraiva 2007; Quarteroni 2009). Sá (2012) reports the use of mathematical modelling to help answer complex biosciences questions, that is, mathematics has been helping answer an increasing number of questions from the biological world. Biembengut (2009) outlines how mathematical modelling performed at various levels of education has also been improving since its initial proposal. This occurs through a significant rise in research and experience reports in mathematical education, focusing on mathematical modelling. There is also a significant academic production of monographs, dissertations, theses and articles, as well as its insertion in official documents and educational guidelines.

Mathematical models consist of sets of equations (algebraic and / or differential) that represent the relationships between the independent variables (controllable) and the dependent ones (observable). This model serves as a framework that attempts to explain a system or phenomenon as closely as possible to reality, without reaching reality itself, which is only possible by direct observation (Almeida et al. 2012).

Mathematical modelling, which is an interactive process between reality and mathematics and aims to build a model between them, consists of several procedures, divided into six subparts (Biembengut and Hein 2011):

(a) *Interaction,* through the direct or indirect study of the issue

- Problem-situation recognition
- Familiarization with the issue being modelled, determining a theoretical framework

(b) *Mathematization* or translation of problem-situation to mathematical language

- Problem formulation, hypotheses formulation
- Problem solving in terms of the built model

(c) *Mathematical model*

- Solution interpretation
- Model validation, process assessment

In the development of these procedures, the better mathematical knowledge the students have, the more complex the situations that can be understood and resolved by them. The value of existing models is not restricted to their mathematical sophistication, but it is related to personal engagement in the process and its results.

From a review of the literature with regard to research into modelling in Brazil, Malheiros (2012) points to the consolidation of this line of research while maintaining intersections with various educational theoretical trends, whether in mathematics education or other areas of knowledge. Of these, Critical Mathematics Education, interest, interdisciplinarity and contextualization stand out. Among the elements that constitute the common theoretical support to the work just mentioned, is the perspective in the "interest" that keeps students engaged in the proposed learning.

A common hypothesis related to mathematical modelling education is the real issues approach, which arise from the students' interest, and how it can be motivated and support given for the acquisition and understanding of school mathematics methods and content (Brito and Almeida 2005; Carrejo and Marshall 2007). Thus, interdisciplinarity and contextualization loom large as privileged spaces that arouse and maintain interest in learning. Particularly with respect to contextualization, some careful observations are a necessary precaution because "when working with modelling in the classroom, [it does] not contextualize the Mathematics, ie Mathematics already has its own context. (...) However, (...) to develop modelling activities, you can discuss mathematical questions in a different context than the Mathematics itself" (Malheiros 2012, pp. 874–875).

34.3 Methods for Evaluation

Through an intervention, observation and interpretive description, we investigated our didactic activities on the subject by looking at the broader context of vocational training, interpreting what occurred in the performance of the participants involved. Using more descriptive data allowed a priority to be pointed out so as to make sense of, or interpret, phenomena in terms of the meanings that the participants brought. The identification of these meanings is another focal point of the evaluation. As knowledge of reality is perspectival, that is, it is given by different points of view, it was important to bring these meanings to the understanding of reality (Denzin and Lincoln 1994).

The students participating in this evaluation study were taught this class in the first semester of the following courses: Management, Biological Sciences, Accounting and Physics. The groups consisted of 45 students on average. Even though in some cases there were more advanced students, most of them had just finished high school, and mostly came from public schools. The aim of the evaluation was to analyse the main difficulties these students in the Fundamentals of Calculus classes in the four targeted areas had regarding mathematical modelling with functions, as well as the relationship of these difficulties with their fields of study.

Data were obtained through class experiences of the situations and their record in the researcher diaries. We also used an in-class questionnaire for the Physics group. In the extension course, after the Physics students were attending the Calculus class, we administered a questionnaire with five questions on mathematical modelling. We analysed responses from 11 questionnaires that were returned. The qualitative approach to analysis was based on content analysis (Bardin 2011), and sought to establish a set of categories for understanding the investigated subject.

In Biological Sciences and Accounting the usual name for the class is "Mathematics", whereas in Management it is called "Fundamentals of Mathematics". In the Physics course we developed a parallel intervention with Differential and Integral Calculus I, which consisted of an extension course of 20 h, entitled "Oriented Studies in Fundamentals of Mathematics". In this course we taught the concepts of limits, derivatives and integrals based on the discussion of functions related to motion. These classes, except for the extension course, comprise 60 semester hours, divided into four classes of an hour's duration each every week. In Accounting, the classes are at night, whereas the other ones are in the morning.

Fundamentals of Mathematics in Management aims to enable the student to determine the domain, image and graphic of the main functions of a real variable regarding real values. Mathematics in Biological Sciences aims to provide mastery of basic concepts of elementary mathematics and specific problem solving for the other disciplines, seeking to apply mathematics in biological-related situations. Mathematics for Accounting aims to provide the students with tools that will serve as subsidies for understanding the mathematical treatment in economic theories, operational research, as well as improve their reasoning.

The content of functions for the groups had already been studied in high school following a specific order. Thus, we began with an initial example of an area from which more general cases in the subject were explained. Afterwards, we gave them a general characterization of functions, with the deepening of the concepts of limits, derivatives and integrals. In the end of the disciplines, the students should be able to choose situations where the content would be applied to their training areas, besides developing a study and presenting it as a course completion assignment. Our systematic monitoring within the study lasted for 40 h of the overall hours taught in each field of study. Due to its didactic proposal, the monitoring in the extension course was also fully planned and analysed from the perspective of the study.

34.4 Results

In this section, we point out some of the key moments we had in 2012–2013 during both the classes and the extension course on the characteristics of mathematical modelling. We chose these moments to characterize the categories that were identified regarding the students' learning difficulties. Due to the students' difficulties in their first systematic contact with mathematical modelling, we concentrate on the initial moment to work with it in the related disciplines. Regarding this research, it has highlighted more clearly the categories of analysis, which will be presented below.

On the other hand, the data from the extension course refer to the last moments when the students were almost finishing the Calculus discipline, which was studied with the other investigated and extension activities under the responsibility of a different mathematics teacher. It is also important to note that the proposal of the extension course arose from our observation that Physics students had trouble in Calculus, which is taught in the first half of the course and has very high failure rates and evasion, since it impairs their progression in the curriculum.

In Management we highlighted an in-class example on shirts examining the manufacturing costs of a factory. Since the problem was how to calculate their manufacturing costs, as well as the revenue and profit function, we questioned the meaning of each term of the first degree functions:

$$R = p \, . \, q \quad and \quad L = R - C$$

where R = revenue, p = unit price, q = quantity, L = profit, C = cost.
It was difficult for the students to understand both the modelling process itself and its purpose. This was evident in some of the students' responses as follows:

Student A1: What's this for?
Student A2: Why should we have trouble if we have all this technology nowadays?
Student A3: In what part of the course will we use it?
Student A4: It is a lot of stuff!

In the Biological Sciences class students had great difficulty interpreting and building graphical representations of functions. When we asked them to find the practical uses of this mathematics in their area, they just came up with simple definitions. We showed the students a table with the values of dependent and independent variables for the concentration of y mg/kg of aluminium due to phosphorus accumulation in x mg/kg of a specific species of rice (Batschelet 1978), and facilitated a discussion on how to obtain the corresponding value for each term of a quadratic function:

$$y = 0.0065\,x^2 - 0.6213\,x + 14.5151$$

As the next example shows, not all students were convinced of the relevance of this application of mathematics to their future career:

Student CB1: I will work with animals! Where will I apply functions working with animals?

In Accounting to give the functions a context we used a set of examples taken from the area – revenue function, cost and production function. The students found the class tiring, and they wanted to go straight to the function definition. Some student responses in this class follow:

Student CC1: The direction (of the exercise) is very big . . . Give us just the formula.
Student CC2: I'll take a Philosophy course. . . with no math.
Student CC3: You must have a PhD to understand mathematics.
Student CC4: Only the equations . . . it is very easy! When practice starts, we get confused.

In the extension course for Physics students ($n = 12$) who had already completed the Calculus class, we administered a questionnaire with five questions about mathematical modelling. We analysed 11 questionnaires after they were returned. The first question was about the importance of calculus. All students considered it important for their academic life, since it provides tools to solve problems, understand phenomena and answer questions. Only one student said that it assisted their reasoning development. When asked if they had solved any mathematical modelling exercise in Calculus applied to Physics classes, all the students said they had not done this nor did they remember this kind of exercise. Regarding context, the students pointed out that it is associated with the exercise directions or theory, the Physics content, its practical uses, the historical context, or even that it is a methodological resource the teacher uses to start a new content topic. Regarding an exercise where the instructions call for calculating the derivative of a second degree function, most students displayed a feeling of low confidence in being able to solve it, because they did not know this. When they were asked whether the exercise was Physics or not, only three students related it to motion. With regards to another "contextualized" exercise on accelerated motion, students showed more confidence in its resolution, even though only three of them could solve it correctly.

Data analysis, considering the emphasis on the contextualization in modelling developed with the different classes, allowed the identification of the following categories from the analysis:

- *relevance* of the problematized situation;
- *translation* of the situation through modelling;
- *personal context.*

The relevance is mainly associated with the recognition of bonds between the problematized situation, the training area and belonging to the students' everyday life. Translation refers to the field of mathematical concepts that enables proper modelling. The personal context involves factors associated directly with students.

34.5 Discussion and Considerations for Future Teaching

Our teaching approach to mathematical modelling situations in the four classes consisted of working with problems for modelling related to each professional area. Although the students showed more confidence regarding these issues, their level of engagement was weak, since they questioned even the relevance of the approach for their course. The same holds true in the emphasis they expected from the taught subject, that is, they did not want to deepen too much the meaning of the mathematical concepts. Many of them even claimed they chose the course because it did not require much mathematics. There was a commonplace sense among the students that the mathematical concepts would not be very useful in future practice and therefore it would not be worth their cognitive and time efforts to learn it. However, as more specific situations related to the course were presented, the students had several difficulties understanding the mathematics in the problems presented. We attribute these difficulties mainly to the teaching approach that was used in both Elementary and High School.

This questioning of the relevance of modelling activities for vocational training is also related to the isolation of the disciplines of mathematics for non-mathematical students in the curriculum of the courses considered, which do not have a consistent articulation that justifies its importance (Silva 2007). This curriculum isolation reduces the contributions of modelling practices for vocational training in the courses considered, since fragmentation in the mathematics approach has increased from the first years of schooling (Biembengut and Hein 2007).

It is a vicious circle, as these lacking links contribute to increasing the students' difficulties in understanding and applying mathematical theoretical concepts in everyday situations, as well as in situations concerning their future profession. Thus, it is necessary that training situations be established, which will enable us to:

> understand mathematics as a science that can be practiced by living so that some issues arising from the student's everyday life can be worked in the classroom, in a way which enables the exploration and construction of mathematical concepts through activities that

> are meaningful to the students, and help them build their knowledge. (Ferruzzi and Almeida 2013, p. 157)

The results of our evaluation of our teaching matched other results on the merely instrumental value of mathematical concepts, through the establishment of relationships between the modelled variables and other elements, which are commonplace for the students. Thus, modelling strategies can favour a broader understanding of the relationship between knowledge and formative (new learning), translation (of the real situation), structuring (of reasoning) and instrumental mathematics (i.e., mathematics used as a tool); therefore it is most significant (Martini 2006).

It is not enough simply to associate the content involved in mathematical modelling with the students' daily lives; but it is also necessary to ensure a relevant context for them (Bennett 2003). With respect to the notion of task context our results indicate it should be greatly expanded to include other areas within the mathematics. The contextualization comprises areas, spheres or dimensions from personal, institutional (at the undergraduate level as well), social and cultural life, which accounts for knowledge and skills, and allows the students to switch from their passive spectator condition to a more active one (Kato and Kawasaki 2011). Nonetheless, this familiarity of task context by itself is also not enough to cover a broader understanding of modelling. It is also necessary to involve the contextualization categories identified, which involve social reality itself, as well as its understanding and transformation.

We believe category discrimination can really contribute for the contextualization. Therefore, we suggest the same categories as those proposed by Silva and Marcondes (2010): mathematical knowledge application to problematize situations (KA), scientific description of commonplace facts and procedures (SD), social reality understanding (SRU), and social reality transformation (SRT). In the last two categories we promoted a discussion on social problem situations so that we could make the students take a position and effectively act (SRT case) in the problematic social reality, using the content to solve the problem being studied.

Social issues play a guiding role for mathematical, scientific and technological content, so that they contribute to their understanding in a dialectical movement that allows a decision-making focusing on changing the social reality (Aikenhead 2005). This way we may establish comprehensive relationships between the different contexts that lie in the modelled problematic situation, which includes the students' vocational training context (and its various individual and collective interests) in a cyclic movement (or spiralling) of comings and goings related to the presented problem and its broader background panoramic. Based on the three categories we established in our evaluation, we think that these broader relations should be considered not only as expected or possible outcomes, but also as possible obstacles to their own modelling work, especially in non-explicit situations.

The valorisation of social aspects, with a view to understanding and transformation refers to own origins of modelling in mathematics education, associated with the ideas of Paulo Freire and Ubiratan D'Ambrosio in the late 1970s and early 1980s (Malheiros 2012). As Biembengut (2009) states:

> Models are tools that help the person to process information, and stimulate new ideas and insights, providing a structured and comprehensive view, which includes abstract relations. They also enable us to observe and reflect on complex phenomena, and even to communicate our ideas to others. It is an important tool not only to help people's everyday lives but also to stimulate the mental process, helping us to think productively, once on the cornerstone of technology or productions there is a model, i.e., a phenomenon and ideas representation (p. 20).

The results of the evaluation allowed us to understand some of the teaching difficulties regarding mathematical modelling, as well as teaching of mathematical concepts for non-mathematical courses. Finally, we should consider the different contexts presented and their similarities with the classroom situations, where everything is worth attention and should be addressed.

References

Aikenhead, G. (2005). Educación Ciencia-Tecnología-Sociedad (CTS): una buena idea como quiera que se le llame. *Educacion Quimica, 16*(2), 114–124.

Almeida, L. W., Silva, K. P., & Vertuan, R. E. (2012). *Modelagem matemática na educação básica*. São Paulo: Contexto.

Bardin, L. (2011). *Análise de conteúdo*. São Paulo: Edições 70.

Batschelet, E. (1978). *Introdução à matemática para biocientistas*. Rio de Janeiro: Interciência.

Bennett, J. (2003). *Teaching and learning science: A guide to research and its applications*. New York: Continuum.

Biembengut, M. S. (2009). 30 Anos de Modelagem Matemática na Educação Brasileira: das Propostas primeiras às propostas atuais. *Alexandria – Revista de Educação em Ciência e Tecnologia, 2*, 7–32.

Biembengut, M. S., & Hein, N. (2007). Modelling in engineering: Advantages and difficulties. In C. Haines, P. Galbraith, W. Blum, & S. Khan (Eds.), *Mathematical modelling (ICTMA12): Education, engineering and economics* (pp. 415–423). Chichester: Horwood.

Biembengut, M. S., & Hein, N. (2011). *Modelagem matemática no ensino*. São Paulo: Contexto.

Brito, D. S., & Almeida, L. W. (2005). O conceito de função em situações de modelagem matemática. *Zetetike, 13*(23), 63–83.

Carrejo, D. J., & Marshall, J. (2007). What is mathematical modelling? Exploring prospective teachers' use of experiments to connect mathematics to the study of motion. *Mathematics Education Research Journal, 19*(1), 45–76.

Denzin, N. K., & Lincoln, Y. S. (1994). Introduction: Entering the field of qualitative research. In N. K. Denzin & Y. S. Lincoln (Eds.), *Handbook of qualitative research* (pp. 1–22). Thousand Oaks: Sage.

Ferruzzi, E. C., & Almeida, L. M. W. (2013). Modelagem matemática no ensino de matemática para engenharia. *Revista Brasileira de Ensino de Ciência e Tecnologia, 6*(1), 153–172.

Furner, J., & Berman, B. (2005). Confidence in their ability to do mathematics: The need to eradicate math anxiety so our future students can successfully compete in a high-tech globally competitive world. *Dimensions in Mathematics, 18*(1), 28–31.

Kato, D. S., & Kawasaki, C. S. (2011). As concepções de contextualização do ensino em documentos curriculares oficiais e de professores de ciências. *Ciência e Educação, 17*, 35–50.

Laudares, J. B., & Lachini, J. (2005). O uso da matemática em cursos de engenharia na perspectiva dos docentes de disciplinas técnicas. *Revista de Ensino de Engenharia, 24*(1), 39–45.

Lima, P. V., & Saraiva, P. M. (2007). A semi-mechanistic model building framework based on selective and localized model extensions. *Computers and Chemical Engineering, 31*, 361–373.

Malheiros, A. P. S. (2012). Pesquisas em modelagem matemática e diferentes tendências em educação e em educação matemática. *Bolema, 26*(43), 861–882.

Martini, M. G. A. (2006). *O conhecimento físico e sua relação com a matemática: um olhar voltado para o ensino médio*. Masters dissertation, FEUSP, São Paulo.

Neunzert, H. (2013). Models for industrial problems: How to find and how to use them – in industry and in education. In A. Damlamian, J. F. Rodrigues, & R. Sträßer (Eds.), *Educational interfaces between mathematics and industry* (pp. 59–76). Cham: Springer.

OECD. (2014). *PISA 2012 results: What students know and can do* (Vol. 1, rev. ed.). Paris: OECD.

Quarteroni, A. (2009). Mathematical models is science and engineering. *Notices of the AMS, 56* (1), 10–19.

Sá, V. (2012). Equações da vida. *UNESPCiência, 3*(28), 32–34.

Silva, M. D. F. (2007). *O uso da modelagem matemática no curso de elementos de cálculo para engenharia florestal* (pp. 706–721). Ouro Preto: Anais da V Conferência Nacional sobre Modelagem e Educação Matemática.

Silva, E. L., & Marcondes, M. E. R. (2010). Visões de contextualização de professores de química na elaboração de seus próprios materiais didáticos. *Ensaio – Pesquisa em Educação em Ciências, 12*(1), 101–117.

Chapter 35
Applied Mathematical Problem Solving: Principles for Designing Small Realistic Problems

Dag Wedelin and Tom Adawi

Abstract We discuss and propose principles for designing problems that let engineering students practice applied mathematical problem solving. The main idea is to simplify real-world problems to make them smaller, while retaining important characteristics such that the solution to the problem is still of practical or theoretical interest, and that the problem should invoke non-trivial modelling and problem solving activities. We formalize our analysis in three dimensions of learning, which provide a basis for reflection beyond just solving the problem. We further discuss the benefits of being able to consider a large and highly varied set of smaller problems for discerning problem solving patterns, and give examples of such problems. We finally discuss the relationship with other proposed ways of designing problems.

35.1 Introduction

If we wish to teach students to solve real-world problems, it is reasonable to assume that real-world problems will serve as good exercise problems. This is, for example, the starting point in both *project-based* and *problem-based learning* (Kolmos et al. 2009; Mills and Treagust 2003). However, since real-world problems – especially in engineering – are often large and complex, students may encounter only a few such problems during their studies, which may provide insufficient

D. Wedelin (✉)
Department of Computer Science and Engineering, Chalmers University of Technology, Gothenburg SE-412 96, Sweden
e-mail: dag@chalmers.se

T. Adawi
Division of Engineering Education Research, Chalmers University of Technology, Gothenburg SE-412 96, Sweden
e-mail: adawi@chalmers.se

G.A. Stillman et al. (eds.), *Mathematical Modelling in Education Research and Practice*, International Perspectives on the Teaching and Learning of Mathematical Modelling, DOI 10.1007/978-3-319-18272-8_35

variation to learn problem solving and to be able to effectively handle unknown future problems (Bowden and Marton 1997).

In this theoretical chapter, we therefore present, discuss and illustrate a set of design principles for smaller problems that preserve important characteristics of real-world problems. We motivate the design principles by drawing on the literature on problem solving and mathematical modelling, as well as our own previous work in the area. These principles extend – and in some aspects go against – related design principles that have been formulated for pure mathematical problem solving (Schoenfeld 1991; Taflin 2003).

The set of principles that we propose have been successfully implemented in a course in mathematical modelling and problem solving, developed by the first author and offered to second-year engineering students at Chalmers University of Technology (for a description of the course, see Wedelin and Adawi 2014). We therefore expect that the work described in this chapter is useful to anyone who wants to include similar problems in their courses, in order to develop their students' ability to apply mathematics in practice, and we expect the main principles to be applicable also in other domains.

35.2 Real-World Problems Have Solutions That Are of Interest

Why is a real-world problem a *real-world* problem? One answer is that the problem exists because someone is interested in its solution (for a similar characterization, see Jonassen 2011). This can be for different reasons, mainly: (1) The solution is directly needed in practice. (2) The solution contributes to the understanding of some topic. In the second case, the problem may be a theoretical problem whose solution is a useful result or insight within the context in which the problem is posed, for example for the engineer, applied mathematician or other specialist. Note that many school problems, designed just to practice the application of some method, will not fall under either category.

Principle 1 The solution to the problem should be of practical or theoretical interest.

For problems of this kind, it becomes natural not only to solve the problem itself, but also to consider the context in which it is posed, and how its solution can be interpreted and contribute to this context. The problem further acts as a reasonably truthful representative of what you can expect in practice, helping students to recognize the character of fruitful investigations, and what they typically lead to.

An obvious way to create a small problem with this property is to begin with a real-world problem, in either of the two categories above. For the first category, we may then *focus on and simplify a critical aspect of a known applied problem* that we may know about in general terms (such as predicting the weather from past data), a problem we may know from own experience, or for example a problem from a thesis project. The challenge is then to keep the *essence* of the real-world problem: the simplified problem should still be a reasonably truthful model of a real problem that is actually out there, and any solution and its derivation should be similar. This can be seen as a modelling exercise for the teacher. A meaningful textbook problem in some applied subject can possibly be used more or less as it is, provided that it is used in a way that also satisfies the considerations of Sect. 35.3.

For the second category, we may *focus on and simplify a critical aspect of a known historical research task*. This can be done by first thinking about a central concept or result in some area – and then considering a particular realistic situation where it is natural to use, explore or discover this concept or result. Or we may consider some highly simplified "model" problem if its solution illustrates some meaningful phenomenon or effect in the context in which it is posed; such problems are often used in practice as vehicles for improved understanding. We may situate the exploration directly in the theoretical context, such as defining a suitable concept, or performing some derivation or generalization. The result is then of interest as a part of the theory. Tasks of this kind are important since we want our students not just to be able to apply given knowledge, but to develop new specific or general knowledge as needed.

35.3 Real-World Problems Require Exploration

Real-world problems are often ill-defined (Mayer and Wittrock 2006; Jonassen 2011), and difficult to fully understand. They require finding a relevant point of view, using the context and making relevant assumptions to create a model of the real problem in terms of a simpler well-defined problem, which is actually solved, and an interpretation of any solution in the real situation. This is known as *modelling*, which can be seen as a key step in applied problem solving.

Then, in the modelling as well as in the theoretical analysis, real-world problems typically give rise to situations where the method of solving the problem – or some sub-problem – is not known, so investigating and exploring different ways to see and to solve the problem, becomes an integral part of the process in order to successfully proceed and not get stuck. This is traditionally known as *problem solving*. See, for example, Lesh and Zawojewski (2007), Schoenfeld (1992) and Jonassen (2011), for different perspectives on modelling and problem solving.

We can contrast these observations with the common practice in both mathematics and engineering classes (especially when mathematics is involved), to solve well-defined problems with a given method (Jonassen et al. 2006), which gives little room for exploration. Models are present everywhere in engineering, but are often perceived as truths, and many students remain unaware of the concept of a model (Wedelin et al. 2015). Additionally, with more or less given methods, there is little focus on developing problem solving skills.

So, we must ensure that the challenge in solving the small problem is similar to that of the real-world problem, and take care not to simplify too much. The problem then provides an opportunity to learn to explore, and to develop modelling and problem solving skills. Since the problem is not easily solved, it becomes natural to talk to others as a means of moving forward, which is useful in itself and for the supervision.

Principle 2 The problem should be challenging to understand in order to stimulate modelling skills and communication.

Principle 3 The problem should be challenging to solve in order to stimulate problem solving skills and communication.

Of course, the challenge is also a function of the students themselves, and how much theory the teacher provides in advance. A routine problem can become a challenging exploration simply by refraining to first introduce any theory for solving it, creating an opportunity to learn to explore. The scope of the challenge can also be controlled by formulating the question appropriately, possibly in a progression where most students will be able to find something, and where the full solution is within reach at least for some.

We note that a real problem sometimes requires learning and searching for existing theory, and this can certainly be sensible to practice. However, there is a risk that students then focus on the highly visible new theory and methods, and less on the invisible skills, creating an imbalance in the long run. In fact, many students believe that if they have difficulties in solving a problem they need to learn more theory, expecting that new given problems will always be solved with new given methods (Wedelin and Adawi 2014). If the learning objective is to especially improve students' own skills we may consider the following recommendation, although it has little to do with the realism of the problem itself:

Principle 4 The problem should not require extensive new theory to be learned before the problem is attempted.

35.4 Problems as Cases and Three Dimensions of Learning

A realistic problem designed along the lines we have discussed, can be seen as offering *learning opportunities in three dimensions*:

- *Familiarity with real-world problems.* A realistic problem and its solution (including any necessary derivation), acts as a representative example and contributes to a familiarity with real-world problems in the domain of interest.
- *Supporting knowledge.* The concepts and methods needed to solve the problem, and how they are used for this purpose (known in advance or created as a part of the solution process).

- *Processes and skills*. The particular way in which the solution (and its derivation) was found, among many different imaginable ways to approach the problem, and the modelling and problem solving techniques involved.

The first and last dimensions relate directly to the previous discussion, while the second is the conventional dimension relating to the knowledge required to solve a problem. We note that there is a rough relationship between these dimensions of learning and a framework for mathematical modelling competency by Blomhøj and Jensen (2007), although we are here concerned directly with properties of problems rather than competencies.

We note that while these dimensions are important aspects of a problem, an actual problem connects a particular real-world problem from which is has been created, particular knowledge needed to solve it, and some particular approach for solving it. It therefore contains more information than what can be seen in each dimension separately. The problem may also contain other potentially important aspects that we are unaware of.

So remembering the problem as a *case* makes it possible at a later time to discern relevant aspects of the problem, which may not have been of interest when the problem was encountered. This can be important for seeing problem solving patterns across problems, and to constructively relate to old problems when approaching new ones. The importance of cases is widely supported in the literature on applied problem solving (Jonassen 2011; Kolodner et al. 1996).

Even though the principles we have already suggested are likely to ensure that a problem cannot become too small or insignificant, we still – considering the abundance of very small and repetitive problems especially in mathematics – suggest the following principle as a safety precaution:

Principle 5 The problem should be easy to remember as a case.

35.5 Extending the Problem Solving Experience: Perspective and Variation

Given the three dimensions of learning, it makes sense for the students and the teacher to discuss and reflect on the problems especially from these three points of view. This includes the specifics of the problem in each dimension, as well as what the problem as a representative example can convey about each dimension in general. We may tell a story about the corresponding real-world problem, or show a large-scale version of a related problem. Some results, observations or methods can be explained to be generally important. When seeing the solution of the teacher, and alternative solutions, students can reflect on their own difficulties in solving the problem, and why they occurred. Overall aspects, such as strategies in modelling and problem solving, and other considerations, can be discussed based

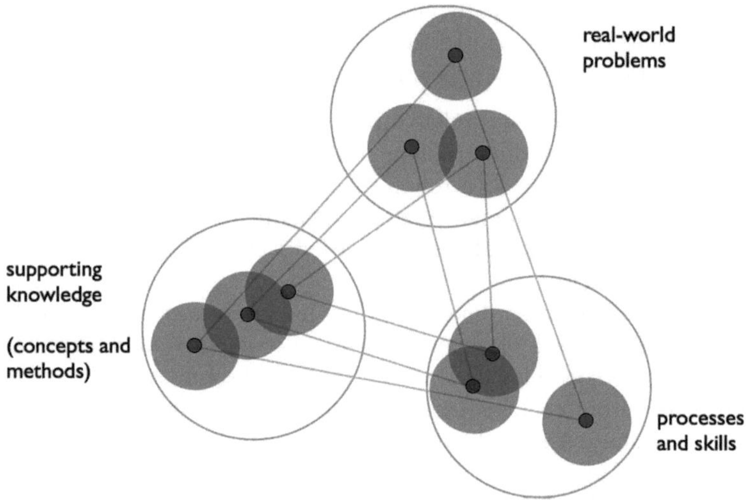

Fig. 35.1 Illustration of how several problems and their perspective span the three dimensions of learning (each *triangle* represents one problem)

on the students' own attempts to solve the problems. We refer to this entire reflection as a *perspective* on the problem.

In the perspective we include relevant *insights, attitudes, beliefs and expectations*. Such aspects are important to convey the views and ways of an experienced teacher, and are known to be important in problem solving (Schoenfeld 1992).

Principle 6 Provide a perspective on the problem in all three dimensions of learning.

We think of the perspective as an integral part of the problem, although it is not presented in the actual problem text, but provided to, and learned by, the students in other ways.

Finally, smaller problems allow us to combine many different problems to create *variation* in all three dimensions of learning, as illustrated in Fig. 35.1. Variation is essential for being able to discern critical aspects of a problem (Bowden and Marton 1997; Marton and Trigwell 2000). Through the variation, students get a chance to experience differences and similarities between problems, including higher-order patterns, and we create a basis for conveying different forms of experience in the language of an expert. We note especially that modelling and problem solving strategies are difficult to fully formalize, and have meaning mainly if the students are able to experience patterns in their work across different problems.

Principle 7 Create a problem set with variation in all three dimensions of learning.

One way to do this is to more or less arbitrarily collect a number of problems, and then iteratively build a subset of these problems under the constraint that they should cover a number of aspects that we have defined in each dimension.

35.6 Examples of Mathematical Modelling Problems

In order to give an impression of how we implement the principles in practice – including the idea of variation – we briefly give five examples of mathematical modelling problems from our course. All of the problems place students in a mode of exploration, where they have to spend time to understand the problem and explore alternatives, and where important metacognitive aspects are trained. (For a detailed analysis of how the students deal with problems of this kind, see Wedelin et al. 2015.) A more exact problem formulation for these and other problems is available on the course homepage (Wedelin 2014); these formulations have been calibrated based on how students typically approach the problems.

Kepler Curve-Fitting Problem We ask the students to suggest a function for fitting a number of given points (time and distance). No method for curve fitting, or the least squares method, is given in advance. The table contains planet data, and the best solution to this problem is Kepler's third law, making it possible to extend the perspective to a historical context. The problem requires some informal judgement about what a "good" solution to this problem is, and invokes real problem solving where students need to explore different functions, discover the so called *modelling cycle*, and so on.

Telephone Network Problem A Swedish mobile phone operator wants to rent communication lines from the national fixed network operator to connect its base stations to their central switch. Given the character of these communication lines, and the prices, how can we decide which lines to rent for a low total cost? The problem is based on a Master's thesis, and many complicating details have been removed. The problem is given as a theoretical modelling exercise, in a progression, where students are first asked to solve an even simpler version of the problem, providing some insight into how varying the problem can influence the way you solve it. The problem can be modelled as a mathematical programming problem, but can also be solved heuristically with a modified spanning tree algorithm, which is well suited to the natural dynamicity of the problem.

Bridge Problem A simple road network including two cities, and some assumptions about how speed changes with traffic intensity, is given. What is the expected travel time between the cities, and how it might improve if a bridge is built? The problem requires additional assumptions about driver behaviour and a precise formulation of equilibrium conditions. It turns out that the travel time *increases* with the new bridge. This is known as Braess' paradox; it can happen in practice, and is an instance of a Nash equilibrium (See Wu et at., Chap. 9, Sect. 9.2).

Drug Dosage Problem How can we calculate the time interval and dosage for a drug? No specific details are given. Many assumptions are required, as well as a combined qualitative and quantitative understanding of the real-world problem and different models, and it is important to split the problem solving process in steps. The problem also shows how you can reach further than what most students expect without substantial subject knowledge, by making significant assumptions and seeing what they lead to, in a kind of thought experiment.

Project Planning Problem How can large projects consisting of many subprojects be represented mathematically and how can the total project time be estimated? We further ask how this time can be computed with the help of a shortest path algorithm. How can the model be extended if we have uncertainty in the estimated times? The well-known methods CPM (Critical Path Method) and PERT (Project Evaluation and Review Technique) have been the source of this problem. The problem includes modelling with graphs, the idea of modelling one problem (the longest path problem) in terms of another (the shortest path problem), and thinking about the modelling of uncertainty.

Insights and patterns that can be illustrated and discerned from this varied ensemble of problems include the usefulness of changing the representation, the importance of a qualitative understanding, how the problem solving can be split in simpler steps, the exploratory nature of problem solving, creating examples, considering extreme cases, the power of making assumptions and drawing their logical consequences to the limit, and so on.

35.7 Discussion of Related Work

Our work was inspired by similar work focusing on *rich problems* (for an overview, see Taflin 2003). Drawing, in particular, on work by Schoenfeld (1991), Taflin (2003) developed a list of design principles for rich problems in mathematics education. These recommendations for designing good problems are similar to ours in that the *problem solving* aspect is emphasized, and in that the problem should be challenging to solve. However, our recommendation that the solution to the problem should be of interest is for rich problems restricted to the *mathematical* domain. Moreover, our recommendation that the problem should be *challenging to understand* is contradicted for rich problems. This is because the notion of rich problems does not take *models and modelling* into account. Our problems offer significant learning opportunities in applied mathematical problem solving, in addition to purely mathematical problem solving and mathematical content knowledge.

The kind of problems we propose bear a strong resemblance to a class of problems known as *model-eliciting activities* (MEAs), which is increasingly being used in engineering education (Diefes-Dux et al. 2008; Hamilton et al. 2008). MEAs are *scaled-down*, real-world problems that require students to develop or adapt a mathematical model for a given situation. With roots in mathematics education research, MEAs were originally developed to serve as instruments for investigating student and teacher thinking at school level (Lesh et al. 2000). A difference is that we start by looking at properties of real-world problems, rather than first considering the kinds of problems that most effectively reveal student thinking. Moreover, we emphasize what we have called a *perspective* for the problems, and the use of many small and *varied* problems in order to help students to discern higher order patterns, such as problem solving strategies. In other words, we take advantage of the opportunities that come with using many small problems. On a more general level, MEAs appear to focus on most phases/ roles in a project, whereas our focus is more specifically on the kind of work undertaken by an applied mathematician involved in the project.

Regarding the notion of *authenticity* (Vos 2011, 2015), we consider our problems to be authentic with respect to the real-world problem characteristics that we intentionally retain, and in some way in how we work with the problems. It is, however, not our goal to be as authentic as possible – the simplification itself is clearly for educational purposes, in order to create smaller problems to highlight important aspects of interest, and to exploit variation. And our point of view is that of a specialist engaged in a particular role in several different projects, rather than a person engaging in all aspects of a single project. We note that definitions of authenticity do not generally consider the *difficulty* of the problems, since real-world problems can be both easy and difficult.

35.8 Conclusions

The proposed way of creating smaller problems from real-world problems underlines aspects of real-world problems that we consider to be especially important for developing applied problem solving skills. The smaller problems have the potential to help students to develop in all of the *three dimensions of learning*, and they provide the students with *a case library* to draw on in future courses and in the workplace. We have also emphasized the importance of a *perspective* for the problems – a reflection on the problems in all three dimensions of learning – as an important complementary part of the problem. A prepared perspective also makes it easier for other teachers to fully understand and use the problems.

Importantly, the use of smaller problems enables *repeated and continuous feedback on the entire problem solving process*. Moreover, working with a set of varied problems opens up for *reflections on higher order patterns*, for example related to problem solving, and allows the teacher to talk about and convey his or her general experience in a meaningful way. Due to their limited size the problems are relatively easy to supervise.

The challenging nature of the problems is, in our experience, very effective for making students' own thinking visible, enabling more pointed feedback for developing complex skills and constructive attitudes. The problems encourage the students to engage in creative thinking, and convey the message that students are expected to do more than just applying given or known methods, including developing new theoretical concepts, models and methods as needed and that they are able to do so.

How do students in our course experience this approach to designing and using problems then? Most students find it extremely motivating, yet quite frustrating. At the beginning of the course they struggle, in particular, to develop effective problem solving skills, and they are hampered by unsuitable attitudes and expectations. However, the response of the students has been exceptionally positive, and after taking the course most students express and demonstrate a fundamental change in their abilities to "think mathematically", in their understanding of the nature of mathematics and its role in their future profession (Wedelin et al. 2015).

Compared to full real-world problems, the small realistic problems we propose have a bias towards being more condensed and simplified, with potential discoveries just around the corner, and with less time-consuming (and possibly boring) work. However, we have found these smaller problems very useful, and see them as a stepping-stone towards a full ability to handle real and larger projects. A course like the one we give therefore acts as an intermediate step between traditional engineering courses and full-scale projects of the kind our students will meet in their profession.

References

Blomhøj, M., & Jensen, T. H. (2007). What's all the fuss about competencies? In W. Blum, P. Galbraith, H. W. Henn, & M. Niss (Eds.), *Modelling and applications in mathematics education: The 14th ICMI study* (pp. 45–56). New York: Springer.

Bowden, J., & Marton, F. (1997). *The university of learning*. London: Kogan Page.

Diefes-Dux, H. A., Hjalmarson, M. A., Miller, T. K., & Lesh, R. (2008). Model-eliciting activities for engineering education. In J. S. Zawojewski, H. A. Diefes-Dux, & K. J. Bowman (Eds.), *Models and modeling in engineering education* (pp. 17–35). Rotterdam: Sense.

Hamilton, E., Lesh, R., Lester, F., & Brilleslyper, M. (2008). Model-eliciting activities (MEAs) as a bridge between engineering education research and mathematics education research. *Advances in Engineering Education, 1*(2), 1–25.

Jonassen, D. (2011). *Learning to solve problems*. New York: Routledge.

Jonassen, D., Strobel, J., & Lee, C. B. (2006). Everyday problem solving in engineering: Lessons for engineering educators. *Journal of Engineering Education, 95*(2), 139–151.

Kolmos, A., de Graaff, E., & Du, X. (2009). Diversity of PBL – PBL learning principles and models. In X. Du, E. de Graaff, & A. Kolmos (Eds.), *Research on PBL practice in engineering education* (pp. 9–21). Rotterdam: Sense.

Kolodner, J. L., Hmelo, C. E., & Narayanan, N. H. (1996). Problem-based learning meets case-based reasoning. *Proceedings of the International Conference on Learning Sciences* (pp. 188–195). Evanston: International Society of the Learning Sciences.

Lesh, R., & Zawojewski, J. (2007). Problem solving and modeling. In F. Lester (Ed.), *The second handbook of research on mathematics teaching and learning* (pp. 763–804). Charlotte: Information Age Publishing.

Lesh, R., Hoover, M., Hole, B., Kelly, A., & Post, T. (2000). Principles for developing thought-revealing activities for students and teachers. In A. Kelly & R. Lesh (Eds.), *Research design in mathematics and science education* (pp. 591–646). Mahwah: Erlbaum.

Marton, F., & Trigwell, K. (2000). Variatio est mater studiorum. *Higher Education Research and Development, 19*(3), 381–395.

Mayer, R., & Wittrock, M. (2006). Problem solving. In P. A. Alexander & P. H. Winne (Eds.), *Handbook of educational psychology* (pp. 287–303). Mahwah: Erlbaum.
Mills, J. E., & Treagust, D. F. (2003). Engineering education: Is problem-based or project-based learning the answer? *Australasian Journal of Engineering Education, 3*, 2–16.
Schoenfeld, A. H. (1991). What's all the fuss about problem-solving? *Zentralblatt für Didaktik der Mathematik, 23*(1), 4–8.
Schoenfeld, A. (1992). Learning to think mathematically: Problem solving, metacognition and sense making. In D. Grouws (Ed.), *Handbook of research in mathematics teaching and learning* (pp. 334–370). New York: Macmillan.
Taflin, E. (2003). *Problem solving – An analysis of rich mathematical problems* (in Swedish). Licenciate thesis, Department of Mathematics, Umeå University.
Vos, P. (2011). What is "authentic" in the teaching and learning of mathematical modelling? In G. Kaiser, W. Blum, R. Borromeo-Ferri, & G. Stillman (Eds.), *Trends in teaching and learning of mathematical modelling* (pp. 713–722). New York: Springer.
Vos, P. (2015). Authenticity in extra-curricular mathematics activities: Researching authenticity as a social construct. In G. A. Stillman, W. Blum, & M. S. Biembengut (Eds.), *Mathematical modelling in education research and practice: Cultural, social and cognitive influences* (pp. 105–113). Cham: Springer.
Wedelin, D. (2014). Course home page at www.cse.chalmers.se/edu/course/DAT026 (or via the home page of Dag Wedelin).
Wedelin, D., & Adawi, T. (2014). Teaching mathematical modelling and problem solving: A cognitive apprenticeship approach to mathematics and engineering education. *International Journal of Engineering Pedagogy, 4*(5), 49–55.
Wedelin, D., Adawi, T., Jahan, T., & Andersson, S. (2015). Investigating and developing engineering students' mathematical modelling and problem solving skills. Accepted for publication in *European Journal of Engineering Education*. doi: 10.1080/03043797.2014.987648

Part IV
Influences of Technologies

Chapter 36
Visualisation Tactics for Solving Real World Tasks

Jill P. Brown

Abstract Many affordances useful in facilitating the solution of real world tasks are available in Technology-Rich Teaching and Learning Environments (TRTLE's). Of particular use are those allowing visual image generation by technology. Whilst the TRTLE provides additional opportunities and approaches to engaging with the real world, additional complexities also exist. One of the transformational powers of the technology is to produce technology-generated images to clarify and refine students' mental models of the situation, but is this power being realised? Following a grounded theory approach, this study showed that students often did not take up the opportunities, such as the usefulness of the data plot informing their choice of function model or comparing models with data or each other, even though they had the technological and mathematical knowledge to do so.

36.1 Background

A *Technology-Rich Teaching and Learning Environment* (TRTLE) exists when teachers and students in the classrooms potentially have access to, and teachers professional development support for, a wide range of electronic technologies. To qualify as 'rich', with respect to a TRTLE, the environment would need to include unfettered access to electronic technologies that enable particular mathematical explorations. Building on the views of Pollak (1986) at ICME 5, Tikhomirov (1981) and others in more recent times (e.g., Borba 2005, 2012), claim the use of electronic technologies can transform mathematics, mathematical activity, and mathematical thinking. Tikhomirov argues, "a transformation of human activity occurs, and new forms of activity emerge" (p. 271) through the use of electronic technologies, that

Data were collected in an ARC funded Linkage Project – LP0453701 at The University of Melbourne.

J.P. Brown (✉)
Education VIC, Faculty of Education & Arts, Australian Catholic University, Level 1, 250 Victoria Parade, Melbourne, VIC 3065, Australia
e-mail: jill.brown@acu.edu.au

G.A. Stillman et al. (eds.), *Mathematical Modelling in Education Research and Practice*, International Perspectives on the Teaching and Learning of Mathematical Modelling, DOI 10.1007/978-3-319-18272-8_36

is, human activity is mediated by technology use (p. 277). Extending this, Borba (2005) introduced the metaphors humans-with-media and students-with-technology to describe the perspective where technology and humans are "completely intertwined" (p. 56). He claims "knowledge is shaped by technologies of intelligence and that such technologies are actual actors in producing knowledge and communication even though there are many other factors that shape such processes" (p. 59). Thus, in a TRTLE human mathematical modelling activity can be transformed although this transformation can bring added complexity to task-solving (Galbraith et al. 2007).

"Rapidly advancing modelling technologies . . . are making modelling accessible to an ever-broadening cross-section of knowledge workers" (Waisel et al. 1999, p. 1). Whilst TRTLE's offer many opportunities, one critical to modelling is the opportunity to use visual images generated by technological tools. However, Waisel et al. (2008 p. 360) note that "student modellers drastically underuse visualization". Where students have access to graphing technology this allows links to be made between different representations (algebraic, graphical, numerical). Such calculators make "plotting and dynamically changing graphs" simple. Thus, technology assists the development of students' modelling abilities as these two enablers are identified by Pead and Ralph as "central to the modelling process" (2007, p. 311).

In terms of mental models theory, Waisel et al. (2008), building on the work of Johnson-Laird (1988), note there are (at least) three types of mental representation: mental models, propositional representations, and images. Mental models, refer to how we imagine a situation to be or could be as we can have multiple mental models of a situation. Propositional representations are derived from information presented as text to describe the situation (i.e., linguistic comprehension). Mental images are views of some aspect of a mental model and are formed as we pose questions about the situation (i.e., internal manipulations). Externalising these mental representations, using technological (and other) tools, facilitates the development and refinement of the mental model(s).

Real world situations are by their very nature complex, hence visualisation is a critical component for successful modelling. This visualisation (both mental images and technology-generated external images) includes the comparison of a model with data (i.e., plot with function graph) and comparing models (i.e., visual comparison of models for different situations and/or to consider multiple models for the same data). Visualisation enables modellers to develop a 'good sense' of the appropriateness of their models within the real situations hence positioning themselves to undertake 'appropriate and foreshadowed' analysis of the situation under investigation. Visualisation may lead to the creation of better models and the use of these models in subsequent analysis through increased understanding of the relationship between the model and the real world. The study to be reported here investigates technology-generated visualisation practices of senior secondary mathematics students. However, Stillman (2004) found the need for such students when attempting real world application or modelling tasks to have "a well developed repertoire of cognitive and metacognitive strategies.. . .[with the latter including] strategies for monitoring, regulating and coordinating the use of these cognitive strategies" (p. 63) in order to benefit from facilitating conditions such as task features which encourage visualisation especially where there is access to visualisations tools. The coordination of these multiple mental representations and external models (e.g., a propositional model from text, an imagined mental model, a view of either as we query them and an external representation on a calculator screen) are of vital importance for efficient functioning of working memory when attempting applications and modelling tasks (Stillman 2004).

36.2 Affordances

Affordances (Gibson 1966, 1977) of the environment are taken to be the offerings of such an environment for both facilitating and impeding learning, describing allowable actions between the technology and the student. However, the existence of an affordance does not necessarily imply that activity will occur. To take advantage of the opportunities arising, students need to perceive affordances and act on them (e.g., Doerr and Zangor 2000).

One affordance related to visualisation that is particularly useful when solving real world tasks is *Function View-ability* (Brown 2013). Function View-ability is an umbrella term for affordances allowing particular views of a function to be observed. Affordance bearers (Scarantino 2003) are the objects in the environment that offer the affordance. Manifestations of this affordance include both *global function viewing* (focus on 'overall' view) and *local function viewing* (focus on a particular aspect or feature). Both involve the entering of a function in the function window (y=) and using one of several affordance bearers (e.g., GRAPH, WINDOW, ZOOM on a TI-84 graphing calculator) to facilitate the finding of a Viewing Window that allows aspects of the function of interest to be observed.

36.3 The Study

The study reported here examines two Year 11 classes that could be considered TRTLE's in the sense indicated above. The research question is: *To what extent is the expected transformative power of the technology, specifically the usefulness of technology-generated visualisation, being realised in typical classrooms when task solving?* This visualisation relates to the perception of affordances that would be useful during task solving, as well as the strategies undertaken in the enactment of these affordances. The 16–17 years old students were solving a real world function task, *The Platypus Task.*[1]

[1] See Brown and Edwards (2011) for further details of the task.

Platypus Task
The platypus is an endangered species that may become extinct unless action is taken to save it. An annual survey held in a nearby national park showed an alarming decrease in platypus numbers over the years 1993–1998. Two sets of data representing a platypus population, before and after an intervention project, were presented. Find a model to represent platypus numbers over time for both data sets. Questions then considered included: did the intervention improve the situation, what was the predicted population a decade later, and when would the population return to the initial value?

Affordances that would be useful in determining and subsequently using models for the platypus populations included: *Data Display-ability, Function View-ability, Represent-ability*, and *Check-ability*. These will be elaborated in the following sections. Technology-generated visualisation plays a critical role in each of these affordances.

36.3.1 The Participants and TRTLE's

The analysis reported here is based on deep examination of the two focus TRTLE's. These included two teachers, Peter and James [pseudonyms], and a total of 36 students. The students in both TRTLEs had just completed a unit on polynomial functions (linear, quadratic, cubic) in Year 11 Mathematical Methods and were of average age 16 years. The first TRTLE [P11] included Peter and 16 students (nine female, seven male). The second TRTLE [J11] included James and 20 students (2 female, 18 male).

The educational context in Victoria, Australia, where the study was set has included the expectation of technology use (including but not limited to graphing calculators) in senior secondary mathematics for over two decades. Graphing calculators include function graphing capabilities, data plotting capabilities, the capability to access these simultaneously (i.e., function graph can be 'overlaid' on a plot of data); table generating capabilities (ability to produce ordered pairs given algebraic representation of a function is entered); function tracing capabilities (ability to trace along function graph and have numerical coordinates displayed); and function calculation capabilities (ability to calculate particular function values including zeroes and local optimal values). All of these are often useful in solving modelling and application tasks. Extended lesson observation prior to task implementation provided evidence the students were technologically and mathematically capable to engage with the task. The students in this study all owned such calculators and made substantial use of them in class prior to task implementation.

36.3.2 Data Collection

Data collection included audio and video recording of students working on the task which allowed the researcher to "capture rich behaviour and complex interactions" (Powell, Francisco, and Maher 2003, p. 407)—two key elements of TRTLE's. Other data sources were individual student scripts, question and answer strips for recording questions and responses of others, post-task sheets and key recordings of students' graphing calculator keystrokes. The task sheet, question and answer strips, and the post task record sheet were all intended to encourage students to record as many details as possible about their thinking during task solution. Following task implementation, post-task interviews were undertaken with six students from P11 and ten students from J11.

36.4 Analysis

Data analysis followed a grounded theory approach with transcriptions including key recordings and student written work subject to open and axial coding (Strauss and Corbin 1998). This showed that many appropriate affordances were perceived and enacted in choosing various models. Several different yet appropriate approaches were taken. The analysis reported here focuses on the use of technology-generated visualisation particularly during the model finding phase of the task. This included plotting the data (*Data Display-ability* and multiple *Data Display-ability* where both data sets were viewed together), viewing the graphical representation of a function used to model the data (*Function View-ability*), comparing a function model to the data (*Data Display-ability* and *Function View-ability*), and viewing multiple models of the situation (multiple *Function View-ability*). The latter included multiple models for the same data or viewing models for both data sets.

36.4.1 Data Display-Ability

Whilst engaging with a task such as *The Platypus Task*, it was expected most modellers would begin by viewing a plot of the data (*Data Display-ability*) as they began their consideration of a suitable model, but this was not necessarily the case here. Table 36.1 shows that only 28 of the 36 students enacted the affordance *Data Display-ability*. In enacting this affordance three different strategies were used. These strategies are related to whether the data were displayed (a) prior to finding a model, (b) after a function model had been found and viewed, or (c) a plot of the data was viewed simultaneously with a function model viewed graphically. The first two strategies were additionally undertaken using a variety of sets of

Table 36.1 Enactment of the affordance *Data Display-ability* in solving the real world task

Strategy	Affordance bearers	Number of students
View plot of data to consider appropriate function type given shape	LIST, StatPlot, ZoomStat	13
	LIST, StatPlot, WINDOW	7
	LIST, StatPlot, WINDOW + ZoomOut	1
View plot of data after function graph has been viewed, in already set up window	LIST, StatPlot, GRAPH [WINDOW used previously]	4
	LIST, StatPlot, ZOOM Standard, ZoomStat	1
View plot of data *simultaneously with function graph*	LIST, StatPlot, ZoomStat	2

affordance bearers as shown (Table 36.1), illustrating the additional complexities of solving real world tasks in a TRTLE as multiple enactment pathways are available to the modeller. The majority of students viewed a plot of data thus providing technology-generated visualisation in order to support subsequent consideration of an appropriate function type given the shape of the data displayed. These students looked at the plot to inform their model construction before finding a model. A further five students enacted *Data Display-ability* after having already viewed a function model. An additional two students generated the external visualisation of the data simultaneously with their function model. These seven students were using the external visualisation for purposes other than for supporting the mental development of a model for the data such as checking.

36.4.2 Function View-Aability

Surprisingly two students although they plotted the data, did not consider that finding a function to model the data would be useful. The remaining 34 students all used the regression capabilities of the technology (*Function Identify-ability*) to do so. Thirty-one of these successfully enacted *Function View-ability*, producing an external visualisation of their model of the situation. For the three unsuccessful students all three perceived *Function View-ability* but were thwarted by a combination of a lack of technological knowledge and an apparent lack of connection between their mathematical knowledge and technological knowledge in enacting this affordance.

36.4.3 Data Display-Ability and Function View-Ability

As shown in Table 36.2, 23 of the students enacted both *Data Display-ability* and *Function View-ability*, thus viewing at least one model simultaneously

Table 36.2 Affordances perceived and enacted: *Data Display-ability* and *Function View-ability*

		Data display-ability (DD)	
		No	Yes
Function view-ability (FV)	No	2	3
	Yes	6	25

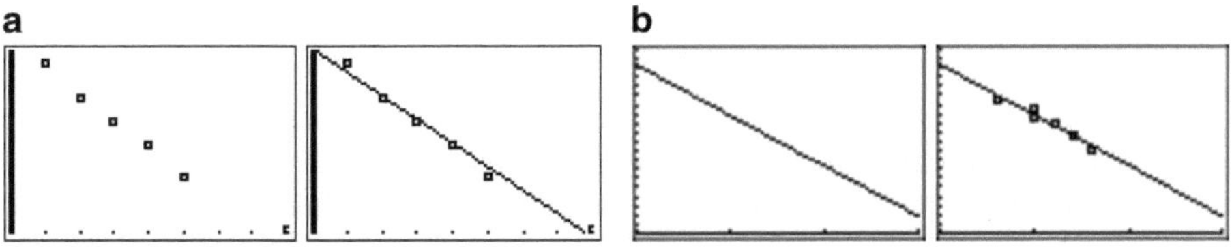

Fig. 36.1 Using (**a**) *Data Display-ability* followed by *Function View-ability* or (**b**) the reverse

with the data. Of the remaining students, two enacted neither affordance and five enacted *Data Display-ability* only – as noted earlier three of these perceived *Function View-ability* but were unable to enact this. The remaining six students enacted *Function View-ability* only. As illustrated in Fig. 36.1, at times students enacted *Data Display-ability* prior to *Function View-ability* and at other times the converse was true. Furthermore, sometimes they were enacted simultaneously. These differences tended to relate to the purpose of enactment, for example, considering the type of function that might be a useful model (Fig. 36.1a) or checking the fit of the algebraic model to the data (Fig. 36.1b).

36.4.3.1 Multiple Data Display-Ability and Multiple Function View-Ability

In a task such as *The Platypus Task*, affordances of the technology include opportunities to view the data before and after intervention simultaneously. If one's working memory is challenged by considering one mental model, then no doubt considering two related mental models adds to the cognitive demand. Thus the opportunities offered by the TRTLE for technology-generated visualisations of both data sets and of multiple models would seem to be particularly useful for mediating cognitive demand of the task (Stillman et al. 2004). In this case, multiple *Data Display-ability* from multiple data storage lists allows the modeller to produce multiple external visual representations of the data. In addition, multiple *Function View-ability* from multiple entries in a function window allows the modeller to produce multiple external visual models. In terms of the transformational power of the technology the enactment of multiple *Data Display-ability* simultaneously with multiple *Function View-ability* has the greatest potential in the model finding phase of this task.

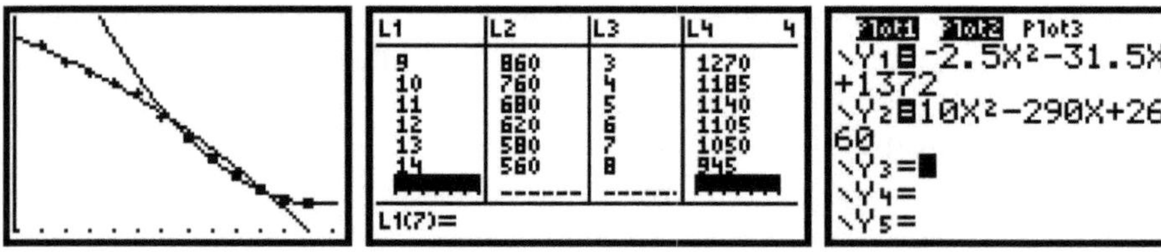

Fig. 36.2 Visually representing models before and after intervention models simultaneously with the data

Surprisingly 24 of the students restricted their use of the lists to only two of the many available. Firstly, this was time inefficient as they needed to delete one set of population data on several occasions in order to view the other set, as they moved back and forth in considering the data sets. Secondly, it was restrictive from a modelling sense as they could not compare before and after intervention plots of the data or models for these data using the technology. These 24 students were thus unable to enact multiple *Data Display-ability* and thus produce technology-generated visualisations of the two data sets simultaneously. Of the 12 students using at least four data storage lists and thus able to view multiple plots (i.e., of population data before and after the intervention project), all but one also used multiple function entries, allowing the possibility of viewing multiple functions simultaneously. Figure 36.2 shows the visualisations produced by one of these 12 students.

36.4.4 Consideration of Multiple Models

In addition to expecting all students to enact *Data Display-ability*, it was expected that some students might also compare different graphical models for the same data set. It was certainly expected that all students would view a plot of their data. The limited enactment of these affordances was unexpected. Key recordings provided evidence that only 23 (of 36) students actively considered multiple models for one or both data sets, with only nine students considering this for both data sets. For four of these nine, considering multiple models for both data sets - consideration stayed at the mental model stage only for the before intervention data. A fifth student found alternative models but did not view these. Hence, 22 students successfully found multiple models (i.e., different function types) and produced technology-generated visualisations of these, thus enacting multiple *Function View-ability*. Four of these students found multiple models for the data both before and after intervention. The remaining students did so for one data set only. The majority of occurrences of viewing multiple models occurred sequentially, rather than simultaneously.

Only three students viewed their multiple models for the same data simultaneously (two doing so for both data sets). Figure 36.3 provides an illustration of the

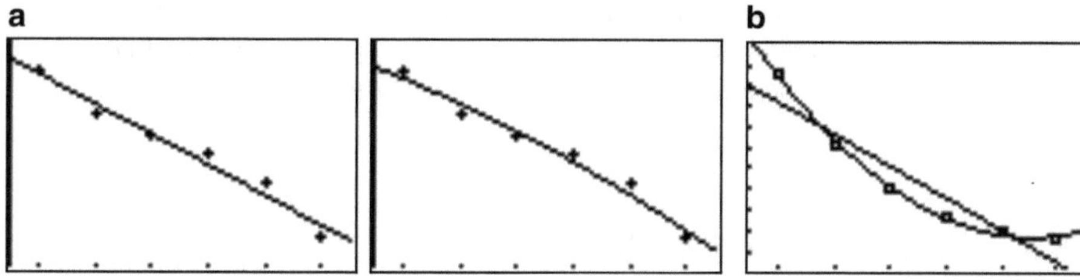

Fig. 36.3 Using *Data Display-ability* and *Function View-ability* (**a**) sequentially (**b**) simultaneously in considering multiple models for a data set

two ways of viewing multiple models, however Fig. 36.3a may be misleading, as for the students when one view was visible, the other was not. Figure 36.3b gives an indication of the advantages of the technology-generated visualisation being viewed simultaneously. These examples show concurrent use of *Data Display-ability*, however, concurrent use was not always enacted. Surprisingly, 20 students did not record at least one of the models they considered. The key recordings were invaluable in establishing what models were considered in these cases.

Most students who considered multiple models did so at the model finding stage of task solution. As noted previously, some of these considered multiple models but for various reasons were unable to pursue these. One student, May, considered an exponential function. Using the Question/answer strips she asked: 'I've forgotten everything about exponentials. How do you predict the population in *x* amount of years?' (No response was given.) Her unsuccessful tactic of asking a knowledgeable other suggests May is using prior knowledge of the task context yet with all recent work being with polynomials, she seems to have forgotten how to enact *Function Identify-ability* if an exponential function is required.

However, on occasions it was during the subsequent analysis that students reconsidered their initial models. One student, Ada, initially used mathematical knowledge to facilitate affordance enactment in selecting a quadratic model for the data. Later she changed her mind apparently on the basis of her interpretation of the constraints, in this case using non-mathematical cues to select a model, illustrated using post-task interview data.

Ada: I think I changed my mind half way through.
I: So you actually did think about it being linear?
Ada: Yeah.
I: And it wasn't because of the graph itself?
Ada: It was because I didn't think it would go back up again. I don't think it would. ...
I: What made you change? You changed your mind because it went up?
Ada: Because I started with a quadratic but I didn't know how it could go up. I think I was thinking too literally about the platypuses. I didn't think it would go up again and I thought it might be more of a linear.

Ada took the question: Would the platypus become extinct? to mean extinction was inevitable and so became dissatisfied with her original quadratic model that did not predict extinction and hence used a linear model that did.

36.5 Discussion and Conclusion

The transformational power of the technology was not realised by many students. An apparent lack of perception of several key affordances related to visualisation (i.e., *Data Display-ability, Function View-ability*) restricted opportunities for students to be provided with additional insight that would have been helpful during task solving. Ultimately it is the student who must bring their resources and competence in enacting affordances to the given enactment context. More attention needs to be given to developing both of these.

Waisel et al. (1999) suggest "in order for potential modellers to make the most efficient and effective use of these new technologies, the software must incorporate visualization in such a way as to facilitate the process of modeling" (p. *i*). Graphing calculators incorporate such visualization opportunities; however, we are not seeing evidence of the actualisation of this visualization to the extent expected for a TRTLE. Waisel et al. further argue that "software designers and developers, who have tended to focus more on perceptual than cognitive issues, need to better understand the cognitive role of visualization in modeling" (p. *i*) when designing visualisation enabling software. Based on students' reluctance to visualise, there is little doubt that teachers also need to better understand the cognitive role of visualization in modelling as students are not appreciating how the visualisation can support their mathematisation, through concrete supports for their mental models.

Making a mental model for the situation and refining this through technology-generated external visualisations follows Borba's (2005, 2012) notion that it is the two way to-ing and fro-ing between the human and the technology, that is the seamless interplay between the mental model → technology-generated model → refined mental model. When this occurs a direct consequence will be this facilitation of the modelling process. The calculator offers this facilitation, however, it appears the students are not appreciating this potential.

To form a mental model, one begins with comprehension where

> the information necessary for forming a mental model is gathered. During description a mental model is formed based on the preliminary knowledge gathered during the comprehension phase. The contents of a mental model . . . are inaccessible until the mental model . . . is queried and views of it are extracted (Waisel, Wallace, & Willemain 1999 p. 3).

Up until this point the mental model is manipulated in the modeller's mind. Limitations of one's working memory foreshadow the benefits of using external aids such as technology-generated visualisations (Stillman 2004; Stillman et al. 2004). Thus the mental representation is formed from the text-based information and other images provided in the complex real world situation. The coordination of these is able to be enhanced and modified through 'deliberate/ thoughtful' use of technology-generated supports. Stillman argues it is the coordination of these (the mental model and the image on the screen, for example, and the integration of the two) that is difficult for students, therefore they do not use the external aids to the extent we would expect them to do. However, Brown (2013) notes in other cases, students do not even attempt to use them (the technology-generated images) as they do not perceive the necessity for their use. So despite the potential for technology to have a transformational impact on what students do in classes where modelling is conducted this is not necessarily being realised. The research reported here thus suggests several future lines of inquiry not the least of which is the anticipation of use of digital tools in a modelling context (See Stillman et al. Sect. 7.2).

References

Borba, M. (2005). Humans-with-media: Transforming communication in the classroom. In A. Chronaki & I. M. Christiansen (Eds.), *Challenging perspectives on mathematics classroom communication* (pp. 51–77). Greenwich: Information Age.

Borba, M. (2012). Humans-with-media and continuing education for mathematics teachers in online environments. *ZDM_The International Journal on Mathematics Education, 44*(6), 801–814.

Brown, J. (2013). *Perceiving and enacting the affordances of technology-rich teaching and learning environments (TRTLE's) for student understanding of function.* Unpublished doctor of philosophy thesis, University of Melbourne, Melbourne.

Brown, J., & Edwards, I. (2011). Modelling tasks: Insights into mathematical understanding. In G. Kaiser, W. Blum, R. Borromeo Ferri, & G. Stillman (Eds.), *Trends in teaching and learning of mathematical modelling* (pp. 187–197). New York: Springer.

Doerr, H., & Zangor, R. (2000). Creating meaning for and with a graphing calculator. *Educational Studies in Mathematics, 41*(2), 143–163.

Galbraith, P., Stillman, G., Brown, J., & Edwards, I. (2007). Facilitating middle secondary modelling competencies. In C. Haines, P. Galbraith, W. Blum, & S. Khan (Eds.), *Mathematical modelling (ICTMA12): Education, engineering and economics* (pp. 130–140). Chichester: Horwood Press.

Gibson, J. J. (1966). *The senses considered as perceptual systems.* Boston: Houghton Mifflin.

Gibson, J. J. (1977). The theory of affordances. In R. Shaw & J. Bransford (Eds.), *Perceiving, acting and knowing: Toward an ecological psychology* (pp. 67–82). Hillsdale: Erlbaum.

Johnson-Laird, P. N. (1988). *The computer and the mind.* Cambridge, MA: Harvard University Press.

Pead, D., & Ralph, B. (2007). Uses of technologies in learning mathematics through modelling. In W. Blum, P. Galbraith, H.-W. Henn, & M. Niss (Eds.), *Modelling and applications in mathematics education (14th ICMI Study)* (pp. 309–318). New York: Springer.

Pollak, H. O. (1986). The effects of technology on the mathematics curriculum. In M. Carss (Ed.), *Proceedings of the fifth international congress on mathematical education* (pp. 346–351). Boston: Birkhäuser.

Powell, A., Francisco, J., & Maher, C. (2003). An analytical model for studying the development of learners' mathematical ideas and reasoning using videotape data. *Journal of Mathematical Behavior, 22*(4), 405–435.

Scarantino, A. (2003). Affordances explained. *Philosophy of Science, 70*, 949–961.
Stillman, G. (2004). Strategies employed by upper secondary students for overcoming or exploiting conditions affecting accessibility of applications tasks. *Mathematics Education Research Journal, 16*(1), 41–76.
Stillman, G., Edwards, I., & Brown, J. (2004). Mediating the cognitive demand of lessons in real-world settings. In B. Tadich et al. (Eds.), *Towards excellence in mathematics* (pp. 489–500). Melbourne: MAV.
Strauss, A., & Corbin, J. (1998). *Basics of qualitative research: Techniques and procedures for developing grounded theory* (2nd ed.). Thousand Oaks: Sage.
Tikhomirov, O. K. (1981). The psychological consequences of computerization. In J. V. Wertsch (Ed.), *The concept of activity in Soviet psychology* (pp. 256–278). Armonk: M. E. Sharpe.
Waisel, L. B., Wallace, W. A., & Willemain, T. R. (1999). Visualizing modelling heuristics: An exploratory study. In *ICIS 1999, proceedings of the 20th international conference on information systems* (ICIS), Paper 16. http://aisel.aisnet.org/icis1999/16
Waisel, L. B., Wallace, W. A., & Willemain, T. R. (2008). Visualization and model formulation: An analysis of the sketches of expert modelers. *Journal of the Operational Research Society, 59*(3), 353–361.

Chapter 37
Developing Modelling Competencies Through the Use of Technology

Ruth Rodríguez Gallegos and Samantha Quiroz Rivera

Abstract This chapter focuses on the study of the development of modelling competencies in a Differential Equations class where diverse technology is used. First, it discusses how modelling can be used for the design of didactic sequences for students of engineering. Secondly, it shows an innovative proposal that includes the incorporation of diverse technology for teaching Differential Equations in the context of RC (Resistor-Capacitor) and RL (Resistor-Inductor) Circuits. Finally, it details qualitatively the way technology promotes the development of mathematical modelling in students of engineering.

37.1 Introduction

Incorporating modelling into the school curriculum has shown encouraging results such as the development of competencies for problem solving that promote the relation between school mathematics and everyday mathematics (Rivera et al. 2015). In universities, solving problems based on professional contexts is now a priority objective according to international and national projects such as the Programme for International Student Assessment (PISA) developed by the Organization for Economic Co-operation and Development (OECD) (2003, 2010) and the Latinoamerican Tuning Project (Beneitone et al. 2007). Recognizing that the development of technology generates more possibilities for using mathematical modelling in classrooms (Galbraith et al. 2007), modelling activities are being used along with technological devices in order to promote increased understanding based on this strategy. This chapter shows the design of innovative didactic sequences for teaching Differential Equations (DE) that involves modelling in Resistor-Capacitor (RC) and Resistor-Inductor (RL) circuits contexts. Also, it describes how the

R. Rodríguez Gallegos (✉) • S. Quiroz Rivera
Department of Mathematics, Tecnológico de Monterrey, Av. Eugenio Garza Sada, 2501, Col. Tecnológico Aulas 7, 227, Monterrey C.P. 64849, Mexico
e-mail: ruthrdz@itesm.mx; samanthaq.rivera@gmail.com

G.A. Stillman et al. (eds.), *Mathematical Modelling in Education Research and Practice*, International Perspectives on the Teaching and Learning of Mathematical Modelling, DOI 10.1007/978-3-319-18272-8_37

technology was chosen in order to support the lessons and an analysis which identifies the modelling competencies that were promoted for students of engineering.

37.2 Mathematical Modelling and the Development of Competencies

The main reason for teaching mathematics in schools is so that students can apply the mathematical content for solving problems in a variety of contexts and daily situations out of schools (Alsina 2007). In order to accomplish this purpose, mathematical modelling provides a way that relates these worlds: that of the scholar and the real world.

The representation of mathematical modelling in a scheme has been a central aspect in much research. Student work has been analyzed using different points of view and descriptions of the stages of the modelling cycle. Some researchers that focus on that purpose are Blum (2002), Borromeo Ferri (2006) and Confrey and Maloney (2007). In our research about the teaching of mathematics at university level, we analyzed the modelling cycle through a scheme proposed by Rodríguez (2007, 2010). This scheme is special for the kind of physical-technical examples we use. It is shown in eight stages of four domains: Real, Pseudo-Concrete, Physical and Mathematical (see Fig. 37.1).

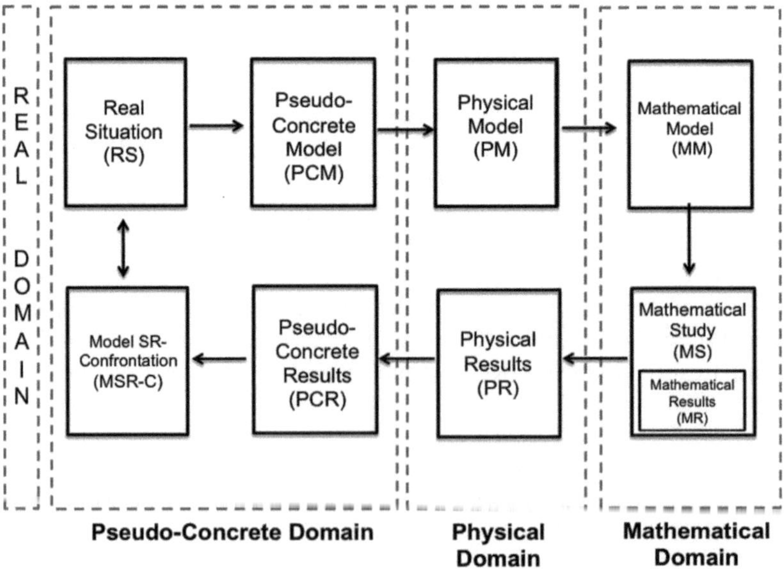

Fig. 37.1 Mathematical modelling cycle (Rodríguez 2007, 2010)

On certain occasions, the modelling situations that are presented to students, can be considered to be in the Real Domain. Perhaps these situations have been modified for their teaching in schools through a didactic transposition, a term proposed by Chevallard (1985). The approach of situations that appear in the scholar context, make us consider the existence of another domain named Pseudo-Concrete. The Physical Domain appears when the contexts chosen for modelling mathematics require previous physical knowledge. The existence of that domain permits us to recognise the importance of specific knowledge that generates a Physical Domain and after that, the elaboration of a mathematical model. The mathematical modelling cycle presented in Fig. 37.1 can support the design of didactical sequences for learning mathematics where the teacher can generate inherent activities related to each stage presented in the scheme and eventually achieve better learning of mathematics to obtain better results. This is the case of the examples that will be presented in the next sections.

Some of the first results obtained help us to consider and identify some modelling competencies that were developed by students of engineering in a DE class. Based on previous studies, especially Henning and Keune (2007), Houston (2007), Kaiser (2007), Maaß (2006) and Singer (2007) with the purpose of unifying the results of these, we chose a qualitative study that began a characterisation of the main modelling competencies developed by the students in a DE class designed according to transitions in the mathematical modelling cycle (see Table 37.1). The didactical sequences described, incorporate innovative technology that researchers selected specifically according to the objectives of each lesson. The central question to discuss is:

Table 37.1 Mathematical modelling competencies (Rodríguez et al. 2014)

Transitions between stages of the modelling process	Modelling competencies promoted
PCM → PM	To identify and structure problems
	To understand and analyse the real problem
PM → MM	To determine and manipulate variables
	To create a mathematical model through real terms
MM → MR	To manipulate the variables on the mathematical model
	To use mathematical knowledge to solve the problem
MR → PR	To interpret the model in terms of the physical domain
PR → PCR	To interpret the model in real terms
	To interpret the result in the real situation
PCR → MSR-C	To adapt the model to new situations
	To reflect on and validate the model
	To evaluate the mathematical model
	To communicate the model and its results

How does technology impact in the promotion of specific modelling competencies?

37.3 The Use of Technology and the Differential Equations

For many years now, teaching DE to students of engineering has emphasized producing and memorising the analytical proceedings that are presented to students like "cuisine recipes". The consequences of this practice have been shown to be the poor learning of the DE and a total incomprehension by students of the applications of mathematics in real and professional contexts that they eventually use (Artigue 1996; Blanchard 1994; Rasmussen 2001).

The DE class at Tecnológico de Monterrey, Monterrey campus, began a curricular reform in 2008. These efforts resulted in a redesign of Differential Calculus and Integral Calculus courses. The actual DE class is based on modelling as the principal teaching strategy. We designed didactical sequences to allow students to relate to diverse contexts where physics and empirical phenomena are studied. The majority of the didactical sequences designed incorporates the use of innovative technology available in the university. This allows the students the possibility of interaction with some phenomenon described in experimental ways. Some of the technologies used are: programmable calculators; temperature, voltage and movement sensors; physical material (electric circuits); specialised software and online resources.

In this chapter we present two didactical sequences designed for the learning of RC and RL circuit modelling through a DE by the linear method. This class was selected as it used diverse technology and the sequence has been adapted and improved each semester.

37.4 Experimental Situation: RC and RL Electric Circuits

The experimental situation that we will detail was designed and implemented beginning in August 2011. Both sequences are associated with mathematical content and their teaching requires a technological device as shown in Table 37.2.

The main difference between the didactical sequences is the use of physical materials, calculators and sensors in the first compared to the use of a simulator in the second. The choice of a simulator for experimenting on a RL circuit was made because of the absence of this physical material in the university. Nevertheless, we found in the Project Physics Education Technology (PhET) from Colorado University, a simulator that has the elements of the RL circuit that can be easily manipulated because of the user-friendly interface. The didactical sequences are as follows:

Table 37.2 Didactical sequences chosen

Didactical sequence	Mathematical content	Technology
RC circuits	Analytical linear method	Electric circuits (connectors, capacitor, resistance and four batteries)
		Calculator
		Voltage sensor
RL circuits	Analytical linear method	Simulator

(a) RC Circuit

- Phase 1. Formulation of a DE that models the charge and discharge of the RC circuit.
- Phase 2. Assembly of the electric circuit physically and re-collection of data using voltage sensors.
- Phase 3. Analysis of the graphs given by the sensor, recognizing their behavior.
- Phase 4. Analytical solution of the DE by linear methods and comparison with the graph.

(b) RL Circuit

- Phase 1. Formulation of a DE that models the current of the RL circuit.
- Phase 2. Assembly of the electric circuit using a simulator.
- Phase 3. Analysis of the graphs given by the sensor of the simulator, recognizing their behavior.
- Phase 4. Analytical solution of the DE by linear methods and comparison with the graph.

It is possible to appreciate that Phase 2, "Assembly of the electric circuit physically and re-collection of data using voltage sensors," and Phase 3, "Analysis of the graphs given by the sensor, recognizing their behavior," are where the technology can play an important role. In Fig. 37.2 the stages of the modelling cycle indicate where the phases are shown as A and B, respectively. Both Phases 2 and 3 where the technology is applied are situated in the transitions between the Pseudo-concrete, Physical and Mathematical Domains of the modelling cycle.

This research follows a qualitative paradigm. In keeping with this type of study (Creswell 2007; Stake 2005), participant observation was used and the collection of data was undertaken using a semi-structured observation instrument and the analysis of documents made by the students. All of these were supported by the lesson being videotaped. The participants consisted of 25 third and fourth semester students with different engineering majors who were enrolled in the DE class of August, 2012. A sample of nine students working in three teams was chosen randomly. The two didactic sequences were taught in two separate lessons each of 90 min duration.

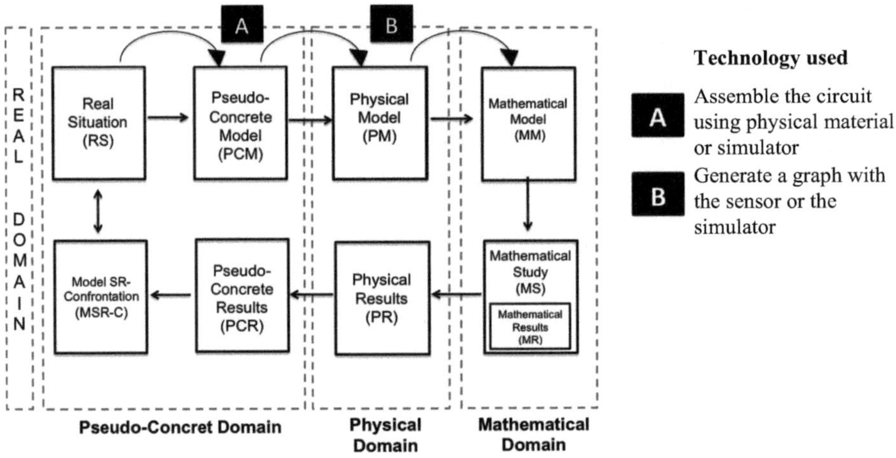

Fig. 37.2 Using technology in the modelling cycle (Rodríguez 2007, 2010)

37.5 Results

We present the results categorised into the two principal uses of the technology indicated on Fig. 37.2.

37.5.1 Assembly of the Electric Circuit

The first activity where the technology was used consisted of the assembling of the RC circuit by each team. We present the results in both didactical sequences:

(a) *RC Circuit*: Observations showed that students in each team could assemble the electric circuit using the given materials. In order to do that, they made use of a tutorial designed by the teacher indicating the way they had to put each element together. The manipulation of the materials was intuitive and after some attempts the students could charge and discharge the capacitor on several occasions. The assembly of the electric circuit generated dialogue and discussion between the members of the different teams that provided explanations about the phenomenon. We registered that they questioned themselves about what was going on in the circuit in order to understand. The dialogue appeared as answers to those questions and we noted that students had conceptions about the phenomenon and they used physical terms to describe them. One team, for example, drew a scheme of a RC circuit in order to explain the phenomenon to their partners.

(b) *RL Circuit*: In the second didactical sequence, referring to the assembly of a RL circuit, students manipulated the PhET simulator. Again, although it was the first time students used the simulator, they did not have any difficulties, and they learned how to use the simulator very quickly. An important difference noted was that, in using the simulator, it was possible to change the magnitudes such as inductance, voltage and resistance. Despite the problem that the software determines the specific values, it was observed that students manipulated the ranges of the magnitudes, and it motivated them to start a discussion about possible answers.

37.5.2 Generation of a Graph Using the Sensor or Simulator

The second activity planned with technology was to generate a graph through data collected using the sensors. Again we present results of both didactical sequences:

(a) *RC Circuit*: The students connected the calculator and the voltage sensor to the RL circuit already assembled. The collection of data was made after several attempts, enabling a better manipulation of the equipment and better measurements. The qualitative analysis of the graph included an explanation of its behaviour, its form and the asymptotes according to the phenomenon studied. The answers described that the capacitor was charged exponentially with an asymptote at 6 V. Some teams collected data again because the graph showed negative values as a result of mistakes in the polarity of the wires. This allowed us to see that students related the mathematics and the problem they were required to solve.

(b) *RL Circuit*: To collect data in the RL circuit the sensor of the simulator was used. As this sensor had to be manipulated in the computer, students had to capture the moment when the current changed and the moment when it was stabilized. The three teams were able to register the graph and drew it for its qualitative analysis. On this occasion, the graph shown was easily read as an exponential function with asymptote at 1A. As this didactical sequence was implemented after the study of the RC circuit, students accomplished a faster comprehension of the variables manipulated and the stabilization of the graph.

37.5.3 Modelling Competencies Promoted

After the analysis and triangulation of the results, and the dialogues between students, it was possible to identify which modelling competencies were promoted in the implementation of the technology previously described. First, the emergence of questions for understanding the phenomenon with physical material or with the simulator, enabled a dialogue between team members that promoted the *competency to identify real problems*. Before the main question of the lesson was proposed, the students of the three teams talked about the possible variables for consideration. The same comments promoted the *competency to understand and*

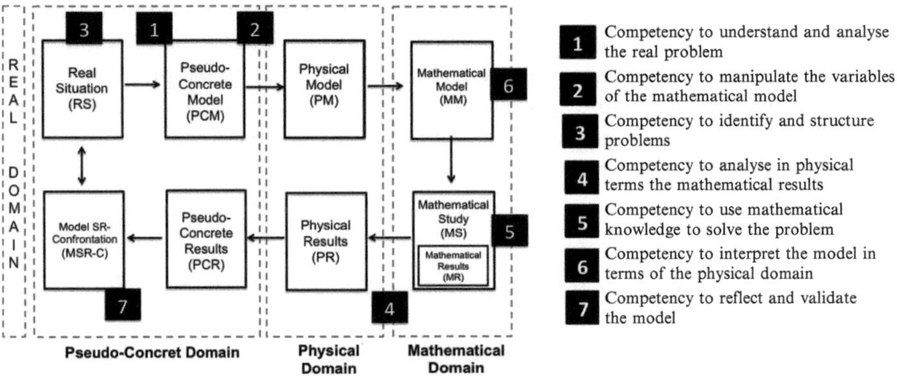

Fig. 37.3 Modelling competencies promoted by the use of technology

analyse the real problem. This competency became possible due to the transition between the Real Problem and the Pseudo-Concrete Model.

Generating graphs and the qualitative analysis provided the opportunity to use physical terms to explain the mathematical results. Also, it promoted discussions about the problem variables. That aspect was important in order to solve analytically the DE, validate the solution and find mistakes as the students already knew the answer in a physical domain. It was possible to promote the *competency to interpret the model in physical terms* and also the *competency to reflect on and criticize the mathematical model*. In Fig. 37.3 the modelling competencies promoted in the modelling cycle are shown.

37.6 Conclusions and Discussion

This investigation concludes that using technological devices can promote the transition between the different stages of the mathematical modelling cycle. Principally, the use of technology can make possible the experience of the phenomenon in the lesson. It permits a better understanding of the context analyzed and promotes the modelling competencies specified previously.

In addition, the use of a simulator in the second didactical sequence was just as effective as the first didactical sequence where concrete materials were used. We believe that it was possible because of the correct selection of a device that has all the elements needed for the experimentation. We recommend for future investigations the design of didactical sequences that include other technological devices and a deep analysis about the promotion of competencies not only mathematical, but also technological and collaborative.

References

Alsina, C. (2007). Less chalk, less words, less symbols... more objects, more context, more actions. In W. Blum, P. L. Galbraith, H.-W. Henn, & M. Niss (Eds.), *Modelling and applications in mathematics education: The 14th ICMI study* (pp. 35–44). New York: Springer.

Artigue, M. (1996). Learning and teaching elementary analysis. In C. Alsina, J. M. Alvarez, M. Niss Aérez, L. Rico, & A. Sfard (Eds.), *8th International congress on mathematics education – Selected lectures* (Vol. 1, pp. 15–29). Sevilla: SAEM.

Beneitone, P., Esquetini, C., González, J., Maletá, M., Siufi, G., & Wagenaar, R. (2007). *Reflexiones y perspectivas de la Educación Superior en América Latina, Informe Final Proyecto Tuning Amética Latina 2004–2007*. Madrid: Universidad de Deusto.

Blanchard, P. (1994). Teaching differential equations with a dynamical systems viewpoint. *The College Mathematics Journal, 25*(1), 385–393.

Blum, W. (2002). ICMI Study 14: Applications and modelling in mathematics education – Discussion document. *ZDM – The International Journal on Mathematics Education, 34*(5), 229–239. doi:10.1023/A:1022435827400.

Borromeo Ferri, R. (2006). Theoretical and empirical differentiations of phases in the modelling process model. *ZDM – The International Journal on Mathematics Education, 38*(2), 1–10.

Chevallard, Y. (1985). *La transposition didactique: du savoir savant au savoir ensigné*. Grenoble: La Pensée Sauvage.

Confrey, J., & Maloney, A. (2007). A theory of mathematical modelling in technological settings. In W. Blum, P. L. Galbraith, H.-W. Henn, & M. Niss (Eds.), *Modelling and applications in mathematics education, the 14th ICMI study* (pp. 57–69). New York: Springer.

Creswell, J. (2007). *Qualitative inquiry and research design*. Thousand Oaks: Sage.

Galbraith, P. L., Stillman, G., Brown, P., & Edwards, I. (2007). Facilitating middle school modelling competencies. In C. Haines, P. Galbraith, W. Blum, & S. Khan (Eds.), *Mathematical modelling (ICTMA 12): Education, engineering and economics* (pp. 130–140). Chichester: Horwood.

Henning, H., & Keune, M. (2007). Levels of modelling competencies. In W. Blum, P. L. Galbraith, H.-W. Henn, & M. Niss (Eds.), *Modelling and applications in mathematics education: The 14th ICMI study* (pp. 225–232). New York: Springer.

Houston, K. (2007). Assessing the "phases" of mathematical modelling. In W. Blum, P. L. Galbraith, H.-W. Henn, & M. Niss (Eds.), *Modelling and applications in mathematics education: The 14th ICMI study* (pp. 249–256). New York: Springer.

Kaiser, G. (2007). Modelling and modelling competencies in school. In C. Haines et al. (Eds.), *Mathematical modelling: Education, engineering and economics* (pp. 110–119). Chichester: Horwood.

Maaß, K. (2006). What are modelling competencies? *ZDM – The International Journal on Mathematics Education, 38*(2), 113–142.

Organization for Economic Co-operation and Developent [OECD]. (2003). *Informe PISA 2003. Learning*. Madrid: OECD Publishing.

Organization for Economic Co-operation and Developent [OECD]. (2010). *PISA 2009 assessment framework. Key competencies in reading, mathematics and science. Assessment*. Paris: OECD Publishing.

Rasmussen, C. (2001). New directions in differential equations: A framework for interpreting students' understandings and difficulties. *The Journal of Mathematical Behavior, 20*(1), 55–87.

Rivera, S., Londoño, S., & López, C. (2015). Measurement of area and volume in an authentic context: An alternative learning experience through mathematical modelling. In G. A. Stillman, W. Blum, & M. S. Biembengut (Eds.), *Mathematical modelling in education research and practice: Cultural, social, and cognitive influences* (pp. 229–240). Cham: Springer.

Rodríguez, R. (2007). *Les équations différentielles comme outil de modélisation mathématique en Classe de Physique et de Mathématiques au lycée: une étude de manuels et de processus de*

modélisation d' élèves en Terminale S. Grenoble: University Joseph Fourier Grenoble. Unpublished doctoral dissertation.
Rodríguez, R. (2010). Aprendizaje y enseñanza de la modelación: el caso de las ecuaciones diferenciales. *Revista Latinoamericana de Matemática Educativa, 13*(4-1), 191–210.
Rodríguez, R., Quiroz, S., & Illanes, L. (2014). Competencias de modelación y uso de tecnología en ecuaciones diferenciales. In E. Flores (Ed.), *Acta Latinoamericana de Matemática Educativa, 27*. México: CLAME.
Singer, M. (2007). Modelling both complexity and abstraction: A paradox? In W. Blum, P. L. Galbraith, H.-W. Henn, & M. Niss (Eds.), *Modelling and applications in mathematics education, the 14th ICMI study* (pp. 233–240). New York: Springer.
Stake, R. (2005). Qualitative case studies. In N. Denzin & Y. Lincoln (Eds.), *The Sage handbook of qualitative research* (3rd ed., pp. 443–466). Thousand Oaks: Sage.

Chapter 38
Model Analysis with Digital Technology: A "Hybrid Approach"

Débora da Silva Soares

Abstract The advent of informatics and its fast development has allowed for new ways of doing, thinking about, and applying mathematics. This chapter will address one role of the software, Modellus, in the development of a teaching approach based on Model Analysis, that proposes the analysis of a mathematical model for a phenomenon to discuss old and new mathematical concepts with students. The software gave access to graphical and numerical representations of model solutions, which allowed students to analyse the model, even though it involved mathematical concepts still advanced for them. Based on the notion of reorganization, the shift in the way of dealing with mathematical models in classrooms suggests that Model Analysis can be understood as a reorganization of modelling and applications.

38.1 Introduction

Applied Mathematics is the only obligatory mathematics discipline in the entire curriculum of the Biology Major course at the State University of São Paulo, Rio Claro, SP, Brazil. Its syllabus includes the study of functions, notions of limits, derivatives and integrals and their applications, distributed in a workload of 4 h per week. In this course it is usual to find students who do not like mathematics and who actually do not understand why they should learn it nor how they could use this content in their area of interest. Based on this scenario, I have developed a teaching approach with the intention of creating an opportunity for these students to think about mathematics in a way related to biology. This teaching approach is based on a situation where the context is as important as the mathematics and it became the background of the research published in Soares (2012).

The focus of this study was the identification and analysis of the roles of the software Modellus during the development of a teaching approach, in which the

D.S. Soares (✉)
Instituto de Matemática, Universidade Federal do Rio Grande do Sul, Av. Bento Gonçalves, 9500 – Prédio 43-111, Porto Alegre, RS 91509-900, Brazil
e-mail: debora.soares@ufrgs.br

G.A. Stillman et al. (eds.), *Mathematical Modelling in Education Research and Practice*, International Perspectives on the Teaching and Learning of Mathematical Modelling, DOI 10.1007/978-3-319-18272-8_38

core idea was to propose the analysis of a mathematical model for a biological phenomenon from the first day of class. This analysis had a qualitative character and was developed in a way that the discussion of some mathematical content from the Applied Mathematics syllabus was intertwined with it. This core idea is referred to, in general, as Model Analysis (Soares 2012; Soares and Javaroni 2013). In this chapter, I will discuss one of the roles of the software, which is to make the model solutions accessible to the students.

38.2 The Teaching Approach

As mentioned before, the core idea of the teaching approach was to analyse a mathematical model for a biological phenomenon from the first day of class in a way intertwined with the discussion of some mathematical concepts from the syllabus[1] of the discipline. The focuses of this analysis were: to comprehend the meaning of the model (including its hypothesis), the meaning of the parameters and the information given by the model's solutions regarding the phenomenon; to analyse the behaviour of the solutions and the influence of the parameters in this behaviour; to discuss some mathematical concepts (such as function, derivatives, tangent line, maxima and minima) in relation to the phenomenon and the model.

The theme for investigation was the transmission of malaria. This is a disease which is quite prevalent in different parts of the world, including Brazil, and which is considered a neglected disease, since research for its cure is quite rare. In this sense, this theme is relevant both biologically and socially, which were important criteria used to choose it. Furthermore, it is a theme with which students could be involved in the future as professional biologists.

A mathematical model already developed for the study of this problem was presented to the students. The intention was that the model was relatively accurate regarding the phenomenon and possible to be understood by the students. Following these criteria, I decided to work with the Ross-Macdonald model (Basañez and Rodríguez 2004), which is one of the first models developed for the study of malarial transmission. Some important simplification assumptions behind this model are: no human mortality by malaria; no incubation period; and the assumption that once infected, mosquitoes will die infected. These assumptions were also the focus of discussion during the work with students.

The model is composed of a system of two non-linear ordinary differential equations (ODE), which is shown below (38.1), and its variables are: $X(t)$, which is the quantity of infected people in the region during time t; and $Y(t)$ which is the

[1] For more details about the teaching approach, refer to Borba and Soares (2011) and Soares (2012).

quantity of infected (female) mosquitoes in the region during time t. Roughly,[2] the main idea of the model is to describe how each one of these populations varies over time, considering aspects that could influence the transmission of the parasite from one person to another, which occurs due to the bite of a (female) mosquito.

$$\left.\begin{aligned}\frac{dX}{dt}&=\left(\frac{a}{N}\cdot p\right)\cdot Y\cdot(M-X)-g\cdot X\\ \frac{dY}{dt}&=\left(\frac{a}{N}\cdot c\right)\cdot X\cdot(N-Y)-v\cdot Y\end{aligned}\right\}\qquad(38.1)$$

Analysing the model and to comprehend what information it gives about the phenomenon are activities that need the study of the model solutions. The idea of the analysis was not to find analytical solutions for this model (even because this kind of system rarely has expressions for its solutions) but to propose a qualitative analysis. Nevertheless, the students still need to access the solutions. In order to achieve that, they worked with Modellus,[3] a free software package that allows working with models involving functions, iterations, difference equations and ODE. Figure 38.1 shows a screen shot of its interface.

The software allows the setting of values for parameters and initial conditions, and presents graphical and tabular representations for the solutions of the model. Likewise the image shows it is possible to see several graphs simultaneously. In Fig. 38.1 there are three graphs of the function $X(t)$ for three different values of one of the parameters. This characteristic of the software allows one to analyse how parameters influence the behaviour of the solutions, which was one of the focuses of the analysis proposed to the students, as mentioned before.

38.3 The Study

Inspired by design research (Doerr and Wood 2006), I developed three versions of the teaching approach, the first as a pilot project, and the other two, full versions applied in regular classes of the discipline. The research followed a qualitative approach, and different sources of data were used, from which I highlight two: an interview with 12 Biology Major students, in six pairs, at the end of the semester; and videos generated by the software Camtasia Studio,[4] which captures all the activities done on the computer screen plus the audio of students.

These data were the basis to search for answers and to understand the following research question: What are the roles of the software Modellus in the development of a teaching approach based on Model Analysis? Three roles of the software were

[2] For more details about the model, refer to Basañez and Rodríguez (2004), Borba and Soares (2011) or Soares (2012).

[3] Website: <http://www.modellus.fct.unl.pt/>. *Accessed on 15 June 2013.*

[4] Website: <http://www.techsmith.com/camtasia.html>. Accessed on 7 Sep 2013.

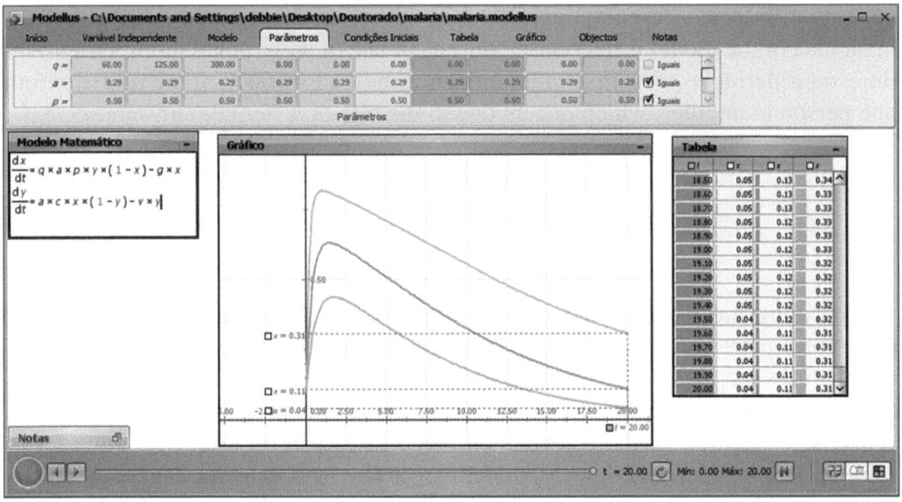

Fig. 38.1 Interface of Modellus

identified (see Soares 2012) and analysed, always taking into account the structure of the teaching approach. In this chapter, I will discuss one of them: to make the model solutions accessible to the students.

38.4 Model Analysis and Mathematical Modelling: Some Reflections

When we talk about mathematical modelling as a teaching approach, despite the diversity of perspectives we can find in the literature, the main idea that comes to mind is that students are invited to create their own models for a problem situation. It is possible to find in literature several schemes that try to represent the different steps involved in doing mathematical modelling. From my point of view, they do not show exactly what happens when one develops a model for a situation, but I think they can help us to develop some reflections.

In Fig. 38.2 there are two schemes: one circular, proposed by Blum and Leiß (2007) and one non-linear proposed by Doerr and Pratt (2008). Although these schemes are different, it is possible to identify some commonalities between them: both include a problem situation, a mathematical model, the determination of solutions for this model, and the interpretation and validation of the model regarding the problem situation. From the scheme proposed by Blum and Leiß (2007) it is possible to infer that one should start the modelling process with a problem situation; but from the scheme proposed by Doerr and Pratt (2008) it is not possible to determine a starting point.

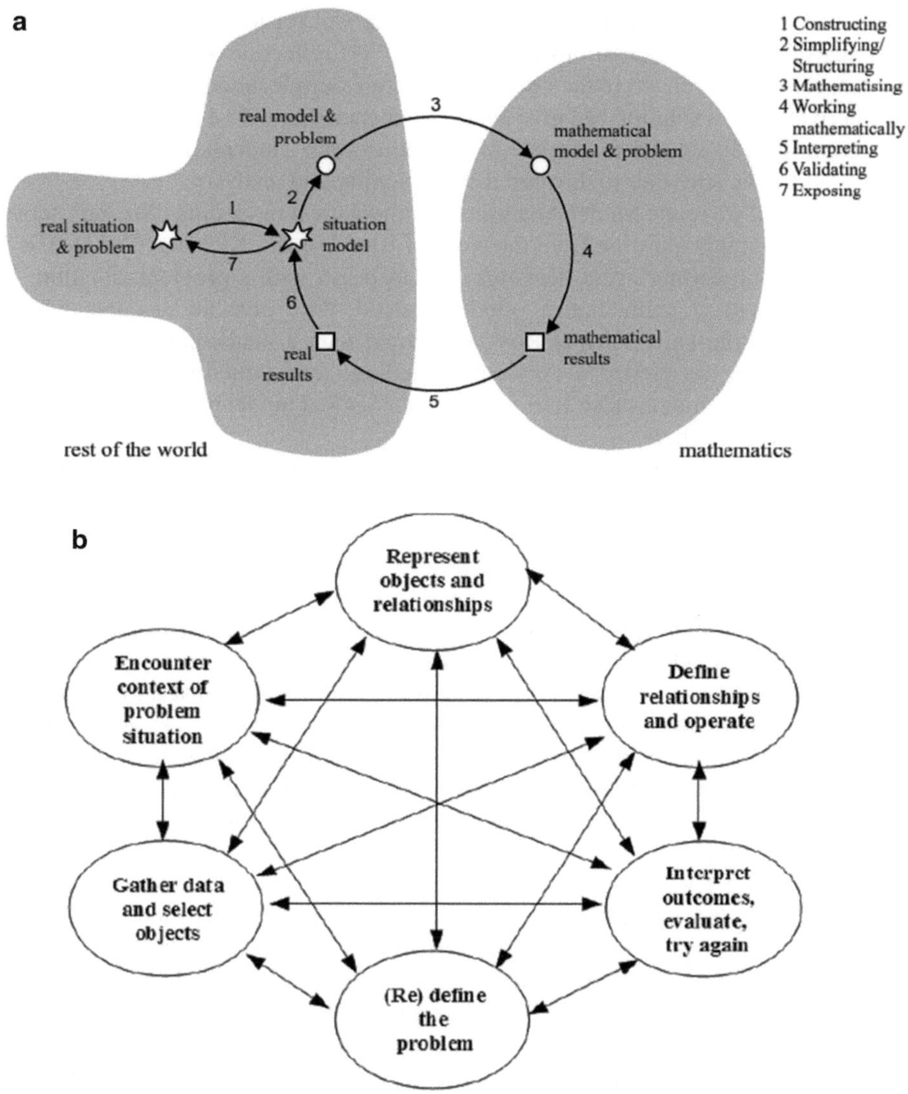

Fig. 38.2 Schemes representing the processes involved in modelling (**a**) Blum and Leiß (2007) (**b**) Doerr and Pratt (2008)

Now when we look at Model Analysis we notice that students analyse a model that already exists. In other words, they do not develop their own model for a situation. Analysing the teaching approach proposed in Soares (2012), it is possible to identify some tasks that were involved in the students' work: (i) study of the phenomenon to understand how the transmission of malaria happens in nature and what factors influence it; (ii) analysis of the model's assumptions of the model; (iii) analysis of the behaviour of the solution and the influence of parameters in this behaviour; (iv) interpretation of results for the model regarding the phenomenon; (v) analysis of some limitations of the model; (vi) reflection about mathematical concepts in a way related to the phenomenon, for example, about what information the derivative of a solution can give about the phenomenon. Tasks (iii), (iv), (v) and (vi) were deeply related to the use of Modellus since students used information provided by the software to develop their reasoning and analysis.

How could we relate Model Analysis to mathematical modelling? We gain some leverage about this question from the work of Blomhøj and Kjeldsen (2011). They argue that a modelling cycle does not need to begin with a problem situation; it could start with a mathematical model. Indeed, they give an example where students from the University of Roskilde started with a mathematical model and from there they developed a cycle of modelling, reconstructing that model. In Model Analysis, students also start with a mathematical model but there is a critical point of difference here.

If we look again at those schemes and we think about the teaching approach developed with the Biology students, we notice that the activities they have worked with focused on the step of interpretation (and maybe validation) of the model. Based on the scheme proposed by Blum and Leiß (2007), we also perceive that students worked in the transition from a real model to a mathematical model, since they started by analysing a framework that described the dynamics of transmission of malaria that already considered the assumptions of the model. However, there are some steps such as collecting data, or simplifying the situation, that they did not do by themselves. In other words, they analysed what other people have done. Therefore, in this sense, we could consider that students did not complete the entire cycle of mathematical modelling. Alternatively, we could consider that students are developing a rudimentary modelling cycle, as suggested by Niss (2015) in his discussion about *prescriptive modelling*.

According to Niss (2015), prescriptive modelling deals with situations such as the body mass index (BMI). The main purpose of this kind of modelling is "to design, prescribe, organise or structure certain aspects of [the world]" (Niss 2015, p. 67). The critical point here is that it does not make sense to validate models such as the BMI; however, it is possible to critique it, developing what Niss calls a meta-validation. This author explains that the modelling cycle as suggested in the schemes above is not completed in the case of prescriptive modelling. For example, the mathematical treatment to the problem of BMI is reduced to simple calculations of the specific index. What eventuates in this case is a rudimentary modelling cycle – but it is still modelling, according to Niss (2015). Similar to prescriptive modelling, what happened in the teaching approach suggested in Soares (2012) was the development of a rudimentary modelling cycle, with focus on the transition from a model situation to a mathematical model and on the interpretation of it. The reduction of the modelling cycle, in this case, is closely related to the use of the software Modellus. As a consequence, in the next section, I bring some data to guide a discussion about the software.

38.5 The Role of Software Modellus

As mentioned before, Modellus had an important role in the development of the teaching approach: it allowed the students to have access to the solutions of a mathematical model quite accurate to the phenomenon, even though they did not know how to solve an ODE system nor the concept of derivative. This possibility offered by the software is one of the factors that causes the reduction of the modelling cycle, since students do not need to solve the model to find its solutions; that is, they do not develop a mathematical treatment of the model. Another possibility offered by the software is to allow for experimentations such as the modification in parameter values to study the influence of different conditions in the evolution of the phenomenon. Both of these possibilities created the opportunity for the students to interpret the model results, which was an important part of their work during the semester. Below, I present an extract from the dialogue of a pair of students, called N and R, while working on a task with a goal to analyze the influence of parameters on the behaviour of the solutions. The students were analyzing the parameter c, which gives the probability of a mosquito being infected by the malaria parasite (*plasmodium*) according to its species, see equation (38.1).

N: Compare the graphs for the three cases [see Fig. 38.3]. In the first species... the graph is decreasing.
R: Is this the first species? [R. points to the third graph from the top to bottom].
N: So... I don't know... I think so, because here... first X, second X, third X... This one that is decreasing [refers to the graph in gray]... look how it was big in the beginning... then it has grown and then it decreased.
[...]
N: What is changing... it only changes the type of mosquito. So, the graph of the first species is decreasing because the probability of this species of being infected by *plasmodium* is small... But... ok, it is small, but why [does] it grow?
R: Why? (Extract from students dialogue, Task 4)

In this extract, it is possible to notice that students, at first, identify which graph represents which species of mosquito. Later on, they describe the behaviour of the solutions, particularly of the graph related with the first species (graph in light gray, Fig. 38.3), and then try to understand this behaviour in terms of the phenomenon (malarial transmission). N suggests that the small probability of the mosquito to be infected would be a possible explanation for this behaviour, but then she asks, "Why [does] it grow?" referring to the first part of the graph. Students engaged in the attempt to understand this behaviour and used a collection of information to arrive at a plausible explanation. Due to space restrictions, it is not possible to show all their dialogue, but this short extract gives the reader an idea of the kind of discussion the students were encouraged to develop in some tasks and is also sufficient for our following discussion.

In my understanding, based on the theoretical construct *humans-with-media* (Borba and Villarreal 2005), the software has a central role in the processes of

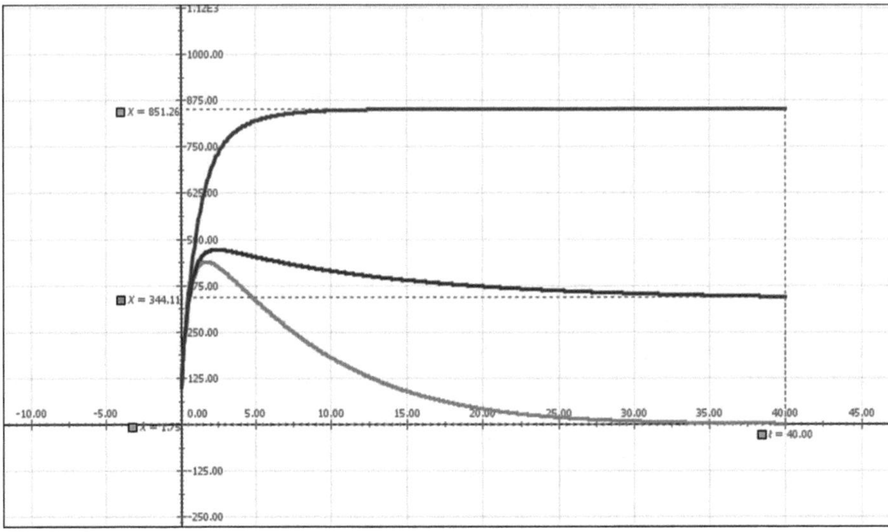

Fig. 38.3 Graph of function *X(t)* showing three cases for parameter *c*

knowledge production. In this sense, the possibilities and restrictions it offers influence the students' reasoning and the kind of interaction with the software. Therefore, if some different software were used, different kinds of interaction and different ways of reasoning with the software would be prompted. According to what I have discussed in Soares (2012), it is possible to perceive in the above extract the reorganization of the processes involved in the analysis of the model. In fact, if using only paper and pencil to develop a study about the influence of the parameters on the behaviour of model solutions, one should at first choose a value for one parameter and then carry out all the procedures involved in qualitative analysis of an ODE system and in the outline of the solution (linearization, fixed points, analysis of stability, etc.). On the other hand, working with Modellus, for example, one can focus completely on the analysis of the behavior of solutions, since the software gives feedback according to the different values chosen for the parameter. In other words, one does not need to focus his/her reasoning on the calculations and procedures involved in the analysis developed only with paper and pencil. This shift in the focus of reasoning characterises the processes of *reorganization of thinking*.

An interesting point to consider is that reorganization of the analysis of the model seems to create a "hybrid approach" between mathematical modelling and applications. The differences between these two approaches on working with models in classrooms have been a focus of discussion in literature for a long time. According to Niss et al. (2007) modelling focuses on the direction "reality → mathematics", emphasizes the *processes* involved, and it could be characterised by the following question: "Where can I find some mathematics to help me with this problem?" (p. 10). On the other hand, applications focus on the direction "mathematics → reality", emphasize the *objects* involved, and it could be characterized by the question: "Where can I use this particular piece of mathematical knowledge?" (Niss et al. 2007, p. 11).

As I have already argued in Soares (2012), although there are some convergences between Model Analysis and applications, they are not the same, since the aim of Model Analysis is neither to apply mathematical content in a real situation to illustrate its use nor to analyse a model by itself. What we can notice in Model Analysis is that, with the advent of appropriate software, students are able to analyze an existing mathematical model for a phenomenon not as a way of applying the mathematical content they have already learned, but as a way to discuss new mathematical concepts, to relate these with a real situation, and to understand their meaning regarding the phenomenon. From my point of view, this "hybrid approach" is a consequence of changes in the nature of modelling activity in particular, and mathematical activity in general, which are engendered by the use of software.

In pedagogical terms, this kind of approach presents some potentialities and limitations, which I will discuss in the following. One of the main gains is that students can deal with mathematical models a bit more realistically earlier in their academic life, since they do not need to know all the mathematical content underlying the model, which can be too advanced for them. The relevance of this fact is that, for students, it is important to understand how the mathematical content can be useful for their area of interest.

Another important gain of this kind of approach is the creation of opportunities for the students to develop some modelling competences (Maaβ 2006) such as to interpret mathematical results in terms of phenomena from other scientific areas. This kind of competence is deeply related to the processes of interpretation of the model and its solutions, which are quite emphasized by Model Analysis. In fact, even modelling competences such as to mathematise relevant quantities could be developed in a teaching approach like this if we start with the analysis of a model, examine its limitations and modify the model to create a new one that is better. In a situation like this, modelling would emerge from Model Analysis.

Regarding limitations of this teaching approach, the main one observed was related to the analytical approach of concepts such as derivatives. Since the solutions $X(t)$ and $Y(t)$ of the model do not have an analytical expression, it is not possible to work with algebraic aspects directly with the solutions. In the Applied Mathematics course these aspects were addressed with the typical functions studied in calculus courses (polynomial functions, exponential functions, etc.). This strategy actually worked very well with Biology students since the qualitative analysis was a pedagogical goal of the teaching approach. However, depending on the pedagogical aims a teacher has, some changes in the phenomenon or the mathematical model may be necessary.

38.6 Final Considerations

The advent of informatics and its fast development has allowed for new ways of doing mathematics and applying mathematics (Napoletani et al. 2014). In the field of mathematics education these new ways of thinking have also been noticed and have been the focus of research (Borba and Villarreal 2005). In this chapter, one role of the software Modellus has been discussed in the development of a teaching approach based on Model Analysis: to give access to graphical and numerical representations of the model solutions, which allowed students to analyse the Ross-Macdonald model, even though it involved mathematical concepts still advanced for them. As a consequence of this role, students did not need to develop a mathematical treatment of the model. In this sense, we can understand that they were engaged in a "smaller" or rudimentary modelling cycle. In addition, this possibility offered by Modellus allowed the formation of a "hybrid approach" where an existing mathematical model is analysed as a base for the discussion of new (to the students) mathematical concepts. Based on the notion of reorganization, the shift in the way of dealing with mathematical models in classrooms suggest that Model Analysis can be understood as a reorganization of modelling and applications. Finally, it is important to mention that the study of the mathematical model was not only an avenue for presenting mathematical concepts. Even though Biology students did not change the equations of the model, they were already preparing themselves to do modelling in the future, since modelling competences such as (a) understanding the real problem, identifying quantities and variables; (b) interpreting mathematical results in extra mathematical contexts; and (c) critically reflecting and analysing the found solutions, were all addressed.

References

Basañez, M.-G., & Rodríguez, D. J. (2004). Dinámica de transmisión y modelos matemáticos em enfermedades transmitidas por vectores. *Entomotropica, 19*(3), 113–134.

Blomhøj, M., & Kjeldsen, T. H. (2011). Students' reflections in mathematical modelling projects. In G. Kaiser, W. Blum, R. Borromeo Ferri, & G. Stillman (Eds.), *Trends in teaching and learning of mathematical modelling* (pp. 385–396). New York: Springer.

Blum, W., & Leiß, D. (2007). How do students and teachers deal with modelling problems? In C. Haines, P. Galbraith, W. Blum, & S. Khan (Eds.), *Mathematical modelling: Education, engineering and economic* (pp. 222–231). Chichester: Horwood.

Borba, M. C., & Soares, D. S. (2011). Modeling in Brazil: A case involving biology. In W. Blum, R. Borromeo Ferri, & K. Maaß (Eds.), *Mathematikunterricht im Komtext von Realität, Kultur und Lehererprofessionalität* (pp. 53–61). New York: Springer.

Borba, M. C., & Villarreal, M. E. (2005). *Humans-with-media and the reorganization of mathematical thinking: Information and communication technologies, modeling, experimentation and visualization.* New York: Springer.

Doerr, H., & Pratt, D. (2008). The learning of mathematics and mathematical modeling. In M. K. Heid & G. W. Blume (Eds.), *Research on technology and the teaching and learning of mathematics* (Research Syntheses, Vol. 1, pp. 18–23). New York: Information Age Publishing.

Doerr, H., & Wood, T. (2006). Pesquisa-Projeto (*design research*): aprendendo a ensinar Matemática. In M. C. Borba (Ed.), *Tendências Internacionais em Formação de Professores de Matemática* (pp. 113–130). Belo Horizonte: Autêntica.

Maaβ, K. (2006). What are modelling competences? *ZDM, 38*(2), 113–142.

Napoletani, D., Panzo, M., & Struppa, D. C. (2014). Is big data enough? A reflection on the changing role of mathematics in applications. *Notices of the AMS, 61*(5), 485–490.

Niss, M. (2015). Prescriptive modelling – challenges and opportunities. In G. A. Stillman, W. Blum, & M. S. Biembengut (Eds.), *Mathematical modelling in education research and practice: Cultural, social, and cognitive influences* (pp. 67–80). Cham: Springer.

Niss, M., Blum, W., & Galbraith, P. (2007). Introduction. In W. Blum, P. Galbraith, H.-W. Henn, & M. Niss (Eds.), *Modelling and applications in mathematics education. The 14th ICMI study* (pp. 3–32). New York: Springer.

Soares, D. S. (2012). *Uma abordagem pedagógica baseada na Análise de Modelos para alunos de Biologia: qual o papel do software?* Unpublished doctoral dissertation, Universidade Estadual Paulista, Rio Claro.

Soares, D. S., & Javaroni, S. L. (2013). Análise de Modelos: possibilidades de trabalho com modelos matemáticos em sala de aula. In M. C. Borba & A. Chiari (Eds.), *Tecnologias Digitais e Educação Matemática* (pp. 195–219). São Paulo: Editora Livraria da Física.

Chapter 39
Collective Production with Mathematical Modelling in Digital Culture

Arlindo José de Souza Júnior, João Frederico da Costa Azevedo Meyer, Deive Barbosa Alves, Fernando da Costa Barbosa, Mário Lucio Alexandre, Douglas Carvalho de Menezes, and Douglas Marin

Abstract In our learning collective, we established the Nucleus for Research in Media in Education including teams from computing, education and science. This research group presents the role of mathematical modelling in the production of Learning Objects, from the interweaving of theoretical and experimental knowledge from individuals implicated in this process. In the production of learning objects involving mathematical models, we note the existence of the following phases of construction: (1) obtaining a mathematical model; (2) describing the reality of the model; (3) the user accesses the application of the information; (4) deciding on how to present the history of the mathematical model; (5) programming the mathematical model; and (6) analysing the interaction between users and their learning via a simulation model. This context shows that the construction of learning objects requires firstly, a process of mathematical modelling.

A.J. de Souza Júnior (✉) • D. Marin
Mathematics, Federal University of Uberlândia, João Naves de Avila, 2121, Campus Santa Mônica – Bloco 1F – Sala 1F120, Uberlândia, MG 38.408-100, Brazil
e-mail: arlindoufu@gmail.com; douglasmarin2007@gmail.com

J.F.d.C.A. Meyer
Mathematics, State University of Campinas, João Naves de Avila, 2121, Campus Santa Mônica – Bloco 1F – Sala 1F120, Uberlândia, MG 38.408-100, Brazil
e-mail: joni@ime.unicamp.br

D.B. Alves
Mathematics, Federal Institute of Education, Science and Technology, João Naves de Avila, 2121, Campus Santa Mônica – Bloco 1F – Sala 1F120, Uberlândia, MG 38.408-100, Brazil
e-mail: deivemat@gmail.com

F.d.C. Barbosa • M.L. Alexandre • D.C. de Menezes
Education in Science and Mathematics, Federal University of Uberlândia, João Naves de Avila, 2121, Campus Santa Mônica – Bloco 1F – Sala 1F120, Uberlândia, MG 38.408-100, Brazil
e-mail: fermat.ufu@gmail.com; mariomla@hotmail.com; douglasmatufu@gmail.com

G.A. Stillman et al. (eds.), *Mathematical Modelling in Education Research and Practice*, International Perspectives on the Teaching and Learning of Mathematical Modelling, DOI 10.1007/978-3-319-18272-8_39

39.1 Introduction

The term *Learning Objects* began to be used by Wayne Hodgins in 1992. He was observing one of his sons playing with Lego blocks while reflecting about teaching strategies. It was then he realized that in this situation it was necessary to build teaching blocks able to connect and express a series of educational content. He used the term Learning Objects to identify such educational blocks (Assis 2005).

Learning Objects can be understood as "Any digital resource that can be reused to support learning" (Wiley 2000, p. 3). Learning Objects can be created in any media or format and can be as simple as an animation or slideshow or complex such as a simulation. It is an activity of handling situations with computer-mediated interaction, it is supplementary to imagining and enables the ability to anticipate the consequences of our actions. Learning Objects make use of images, animations, applets, documents VRML (virtual reality), text files, hypertext and others (Macêdo 2007).

Since 2004, the Brazilian government has been funding a project called Interactive Virtual Education Network (RIVED).[1] The main idea is to break the disciplinary educational content into small snippets which can be reused in various learning environments. Thus any material that comes from electronic information for the construction of knowledge can be considered a Learning Object. This information can be in the form of an image, a page of HTML, an animation or simulation. In other words, Learning Objects must have a defined educational purpose, an element that encourages student reflection and the application is not restricted to a single context.

For the preparation and development of Learning Objects a multidisciplinary team is needed in which pre-service teachers and faculty experts in areas of knowledge work collaboratively with teachers, computer teachers, programmers and web designers. It is essential that there is communication between the multidisciplinary team and the teachers of primary students. "We point out the prospect that academics who work the investigation of initial and continuing training of teachers could form partnerships with teachers of elementary and middle school in order to develop projects designed to work in the school routine" (Souza Júnior and Lopes 2007, p. 8).

The participation of undergraduates in this project enriches their education, however, we must discard the myth that beginning teachers will change the world when they come to school. There are many external factors that prevent teachers in practice from using new technologies. However, that does not mean that student teachers should not participate in such projects, we believe it is important, but you need to see things more realistically.

---

[1] Learn more at <http://rived.mec.gov.br/site_objeto_lis.php>, accessed 05/06/2011.

Some authors believe that teacher training should provide pre-service teachers with the ability to produce their own materials.

> In the training of teachers, for example, there must be concern with pedagogy consistent with the technological development of society in a way that future teachers need not only know how to use technological resources that have been prepared and developed by others, but yes, do they know their own material and even know how to use the new technologies from the perspective of mediation training. (Abreu et al. 2006, p. 337)

In this context, Souza Júnior et al. (2007) stated that the aim of the RIVED project was to attempt to improve teaching and learning and encourage the use of technology in schools, seeking a new vision for education. For these authors the construction and development of learning objects is a social production that fulfills two well defined roles: Firstly, that the pre-service teachers can, in conjunction with a team, create learning objects and secondly, immersion of technology in the educational environment.

The RIVED project began a collective production of knowledge between the areas of Computing, Education, and Science at the Federal University of Uberlândia. These productions, according to Cintra (2010) made the knowledge areas involved knit together with the important support of a core group of researchers, in this case, the Center for Research in Media in Education (NUPEME). In publishing, the three teams that participated in the RIVED-UFU perspective of collective work, integrated the mentioned areas of knowledge. Cintra also states that the project established a culture of analysis, production and use of learning objects in special courses in Mathematics, Chemistry and Computer Science.

Souza Júnior et al. (2010) argue that NUPEME aims to develop technological resources and discussion of the pedagogical use of technology in the educational process. Thus, we are striving to find the interaction between mathematics and Learning Objects. However, for that to happen, we need to find the mathematical model of a problem. According to Skovsmose (2007):

> ... the concept of math[ematical] modelling as a representation of reality is related to dualism, a perspective of two worlds. On one hand, we can work with mathematical concepts as part of the structured world, as suggested by formalism. On the other hand, we can work with the reality of the empirical world. A mathematical model becomes a representation of part of this reality. (p. 107)

However, he cautions that, "We can make good and bad representations. The results of the model can be more or less adequate," and raises questions such as "What part of reality does this model address?" "What mathematics is used in building the model?" and "How well does the model represent reality?" We understand these questions, but we see them as ways make a difference in the development of Learning Objects for mathematics teaching and learning. We now make the case for our argument that it is through the mathematical model that Learning Objects connect education, mathematics and computing.

39.2 Production of Learning Objects: An Approach to Bring the Real World into the Classroom

People engage with digital technologies in different contexts and embedded within these contexts are different expectations about the nature of the interaction. This engagement is closely interconnected to learning to recognise and produce therefore Cortella (2009) understands work as poiesis, a Greek word which he translates as: What we build, and that is to create the subject creates itself as it creates the world, otherwise every time we look at what we did not like being ourselves we are oblivious. Thus, it is natural to allow undergraduate students to participate in the act of producing Learning Objects so that they see if it is so.

This perspective is aimed at finding a solution to the question: *What is the "path" in the production of Learning Objects for teaching and learning of mathematics?* We understand that such questioning seeks "a process that emerges from [our] own reason[ing] and [is] part of our lives as a way [of deconstruction] and expression of knowledge" (Biembengut and Hein 2002, p. 11). This process is mathematical modelling, because the mathematical model represents part of a given reality. The process by which we developed Learning Objects occurs in five phases:

First Phase The first action in building a Learning Object is to develop a mathematical model, since the model is the story of the real world in mathematical symbols. We think this is the hardest part and we should have the greatest concern, because it determines the rest of the construction of Learning Objects. Those responsible for this phase are the pre-service teachers in mathematics, under the guidance of expert faculty in mathematics.

Second Phase In this phase is a description of the real world, of people and the problem that involves the mathematical model. It aims to find the elements of the scenario: time, place and social climate, for the construction of learning objects. Again, pre-service teachers in mathematics, under the guidance of expert faculty in mathematics are responsible for this phase.

Third Phase In this phase, we choose the software for the development of Learning Objects. For the case of objects developed by members of NUPEME that are reported in this chapter these are produced using Adobe Flash. Those responsible for this phase are the undergraduate students in computing, under the guidance of faculty teachers of computer science.

Fourth Phase This involves the audiovisual resources. We think about how we want to present the history of the mathematical model as well as the legal implications of using image, video or music. We define and start working with the construction of the characters, background and other elements. It is at this phase that the interface, the shapes and colours of Learning Objects are determined. Those responsible for this phase are the undergraduate students in computing and the pre-service teachers in mathematics, under the guidance of expert faculty. The success of this phase depends on the interaction between these two groups.

Fifth Phase The activity of this phase is to provide functionality and movement to the mathematical model(s) by means of computer programming. Those responsible for this phase are the undergraduate students in computing, under the guidance of faculty teachers of computer science. The pre-service teachers in mathematics are responsible for testing the Learning Objects in elementary schools. To the extent that errors are found new versions are developed.

It is argued that, to create a learning object, we need to be involved with mathematical modelling, since this is the "backbone" of what is created. Thus, the Learning Object becomes a means of expression, an act of mathematical modelling on the computer itself. This act of producing reveals an interaction between undergraduate mathematics and computer science students, as well as with their teachers. What is evident is a collective reflection on the construction of computational and mathematical knowledge. From this point of view, we illustrate our argument with the production of the Learning Object: *Population Dynamics*, developed by undergraduate students in mathematics and computer science in the RIVED project carried out at the Federal University of Uberlândia in 2010.

39.3 Learning Object: *Population Dynamics*

Population Dynamics[2] is a learning object that has as reference the history of Italian marine biologist Umberto D'Ancona, during the years 1914–1923 when he conducted studies on the fish sold in the markets of Trieste, Fiume and Venice. He noted that because of the First World War, during which fishing was greatly reduced in part of the Adriatic Sea, there was an initial reduction in fish numbers of certain species followed by an increase in number of others. During the time fishing was suspended, shark populations (predators) increased, then decreased when fishing resumed.

The educational objectives of this object are given to answer the questions asked by D'Ancona (1954): *How does the intensity of fishing affect the fish population?*

This case of predators and prey is well-studied by mathematical modellers, so several models have been developed. The model used in our project uses simple difference equations:

$$P_{n+1} = P_n + A \cdot P_n - B \cdot P_n \cdot Q_n - E \cdot P_n \quad (39.1)$$

$$Q_{n+1} = Q_n - C \cdot Q_n + D \cdot P_n \cdot Q_n - E \cdot Q_n \quad (39.2)$$

where

P_n is the initial population of fish (prey).
Q_n is the initial population of sharks (predators).

[2] Learn more at < http://goo.gl/nyFuV8>, accessed 14/09/2013.

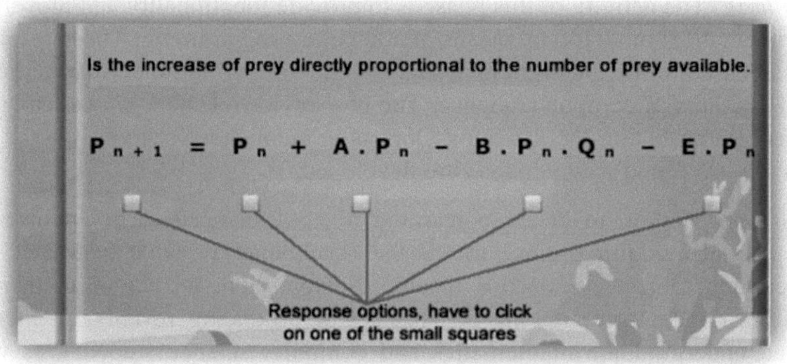

Fig. 39.1 Simulation environment: statement 1

A is a positive real constant, which represents the reproduction rate of prey.
B is a positive real constant, which represents the mortality rate of prey when it encounters with sharks.
C is a positive real constant, which represents the mortality rate of sharks by natural causes.
D is a positive real constant, which represents the birth rate of sharks when feeding on prey. To simplify the model, we consider $B = D$.
E is a positive real constant, which represents the fishing rate. To simplify the model, we consider the same rate of fishing for the two populations.

From this model, three statements guided the Learning Object simulation. The initial statement was posed as a question: Is the increase of prey directly proportional to the number of prey available (see Fig. 39.1).

It is easy to verify that the *increase in prey* $= A \times$ *number of prey*. However, the important thing is that from such questioning one can ask the user to reset the initial population of sharks (i.e., the predators, Q_n) and fishing (i.e., the prey, P_n). In this way, by simulating as shown in Fig. 39.2, it is clear that the evolution of prey over time can be represented by mathematical arguments as follows:

$$P_{n+1} = P_n + A \cdot P_n \tag{39.3}$$

In this case, understanding that (39.3) is a geometric progression to calculate from known values:

$$P_{n+1} = P_n(1 + A)^m \tag{39.4}$$

Thus for P_{n+2} we have

$$P_{n+2} = P_{n+1}(1 + A) = P_n(1 + A)(1 + A) = P_n(1 + A)^2$$

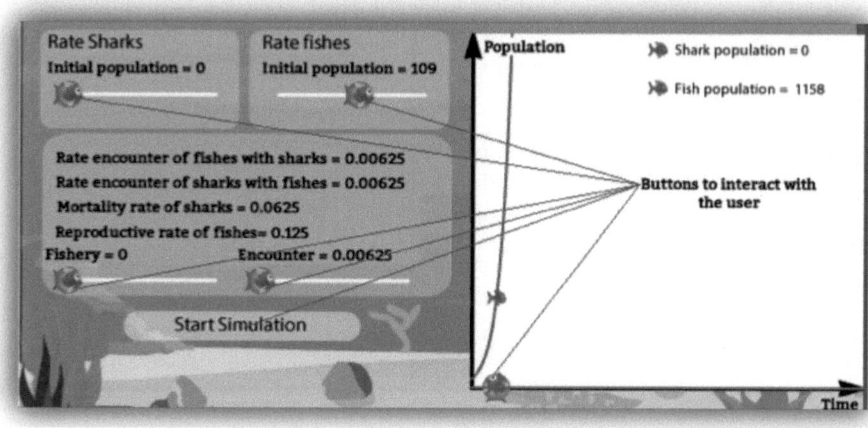

Fig. 39.2 Simulation environment: simulator

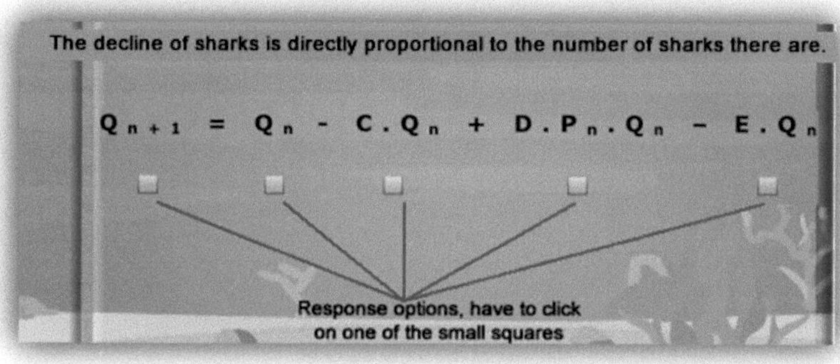

Fig. 39.3 Simulation environment: statement 2

The process is analogous to P_{n+m}, for any n and m belonging to the set of natural numbers. We can generalize this relationship and write it as follows:

$$P_{n+m} = P_n(1+A)^m \tag{39.5}$$

which takes the form of a geometric progression with common ratio $(1+A)$.

Similarly, for the second statement (Fig. 39.3): The decline of sharks is directly proportional to the number of sharks there are.

It is also easy to verify that the *decrease of sharks* $= -C \times$ *number of sharks*. So, because of this you can ask the user to reset the initial population of prey (Pn) and fishing rate (E) for zero. Thus, to simulate, in Fig. 39.4, it is clear that the evolution of predators with time can be represented by mathematical arguments as follows:

$$Q_{n+1} = Q_n - C \cdot Q_n \tag{39.6}$$

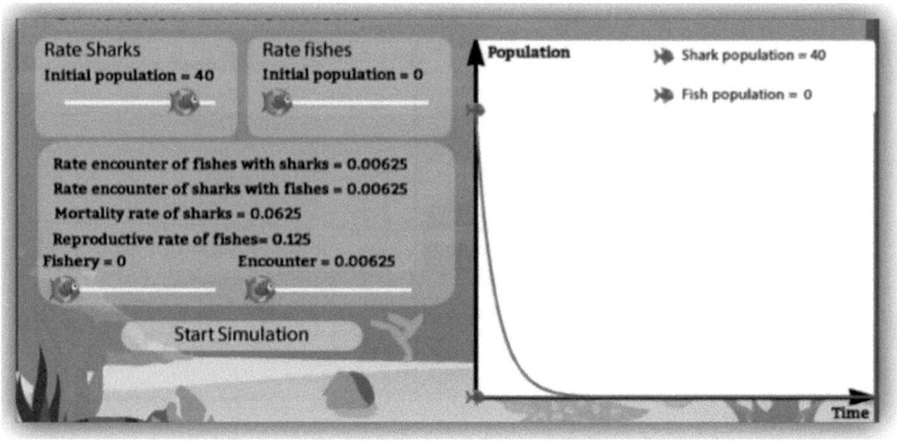

Fig. 39.4 Simulation environment: simulator

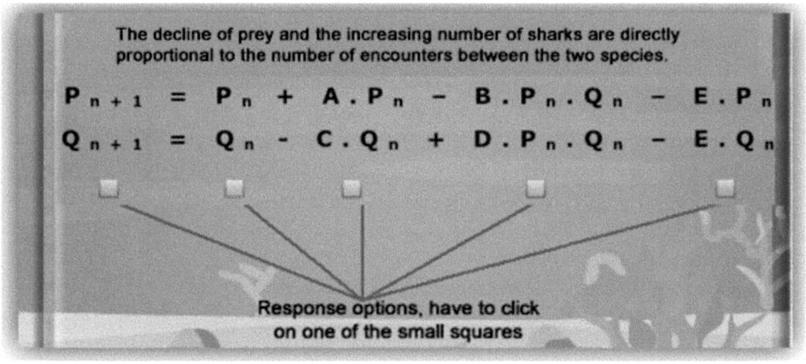

Fig. 39.5 Simulation environment: statement 3

As this is analogous to expression (39.3), we conclude that $Q_{n+m} = Q_n(1-C)^m$ it is the form of a geometric progression with common ratio $(1-C)$.

The third statement was: The decline of prey and the increasing number of sharks are directly proportional to the number of encounters between the two species (Fig. 39.5).

It can be noticed that the *decrease of prey* $= B \times$ *number of prey* $\times$ *number of sharks* and *sharks have increased* $= D \times$ *number of prey* $\times$ *number of sharks*. Bearing this in mind, linking the first two statements we can talk about changes in prey and predator variation, in other words:

$$\begin{aligned} \text{For fish } \Delta P_n &= A \,.\, P_n - B \,.\, P_n.\, Q_n \text{ and} \\ \text{for sharks } \Delta Q_n &= -C \,.\, Q_n + D \,.\, P_n.\, Q_n \end{aligned} \tag{39.7}$$

To answer the initial statements: How does the change in fishing affect fish populations as prey? And can change in fishing facilitate the reduction of fishing

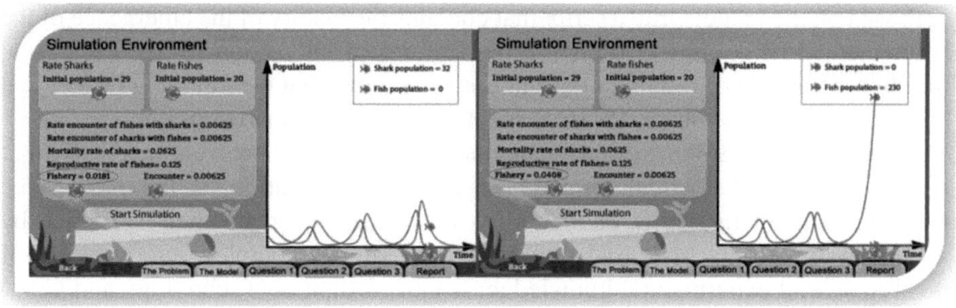

Fig. 39.6 Simulation environment: simulator

sharks in relation to the number of sharks? It is necessary to add the fishing factor for prey ($-E\,.\,P_n$) and sharks ($-E\,.\,Q_n$). This returns us to equations (39.1) and (39.2):

$$\begin{aligned} P_{n+1} &= P_n + A.P_n - B.P_n.Q_n - E.P_n \\ Q_{n+1} &= Q_n - C.Q_n + D.P_n.Q_n - E.Q_n \end{aligned}$$

Then, in the simulation, it is observed that by increasing the intensity of fishing for sharks ($E\,.\,Q_n$) the decrease in sharks may occur to extinction and the number of prey (P_n), that is fish, increases, in the case when the extinction of sharks is increasing exponentially Fig. 39.6.

As seen in Fig. 39.6, the fishing coefficient (E) was 0.0181 (shown as Fishery) had a final population of sharks of 32 and a population of fish as zero. When we increase only fishing rate to 0.0408, the final population of sharks has dropped to zero and prey (fish population) increased to 230. From this we can conclude two things: The lower the fishing rate the better fishing is for sharks and the higher this factor, the better fishing is for the fish. This explains the increase in the percentage reported values of D'Ancona (1954).

39.4 Final Considerations

We believe that teachers can be the authors of their own Learning Objects as Paim (2005) describes a teacher as one who "thinks, plans, defines and implements educational activities" (p. 149). For the concept of authorship, we draw on the writings of Kramer (2002), who states:

> Being author... say the word itself, coining it and your personal brand to brand yourself and to others by word said, yelled, dreamed, spelled... Being author means to rescue the possibility of 'human', to act collectively for what characterizes and distinguishes men... Being author means producing with and for each other. (p. 83)

For pre-service teachers in mathematics to be authors of a Learning Object, we have outlined a process which comprises five phases: (1) Find a mathematical model;

(2) based on the model write a script that contains the history of the emergence of the mathematical model; then (3) choose the features of interaction for the user; then, (4) choose the type of visual, or design, image or video that the Learning Object will have; finally, the last phase, (5) implement in a programming language.

This context shows that the construction of Learning Objects requires firstly that a process of mathematical modeling occur. The formation of the pre-service teachers in mathematics is facilitated as they learn to create their existence by means of three sets of knowledge consisting of the complex interweaving of theoretical and experimental knowledge; from Information Technology and Communication, interacting – simulating with the Thematic and in turn interacting – simulating the mathematical model and this subsequent interaction by simulation-to-Information Technology and Communication.

References

Abreu, M. F., et al. (2006). *Utilizando Objetos de Aprendizagem no processo de ensino e aprendizagem de química no ensino médio: o caso dos óxidos e da poluição atmosférica*. http://goo.gl/VXaqCK. Accessed 10 June 2011.

Assis, L. S. (2005). *Concepções de professores de matemática quanto a utilização de objetos de aprendizagem: um estudo de caso do projeto Rived – Brasil*. São Paulo: Pontifical Catholic University.

Biembengut, M. S., & Hein, N. (2002). *Modelagem Matemática no Ensino*. São Paulo: Contexto.

Cortella, M. S. (2009). *Qual é a tua obra?: Inquietações propositivas sobre gestão, liderança e é tica*. Petrópolis: Vozes.

D'Ancona, U. (1954). *The struggle for existence*. Leiden: Brill.

de Cintra, V. P. (2010). *Projeto Rived: Um estudo de caso de uma equipe de matemática*. Rio Claro: State University Paulista "Júlio de Mesquita Filho".

de Macêdo, L. N., et al. (2007). Desenvolvendo o pensamento proporcional com o uso de um Objeto de Aprendizagem. In C. L. Prata & A. C. A. A. de Nascimento (Eds.), *Objetos de aprendizagem: uma proposta de recurso pedagógico* (pp. 17–26). Brasília: Mec, Seed.

de Souza Júnior, A. J., & Lopes, C. R. (2007). Saberes docentes e o desenvolvimento de Objetos de Aprendizagem. In C. L. Prata & A. C. A. A. de Nascimento (Eds.), *Objetos de Aprendizagem: uma proposta de recurso pedagógico* (pp. 7–15). Brasília: Mec, Seed.

de Souza Júnior, A. J., et al. (2007). *Informática na educação: elaboração de Objetos de Aprendizagem*. Uberlândia: Edufu.

de Souza Júnior, A. J., et al. (2010). *Objetos de Aprendizagem: Aspectos conceituais, empíricos e metodológicos*. Uberlândia: Edufu.

Kramer, S. (2002). *Por entre as pedras: arma e sonho na escola*. São Paulo: Ática.

Paim, E. A. (2005). *Memórias e experiências do fazer-se professor(a) de História*. Campinas: State University of Campinas.

Skovsmose, O. (2007). *Educação crítica: incerteza, matemática, responsabilidade*. São Paulo: Cortez.

Wiley, D. A. (2000). Connecting learning objects to instructional design theory: A definition, a metaphor, and a taxonomy. In D. A. Wiley (Ed.), *The instructional use of learning objects* (pp. 1–35). Bloomington: Association for Education Communications and Technology.

Part V
Assessment in Schools and Universities

Chapter 40
Learners' Dealing with a Financial Applications-Like Problem in a High-Stakes School-Leaving Mathematics Examination

Cyril Julie

Abstract Research of students' ways of working with modelling and applications-like problems in time-restricted examinations is rare. Using ideas and notions from ethnomethodology and the sociological study of work in science, the actual scripts of examinees were analysed to tease out the examinees' ways of working with a modelling and applications-like problem in a high-stakes school-leaving examination. The analysis was anchored around the various agencies exerted by elements present in the context of high-stakes school-leaving examinations. Three ways of working which characterised the candidates' ways of working are focused on. It is demonstrated how the prevailing contexts of writing high-stakes examinations exercised agency for these ways of working. The pragmatic value of analysis of this nature is recommended.

40.1 Introduction

Applications of mathematics are currently an element of school mathematics curricula in most, if not all, countries. It is thus expected that this element will be part of examinations students are required to write to graduate from schools. Most of these examinations are time-restricted examinations. If the common modelling and applications cycle is taken into account, then it is obvious that time-restricted examinations which a student must complete on his or her own, are not conducive for allowing students to demonstrate their competence to deal with the applications of mathematics. However, in these kinds of examinations the form in which questions related to applications appear is of a guided nature. In the South African situation with respect to applications of mathematics to financial

C. Julie (✉)
School of Science and Mathematics Education, University of the Western Cape, Robert Sobukwe Road, Bellville 7530, South Africa
e-mail: cjulie@uwc.ac.za

G.A. Stillman et al. (eds.), *Mathematical Modelling in Education Research and Practice*, International Perspectives on the Teaching and Learning of Mathematical Modelling, DOI 10.1007/978-3-319-18272-8_40

contexts, further guidance is provided by supplying candidates with a formula sheet which contains financial models. The examination problems, however, are, of such a nature that the given models cannot always be directly applied. Examinees have to do certain adaptations to the models in order to use information from the formula sheet for the situation they are presented with.

In time-restricted examinations of the kind mentioned above, candidates also have to particularise a model to suit the situation at hand. Item analysis of the various questions of the mathematical topics in a national examination in South Africa rendered that candidates found the items related to financial situations the most difficult (Jacobs et al. 2014). In this chapter the outcomes of a qualitative analysis of the ways of working with applications problems, as manifested in the actual scripts of learners, are presented and discussed. The focus is one question and the analysis is inspired by ethnomethodological studies of mathematical work.

40.2 Theoretical Machinery

The study reported here is situated within ethnomethodology and the social study of science. Ethnomethodology "is the study of common, everyday methods – of practical action and practical reasoning" (Livingston 1987), p. 4). It focuses on how people go about achieving what they achieve in the contexts within which the achievement is realised. In this sense, with respect to mathematics, it studies the accomplishment seen as the realization/production of solutions (or personal resolutions) to problems which are being pursued in mathematics, preferably in real-time, at its site of pursuit. It seeks "to describe members' accounts of formal structures wherever and by whomever they are done, while abstaining from all judgements of their adequacy, value, importance, necessity, practicality, success, or consequentiality" (Coulon 1995), p. 42). Thus when studying students' ways of doing mathematics from this perspective, the 'correctness' of their answers or compliance to some predetermined methods are treated with indifference.

A seminal ethnomethodological study in mathematics was done by Livingston (1986). He uses his own way of working on proofs as data to open up the lived work of proving in mathematics (Livingston 1986, 2006). He partitions the work into two components: the lived work and an account of the work. For proving he asserts that

> What is written and said is not really the 'whole' proof. It is a proof-account. The proof – as one coherent social object – consists of a pair: [a proof-account/the lived-work of proving to which the proof-account is essentially and irremediably tied]. The pairing – as one integral object, not as two distinct 'parts' circumstantially joined – is the 'proof' in and as the details of its own accomplishment. (1986, p. 111)

Different from Livingston, Pickering (1995) renders a description of Hamilton's construction of quaternions. He proposes disciplinary agency as "the agency of a discipline—elementary algebra, for example—that leads us through a series of manipulations within an established conceptual system" (p. 115). According to him in the pursuance of solution-seeking in mathematics, there are periods when the problem solver is passive and the discipline exerts its agency. In this sense mathematical work is characterised as a dance of agency where human and disciplinary agency intertwine during the solution-seeking process. He refers to this intertwinement of these agencies as "the *dialectics of resistance and accommodation*, where resistance denotes the failure to achieve an intended capture of agency in practice, and accommodation an active human strategy of response to resistance" (p. 22). Resistance occurs when an action does not produce an expected outcome. Pickering demonstrates how Hamilton in his construction of quaternions encountered resistance when the algebraic and geometric approaches rendered contradictory results. Accommodation is the paths embarked upon to resolve resistance.

Merz and Knorr-Cetina (1997) analysed "the work and accomplishments of theoretical physicists" (p. 73). Their work is situated within the application of procedures of the new sociology of scientific practices on mathematics and, to a certain extent, emulate the work of Livingston and Pickering to render an analytic account of the practical work of theoretical physicists since theoretical physics is "like mathematics,...a thinking science [where] work is performed at the desk, instruments are reduced to the pencil and the computer, and processing is realized through writing" (p. 74). They used the emails which the physicists used to communicate their ongoing work and also data they obtained from conducting interviews with the physicists. They assert that theoretical physicists achieve resolutions to problems

> through a first-order deconstruction, and, if this is hard to achieve, through a second round of detours and tricks. [And] those [strategies] which can be deployed toward both first-order difficulties and second-round problems...In a third run [of] theorists' practical work of dealing with computational objects [they employ strategies such as] variation, "doing examples", and model objects. (p. 95)

In previous work Julie (2003) used constructs from some of the above work to analyse how a group of four practising teachers with no prior experience of modelling went about to construct a mathematical model. He shows how a particular table constructed by the teachers drove the mathematical model-building activity at a given time. This table played a crucial, mediating and guidance-giving role. The performative actions they employed were identified as shedding, compensation and variation.

Greiffenhagen (2008), on the other hand, had as data videos of actual lectures to graduate students in Mathematics and interactions between doctoral students in mathematics and their supervisors to illuminate such lived work in mathematics. He found that mathematical work at its site of production is "both retrospective and prospective, i.e., the lecturer is summarizing what he has just done (showing that the two items are less than one), before projecting what he will do next" (Greiffenhagen 2008, p. 17).

The difference between the lived work and the account of such work in mathematics is also foregrounded by Roth (2012). Using the proof of the theorem "the sum of the internal angles of a triangle is 180°", he demonstrates how the lived work of the proof emerges through drawing, seeing and concluding. He insists that "the lived (subjective) work has to be enacted each and every time by the person actually doing or following (observing) the proof" (Roth 2012, pp. 239–240).

The aforementioned studies reveal the possibilities and complexities of analysing the lived work doers of mathematics engage in when constructing resolutions to mathematical problems. It is clear that the constructs being used are diverse and Livingston (2006) calls for "a vocabulary that provides descriptive insight into the lived, perceived details and reasoning of mathematical discovery work" (p. 63). It is not the intention in this chapter, to work towards such a vocabulary. Rather, the study reported here is inspired by, and draws on, the afore-mentioned studies. Where appropriate, some of the constructs are used and others are constructed to tease out the 'lived' work of students' production of responses to an applications-like problem in a high-stakes school mathematics examination.

40.3 The Context

In the above studies the contexts within which the resolution pursuance occurs is accorded primacy. As already alluded to, resolutions of problems in mathematics are accomplished with, as is the case for high-stakes examinations, pen and paper. The temporally-emerging scribblings done on paper form an inextricable part of the context. They exert some agency and, in a sense give guidance on how to (or not) proceed further. In addition to the available pen and paper, the examinee also has access to and can use a modern, non-programmable scientific calculator. The scribblings arise from the examinee's interaction with an available object in the context—the question paper and its attachments. These objects demarcate the disciplinary domain at stake and for a high-stakes examination; it also provides the specific problems that must be solved. The question paper is a source of constant referral. As invigilators of examinations will agree, examinees constantly, physically and/or visually, move between their produced answer text and the question paper with the problem text. Another visible element in the examination environment is a time indicator in the form of a large display clock or some writing of the time available or elapsed on the chalkboard or an examinee's personal time indicator device. Last, but crucial, the mathematically-historicised examinee is present as agent. The examinees bring to the solution pursuance their entire experience of at least 12 years of school mathematical work and the agentic action "begins within the actor…[which] covers behavior ranging from [the] automatic (throwing a ball) to [the] carefully considered (solving a math problem)…(Hitlin and Elder 2007, p. 175).

For this study the data are the answer scripts of examinees, purposefully selected to have a wide range of calculation protocols. The problem that is focussed on was presented as:

QUESTION 7

7.1 At what annual percentage interest rate, compounded quarterly, should a lump sum be invested in order for it to double in 6 years?
(Department of Basic Education 2010), p. 6)

The entire examination consisted of ten questions, carried a total of 150 marks and the maximum allotted time for the examination was 3 h. Rule of thumb advice given to candidates in preparation for the examination is that they should not spend more than 1½ minutes per mark on a question. Thus for a question carrying five marks, as is the case for question 7, they should not spend more than 8 min.

The formulae related to financial mathematics given on the information sheet attached to the question paper were: $A = P(1 + ni)$, $A = P(1 - ni)$, $A = P(1 - i)^n$, $A = P(1 + i)^n$, $F = \frac{x[(1+i)^n - 1]}{i}$, $P = \frac{x[1-(1+i)^{-n}]}{i}$.

From the formulation of the problem and the fact that examinees had experience in constructing these models it is clear that the expectation is that they would select a model, particularise it for use with the given information, do the necessary calculations and perhaps check the reasonableness of their answers. The majority of their work thus has to do with the calculation component of modelling.

In what follows an analysis of how examinees accomplished their results is presented. I do it primarily by focussing on the agencies exerted by different elements of the above context. Three solution paths are focussed on.

40.4 Analysis of the Production of Answer Responses by Examinees

40.4.1 Translation to Familiar Symbolism

Correct solutions were accomplished by basically selecting and transferring the correct model from the formula sheet, doing the particularisations for the compounding period, the doubling of the outcome amount and executing the correct calculations. The doubling of the outcome amount normally resulted in the working model becoming $2P = P(1 + i)^n$. There were slight deviations from this set procedure. Figure 40.1 demonstrates such a deviation. Instead of using the symbol A, as given for the model in the formula sheet for the outcome amount, x is used. This is the familiar habituated symbol used with polynomial algebra and functions in South Africa. In a sense this is a case of simplification by going to "less complex special cases" (Merz and Knorr-Cetina 1997, p. 100). The simplification in this case is something akin to 'let me get it in symbols which I have experience of working with'. Also noticeable of resorting to familiar symbols of polynomial algebra and functions is the explicit writing of the division ($\div\ x$) that was performed in line

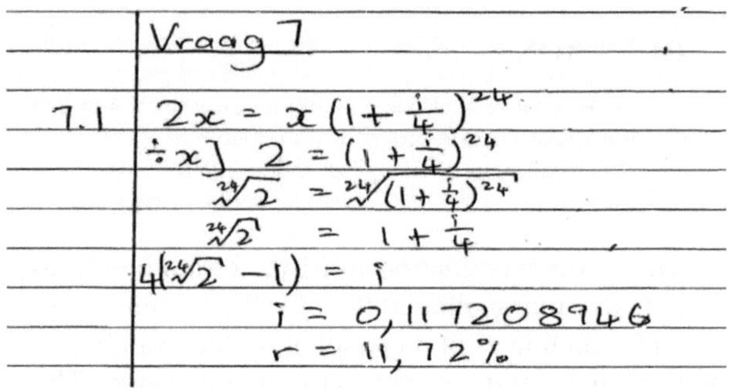

Fig. 40.1 Translation to familiar polynomial and functions symbols

2. Further, in the last line *i* is replaced by *r*, the symbol generally used colloquially for interest in financial matters and particularly its Afrikaans equivalent '*rente*'. Although *i* was written on the right-hand side in line 5, this changed to the left-hand side for the subsequent lines due to the agency exerted by the historicised self of what an expected representation should be for a response in mathematics.

40.4.2 *Anticipation of Adjustments to Be Made*

Another way of working to obtain a correct answer was to start with transforming the model from the formula sheet to make explicit that the compounding period should be adjusted and this adjustment should be to the '*per annum*' connotation of the model. This is indicated by *m* in the first line in Fig. 40.2. It emulates Greiffenhagen's (2008) notion of mathematics lecturers projecting of what will be done next. The "done next" is confirmed be replacing *m* with 4 and 6×4. (It should be noted that the 2 *m*'s represent different entities as the further procedure with the calculation indicates.)

There is also resort to the *x* notation scheme as comes through in line 2. This is discarded as the scratching out in line 2 shows. The work proceeds with the explicit numbers, 2 and 1, (line 3) and the need for working with abstract symbolic manipulation is abandoned since what is accomplished can accountably be accomplished without it.

40.4.3 *U-Turning*

A way of working forthcoming from the analysis is that of 'abandoning a solution path after a final answer was obtained'. The calculation is firstly worked through to its conclusion. Upon obtaining an answer a 'change of course' is taken. A U-turn—coming to a destination, finding that it is not the desired destination, going back and choosing another route—is made. Three kinds of U-turns were identified.

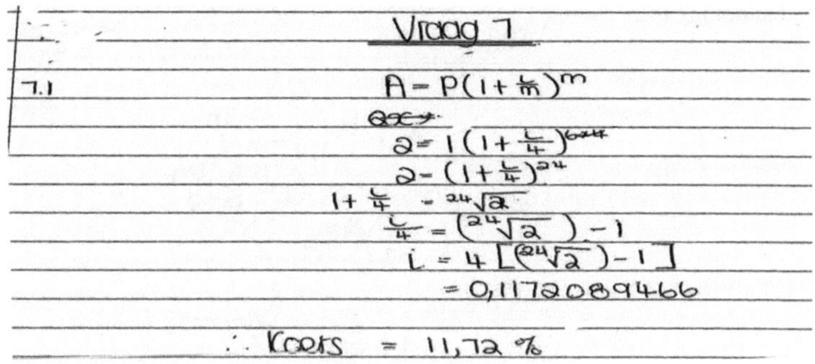

Fig. 40.2 Indication of anticipated adjustment to be made

In Fig. 40.3 a kind of 'go back to the start, restart with a different interpretation of some of the given' strategy is exhibited. Upon obtaining the answer, some scratching out is done, indicating the going-back to start. The new course of pursuit is to substitute the 24 for n with 6, the number of years given in the problem-text. The 4, originally acquired from "compounded quarterly" is replaced by 12, presumably the number of months in the year. It is plausible to conjecture that the examinee re-read the problem after its first round of calculation and the phrases "6 years" and "compounded quarterly" exercised some agency for the impetus to pursue a different path.

In order to accomplish a resolution of the problem, Fig. 40.4 indicates that a complete working through is done. This is marked 'rough work' and deleted (Fig.40.4a). The rough work is very 'detailed' rough work and not of the kind found where 'rough work' is 'rough' indicating a few hints, a possible path to pursue and perhaps some interim calculations. The first three lines of the 'rough work' and the offered 'real' response (Fig. 40.4b) are the same. In line 4 of the 'real' response an equality is set up between an element, $\left(1 + \frac{i}{4}\right)^{24}$, of the previous line, the 2 being replaced by $(1 + i_{\text{eff}})^6$ and the sides being switched. This kind of U-turn is different from the previous one. It is a 'go back to the start and restart with a different strategy'. Noticeable is that the left-hand side expression contains an element, i_{eff}, not part of the surrounding context within which the calculation is being pursued. Its clue lies in the line, 'effektiewe jaarlikse rentekoers (effective annual interest rate)'. The $(1 + i_{\text{eff}})^6$ is near to $r = \left(1 + \frac{i}{n}\right)^n - 1$, the model for effective annual interest rate, with i the nominal interest rate. The historicised self of examinee together with the phrase "annual percentage interest rate" exerted an interpretative agency which resulted in a U-turn being taken and pursuing a different path.

A third kind of U-turn inspiration is demonstrated in Fig. 40.5. It is of the nature 'go to the start, restart because the obtained result does not comply with the dictates of the extra-mathematical context'. The work proceeded until the result 48,98 % was reached. Deletion from line 3 of this result is executed. In the reworked line

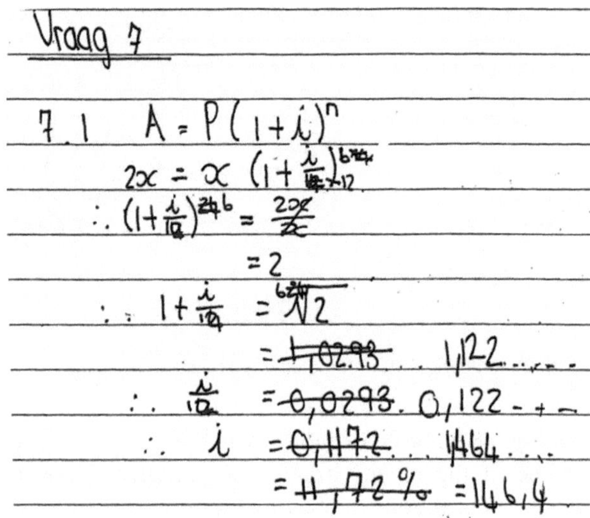

Fig. 40.3 Abandoning the 'correct'

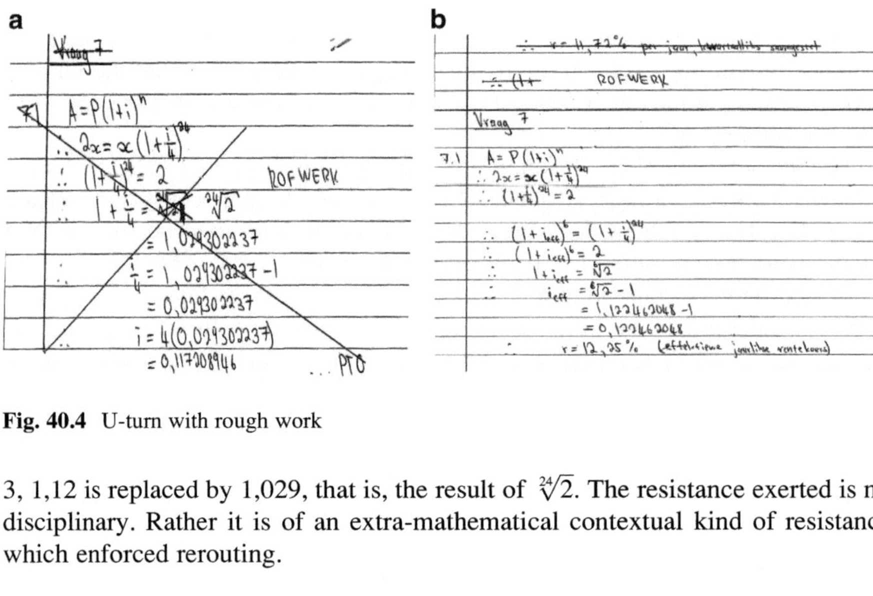

Fig. 40.4 U-turn with rough work

3, 1,12 is replaced by 1,029, that is, the result of $\sqrt[24]{2}$. The resistance exerted is not disciplinary. Rather it is of an extra-mathematical contextual kind of resistance, which enforced rerouting.

40.5 Discussion and Conclusion

The analysis of the examinees' ways of work from a quasi-ethnomethodological perspective and work forthcoming from the new sociology of scientific practices is deemed important for a variety of reasons. Firstly, given that high-stakes examinations place particular constraints on the way problems related to mathematical

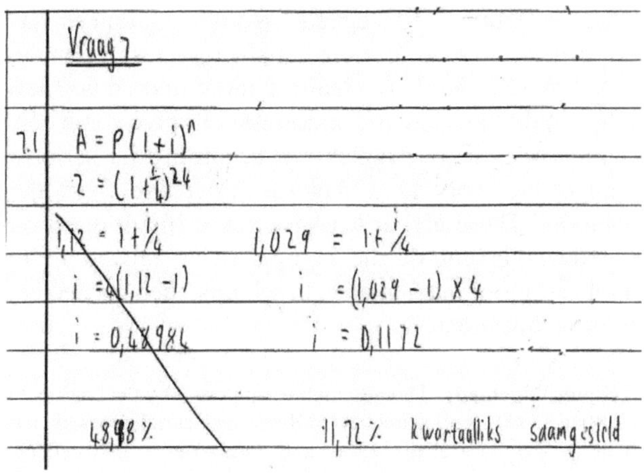

Fig. 40.5 U-turn resulting from exertion of extra-mathematical contextual agency

modelling can be posed, it is incumbent upon researchers to find ways to tease out the different paths followed by examinees in their solution-seeking endeavours. Although, as has been indicated above, research of this nature has been done on the production of mathematical work by experienced mathematical workers at the site where the mathematical work is done, it has not been the case for mathematical work of school learners. Particularly, the analysis of learners' mathematical work produced in an examination setting is virtually non-existent. Secondly, such analysis has pragmatic value. It can be made available to and discussed with teachers for them to decide how they will adapt (or not) their teaching taking into account the results emanating from analyses of the kind above. Lastly, linked to the pragmatic value, is the development of some fairly localised language with which to engage in productive conversations with teachers about applications and mathematical modelling work of their learners. From experience in working with teachers, I have found that their preference for speaking about the work of their learners is one that is less disciplinary-loaded. For example, they would speak about "problems that are turned around" when referring to examination items which are not posed in the standard form that it was taught. From the accounts of ethnomethodological work above, the approach offers the possibility for the development of such localised terminology exemplified by learners' real work. This can lead to having productive conversations with teachers to enhance their ways of dealing with mathematical modelling and applications in their classrooms.

An issue that is normally raised when attempts are made to tease out the 'lived' work from the responses they produce in high-stakes school examinations is whether the learners should be interviewed. The impracticability of this is obvious. Furthermore, Roth (2012) draws attention to that if one asks someone to explicate what he/she was doing when she/he was doing say, a proof of a theorem, he/she rather gives an account of the doing and not the doing.

The above analyses show the struggles examinees go through to find satisfactory resolutions to the adaptation and calculation components of the modelling process.

Three questions are begged. Would examinees have used what they initially constructed to lead to more satisfactory resolutions if they were not pressured by time? Would they have learned more and some 'new' tricks if a knowledgeable other could have had the opportunity to give some commentary on their work? Is it fair to assess their struggles only according to a script such as a memorandum which depicts the work as a flawless procedure from the problem statement to the solution? These are, at least for me, difficult questions but if we are interested in the development of the competencies related to mathematical modelling and applications of mathematics in schools, then we need to search for practical answers to these questions.

Acknowledgement This research is supported by the National Research Foundation under Grant number 77941. Any opinions, findings and conclusions or recommendations expressed in this material are those of the author and do not necessarily reflect the views the National Research Foundation.

References

Coulon, A. (1995). *Ethnomethodology*. Thousand Oaks: Sage.

Department of Basic Education. (2010). *National senior certificate: Grade 12 – mathematics P1, November 2010*. Pretoria: Republic of South Africa.

Greiffenhagen, C. (2008). Video analysis of mathematical practice? Different attempts to "open up" mathematics for sociological investigation [73 paragraphs]. *Forum Qualitative Sozialforschung/Forum: Qualitative Social Research*, *9*(3), Art. 32. http://nbnresolving.de/urn:nbn:de:0114-fqs0803323

Hitlin, S., & Elder, G. H., Jr. (2007). Time, self, and the curiously abstract concept of agency. *Sociological Theory, 25*(2), 170–190.

Jacobs, M., Mhakure D., Fray, R.L., Holtman, L., & Julie, C. (2014). Item difficulty analysis of a high-stakes mathematics examination using Rasch analysis. *Pythagoras*, *35*(1), Art. 220, 7 pp. http://dx.doi.org/10.4102/pythagoras.v35i1.220

Julie, C. (2003). Work moments in mathematical modelling by practising mathematics teachers with no prior experience of mathematical modelling and applications. *New Zealand Journal of Mathematics*, *32*(Supplementary Issue), 117–124.

Livingston, E. (1986). *The ethnomethodological foundations of mathematics*. New York: Routledge and Kegan Paul.

Livingston, E. (1987). *Making sense of ethnomethodology*. London: Routledge & Kegan Paul.

Livingston, E. (2006). The context of proving. *Social Studies of Science, 36*(1), 39–68.

Merz, M., & Knorr-Cetina, K. (1997). Deconstruction in a "thinking" science: Theoretical physicists at work. *Social Studies of Science, 27*(1), 73–111.

Pickering, A. (1995). *The mangle of practice: Time, agency and science*. Chicago: The University of Chicago Press.

Roth, W.-M. (2012). The work of seeing mathematically. In S. Mukhopadhyay & W.-M. Roth (Eds.), *Alternative forms of knowing (in) mathematics: Celebrations of diversity of mathematical practices* (pp. 227–246). Rotterdam: Sense.

Chapter 41
Evidence of Reformulation of Situation Models: Modelling Tests Before and After a Modelling Class for Lower Secondary School Students

Akio Matsuzaki and Masafumi Kaneko

Abstract In this chapter we discuss the reformulation of situation models. In previous modelling teaching and tests, the modelling tasks related to a real situation were reported for elementary school students. Here we administer a test before and after a modelling class for lower secondary school students. The original situation at the pre-test and the post-test were not changed. The students were required to set up a problem from a real situation, and draw pictures to explain how they setup the problem. We focus on pictures to explain the problems as evidence of situation models. As results, we can confirm that the reformulation of situation models occurred when simplifying the task, clearly explaining the task for others, or structuring the task mathematically.

41.1 Situation Models Based on Individual Modelling

Blum and Leiß (2007) indicated that there are reciprocal processes between the real situation and the situation model in a modelling cycle, and explained that this process is most important as a phase in understanding a modelling task. Borromeo Ferri (2006) focused on differences in individual mathematical thinking styles, and explained the presence of these processes through distinctions between the situation model/mental representation of the situation and the real model. In this chapter, we

A. Matsuzaki (✉)
Faculty of Education, Saitama University, Shimo-Okubo 255, Sakura-ku, Saitama-shi, Saitama, Japan
e-mail: makio@mail.saitama-u.ac.jp

M. Kaneko
Graduate School of Education, Naruto University of Education, Nakashima 748, Takashima, Naruto-cho, Naruto-shi, Tokushima, Japan
e-mail: mkaneko@naruto-u.ac.jp

G.A. Stillman et al. (eds.), *Mathematical Modelling in Education Research and Practice*, International Perspectives on the Teaching and Learning of Mathematical Modelling, DOI 10.1007/978-3-319-18272-8_41

use two modelling tests before and after an experimental modelling class, report on them, and discuss evidence of reformulations of situation models. In our previous work on modelling teaching and tests, modelling tasks related to the real situation were reported for elementary school students (Matsuzaki and Kawakami 2010) and for upper secondary school students (Matsuzaki 2007). In addition, Matsuzaki (2011) compared the modelling of a graduate school student with that of a working adult. Findings indicate that modellers require a variety of experiences for solving tasks successfully.

There is the possibility that modellers imagine situations from modelling tasks, and the situations might be able to change. Here we distinguish two kinds of representations of the situation model: First, one is a situation model formulated from the initial imagined situation, and formulation of a situation model might have occurred. Second, a situation model could be reformulated from the initial situation model through processing modelling or it does not change in the process. In previous studies we adopted drawing pictures as the method for gaining insight into these experiences in setting up the problem from the task situation in the real world (Kawakami and Matsuzaki 2012; Matsuzaki and Kawakami 2010). We also focussed on formulation and reformulation of situation models from modelling tasks. Here, modelling tasks to capture situation models are prepared to test different situation models based on each modeller's individual experiences.

In this chapter we try a similar research setting to the previous studies (Kawakami and Matsuzaki 2012; Matsuzaki and Kawakami 2010) but this time for lower secondary school students. We administer a test before and after a modelling class. On both tests, students were required to set up a problem from a real situation, and draw pictures to explain how they set up the problems for other students. We focus on differences in pictures because changes in these we believe are aspects of products of actualizing situation models. Of course, only these artefacts are not enough for us to interpret situation models drawn on worksheets. In addition, for explaining formulation or reformulation of situation models, we interview selected students after each test, and try to interpret their responses with protocols from interviews. We analyse situation models that these students imagined for empirical evidence of situation model reformulation actions.

41.2 Case Study

Matsuzaki and Kawakami (2010) reported on an experimental modelling class in an elementary school. In that case, the students were required to create models through talking or discussing with each other, and they drew pictures for explaining the problem to their classmates. In this chapter, we adopt a similar approach for lower secondary school students. An experimental class was implemented with a seventh grade class of 41 female students at a private lower secondary school in the Tokyo metropolitan area. The teacher and the interviewer was the second author. All the students were presented with the same problem situations and required to tackle some tasks through setting up problems.

41.2.1 The Modelling Pre-test

We prepared the following *Brightness Situation Task* and implemented it for 30 min with the class. The aim of these tasks is to formulate an initial situation model. In particular, task (1a) requires verbal expression whereas (1e) requires a pictorial response. Although we presented the following five tasks, the students to this point had not been instructed in modelling at that time.

The Brightness Situation

You read a book for 10 min every morning. Today's weather is cloudy, and sunlight that is enough for reading doesn't reach the classroom. So you think that you would like to read a book continuously in a bright place.

(1a) What kind of things did you imagine from this situation?
(1b) What kind of things should you research to read? Describe findings, what is already-known, or what you would like to find out.
(1c) Set up a problem by using all or some of (1b).
(1d) Does the problem that you set up in (1c) seem able to be solved? Answer 'Yes' or 'No'. If your answer is 'Yes', how do you solve the problem? If your answer is 'No', why did you think that you could not solve the problem? What is necessary for you to solve the problem?
(1e) You will explain the problem to your classmates. Draw a picture for explaining the problem in detail.

Our intention was to have students set up problems through the tasks (1a) to (1c). In other words, we expected students to achieve this through the processes of understanding and/or simplifying/structuring the task to create a real model or mathematical model via a situation model, with the questions or the demand for extra-mathematical knowledge arising (Borromeo Ferri 2006, p. 92). In this chapter we hope to show evidence of the formulation or reformulation of situation models. After the pre-test, we interviewed five students on the following day.

41.2.2 Experimental Class

An experimental class was implemented for 50 min two days later. The teacher provided two problems and two pictures to the students. Problems were copied and pasted from student work for task (1c) and pictures were copied and pasted from student work for task (1e). During this time, all the students wrote down their own opinions about each problem and each picture.

The teacher selected two types of problems. The first one (see Fig. 41.1) was the problem of the relationship between velocity and distance, and it was extremely

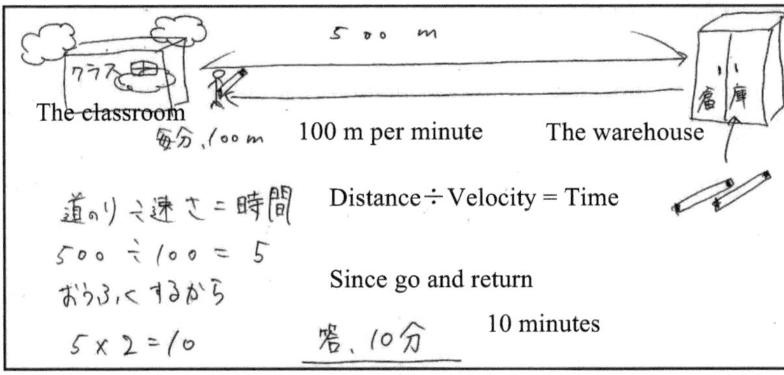

Fig. 41.1 A picture drawn by H.M

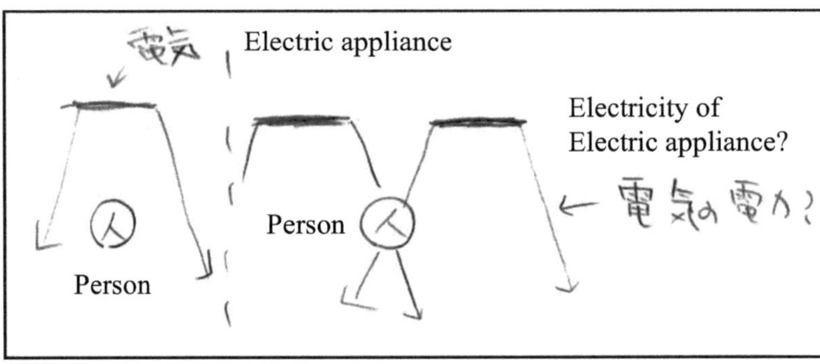

Fig. 41.2 A picture drawn by S.K

easy for the students to generate an answer using calculations. This problem was a type of mathematical problem and a picture for explaining the problem was considered a *mathematical model*. The second one (see Fig. 41.2) was easy for the students to answer without using mathematics because they could answer based on their own daily-life experiences. This problem was a type of non-mathematical problem and a picture for explaining the problem was considered a *real model*.

The Problem of Brightness Set Up by H.M.
A fluorescent tube is not working in the classroom. The teacher goes for a replacement tube in the warehouse which is 500 m from the classroom. The teacher walks 100 m per minute. How long until the teacher goes for a fluorescent tube and back to the classroom?

The Problem of Brightness Set Up by S.K.
One person sits down right under the electric appliance, and another one sits down between two electric appliances. Which place is bright?

Each student explained the problem in terms of tasks (1a), (1b) and (1d). After hearing their explanations, the other students asked the questions. Then the teacher asked all the students which problem and picture was better. Responses from the class about the better problem gave the following characteristics: include required information to solve a problem; make a problem sentence short; the situation should be fine to share, as we have to share setting up of the problem; and do not include extra information. Responses about the characteristics of the better picture were: it is easy to find problem situations, add explanations to picture and draw a picture to help comprehend situations.

Following this, all the students wrote down what they found and stated their own explanations on worksheets.

41.2.3 The Modelling Post-test

The post-test was implemented for 30 min three days after this class. The students were presented with the same brightness situation as in the pre-test. The aim of these tasks is to reformulate situation models. The students tackled the following five tasks:

(2a) You will set up a problem referring to data. What kind of things did you imagine from this situation for making a problem?
(2b) What kind of things should you research to read other than data? Describe your findings and knowledge, or the things that you would like to find out.
(2c) Set up a problem by using all or some of (2b).
(2d) You will explain the problem you set to your classmates. Draw a picture for explaining the problem in detail.
(2e) Solve the problem that you set up in (2c).

After the post-test, we interviewed the same five students who were interviewed after the pre-test.

41.3 Reformulation of Situation Models

At both tests, the students were asked to draw pictures for explaining their own setting up of the problem to others. Here we focus on differences in pictures drawn by two students who were selected from interviewees after each test. We analyse reformulation of situation models that these two students, A.S. & Y.R., imagine for the situations by referring to their worksheets and their interview protocols.

41.3.1 Responses from A.S. Through Modelling Tests and Experimental Class

41.3.1.1 Responses from A.S. at the Pre-test

For the first brightness situation [task (1a)], A.S. described "turn on light, go to the toilet, move to bright place (especially, lavatory etc...), go to the library, and change a big book which is easy to read" on her worksheet. For task (1b), she wrote "Morning reading time is basically held in the classroom" and "Sunlight is not hoped for" as her findings and already-known. Continuing, she listed: "Where is the brightest spot in the classroom?", "Is it possible to implement morning reading except in the classroom?" and "Is it possible to turn on light at hand?" as the things that she would like to find out. In task (1c), she set up the following problem considered a *real model*.

A Problem Set Up by A.S. in Task (1c)
I would like to implement morning reading in a bright place. Outside is cloudy, and sunlight that is enough for reading doesn't reach the classroom. In this case, it is possible to go out or to use some tools. How do you action this?

In task (1d), she responded "Yes" to whether she thought the problem was able to be solved and wrote reasons on her worksheet as follows: "I suppose that I go out from the classroom or to use some tools." Additionally, she pointed out two possible solutions "to know bright spot and to use light". In task (1e), she drew a picture on her worksheet (Fig. 41.3). This picture reflects the alternatives that she pointed out in task (1d).

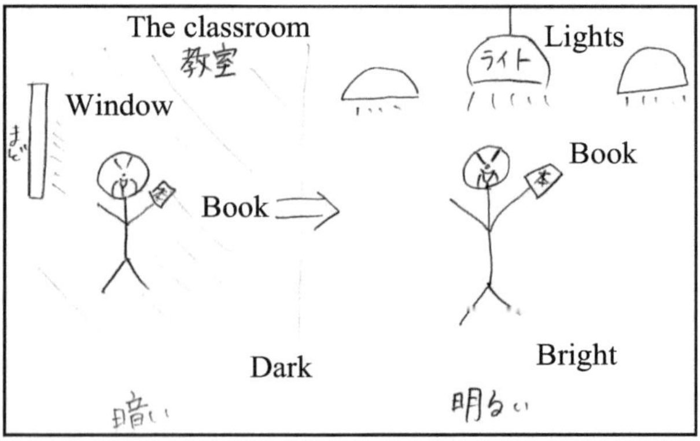

Fig. 41.3 A picture drawn by A.S. in task (1e)

41.3.1.2 Responses from A.S. at the Experimental Class

During the experimental class, A.S. wrote: "Another teacher may walk 200 m per minute" in response to the problem set up by H.M. She described "It is so clear for me, as adding explanations" to a picture figure drawn by H.M. In addition, she noted: "the figure is easy for elementary school pupils who just learn 'velocity' to understand the problem" as a result of discussing with classmates. A.S. commented that 'I think that positions of under electric appliance etc... have no relations with brightness' in response to the problem set up by S.K. In addition, she added the annotation: "Generated electricity radiates..." to a picture drawn by S.K. and described an alternative explanation (Fig. 41.4).

41.3.1.3 Responses from A.S. at the Post-test

In task (2a), A.S. noted, "The seat near the window is brighter than the seat at the centre of the classroom" and drew a supplementary picture (Fig. 41.5) on her worksheet. In task (2b), she commented: "It is better to go to the seat near the window when possible" as a finding and already-known. She asked, "Is it possible to move seats in the morning reading time?" and "Are there differences in brightness right under the light and between lights in the case of window seats?" as the things that she would like to find out. In task (2c), she set up the following problem considered a real model.

A cross section is

In a case of slit light, it becomes

but I think that normal light becomes like the figure above.

Fig. 41.4 Alternative explanation and pictures described by A.S

Window

Bright

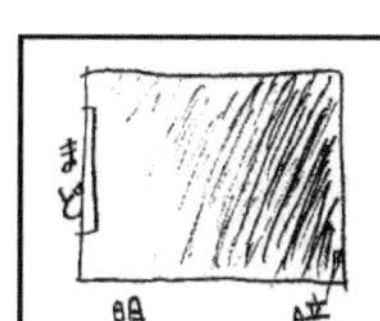

Dark

Fig. 41.5 A picture drawn by A.S. in task (2a)

A Problem Set Up by A.S. in Task (2c)
It is supposed that you sit down beside the corridor side seat in the classroom. Some students say that it is possible to change their seat. Which seat do you want to change your seat to?

In task (2d), she drew pictures like a four-frame comic book page on her worksheet (Fig. 41.6). To clarify her thinking, the interviewer asked her:

Interviewer [00:59]: Why did you draw pictures within a four-frame?
A.S. [01:07]: I think that four-frame comics are fine for classmates to explain.

In task (2e), she solved the problem she set up by herself in task (2c). Her answer was "we can find that the window seat is the brightest, and I change B's seat which is most near to windows there".

41.3.2 *Responses from Y.R. Through Modelling Tests and Experimental Class*

41.3.2.1 Responses from Y.R. at the Pre-test

For the brightness situation [task (1a)], Y.R. wrote "If curtains in the classroom are closed, I open the curtains to their full width and adopt sunlight. Now the curtains are fully opened, but it is not possible to get enough sunlight; because of this, when the classroom is dark, I turn on the lights and close the curtains. As a result, lights are shut by the curtains and the classroom becomes bright" and "I want to go to the biotope[1] and read a book. Outside is brighter and I want to feel bright and nature".

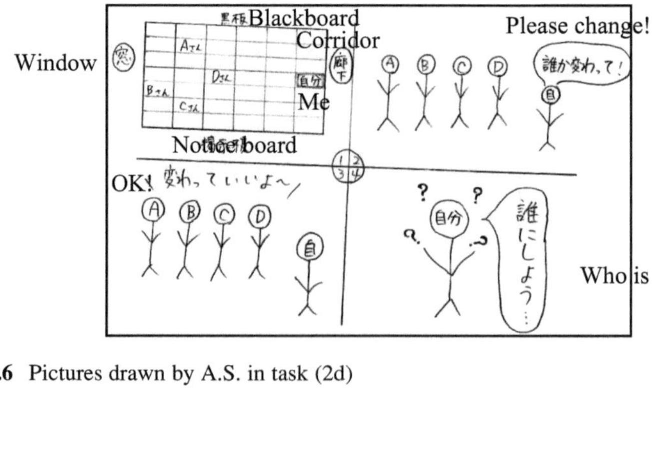

Fig. 41.6 Pictures drawn by A.S. in task (2d)

[1] The biotope is a small environmental natural habitat in the school grounds.

For task (1b), she described "Only children without adults cannot enter the biotope" and "Biotope is usually locked" as her findings and already known. She noted that: "I ask the teacher whether it is possible to go to biotope in morning reading time" as what she would like to find out. In task (1c), she set up the following problem considered a real model.

The Problem Set Up by Y.R. in Task (1c)
You and your friends reached the biotope for playing. Now the biotope is not locked. But children alone cannot enter the biotope and you have to get permission from the teacher. At that time, there are birds in the biotope. Do you enter or not enter?

In task (1d), she responded "Yes" to her problem being solvable but she did not write reasons and did not draw a picture in task (1e). To clarify the interviewer asked her reasons for no responses:

Interviewer [03:24]: I want to ask why you did not draw a picture in your worksheet. Perhaps did you have no time?

Y.R. [03:27YR]: Yes.

41.3.2.2 Responses from Y.R. at Experimental Class

In the experimental class, Y.R. noted that, "This problem was different from what I imagined as 'the problem'" in response to the problem set up by H.M. She added, "We find that the problem can be solved. But, we wonder that a mathematical problem is possible or not" as the results of her group discussion. To a picture drawn by H.M., Y.R. commented, "I think it is a very clear picture and expressions". Y.R. wrote "this problem was set up based on pondering in daily-life" in response to the problem set up by S.K. In addition, she recorded, "We want to know the kinds of electric appliance. For example, electric radiation from a fluorescent tube or a spotlight is different", as a result of their group discussion. To a picture drawn by S.K. (see Fig. 41.2), Y.R. responded that, "I think that the person who sits down right under the electric appliance is brighter than, but I reconsider the person who sits down between two electric appliances after seeing this picture". She added: "We cannot understand the wording 'electricity of electric appliance'". Using more complicated wording, "area of lighting" etc....", was taken as the result of their group discussion.

41.3.2.3 Responses from Y.R. at the Post-test

In task (2a), Y.R. wrote “I want to go near the windows to read a book” and “I want to go to the biotope to read a book” on her worksheet. For task (2b), she noted: “Morning reading time is basically held on my seat” and “My current seat is the third seat from the window and the third seat from the front” as her findings and already known. She asked: “Is it possible to walk around in morning reading time?” and “Is it possible to change window seats?” as the things that she would like to find out. For task (2c), she set up the following problem considered to be a real model.

The Problem Set Up by Y.R. in Task (2c)
It is possible to change seats only in morning reading. Six window seats have already changed when you want to sit down at a seat at a bright place. Which place do you sit down?

In task (2d), she drew a picture on her worksheet (Fig. 41.7). In task (2e), she solved the problem set up in task (2c). Her answer was as follows: “Seats under the fluorescent tubes are in the first, fourth and sixth rows; but six rows have already changed. In this time, brightness is not changed in the case of seats right under the fluorescent tubes and in the case of ones between the fluorescent tubes. As a result, seats near the windows are bright. The answer is the fifth row which is the second nearest seat from the windows.” Additionally, she explained this reasoning in interview:

Y.R. [00:16]: Well. . .I can find that the window side is bright from data. I don’t measure brightness in the biotope using luxmeter; but, the biotope is brighter since the biotope is outside.

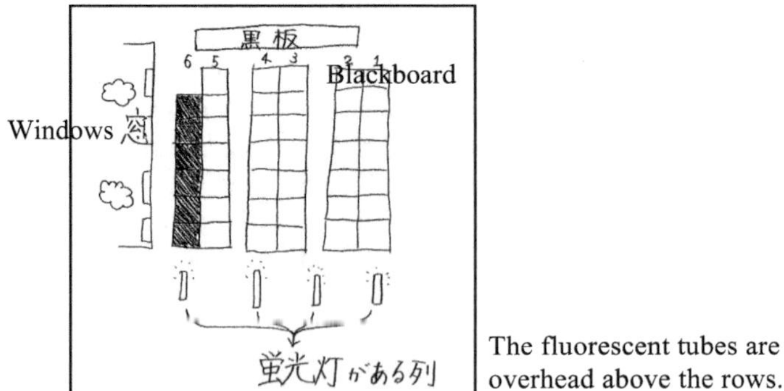

Fig. 41.7 A picture drawn by Y.R. in task (2d)

41.4 Discussion and Conclusion

In this chapter we discuss reformulation of situation models, and administer a test before and after a modelling class for lower secondary school students. The original situation at the pre-test and the post-test were not changed, the students were required to set up problems and explain them to their classmates. We focus on pictures to explain the problems as being evidence of situation models because processes of understanding and/or simplifying/structuring the task to create real models or mathematical models via situation models (Borromeo Ferri 2006). Both instances of setting up the problems by the two students, A.S. & Y.R., are real models.

In the case of the student A.S., both setting up the problems in task (1c) and in task (2c) are based on the original real situation, and reflect the real classroom situation. At the pre-test, she did not set the limit of possible places. In task (1c) she described two ways "to go out or to use some tools" in the third sentence, however she did not show the former way on the picture in task (1e). She responded that it did not matter if the situation was inside or outside in task (1d). From the picture at the pre-test we can confirm that reformulation of situation models has occurred to simplify the task. At the post-test, she limited the places to within the classroom based on her personally belonging to the classroom situation. Furthermore, she devised pictures like a four-frame comic (Fig. 41.6) to explain the problem in task (2d). The first scene shows the real classroom situation, however she explained for "some students" with the supplementary three remaining scenes. Here the situation model has occurred to explain the task for her classmates clearly.

In the case of student Y.R., setting up of the problem in task (1c) and the problem in task (2c) are different, however biotope or placement of the seats are able to be understood by all the students. At the pre-test, she did not draw explanatory images in tasks (1d) and (1e) as she had no time to draw. At the post-test, she raised two images in task (2a), that were continuous with her thinking for the pre-test, although setting up the problem was limited to being inside the classroom. In addition, this classroom situation was easy to share with her classmates, since the situation of brightness was constructed based on her school. In other words, we can say that the problem in task (2c) are based on the original real situation, and reflects the real classroom situation: There are 41 seats consisting of six lines in a classroom. In Fig. 41.7 she excluded six window seats. So she set up the problem by referring to data collected by use of a luxmeter, as the window side is bright. After the experimental class, we had provided the luxmeter to measure brightness and some students tried to gather real data at various spots or places. Some of them set up the problem using these data. Thus a situation model has occurred to structure the task mathematically.

We re-iterate we have been able to confirm empirically the occurrence of situation model reformulation processes occurring through our modelling tests. In the future we will attempt to classify all the problems set up by the students (cf Kawakami and Matsuzaki 2012) and characteristics of sharing types of models as our next studies.

References

Blum, W., & Leiß, D. (2007). How do students and teachers deal with modelling problems? In C. Haines, P. Galbraith, W. Blum, & S. Khan (Eds.), *Mathematical modelling: Education, engineering and economics* (pp. 222–231). Chichester: Horwood.

Borromeo Ferri, R. (2006). Theoretical and empirical differentiations of phases in the modelling process. *ZDM – The International Journal on Mathematics Education, 38*(2), 86–95.

Kawakami, T., & Matsuzaki, A. (2012). A new approach for teaching mathematical modelling in elementary school: Focusing on setting up problems from the task situation in the real world. *Journal of Japan Society of Mathematical Education, 94*(6), 2–12 (in Japanese).

Matsuzaki, A. (2007). How might we share models through cooperative mathematical modelling? Focus on situations based on individual experiences. In W. Blum, P. Galbraith, H. Hans-Wolfgang, & M. Niss (Eds.), *Modelling and applications in mathematics education* (pp. 357–364). New York: Springer.

Matsuzaki, A. (2011). Using response analysis mapping to display modellers' mathematical modelling progress. In G. Kaiser, W. Blum, R. Borromeo Ferri, & G. Stillman (Eds.), *Trends in teaching and learning of mathematical modelling* (pp. 499–508). New York: Springer.

Matsuzaki, A., & Kawakami, T. (2010). Situation models reformulation in mathematical modelling: The case of modelling tasks based on real situations for elementary school pupils. *Proceedings of the 5th East Asia Regional Conference on Mathematics Education*, *2*, 164–171.

Part VI
Applicability at Different Levels of Schooling, Vocational Education, and in Tertiary Education

Chapter 42
Mathematical Modelling in the Teaching of Statistics in Undergraduate Courses

Celso Ribeiro Campos, Denise Helena Lombardo Ferreira, Otávio Roberto Jacobini, and Maria Lúcia Lorenzetti Wodewotzki

Abstract In agreement with the elements of the statistics education theoretical framework, as well as critical education, mathematical modelling allows the possibility of pedagogical projects that value interdisciplinarity and active participation of the student in knowledge construction. In this chapter, we present a scenario that occurred in the statistics discipline of an undergraduate course. We adopt Critical Education practices to discuss global warming, its causes and consequences for society.

42.1 Introduction

At any grade level, Critical Education aims to promote political awareness and the discussion of social issues related to students' reality. In our view, as in the opinion of the main articulators of this theory (Freire 1974, 1998, 2000; Giroux 1988, 2010; McLaren 2000), such a goal can be pursued independently from the syllabus of a discipline. We believe that educators can build adaptations to the content in order to cover topics that facilitate the discussion and the debate of political and social issues relevant to students' reality. Nevertheless, it is important to clarify that, as we follow the Critical Education ideas, we do not search for a method or a rule, because

C.R. Campos (✉)
FEA, Pontificia Universidade Católica de São Paulo, Rua Padre Mariano Ronchi, 1090, apt. 74, São Paulo 02932-000, Brazil
e-mail: crcampos@pucsp.br

D.H.L. Ferreira • O.R. Jacobini
Matemática, Pontifícia Universidade Católica de Campinas, Rua Padre Mariano Ronchi, 1090, apt. 74, São Paulo 02932-000, Brazil
e-mail: lombardo@puc-campinas.edu.br; otavio@puc-campinas.edu.br

M.L.L. Wodewotzki
IGCE, UNESP, Rua Padre Mariano Ronchi, 1090, apt. 74, São Paulo 02932-000, Brazil
e-mail: mariallwode@gmail.com

G.A. Stillman et al. (eds.), *Mathematical Modelling in Education Research and Practice*, International Perspectives on the Teaching and Learning of Mathematical Modelling, DOI 10.1007/978-3-319-18272-8_42

"individuals who reduce [the] learning process to the implementation of methods should be dissuaded from entering the teaching profession" (Giroux 1988).

The evolution of Critical Education led to the construction of Critical Mathematics Education, articulated by Ole Skovsmose, in particular (Skovsmose 1994). In Campos (2007) and especially in Campos et al. (2011), we show how Critical Education can relate to Statistics Education and we build the foundations of Critical Statistics Education. In this context, we take these foundations to highlight and practice the valuing of socio-political interfaces within the teaching of Statistics in undergraduate courses. In this way, in this chapter we present a fragment of the theoretical basis of Critical Statistics Education and show, through a mathematical modelling project, how it is possible to obtain positive results under Statistics Education and Critical Education. This project was developed and applied in the statistics discipline of an undergraduate course in Economic Sciences. One aspect that seems worth noting is that, as recommended by Campos (2007), mathematical modelling and working with projects that value the use of technological tools on the one hand help to motivate students to practice research and, on the other hand, extend and diversify the universe of opinions and insights on the topics addressed, enriching and stimulating discussions in the classroom.

42.2 Critical Education, Statistics Education and Critical Statistics Education

We understand Critical Education as an evolution of critical thinking, having emerged as a counterpoint to traditionalism in the educational system. Its foundations can be credited mainly to Jurgen Habermas in Germany (Habermas 1971) and to Paulo Freire in Latin America. Freire's emancipatory vision of education (Freire 1974, 1998, 2000) inspired Giroux (1988, 2010), who advanced the idea of democratisation and politicisation of pedagogy, in a vision of the teacher as a transformative intellectual. "Essential to the category of transformative intellectual is the need to make the pedagogical more political and the political more pedagogical" (Giroux 1988, p. 163). Thus, Giroux defended the use of forms of pedagogy that treat students as critical agents, use an affirmative dialogue and argue for a qualitatively better world for all.

Incorporating these concepts, Skovsmose (1985) progressed the development of Critical Education and stated that "it is essential that the problems must be related to fundamental social conflicts and it is important that students can recognize problems as their own problems" (p. 345). Focusing on the democracy issue, Skovsmose highlighted that it must be present in mathematics education and, in this way, he built a Critical Mathematics Education (1994), in which he valorised working with modelling projects.

Although the American Statistics Association had developed a section on Statistics Education in 1973, it was only in 1982 that the first International Conference on Teaching Statistics (ICOTS) was convened, improving the research in the Statistics Education field. Therefore, in the 1990s, there was an increasingly strong call for Statistics Education (Garfield and Ben-Zvi 2008) and the *Journal of Statistics Education* was first published in 1993. Since its beginning, Statistics Education has been designed in a context of concern, questioning and reflection on the problems of teaching and learning statistics. This is revealed primarily through the difficulties that students have to think or reason statistically (even if they demonstrate calculation ability). This context led to the development of Statistics Education, which seeks to differentiate the pedagogical issues presented by this discipline from those experienced by mathematics. Thus, several authors (e.g., Chance 2002; Gal 2004; Garfield 2002; Pfannkuch and Wild 2004) converged on the idea that the teaching of statistics should focus on developing three specific competences: *statistical literacy*, *statistical thinking*, and *statistical reasoning*.

Statistical literacy, according to Gal (2004), occurs primarily by two interrelated components: (a) people's ability to interpret and critically evaluate statistical information, arguments relating to research data and stochastic phenomena that can be found in different contexts; and (b) people's ability to discuss or communicate their reactions to such statistical data, their interpretations, opinions and understandings. To Gal (2004), the understanding and interpretation of statistical data require the student to have the attitude of making inquiries, neither passively treating the data available to them nor the results that are obtained. Also according to Gal (2004), to develop statistical literacy, educators should encourage dialogue, discussion, valorising the students and their ideas and interpretations, when faced with real-world messages containing statistical elements and arguments.

Statistical thinking is related to the ability of identifying the statistical concepts involved in investigations and in problems, including data variability, uncertainty, how and when to properly use the analysis and estimation methods, and evaluating the statistical problem in a global way. This capacity, according to Chance (2002), foresees that the student has the ability to explore the data in order to extrapolate what is prescribed in the texts, and generate new questions outside those indicated in the investigation. Pfannkuch and Wild (2004) went deeper studying this capacity and highlighting its importance.

The way in which people reason with statistical concepts comprises what is generally called *statistical reasoning*. According to Garfield (2002), statistical reasoning means to make appropriate interpretations on a certain data set, to represent the data properly and make connections between the concepts involved in a problem, highlighting variability, uncertainty and probability, for example. The development of statistical reasoning should lead the student to be able to understand, interpret and explain a statistical process based on real data. Based on this, Ben-Zvi (2008) highlights the importance of this competence, stating that all citizens should have it and that it should be a standard component in education.

Several researchers (e.g., Rumsey 2002; delMas 2002) noted many similarities between the three competences, especially between statistical thinking and reasoning. We perceive the thinking and reasoning characterization in a similar way and understand that there is a convergence of cognitive and conceptual aspects between them. In our view, it is more important than observing possible differences among these competences, for statistical educators to undertake further research on how to develop them in students. In this direction, Campos (2007) proposes, briefly, some teaching practices (or pedagogical actions) for working in the classroom: (a) work with real data and relate them to the context in which they are involved; and (b) encourage students to interpret, explain, criticize, justify and evaluate the results, preferably working in groups, discussing and sharing opinions. Additionally, to address the major aspects of Critical Education, Campos et al. (2011) suggest the following actions:

(a) problematise the teaching, working Statistics through projects contextualised within a reality consistent with the student's reality;
(b) foster debate and dialogue among students and between them and the teacher, assuming a democratic pedagogical attitude;
(c) thematize the teaching, privileging activities that enable the discussion of important social and political issues;
(d) use technological devices to make calculations, draw graphs, tables, access information, databases, etcetera, as students live in a highly technological society;
(e) adopt a flexible pace for the development of themes; and
(f) discuss the curriculum and pedagogical framework adopted with the students.

By adopting these actions and teaching practices in the educational process, we will be practising the Critical Statistics Education that goes against the traditional teaching model, in which, according to Giroux (1988), education takes place in an alienating form by assuming a false politically neutral posture.

In this context, we have defended (Campos 2007; Campos et al. 2011) that work with mathematical modelling projects composes an appropriate pedagogical strategy to carry out Critical Statistics Education. Niss (1983) has argued that "the mathematical education goal should be to enable students to notice, understand, judge, use and apply mathematics in society, mainly in significant situations for particular and professional life of each one" (p. 248). We thoroughly believe that through modelling, we create motivation, facilitate learning, give meaning to the content worked, value the applicability of concepts and develop students' critical spirit. This way, we stimulate students to transform their reality whilst we promote the understanding of the socio-political role of statistics. We understand that Critical Statistics Education worked through mathematical modelling consists of an efficient articulation between theory and practice. It also favours the disruption of arbitrary boundaries between disciplines, enabling more widespread and effective ranges, as long as students can experience the possibility of making connections among different content from distinct disciplines and between the scholar and the real world.

It is also noteworthy to point out that statistical literacy has been well linked to modelling competencies by Engel and Kuntze (2011), whereas previously Kuntze et al. (2008) had connected the domain of statistical literacy to the competency of using models and representations. We also note that Skovsmose (1992) had pointed out the connection between Critical Education and mathematical modelling, a point taken up by others such as Mukhopadhyay and Greer (2001) when they proposed the use of modelling as a critical tool for mathematics.

42.3 Environment Description

In the Economic Statistics discipline, taught in an Economics Sciences course by the first author of this work in the second half of 2011, one topic in the program is Correlation and Regression. In an introductory class, the teacher mentioned that regression can be useful for modelling phenomena of several kinds, even outside the economics context. The students were interested and asked for an application example. The professor cited the study of global warming, and since this subject is very controversial and has generated much debate, it was proposed to carry out an activity related to the subject, organising groups, selecting topics to research and making a presentation. Joining would be voluntary and the meetings would be held outside class hours. It should be noted that Brazil would host (in 2012) the United Nations Conference on Sustainable Development, known as Rio +20.

At the first meeting, 20 students attended and were interested in participating. They were divided into four groups. It was decided that the groups would research the following topics:

(i) rainfall in the city of São Paulo;
(ii) temperature in the city of São Paulo;
(iii) rainfall in the city of Rio de Janeiro; and
(iv) temperature in the city of Rio de Janeiro.

The choice of cities was made for convenience, because it might be difficult to obtain historical data for smaller cities. The first task for each group would be to obtain historical data in a series as long as possible for the temperature and precipitation for the cities mentioned. The groups who investigated the temperature obtained their data from the National Institute for Space Research (INPE) site: http://clima1.cptec.inpe.br/. At this site, the temperature data can be obtained at the Database option – climatological city database. The groups who investigated the precipitation obtained their data from the National Aeronautics and Space Administration (NASA) site: http://disc2.nascom.nasa.gov/Giovanni/tovas/. At this site, rainfall data are obtained in the ASCII option, by providing the location latitude and longitude. The precipitation time series obtained by the students referred to the period 1979–2009. The temperature series was from 1980 to 2009.

With the temperature data arranged in a time series, students could carry out a regression over the deseasonalised data, using time as an explanatory variable ($x = 1, 2, 3$, etc.). The aim was to check whether there was a demonstrable upward trend in the series. With rainfall data, the students could make a graph using an Excel spreadsheet in order to visually observe the data volatility. Through the variance statistic of the data, the goal was to see whether there was an increase in the precipitation volatility, that is, whether the rainfall variation was constant throughout the series or whether it had been accentuated in more recent times.

The idea of this study is to determine from empirical evidence if the temperature studied over time was increasing. Our hypothesis, guided by global warming literature, was that the temperature was increasing. Moreover, we knew that one of the global warming consequences is the rainfall patterns change, accentuating the rainfall concentration over time.

Not all the students in the class agreed to participate in the project. Thus, the class was broken into two larger groups. In line with a democratic vision of the pedagogical environment (Giroux 1988; Skovsmose 1994), it turns out this was normal and after a brief dialogue, it was agreed that the groups engaged in the project would make a presentation of their results to the class and students who did not participate would make a presentation on the global warming topic.

42.4 Presentations and Results of Student Investigations

At an appointed date, first there was a presentation by the students who did not participate in the regression project. They showed images related to global warming and an excerpt from the movie *An Inconvenient Truth* (Guggenheim et al. 2006), presented by Al Gore. With no empirical study made by students, there was no arguments against the global warming phenomena; only Gore's discourse, which had convinced the students that made the presentation.

Secondly, groups (i)–(iv) presented, in this order. Groups (i) and (iii), who investigated the rainfall, showed similar results. In both cases, the variance was constant throughout the historical series, as indicated by a visual examination of the graphs and the calculated heteroscedasticity. It should be noted that heteroscedasticity refers to the violation of the homoscedasticity assumption in the least squares method, that is, the variance of each disturbance term u_i, conditional on the chosen values of the explanatory variables, is some constant number (Gujarati 2004). This verification was carried out by a hypothesis test, known as the Park test.[1] The results indicate that there was no verifiable evidence that there has been a significant variation in rainfall over time in both samples studied by the pupils. Figure 42.1 shows the students' work on rainfall data.

Groups (ii) and (iv), who had investigated the temperatures, also obtained similar results. In both cases, the calculated regressions showed a positive value for the *X* variable parameter (time), indicating a slightly upward trend in the series. However, although positive, this parameter showed a close to zero result, and through a hypothesis test, student *t* test at a 95 % confidence level, the parameter

[1] Park test consists in making the following regression: $\ln u^2_i = \alpha + \beta . \ln X_i + \nu_i$. If β turns out to be statistically significant, it would suggest that heteroscedasticity is present in the data (Gujarati 2004).

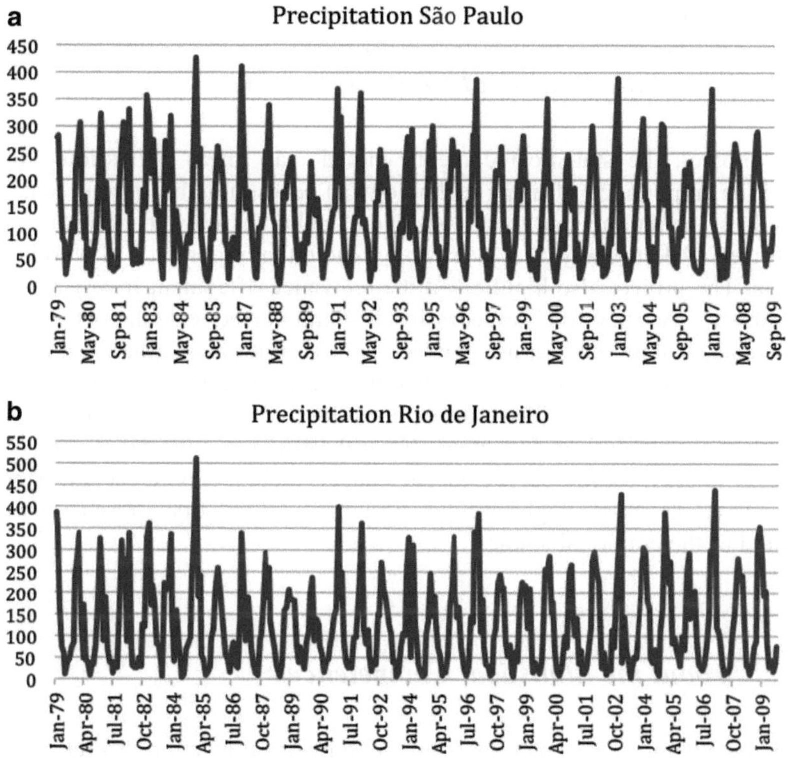

Fig. 42.1 (**a**) Graph of rainfall data made by group (i) using Excel (**b**) by group (iii) using Excel

was statistically null. This result indicates that there is no verifiable empirical evidence in the sample considered that there is a growing trend in the analysed historical temperature series.

After the presentations, the students discussed, along with the teacher, the results obtained. The students themselves mentioned the limitations of the results, noting that they do not refute the global warming hypothesis, mainly due to the fact that only two regions were analysed and the time series obtained might not be as extensive as necessary to observe trends. This discussion also addressed issues relating to people's behaviour and attitudes which they should have in order to not aggravate the global warming problem. Another issue that emerged from the discussion was public policies adopted by governments in relation to the global warming issue. On this point, the students criticised the Brazilian government and also those of other countries that did not adopt suitable policies in order to fight the problem. In line with this, they cited the Kyoto Protocol, which is an international treaty that commits State Parties to reduce greenhouse gas emissions. It was pointed out that some of the major polluter countries did not sign the protocol and others, who had signed, did not make enough efforts to achieve the agreement.

42.5 Analysis

42.5.1 About the Statistical Content

In this project, the statistical content worked was linear regression. Students realised the importance of using technology to access reliable results (data were obtained via the internet and the graphs and calculations were made with an Excel spreadsheet). Additionally, they were able to note the importance of working with real data. The project also highlighted the relevance of content in order to model real problems. As they felt the necessity of doing correct calculations, collecting reliable data, making correct assumptions and interpretations, they had discussed among themselves and thoroughly studied the content before writing the report and preparing the presentation. Thus, we realized that the students developed a deeper knowledge of the statistical content. Nevertheless, we must point out that the students who did not carry out regression analysis did not experience this deepness. In this way, we must emphasise the weakness of the first presentation when compared to the empirical studies.

42.5.2 About the Statistical Competences

For the three competences mentioned in the Statistics Education theoretical framework, we found that working with real situations involved in the regression calculations allowed students a global view of the problem. Students could realise the difficulties surrounding the complexity of the topic and experienced the usage of some statistical tools used in the regression universe. We understand that this work tends to help the development of statistical thinking and reasoning about data and measures. As for literacy, we believe that the preparation of reports, the use of expressions and terminology typical of statistics, the construction of charts and tables, in addition to hypothesis testing performed, and discussions involving the regression theme tend to assist the development of competence with critical thinking and reasoning with statistics. It is important to note that there is no quantitative measure of the development of the competences. Nevertheless, we pay attention to the competences to ensure we are progressing towards them.

Concerning *statistical literacy*, we mentioned Gal (2004), who pointed out that this emerges when students are able to interpret and critically evaluate statistical information, arguments relating to research data and stochastic phenomena. He also noted that literacy appears when people discuss or communicate their reactions to such statistical data, their interpretations, opinions and understandings. We believe that our students achieved this goal as they showed, discussed and communicated their interpretations, critical evaluations and understandings of the regression results.

The interpretation of the statistical results made by students, that is, the interpretation for the hypothesis tests that were carried out, can be seen as an ability to explore the data, extrapolating what is prescribed in the books. Focusing on the idea of verifying an empirical proof of the global warming phenomena, students were able to carry out some interpretations of the results that are non-trivial. In line with this, as was pointed out by Chance (2002), we can say that the students progressed in the development of *statistical thinking*.

Garfield (2002) suggested that the development of *statistical reasoning* should lead the student to be able to understand, interpret and explain a statistical process based on real data. This was completely done by the students when they made correct interpretations of the results and explained their conclusions for the class.

42.5.3 About Critical Education and Critical Statistics Education

As we mentioned previously, according to Giroux (1988), Critical Education is not a method. Thus, it is not a question of measuring its deepness or checking some steps to be followed. Despite this, we understand that in many ways we highlighted Critical Education in this project. Both in the presentations and discussions, many opportunities occurred where the students came face to face with the global warming problem, its causes and consequences. The debate sparked in the students a sense of outrage at the attitudes of people and governments towards a problem that affects everyone, rich and poor, but where the main consequence is for those who are socially disadvantaged, who live in risk areas subject to flooding and landslides. The social problems of irregular occupation of areas that should be preserved as well as the authorities' indifference towards environmental degradation were highlighted in students' discussions. Moreover, students discussed actions they could do in order to change attitudes and raise awareness of the importance of environmental preservation.

As for Critical Statistics Education, in a different way, we realised that during the project we were on the path traced by the theoretical considerations provided by Campos et al. (2011) as:

(i) we had problematised the teaching, we had worked topics related to statistics through contextualised projects linked to a reality consistent with that of the students;
(ii) we had encouraged debate and discussions among students and between them and the teacher, thus taking a democratic pedagogical position;
(iii) we had thematized the teaching and focused on activities emphasising the debate of several important social and political issues;
(iv) we had used technology in teaching and valorised technical skills to students; and
(v) we adopted a flexible pace in implementing the work presentations.

42.5.4 About the Mathematical Modelling

The question remains: Did we (or the students) really work on a mathematical modelling problem? Beginning with a real world problem (global warming), students had built a mathematical model, represented by a function. This model was used to evaluate whether there is empirical evidence of global warming or not. The results were analysed, debated and compared with the real evidence that we feel in our ordinary life, that is, there was a discussion on the validation and limitation of the model.

The model construction allowed us to work on statistical content, to promote the development of some statistical capacities, which are important to learn the discipline, and to implement a critical education class. Actually, students who used regression analysis by means of mathematical modelling revealed a much more powerful discourse, than the others, who just explored Gore's arguments and remained at a weaker level. The importance of the mathematical models to explore the phenomena is clear, so we can conclude that mathematical modelling was a key to making our goals successful.

42.6 Final Considerations

In the execution of the pedagogical activities related to the project described here, we aimed to show the possibility of Critical Statistics Education insertion within a statistics topic in an undergraduate course. In this context, we highlight the socio-political interfaces involved in the proposed thematisation, which emerged from the educational environment experienced by the teacher. Our interest in reporting this experience was to show that the opportunities to insert themes related to social and political problems occur several times in the pedagogical action. We believe that educators must take advantage of these situations to encourage students' critical, investigative and advocator spirit, which has the potential to excel when they are confronted with a social problem that involves their reality.

It also should be emphasized that throughout the project execution reported here, there were two important keys to our success. One was the use of technology in the project development. It brought the realisation of the potential of statistical content and facilitated the students' engagement in the activities executed to the extent that they could realise the power of the tool that they were handling. The other key was use of mathematical modelling. It had permitted us to focus on teaching and learning of statistics within a critical vision of education, helping students develop important capacities listed here.

We believe that, without losing the focus of statistical content, adopting Critical Statistics Education within a technological mathematical modelling environment can greatly enrich the educational process. It had given students the opportunity to better understand their own reality and to find the paths that can lead them to actions that actually represent reactions against the social and political reality in which they live. Thus, we understand that the teacher performs a much more comprehensive role and makes education more meaningful, more interesting and more real.

References

Ben-Zvi, D. (2008). Research on developing statistical reasoning: Reflections, lessons learned, and challenges. *11th International Congress on Mathematical Education (ICME-11) Annals*. Monterrey: Universidad Autonoma de Nuevo Leon. http://www.icme11.org/node/1530.html

Campos, C. R. (2007). *A educação estatística: uma investigação acerca dos aspectos relevantes à didática da estatística em cursos de graduação*. Doctoral thesis. Rio Claro: UNESP-IGCE.

Campos, C. R., Wodewotzki, M. L. L., & Jacobini, O. R. (2011). *Educação Estatística – teoria e prática em ambientes de Modelagem Matemática*. Belo Horizonte: Autêntica.

Chance, B. L. (2002). Components of statistical thinking and implications for instruction and assessment. *Journal of Statistics Education, 10*(3). http://www.amstat.org/publications/jse/v10n3/chance.html

delMas, R. C. (2002). Statistical literacy, reasoning and learning: A commentary. *Journal of Statistics Education, 10*(3). http://www.amstat.org/publications/jse/v10n3/delmas.html

Engel, J., & Kuntze, S. (2011). From data to functions: Connecting modelling competencies and statistical literacy. In G. Kaiser, W. Blum, R. Borromeo Ferri, & G. Stillman (Eds.), *Trends in teaching and learning of mathematical modelling* (pp. 397–406). New York: Springer.

Freire, P. (1974). *Education for critical consciousness*. New York: Continuum.

Freire, P. (1998). *Pedagogy of freedom: Ethics, democracy, and civic courage* (trans: Clarke, P.). Lanham: Rowman & Littlefield.

Freire, P. (2000). *Pedagogy of the oppressed*. New York: Continuum.

Gal, I. (2004). Statistical literacy: Meanings, components, responsibilities. In D. Ben-Zvi & J. Garfield (Eds.), *The challenge of developing statistical literacy, reasoning and thinking* (pp. 47–78). Dordrecht: Kluwer.

Garfield, J. B. (2002). The challenge of developing statistical reasoning. *Journal of Statistics Education, 10*(3). (http://www.amstat.org/publications/jse/v10n3/garfield.html)

Garfield, J. B., & Ben-Zvi, D. (2008). *Developing students' statistical reasoning – Connecting research and teaching practice*. New York: Springer.

Giroux, H. (1988). *Teachers as intellectuals: Toward a critical pedagogy of learning*. Westport: Bergin and Garvey Press.

Giroux, H. (2010). Lessons from Paulo Freire. *Chronicle of Higher Education*. Washington, DC. Retrieved October 20, 2010, from http://chronicle.com/article/Lessons-From-Paulo-Freire/124910/

Guggenheim, D., Bender, L., Buena, S. Z., & David, L. (2006). *An inconvenient truth*. [movie]. Directed by David Guggenheim and produced by Lawrence Bender, Scott Z. Burns and Laurie David. USA, 94 min, digital (computer).

Gujarati, D. N. (2004). *Basic econometrics* (4th ed.). New York: McGraw-Hill.

Habermas, J. (1971). *Knowledge and human interests*. Boston: Beacon Press.

Kuntze, S., Lindmeier, A., & Reiss, K. (2008). Using models and representations in statistical contexts as a sub-competency of statistical literacy – Results from three empirical studies. *11th international congress on mathematical education (ICME-11) annals*. Monterrey: Universidad Autonoma de Nuevo Leon. http://tsg.icme11.org/document/get/474

McLaren, P. (2000). Paulo Freire's pedagogy of possibility. In S. Steiner, H. Krank, P. McLaren, & R. Bahruth (Eds.), *Freirean pedagogy, praxis and possibilities: Projects for the new millennium* (pp. 1–22). New York: Falmer Press.

Mukhopadhyay, S., & Greer, B. (2001). Modeling with purpose: Mathematics as a critical tool. In B. Atweh, H. Forgasz, & B. Nebres (Eds.), *Sociocultural research on mathematics education: An international perspective* (pp. 295–311). Mahwah: Erlbaum.

Niss, M. (1983). Considerations and experiences concerning integrated courses in mathematics and other subjects. In M. Zweng et al. (Eds.), *Proceedings of the fourth international congress on mathematical education* (pp. 247–249). Boston: Birkhäuser.

Pfannkuch, M., & Wild, C. (2004). Towards an understanding of statistical thinking. In D. Ben-Zvi & J. Garfield (Eds.), *The challenge of developing statistical literacy, reasoning and thinking* (pp. 17–46). Dordrecht: Kluwer.

Rumsey, D. J. (2002). Statistical literacy as a goal for introductory statistics courses. *Journal of Statistics Education, 10*(3). www.amstat.org/publications/jse/v10n3/rumsey2.htm

Skovsmose, O. (1985). Mathematical education versus critical education. *Educational Studies in Mathematics, 16*(4), 337–354.

Skovsmose, O. (1992). Democratic competence and reflective knowing in mathematics. *For the Learning of Mathematics, 12*(2), 2–11.

Skovsmose, O. (1994). *Towards a philosophy of critical mathematics education*. Dordrecht: Kluwer.

Chapter 43
Models and Modelling in an Integrated Physics and Mathematics Course

Angeles Domínguez, Jorge de la Garza, and Genaro Zavala

Abstract This is an on-going study of an integrated first-year engineering course of mathematics and physics using models and modelling instruction. This innovation involves: redesigning the course content, combining teaching strategies, reshaping the classroom setting, and the use of technology. The experimental course was taught at a large private university in northern Mexico. This study analyses the students' final projects of this integrated course. The general comments made by the students were positive and affirmed this as a valuable learning experience. Students noted that the course reduced boundaries between physics and mathematics, helping them to better understand the application and need for the mathematics content.

43.1 Introduction

This chapter presents an attempt to close the gap between calculus and physics by teaching an integrated college level course of calculus and physics using models and modelling instruction. This innovation involves redesigning course content, combining teaching strategies, reshaping the classroom setting, and the use of technology. The course was taught at a large private university in northern Mexico in an integrated Physics and Mathematics course for first-year engineering students.

Physics is a science in which all topics are interconnected with a small number of physical laws (Chabay and Sherwood 2010) in which mathematics plays an

A. Domínguez (✉)
Department of Mathematics, Tecnologico de Monterrey, Av. E. Garza Sada 2501 Sur, Col Tecnologico, Monterrey, NL 64849, Mexico

School of Engineering, Universidad Andres Bello, Sazié, 2315, Santiago, Chile
e-mail: angeles.dominguez@itesm.mx

J. de la Garza • G. Zavala
Department of Physics, Tecnologico de Monterrey, Av. E. Garza Sada 2501 Sur, Col Tecnologico, Monterrey, NL 64849, Mexico

School of Engineering, Universidad Andres Bello, Sazié, 2315, Santiago, Chile
e-mail: jdelagarza@itesm.mx; genaro.zavala@itesm.mx

G.A. Stillman et al. (eds.), *Mathematical Modelling in Education Research and Practice*, International Perspectives on the Teaching and Learning of Mathematical Modelling, DOI 10.1007/978-3-319-18272-8_43

important role (Dirac 1938). However, students often lack this connected view of physics and the relationship between the mathematics and physics they are learning at school (Adams et al. 2006; Dray et al. 2008; Martínez-Torregos et al. 2006). There have been efforts both in physics and mathematics not only to increase students' learning, but also to help students to see the connection between physics and mathematics. For instance, in physics, Modelling Instruction (MI) by Hestenes (2010) is based on students' own building of the concepts with activities that are more into what they will encounter in real life. The use of mathematics in the strategy comes naturally in order to try to solve a situation. In mathematics, Michelsen (2006) presents a framework for interdisciplinary instruction involving the interplay between mathematics and science; whilst Lesh and Sriraman (2005) propose mathematics education as a design science which provides a framework structured to promote testing, communication with relevant communities and progression.

The need for bridging the gap between mathematics and sciences is well documented. Several universities have designed integrated courses that involve two or more areas. Some of these are recent such as a course that integrates calculus and introductory science (Carpenter 2007) and a STEM course that integrates chemistry, biology, computer science, physics, and mathematics (Gentile et al. 2012). Our attempt, namely Fis-Mat, focuses on integrating calculus and physics for first year engineering students (Domínguez et al. 2013).

At our university, the first attempt to close the gap between calculus and physics was a redesign of the calculus curriculum (Alanís 2000; Salinas et al. 2011). This redesign uses physics contexts to trigger mathematics concepts (Salinas et al. 2001). However, the mathematics and physics courses were still taught separately, and frequently mathematics concepts that had not yet been taught in the calculus course were needed for application in physics. Fis-Mat fully integrates the first semester calculus course with the first semester physics course and its corresponding physics laboratory. This chapter provides a description of the course, its teaching strategies, the classroom setting, the characteristics of the participants and the academic results. It also offers some concluding remarks and proposed steps for the future.

43.2 Teaching Strategies

Richard Feynman (1998) has stated that "The rules that describe nature seem to be mathematical.... It just turns out that you can state mathematical laws, in physics at least, which work to make powerful predictions" (p. 24). This is what we do in Fis-Mat: we use mathematical laws to study physical phenomena so students can make strong predictions and construct complete models using a variety of representations (Hestenes 1987, 2010; Wells et al. 1995). The main pedagogical approach of Fis-Mat is Modelling Instruction (MI), which focuses on qualitative and quantitative model development in an explicit cycle that allows students to conjecture, test and refine their models (Brewe 2008).

The basic building blocks in science are the models, and science is a modelling process. MI contends that science instruction should teach students the basic rules of modelling and should organise the course content around a small set of scientific conceptual models (Halloun 2004). The conceptual models that structure the course content in MI are shared among members of the learning environment, and the validity, deployment and interpretation of the models are established through classroom activities and discourse. In this approach, models serve as conceptual resources that can be used to develop an understanding of a variety of phenomena. MI "helps students develop model-centred knowledge bases that resemble those of practicing [sic] scientists" (Brewe 2008, p. 1156).

In a MI class, students work in groups of three to solve the given problems; they record their analyses on portable whiteboards. Then, the entire class sits in a circle to discuss their findings. All students are able to see the boards of every other group and are encouraged to ask questions of their peers. The instructor facilitates the discussion to reach a consensus. It is during these interactions that students negotiate the use of conceptual models. The key for success is the design of the problems. In Fis-Mat, most of the physics worksheets were adopted from the material designed by Eric Brewe's group at the Florida International University.

Two other teaching strategies implemented in the Fis-Mat course are based on educational research in the disciplines, tutorials for introductory physics (McDermott et al. 2002) and modelling (Niss et al. 2007). Students make predictions or conjectures about a physical situation and then they are led through scientific reasoning or data collection to verify whether or not their predictions or conjectures are correct. In Fis-Mat, modelling the situation using data analysis involves the use of mathematical concepts. Often, students' predictions or conjectures contradict their results, fostering cognitive conflict (Festinger 1957) that promotes learning.

Another relevant element of this course is the classroom setting. It consists of round tables that accommodate nine students arranged in groups of three. This setting fosters group interaction, promotes communication, empowers students, and in turn facilitates the development of learning skills such as argumentation and self-regulation. The variety of technological tools and equipment available in the room facilitated students' investigation of various models that were constructed based on their own observations and data collection (Zavala et al. 2013).

43.3 Methodology

For this first experience, the participants in the Fis-Mat course were 20 first-year engineering students. The research questions that guided this study are:

How do models reflect students' understanding of the physical and mathematical concepts? How does modelling work as a core for an integrated physics and mathematics course?

43.3.1 Course Description

Fis-Mat uses the physics curriculum as its backbone, while the mathematics brings the support for idea-building and operations. In developing this course, we considered previous research (Dray et al. 2008; Martínez-Torregos et al. 2006) and added modelling as a principal teaching strategy (Blum and Borromeo Ferri 2009; Brewe 2008), along with an innovative classroom design that also serves as a physics laboratory to conduct experiments (Zavala et al. 2013). The primary goals of the Fis-Mat Project are: (a) to improve students' abilities to make connections between physics and mathematics, (b) to increase students' motivation to advance in their engineering studies, and (c) to develop diverse competencies, such as critical thinking and the ability to work collaboratively.

There were four faculty members from the School of Engineering involved in the development of the Fis-Mat course: two from Physics and two from the Mathematics Department. All four met regularly to discuss and make decisions on the implementation. Joint teaching of the course began in the fall of 2012 with one Physics and one Mathematics faculty member. The major topics covered in the Fis-Mat course included at least all the topics of a regular first year physics course and calculus course for engineering majors (Domínguez et al. 2013).

43.3.2 Final Project of the Course

The implementation strategy of using projects in engineering education has proven successful, as it fosters individual responsibility in students and creates an environment that is similar to that of professional engineering. Based on this idea, students had two weeks to work on the *Spring-Mass System Project* as a final assignment, in which they had to design, implement, document, and (orally) present to the entire class on the last day of classes. Seven teams of three students were formed, and each turned in a final report and gave a presentation about their approach and solution to the problem.

Spring-Mass System Project

On a table there is a block attached to a spring, and the spring is attached to a wall. A person slides the block onto the table, stretching the spring. Then, the block is released (still attached to the spring) and it slides onto the table.

Use the modelling tools to analyze the situation and complete the following tasks:

1. Determine the friction coefficient between the block and the table.
2. Determine the spring constant.
3. Construct a mathematical model that describes the motion of the object at any time from the moment the block is released.

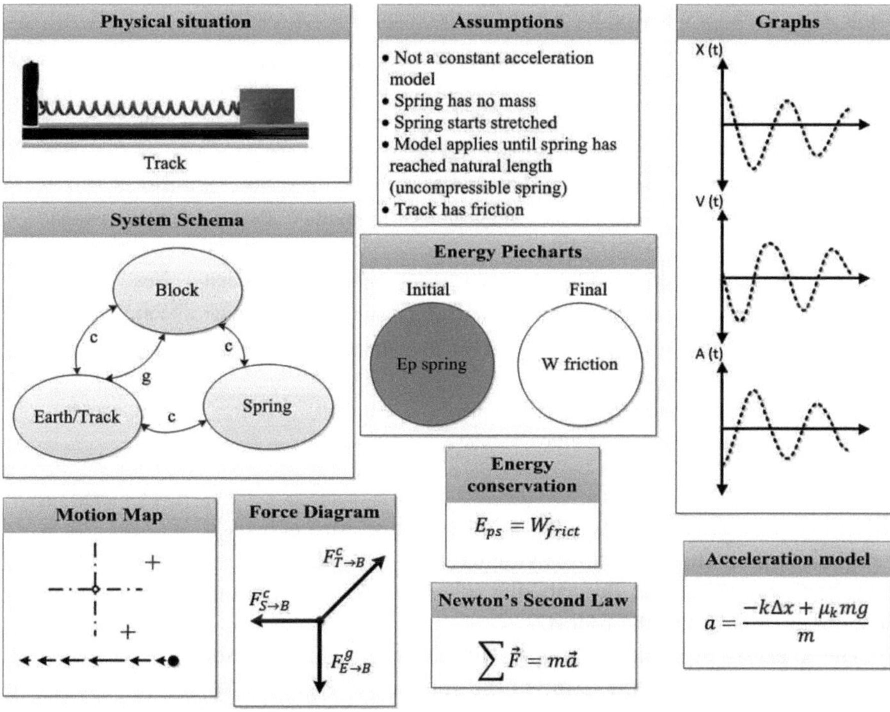

Fig. 43.1 Complete model for calculating the acceleration of the object

The *Spring-Mass System Project* was designed to serve as a complex problem that involves different concepts dealt with during the course, like friction and forces on a spring. Students need to take what they have learned and broaden their understanding of these concepts using numerical methods to solve the problem. The more representations (diagrams, drawings, principles, and relationships that complement one another) used in the model, the more robust the model is (see Fig. 43.1).

The criteria that students had to meet were: (1) establish the connections between physical and mathematical concepts and procedures; (2) present a project design before setting the experiment and collecting the data; and (3) document their work, and prepare a written report and an oral presentation. The report was graded on the results, the consistency of their model, and creativity of the design project; the oral presentation was graded on content and communication skills.

43.4 Results

Students presented reports in which they explained the three different tasks they were asked to work on in the *Spring-Mass System Project*. In this section, tables summarise the collaborative work of the students. Each table corresponds to a

Table 43.1 Model completion for calculating the kinetic friction coefficient

Group	Assumptions	Principle based on	Physical situation	System schema	Force diagram	$\mu_k = \frac{F^C_{W\to B} - ma}{mg}$
Alpha	No friction on pulley	Second law	Verbal	Implicit	Implicit	Yes
Beta	Implicit	Second law	Verbal	No	No	No
Gamma	Implicit	Energy conservation	Yes	Implicit	Yes	Yes, energy
Delta	Implicit	Second law	Implicit	Implicit	Implicit	Yes
Epsilon	Implicit	Second law	Verbal	Implicit	Implicit	Yes
Kappa	Implicit	Energy conservation	Verbal	Implicit	Energy	No, static friction
Lambda	Implicit	Energy conservation	Implicit	Implicit	Energy	Yes, energy

different task. There are two forms of evidence for each task. We use the word "explicit" when students show evidence in the report; however, we use the word "implicit" when students describe the schema or situation but they do not show the schema or drawing in the report.

The analysis of students' models for the first task is presented in Table 43.1. The first column shows the name of the collaborative group, the second column the assumptions made explicitly by the group in their model, and the third column the physical principle students used for their model (Newton's Second Law or Energy Conservation Principle). The following columns indicate whether students made explicit drawings of the physical experiment, the abstraction of the system (system schema) and the force diagrams. The last column indicates whether the students were able to construct the function to calculate the kinetic friction coefficient, μ_k.

We can observe in Table 43.1 that no matter what model students based their answers on, Conservation of Energy or Newton's Second Law, if they explicitly stated it in their reports, they arrived at the correct model for calculating the friction coefficient. Group Kappa made a conceptual mistake that led them to determine the static friction coefficient. That mistake could have been avoided if the motion map had been included, something none of the groups did. The complete model for this task is shown in Fig. 43.2.

When a group arrived at a correct relationship for calculating the kinetic friction coefficient, we are assuming that they had at least implicitly considered force diagrams and system schema. In that case, we also assumed that since all groups designed and constructed their experiments, they had a clear idea of the physical arrangement of the objects.

On the other hand, when a group was unable to determine the correct relationship, we could not consider those assumptions. This was the case with the second group; Beta had only a verbal representation of the situation and was unable to reach the correct representation. We could not assume that group Beta implicitly considered the other representations. It is worth noting that although most of the

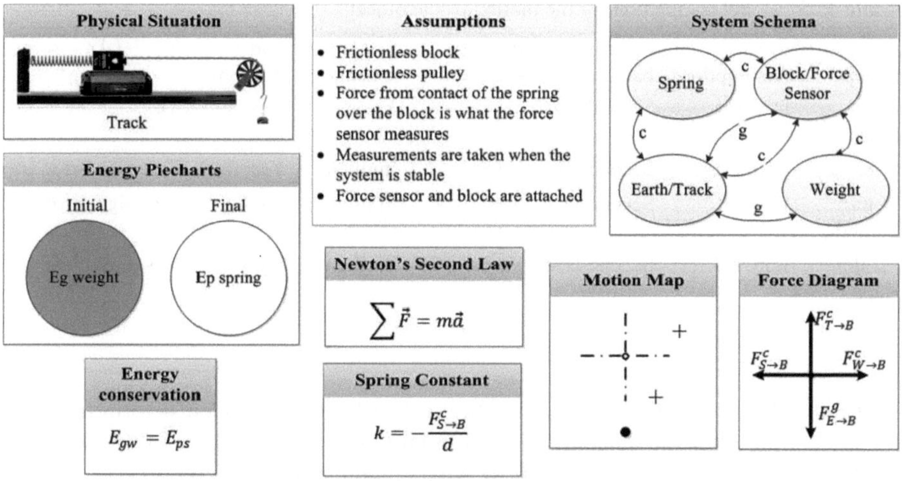

Fig. 43.2 Complete model for calculating the spring's elastic constant

Table 43.2 Model completion for calculating the elastic constant of the spring

Group	Assumptions	Principle based on	Physical situation	System schema	Force diagram	$k = -\frac{F^{C}_{S\to B}}{d}$
Alpha, Beta, Gamma	Implicit	Second law	Verbal	Implicit	Implicit	Yes
Delta, Epsilon, Lambda	Implicit	Second law	Implicit	Implicit	Implicit	Yes
Kappa	Implicit	Energy conservation	Verbal	Implicit	Energy	No

groups recorded no explicit assumptions on the written reports, we know students considered those assumptions based on the results of their work, their oral presentation, or the experiment design submitted before conducting the experiment.

Table 43.2 shows the analysis for the second task in this model, determination of the elastic constant of the spring. The results in this table are presented using the same structure and criteria as in Table 43.1.

The model needed to obtain the spring constant seemed to be easier for these students. This might have been influenced by the fact that students had conducted a similar experiment two weeks before the due date for the final project. As in the first task, students used either Newton's Second Law or the Energy Conservation Principle for their models. However, the group that used energy was not able to construct a model that allowed them to calculate the spring constant. None of the groups recorded explicit assumptions nor used motion maps in their models.

Group Kappa considered friction in their implicit assumptions, but neglected it in their experiment. They might have noticed something was missing, if they had drawn the system schema. Therefore, that representation would have shown that there was one interaction with the track that they had completely ignored.

Table 43.3 Model completion for the motion of the object

Group	Assumptions	Principle based on	Physical situation	System schema	Force diagram	$a = \frac{-k\Delta x + \mu_k mg}{m}$
Alpha, Beta	Implicit	Second law	Yes	Yes	Yes	Yes
Gamma	Implicit	Second law	Verbal	Implicit	Yes	Yes, sign error
Delta, Lambda	Implicit	Second law	Yes	Implicit	Implicit	Yes, sign error
Epsilon, Kappa	Implicit	Second law	Verbal	Implicit	Implicit	Yes, sign error

Table 43.3 shows the analysis for the last task on the model, determining the complete model for the object motion. Similar columns to those in Tables 43.1 and 43.2 are listed in Table 43.3 that show what students did in the report.

Table 43.3 shows that all groups were able to propose a function of how the block would move after it is released. Yet, a sign error appears in most of the work of the groups. That error seems to happen when they explicitly avoid doing force diagrams, leading them to think that the force of the spring is in the same direction as the friction force. A careful use of the force diagram would have helped them avoid such a mistake. Moreover, none of the groups presented a complete model (Fig. 43.1), they only used some representations of the full model as shown in Table 43.3.

43.5 Discussion and Conclusions

Students worked on the project according to the model perspective learnt in class. The project was open enough that students in an implicit or explicit way had to make many assumptions, as well as make decisions on how to solve the problem and on what experiments to perform. These decisions made the project valuable because the process was as important as the results. During the semester, students worked on constructing models which led them to the concepts and principles of mechanics and calculus. The activities had the objective of covering a topic that was in the course content. The final project, on the other hand, had the objective of applying the model-constructing process for a new situation.

Most of the groups succeeded in proposing an acceleration function of the object, although they did not succeed in predicting the block motion. That is, students succeeded in constructing a model, identifying the variables involved and their relationships. In Table 43.3, we presented the evidence that all teams followed the process.

Domínguez et al. (2013) argued that students in this course learned concepts in mechanics better than most students in similar classes in which physics and calculus are taught separately. Their conceptual understanding at the end of the semester was similar to that of students in honours classes in that university. Students' general comments were positive (e.g., "this was a good learning experience"). They commented that the reduced boundaries between physics and mathematics helped them to better understand the application of, and need for, the calculus content.

The evidence collected suggests that the more robust the model developed by a student or group of students, the more likely it is that they will not make mistakes. As a result, they are more likely to have a correct model to present and predict how a physical situation will evolve over time. Also, the evidence shows that students are able to take the modelling skills they have learned and apply them to the constructions of models not seen in class, such as the case of the *Spring-Mass System Project* which corresponds to a non-constant acceleration model.

Our view of modelling brings two elements to the perspective; that is the Modelling Instruction as a main teaching strategy and the model construction process that students face in every class of the Fis-Mat course. In this way, students focus on understanding the relationship among variables and the different representations that can be used as a means to build models that become more robust as more representations are combined. There is, however, much that needs to be done. First, we would like to frame the problem within a context that is a more realistic situation for students. Second, to better understand the model construction process, we would like to analyse students' responses at different stages of the Fis-Mat course. Third, we believe that the modelling approach of the course promotes deeper understanding of the mathematical and physical concepts as well as facilitates the development of critical thinking, problem solving and argumentation competences (Brewe et al 2012). It is desirable to collect evidence of such claims. So far, our findings indicate that, since they actually constructed more robust models, it would appear they found the multiple representations helpful in their problem solving. We believe that the teaching methodologies, activities, classroom setting and evaluation implemented point towards a reasonable direction to integrate physics and mathematics in an engaging way.

Acknowledgements We acknowledge the support received from our Institution through a research chair grant CAT140 and the School of Engineering. We recognize and appreciate Eric Brewe and his Physics Education Research Group's sharing support material developed for teaching a Physics course using MI.

References

Adams, W. K., Perkins, K. K., Podolefsky, N. S., Dubson, M., Finkelstein, N. D., & Wieman, C. E. (2006). A new instrument for measuring student beliefs about physics and learning physics: The Colorado Learning Attitudes about Science Survey. *Physical Review Special Topics – Physics Education Research, 2*, 010101.

Alanís, J. A. (2000). La predicción: un hilo conductor para el desarrollo de un curso de cálculo. In R. Cantoral (Ed.), *El futuro del cálculo infinitesimal, ICME 8, España*. México: Iberoamérica.

Blum, W., & Borromeo Ferri, R. (2009). Mathematical modelling: Can it be taught and learnt? *Journal of Mathematical Modelling and Application, 1*(1), 45–58.

Brewe, E. (2008). Modeling theory applied: Modeling instruction in introductory physics. *American Journal of Physics, 76*(12), 1155–1160.

Brewe, E., Kramer, L., & Sawtelle, V. (2012). Investigating student communities with network analysis of interactions in a physics learning center. *Physical Review Special Topics – Physics Education Research, 8*, 010101. doi:10.1103/PhysRevSTPER.8.010101.

Carpenter, J. (2007). Integrating calculus and introductory science concepts. *Proceedings of the 114th annual ASEE conference and exposition*. Honolulu; Code 70739, 6 p. Washington, DC: American Society for Engineering Education.

Chabay, R. W., & Sherwood, B. A. (2010). *Matter and interactions*. Hoboken: Wiley.

Dirac, P. A. M. (1938–1939). The relationship between mathematics and physics. *Proceedings of the Royal Society (Edinburgh), 59* (Part II), 122–129.

Domínguez, A., Zavala, G., & Alanís, J.A. (2013). Integrated physics and math course for engineering students: A first experience. *Proceedings of the 120th ASEE annual conference and exposition*. Atlanta; Code 99351. Washington, DC: American Society for Engineering Education.

Dray, T., Edwards, B., & Manogue, C. A. (2008). *Bridging the gap between mathematics and physics*. Paper presented at the 11th international congress on mathematics education, Topic Study Group 16. Monterrey.

Festinger, L. (1957). *A theory of cognitive dissonance*. Evanston: Row Peterson.

Feynman, R. (1998). *The meaning of it all: Thoughts of a citizen-scientist*. Reading: Helix.

Gentile, L., Caudill, L., Fetea, M., Hill, A., Hoke, K., Lawson, B., et al. (2012). Challenging disciplinary boundaries in the first year: A new introductory integrated science course for STEM majors. *Journal of College Science Teaching, 41*(5), 44–50.

Halloun, I. A. (2004). *Modeling theory in science education*. Dordrecht: Springer.

Hestenes, D. (1987). Toward a modeling theory of physics instruction. *American Journal of Physics, 55*(5), 440–454. doi:10.1007/978-1-4419-0561-1.

Hestenes, D. (2010). Modeling theory for math and sciences education. In R. Lesh, P. L. Galbraith, C. R. Haines, & A. Hurford (Eds.), *Modeling students' mathematical modeling competencies* (pp. 13–41). New York: Springer.

Lesh, R., & Sriraman, B. (2005). Mathematics education as a design science. *ZDM, 37*(6), 490–505.

Martínez-Torregos, J., López-Gay, R., & Grass-Marti, A. (2006). Mathematics in physics education: Scanning historical evolution of the differential to find a more appropriate model for teaching differential calculus in physics. *Science & Education, 15*, 447–462.

McDermott, L. C., Shaffer, P., & Physics Education Group. (2002). *Tutorials in introductory physics*. Upper Saddle River: Prentice Hall.

Michelsen, C. (2006). Functions: A modelling tool in mathematics and science. *ZDM, 38*(3), 260–280.

Niss, M., Blum, W., & Galbraith, P. (2007). Introduction. In W. Blum (Ed.), *ICMI Study 14: Applications and modelling in mathematics education* (pp. 3–32). New York: Springer.

Salinas, P., Alanís, J. A., Pulido, R., Santos, F., Escobedo, J. C., & Garza, J. L. (2001). *Elements of calculus: Conceptual reconstruction for learning and teaching*. Mexico: Iberoamerica.

Salinas, P., Alanís, J. A., & Pulido, R. (2011). Cálculo de una variable. Reconstrucción para el aprendizaje y la enseñanza. *Didac, 56–57*, 62–69.

Wells, M., Hestenes, D., & Swackhamer, G. (1995). A modeling method for high school physics instruction. *American Journal of Physics, 63*(7), 606–619.

Zavala, G., Domínguez, A., & Rodríguez, R. (2013). ACE: Innovative educational model for teaching physics and mathematics to engineering students. *Proceedings of the 120th ASEE annual conference and exposition*. Atlanta; Code 99351. Washington, DC: American Society for Engineering Education.

Chapter 44
Research-Based Modelling Teaching Activities: A Case of Mathematical Positioning with GNSS

Xiaojun Duan, Dan Wang, and Mengda Wu

Abstract This chapter explores the mechanism of modelling teaching based on research, including modelling problem collection and issuing of problems for selection by students, the feedback of students and consultation between teachers and students, mathematical modelling and problem solving process, discussion and assessment. Based on the Innovation Practice and Internship Base for Mathematical Modelling, research-based problems are issued for senior undergraduates to cultivate modelling competency when solving practical problems. A case of mathematical positioning with GNSS, multipath, is introduced to show the whole process in detail focussing on developing active and critical thinking, in particular the innovative thought of students, by the process. From years of practice, research-based modelling teaching has been shown to be a very effective platform to improve the modelling competency of students.

44.1 Introduction

Teaching and application research activities on mathematical modelling have been carried out for nearly 30 years at the National University of Defence Technology (NUDT) in China. The focus is not only on the mathematical modelling course teaching, but also the training of students' mathematical modelling innovative thinking (Li 2013). The Innovation Practice and Internship Base for Mathematical Modelling is established in our university. One of its tasks is to organize students to carry out mathematical modelling innovative research projects and related activities. Stillman (2015) notes that with mathematical modelling the direction reality to mathematics becomes the focus. The question to ask as a modeller is then: "Where

X. Duan (✉) • D. Wang • M. Wu (✉)
College of Science, National University of Defence Technology, Yanwachi Street 137, Changsha, Hunan 410073, P.R. China
e-mail: xj_duan@163.com; casablanman@hotmail.com; mdwu@nudtwmd.com

G.A. Stillman et al. (eds.), *Mathematical Modelling in Education Research and Practice*, International Perspectives on the Teaching and Learning of Mathematical Modelling, DOI 10.1007/978-3-319-18272-8_44

can I find some mathematics to help me with this problem?" The model has to be built through idealising, specifying and mathematising the real world situation. Based on the Innovation Practice and Internship Base for Mathematical Modelling at NUDT, we built a student innovation training system of mathematical modelling to enable students to experience these processes in a supervised environment.

The system includes not only a generic training for mathematical modelling, but also mathematical modelling training for the professional fields. The training task covers the teaching of mathematical modelling courses; the student's training programs and the scientific research training projects. The support environments of this system involve a mathematical training laboratory, the Innovation Practice and Internship Base for Mathematical Modelling and the professional laboratory of applied mathematics. Unlike many other university programs involving professional/industry training (e.g., Araújo et al. 2013; Cumberbatch and Fitt 2001), the mathematical modelling competency training of undergraduates is an ongoing process in our system involving students from low-level to senior undergraduates. The training goal of the system is to cultivate the student modelling competencies, including not only generic mathematical modelling ability and professional modelling ability, but also an open and cooperative consciousness. In this system, the core is the problem-oriented and the research-based mathematical modelling training. We have designed a program as research-based mathematical modelling, directed at the training of mathematical modelling.

This chapter firstly introduces the mathematical modelling training system, and then the mechanism of modelling teaching based on research is presented in detail, including modelling problem collection and issue, the feedback of students and consultation between teachers and students, and the mathematical modelling and problem solving process. A case of mathematical positioning with the Global Navigation Satellite System (GNSS) is then introduced to show the whole process in detail, focusing on stimulating the interest, recognition and innovative thought of students.

More concretely, in the mathematical modelling related to positioning based on GNSS to be discussed here, there is a need for sophisticated algorithms (Pecchioni et al. 2007) for accurately and reliably processing the GNSS signals for timing and navigation. Mathematical modelling is of critical importance in this. Areas of mathematics that are relevant include linear and non-linear algebra, optimization, filtering, and statistics. This chapter will focus on one of the mathematical issues (i.e., multipath, integrity, resolution of the phase ambiguity, global tomography) that arise in increasing the processing speed, accuracy and reliability of GNSS—the multipath case.

44.2 The Mechanism of Research-Based Mathematical Modelling Activities

During mathematical modelling, modellers are simplifying and structuring a real world problem, transferring it into the world of mathematics, working mathematically and interpreting a mathematical result as well as validating a real result (Maaβ 2006). Thus whilst a mathematical course could provide basic training of mathematical reasoning and logical thinking, mathematical modelling could provide an environment for the overall training for students as a small paradigm of scientific research, including improving their mathematical application ability, innovation, the ability to plan as a whole, synthetic knowledge level, engineering accomplishment, insistence on purpose, expression, and cooperation amongst other things.

44.2.1 The Practical Mathematical Modelling Program in NUDT

If we accept the need for a mathematical modelling course, practical modelling training is also important. Based on the Innovation Practice and Internship Base for Mathematical Modelling at NUDT, we built a student innovation training system. The mathematical modelling training is divided into three levels (see Fig. 44.1). The first level focuses on the teaching of mathematical modelling courses and contest activities (e.g., Li 2013; Xie 2013). It is a generic training and suitable for first and second year. The second level is a middle phase of mathematical modelling training. Student interest and competency in mathematical modelling is much developed by some innovation training programs from real life, such as transportation problems and insurance issues. The third level is a senior phase of mathematical modelling training. In this level, we combine the teacher's research project with the student's training programs and design a program as research-based mathematical modelling, directed at the Junior and Senior undergraduates.

In reality, the middle phase and senior phase are not independent. They are both attempting to train the student's ability in mathematical modelling through solving real world problems, but the problems in the senior phase are more difficult than the former. The problem-oriented research is precisely the important aspects of scientific studies. We call this training research-based mathematical modelling.

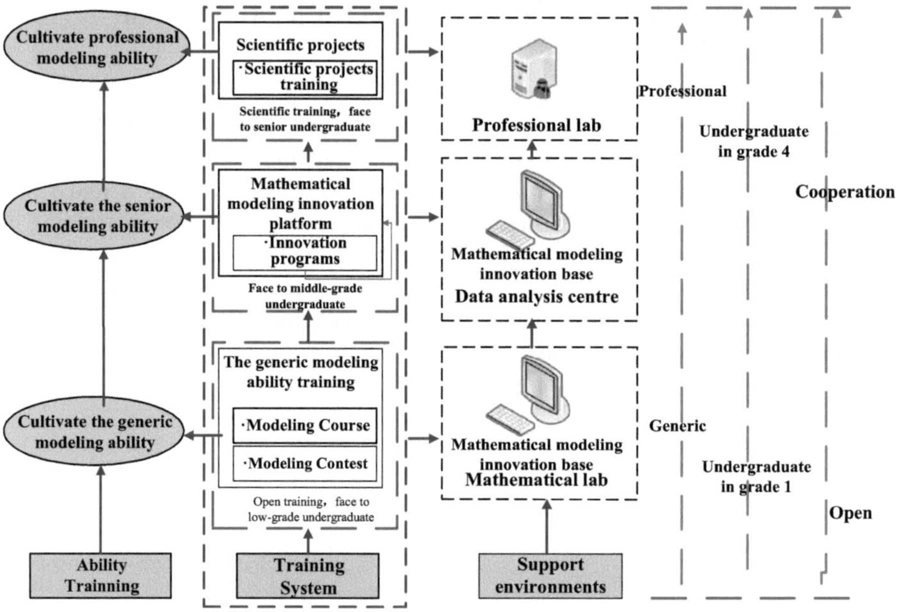

Fig. 44.1 The mathematical modelling training system

44.2.2 The Scheme for Research-based Mathematical Modelling

Now, the scheme of the research-based mathematical modelling in detail is as follows.

1. *Collection of modelling oriented research problems.* Each June, the organising committee of the Innovation Practice and Internship Base for Mathematical Modelling sends notices to tertiary teachers to collect modelling-directed research problems. The basic requirements include: (a) the origin of the problems should be from practical scientific research, (b) the problems are remodified to be fit for the level of undergraduates, (c) they are not too difficult and (d) they do not require too much special background knowledge. Generally, the initial collection of the problems finishes in September. After receiving these problems, the organising committee with experienced modelling teachers helps rewrite or polish the problems to make them easier to understand for undergraduates.
2. *Issuing problems and consultation.* Every early-September, the research-based mathematical modelling problems are issued first on the campus internet for students. Almost half a month later, a consultation conference is held for those students who are interested. Firstly, the teachers introduce the background and the problems, the interested students put questions to the teachers in a

subsequent one-to-one communication process. Several days later, the students are required to submit their choices of problems to the Base.

3. *Learning the background of the problems.* Different from the general mathematical modelling exercises, the research-based mathematical modelling problems are still related to much engineering background. Therefore, in order to solve the problems better, the students need to learn more of the engineering background involved in these problems. In this process, the teachers prepare all needed materials and give several lectures to students, to make sure they understand the problems correctly.
4. *Modelling exploration and solution.* This step is of most importance to cultivate the modelling competency of the students. Since there is no standard solution for the research-based mathematical modelling problems, it is challenging to obtain a better solution satisfying the practical engineering requirements. In many cases, the theoretically best solution may not be the best choice in practice as Pollak (1997) and Neunzert (2013) have pointed out. Consideration might need to be given to the balance for the solving method among the performance, realization cost, and reliability, amongst other things.
5. *Discussion and assessment.* After obtaining a satisfactory solution, summary and feedback are also crucial. Usually, students submit a paper describing the process involved in solving the problems under supervision. The paper is sent to two other teachers for assessment. The students also need to give a defence for their papers, in this process the inquiry and discussion can be intensive.

44.2.3 Cultivation of Modelling Competency

Innovation consciousness and ability are crucial for modelling competency. Without a standard answer for the modelling problems, the inexistence of a unique best solution of the modelling problems provides a capacious creation space for the students (Li 2013). That uncertainty and challenging character makes modelling more attractive.

Students are inspired to use their originality to solve the same problem using different methods from different aspects, which is very helpful for their imagination and open thoughts. The inspiration is based on deep understanding of the knowledge and deep thinking about the problems. In one case, for example, the optimization algorithm is needed to calculate the assignment of resources. Some students chose to use new intelligent algorithms such as neural networks or the genetic algorithm, while others used the classical heuristic algorithm. However, in this case, the classical heuristic algorithm achieved a better result since there existed too many local extrema in the problem. So focusing on only the innovation of the method is not enough, the students need to investigate more rational methods to implement the intrinsic character of practical problems. Especially, in the research-based mathematical modelling process, the practical background is very important, thus the students should consider the convenience to be realised in reality besides other performance of the solution.

We also helped the students to be persistent in solving the problem even when they encounter difficulty, which is a very good character in life. The young are always eager to do well in everything, but cannot keep persisting in many cases. The greatest results are usually attained by simple means. In the training and cooperative research, we encourage the students steadfastly to apply effort and grapple with each difficulty in the process of problem solving. We show examples to demonstrate the different sequels for giving up or insisting on facing difficulties. Since the students have to think more of insisting on overcoming difficulties, many report the training improves their self-confidence by cooperative research experience. They often experience the range of feelings from confusion and anxiety to satisfaction in the process of overcoming difficulties by thinking of many kinds of methods.

It is also very important to improve students' ability to express opinion and understand other people's opinion. It is the first time for many students to take part in a practical program. Some of them are not good at expressing themselves, and also have some problems in communication. Therefore, the supervisors encourage them to show themselves in a better light and make a more active communication atmosphere, identifying some tactics in communication during the training process.

44.3 A Case: Multipath in Mathematical Positioning

The GNSS, including GPS, GALILEO, GLONASS, and Beidou System, utilizes the pseudo-range and carrier phase from a constellation of satellites in earth orbit to accurately locate a receiver antenna position relative to these satellites. Applications include land surveying, autonomous vehicle control, navigation, air traffic control. With the addition of differential or relative signals, the ultra-high precision GNSS is capable of position accuracies of the order of a few centimetres. A user may receive signals coming from several different satellite systems in the future. It is important to make full use of the whole navigation information of all the constellations. There is a need for sophisticated algorithms for accurately and reliably processing the GNSS signals for timing and navigation. Mathematical modelling is of critical importance here. Areas of mathematics that are relevant include linear and non-linear algebra, optimisation, filtering, and statistics.

In this section, the case of multipath is introduced as an example in the Innovation Practice and Internship Base for Mathematical Modelling. Multipath is the propagation phenomenon that results in radio signals reaching the receiving antenna by two or more paths. Causes of multipath include atmospheric ducting, ionospheric' reflection and refraction, and reflection from water bodies and terrestrial objects such as mountains and buildings. It is related to the characteristics of the signal, the processing method in the receiver, the antenna and signal receiving scenario (Satirapod and Rizos 2003; Zhang and Bartone 2004). To solve the real signal from the multipath is an optimization problem. The complex factors make it difficult to eliminate the multipath errors. The mathematical modelling methods are introduced here and an adaptive filter is proposed. We modify the problem to fit for undergraduates, this embodies the idealising, specifying and mathematising processes of modelling.

Through this case we expect to develop the student's modelling ability, including exploring possible solutions to problems, active and critical thinking on exploring the way to solution, scheme selection, data analysis and summarization, and so on.

44.3.1 Guiding Exploring Possible Solution to Problems

In the research-based modelling teaching process, we do not give out the problem-solving thought directly. The students have to seek for the ways to solve the problems themselves. In the multipath mitigation case, the students had to learn more of the engineering background involved in GNSS. In addition, the teacher instructed the students on literature to find, what others have done on GNSS (e.g., Kneissl et al. 2009), and how they have done this (e.g., Pecchioni et al. 2007), what progress has been made, what is the problem which is still not solved, what is the difficulty and hotspot, what is the key, what is valuable, what is meaningless, and so on. Furthermore, the teacher also guided the students in how to improve the efficiency of reading literature.

At the beginning, most of the students were not familiar with GNSS. Through a large number of information inquiries for multipath, they learned that there were two ways for multipath mitigation: Nonparametric and Parametric methods. They gradually understood and compared the two methods. The following is a brief summary of the student design specification concerning the two methods.

Nonparametric method reduces multipath interference based on a phase discriminator function design related to the physical signal. Parametric method is a mathematical modelling approach by considering the received signal as the superposition of the direct signal and multipath signal, it transforms the multipath mitigation problem into a direct signal time delay estimation problem.

When they had any questions about the two methods, the teachers prepared materials and gave several lectures to students on Nonparametric and Parametric methods. In this process, the students felt that they gained a lot. As one student reported, "I really enjoyed expanding my mind. It pushed me to develop my own thoughts".

44.3.2 Directing Solution Selection

The process to guide students to complete mathematical modelling for a practical research question is that of motivating the student to make factor trade-offs and scheme choice. By this training, it promotes the students' ability of macro-control and judgment (see Wu et al. this volume for further details).

Students needed to consider two levels of scheme choice in the process of multipath research. The first level is to select the nonparametric method or parametric method. The second level is to choose the concrete implementation algorithm in the framework of nonparametric method or parametric method. Most of the students decided to take the parametric method in scheme selection in the first stage selection, since the nonparametric method cannot completely solve the problem of multipath mitigation. In the process of the choice of the second level, the parametric methods are mainly concentrated on the Maximum Likelihood estimation (ML) method, and nonlinear filtering method.

The following is a brief summary of the students' design specification concerning the second level.

"ML method, however, does not take advantage of the priori information, it assumes that the parameters in the processing time does not vary over time. In practice, however, due to the existence of the Doppler Effect, the signal delay often varies over time."

"To Nonlinear filtering method, the equation of state reflects the prior information of parameters, measurement equation reflects the observations, which is a blend of a priori knowledge and measurement to estimate the state. Therefore, nonlinear filtering is also a good tool to handle such problems."

At this level, the majority of students chose the ML method which is in more common use, but there were some students who wanted to choose a more challenging solution, namely, the nonlinear filtering method. Supervisors would also advise them that the ML method, which is more popular and easier to realize in practice, would encounter some difficulty in dealing with the time varying delay problem. So, students should make an appropriate choice in this case according to the data they have.

44.3.3 Developing Active and Critical Thinking

Mathematical modelling needs critical thinking. After the mathematical model is built and the solution is obtained, we need to re-evaluate the established mathematical model, and improve the model in time, in order to seek the best results. At the same time, we also need to point out the practical significance of the model, and to promote its performance. In research-based modelling teaching, developing the student's critical spirit and deep thinking habit is always one of the teaching goals.

In this process, we guide the students to think deeply on solving the problem with different methods. The nonparametric method is based on an error control system rather than eliminating system error, so it cannot fundamentally eliminate the interference of multipath error. The ideal multipath error mitigation method is modelling the multipath interference system error, which minimizes the interference of multipath error. The parametric method models the multipath signal processing by considering receiving signals as the superposition of the direct signal and the multipath signal. It estimates the multipath signal delay and direct signal separately. Subsequently, it transforms the multipath mitigation problem into a parameter estimation problem, so as to restrain or eliminate the multipath error.

To accompany the problem solving, we all undertake the training of feedback with probing questions: Is the result right? Is it conforming to the practice? Where are the advantages and disadvantages of this method? Is there a better solution? How can you improve? Can this solution be promoted? In the process of the training, students' critical thinking will be strengthened and improved. As a student noted, "The most attractive aspects were the freedom to create a non-traditional learning method". Another student reflected that, "It was the highlight of my university life. It reminded me of why I went into academia".

44.3.4 Improving the Ability of Data Analysis and Summarising

In this case of Mathematical Positioning with GNSS, the students who chose the parametric method compared the advantages and disadvantages of the different methods, and they made generalizations. The students who chose the ML method and nonlinear filtering method analyzed the characteristics and applicability for the methods in different cases.

Here there are two cases to consider: Case 1 showed that the relative delay of the direct signal and multipath signals remains unchanged within the processing time (Fig. 44.2a). Case 2 showed that the relative delay of the direct signal and multipath signals decreased quickly within the processing time (Fig. 44.2b). The data analysis results are shown in Fig. 44.2, in which ML stands for Maximum Likelihood method, D-EKF stands for EKF (Extended Kalman Filter) method of nonlinear filtering, D-UKF stands for UKF (Unscented Kalman Filter) method of nonlinear filtering.

The data analysis results show that the ML method and nonlinear filtering method have equivalent levels of efficiency in the case of the relative time delay remaining unchanged. However, the ML method is more robust. When delay dynamic change is more obvious, the nonlinear filtering method is significantly better than the ML method in suppressing multipath error. As for nonlinear filtering methods, the UKF is more adaptive for the much stronger nonlinearity problem, so it is better

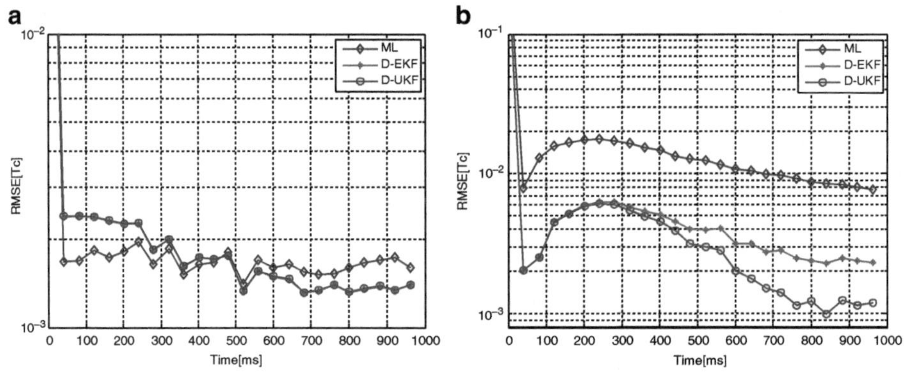

Fig. 44.2 The precision comparison between the ML method and nonlinear filtering within the framework of the Parametric method (**a**) case 1: the relative delay of the direct signal and multipath signals remains unchanged (**b**) case 2: the relative delay of the direct signal and multipath signals decreased quickly

than that of EKF method in this case. Therefore, the methods show different adaptability in different cases.

Thus, based on the practical problem put forward, we improve students' ability of mathematical modelling and help them adapt to solve the practical problem by the process of exploring the solution, critical thinking, scheme selection, data analysis and conclusion.

44.4 Conclusion

We have described the Practical Mathematical Modelling Program in NUDT. We have introduced the mechanism of modelling teaching and shown how to cultivate modelling competency for senior undergraduates. Our main experience is, to create an environment to attract the students' learning desire, cultivate their ability to study independently and enhance their mathematics quality and creative ability. We emphasize the ability to acquire new knowledge and a process of problem solving, rather than knowledge and result. The most important thing is to inspire students to solve the problem.

The solution proposed in this chapter, supported by good practices and experiences from a successful case, is a research-based Modelling Teaching Activity, which provides a capacious creation space for the students. It is the uncertainty and challenging character that makes modelling more attractive to students as noted in the student comment. With years of practice, a lot of achievements have been made in many kinds of student involved modelling activities. We think that research based modelling teaching is very effective and it is a good platform to improve the modelling competency of students. We intend to pursue further research to improve the mechanism.

Acknowledgment This work is supported by the Course Construction Program, and the Innovation Practice and Internship Base for Mathematical Modelling, of NUDT in China.

References

Araújo, A., Barbeiro, S., & Ferraira, J. S. (2013). Laboratory for computational mathematics: An interface between academia and industry. In A. Damlamian, J. F. Rodrigues, & R. Sträßer (Eds.), *Educational interfaces between mathematics and industry: Report on an ICMI-ICIAM-study* (pp. 199–204). Cham: Springer.

Cumberbatch, E., & Fitt, A. (Eds.). (2001). *Mathematical modeling: Case studies from industry*. Cambridge: Cambridge University Press.

Kneissl, F., Stoeber, C., Eissfeller, B. (2009). *Assessment of combined integrity algorithms*. Proceedings of the 22nd International Technical Meeting of the Satellite Division of the Institute of Navigation (ION GNSS 2009), Savannah, pp. 2804–2817.

Li, T. (2013). Mathematical modeling education is the most important educational interface between mathematics and industry. In A. Damlamian, J. F. Rodrigues, & R. Sträßer (Eds.), *Educational interfaces between mathematics and industry* (pp. 51–58). Cham: Springer.

Maaβ, K. (2006). What are modelling competencies? *ZDM, 38*(2), 113–142.

Neunzert, H. (2013). Models for industrial problems: How to find and how to use them—in industry and in education. In A. Damlamian, J. F. Rodrigues, & R. Sträßer (Eds.), *Educational interfaces between mathematics and industry: Report on an ICMI-ICIAM-study* (pp. 59–76). Cham: Springer.

Pecchioni, C., Ciollaro, M., & Calamia, M. (2007). *Combined Galileo and EGNOS integrity signal: A multisystem integrity algorithm*. 2nd Workshop on GNSS Signals & Signal Processing – GNSS SIGNALS 2007, Nordwijk.

Pollak, H. O. (1997). Solving problems in the real world. In L. A. Steen (Ed.), *Why numbers count: Quantitative literacy for tomorrow's America* (pp. 91–105). New York: College Entrance Examination Board.

Satirapod, C., & Rizos, C. (2003). Multipath mitigation by wavelet analysis for GPS base station applications. *Survey Review, 38*, 295–304.

Stillman, G. A. (2015). Applications and modelling research in secondary classrooms: What have we learn? In S. J. Cho (Ed.), *Selected regular lectures from the 12th International Congress on Mathematical Education* (pp. 791–805). Cham: Springer International.

Xie, J. (2013). An introduction to CUMCM: China/contemporary undergraduate mathematical contest in modeling. In A. Damlamian, J. F. Rodrigues, & R. Sträßer (Eds.), *Educational interfaces between mathematics and industry: Report on an ICMI-ICIAM-study* (pp. 435–443). Cham: Springer.

Zhang, Y., & Bartone, C. (2004). Multipath mitigation in the frequency domain. *Proceedings of the IEEE/ION position location and navigation symposium (PLANS'04)* (pp. 486–495). Athens: IEEE/ION.

Chapter 45
Mathematical Texts in a Mathematical Modelling Learning Environment in Primary School

Ana Virgínia de Almeida Luna, Elizabeth Gomes Souza, and Larissa Borges de Souza Lima

Abstract This chapter deals with a study involving mathematical modelling in primary school. It aims at analysing what kinds of text are produced in the learning environment of mathematical modelling. The participants were the teacher and her 22 students, aged 9–10 years old, who attended the fourth grade in a private school in Brazil. According to Bernstein's theory, the principles governing the pedagogical practice of a given context allow the production of different texts for the instruction of school mathematics. In this study, three types of instructional mathematical texts were identified: contextualized, critical and interdisciplinary.

45.1 Introduction

This chapter analyses the development of a mathematical modelling activity which was conducted with primary school students. The main objective was to identify and classify some mathematical texts produced by the teacher and students in the development of modelling activities. At times, in this chapter, the term modelling will be used in substitution for mathematical modelling.

A.V. de Almeida Luna (✉)
Department of Mathematics, State University of Feira of Santana, Transnordestina Avenue, Feira de Santana, Bahia 44036-900, Brazil
e-mail: andrluna@uol.com

E.G. Souza
Scientifics and Mathematics Educational Institute, Federal University of Pará, Augusto Corrêa Road, Km-01-Guamá, Belém, Pará 66075-110, Brazil
e-mail: elizabethmathematics@gmail.com

L.B. de Souza Lima
Despmat Group, Despertar School, Noide Cerqueira Avenue, 4380, Feira de Santana, Bahia, Brazil
e-mail: l.souzabb@hotmail.com

G.A. Stillman et al. (eds.), *Mathematical Modelling in Education Research and Practice*, International Perspectives on the Teaching and Learning of Mathematical Modelling, DOI 10.1007/978-3-319-18272-8_45

Studies analysing the insertion of mathematical modelling into the first years of schooling are not commonly found in the international literature as reported by English and Sriraman (2010) and English (2009). Nevertheless, existing studies highlight that this insertion contributes to the development of skills by young students, to solve problems, argue mathematically, understand and relate mathematical concepts and objects, etcetera (Biembengut 2007; English 2009, 2010; English and Sriraman 2010).

This study presents an analysis of the discursive production of students and a teacher in conducting a modelling activity based on the theoretical concepts of the sociologist Basil Bernstein (2003). Bernstein's theoretical framework (2003) allows us to identify the nature of the participants' discursive production in light of the understanding of social relations formed between the students and the teacher in performing modelling activities.

Among the various conceptions of mathematical modelling (see Lesh et al. 2010), we understand it as a learning environment in which students are invited to question and investigate problems with reference to reality (Barbosa 2010). Modelling activities enable students to discuss daily life problems in mathematics classes. The use of modelling in primary school allows students to be in contact with mathematics at an early age, and by doing so, they can understand the importance of mathematics and its presence in various situations of the real world at the very early stage of their scientific thinking (English 2013).

The teacher and students in primary school (Years 1–5) present different characteristics, as elsewhere, compared with those at other levels of schooling within the Brazilian educational system. In particular, the mathematical content is integrated with other school subject matter content. Apart from that, the students have only one teacher teaching different subject matter, and this teacher may not have specific academic training in mathematics, but just a generalist qualification.

When we refer to modelling in primary school, we aim to provide students with problems and real data from the very beginning of their schooling. According to Lesh and Fennewald (2010), modelling does not work with what is called a textbook problem, which deals with particular mathematical knowledge, in which the student will have to use predetermined mathematical tools. On the contrary, students use their mathematical knowledge to understand the theme chosen and respond to real problems. In addition, the implementation of mathematical modelling in primary school enables students to begin to identify the presence of mathematics in society, and reflect critically about such presence.

Therefore, we suggest that mathematical modelling in primary school should be used in a socio-critical perspective (Barbosa 2006; Kaiser and Sriraman 2006). The socio-critical perspective is viewed as that which focuses on the development of modelling as the analysis of the presence of mathematics in society, through discussion and critical reflection on the assumptions that underpin the social arguments based in mathematics.

45.2 Pedagogical Practice Based on the Theory of Basil Bernstein

In order to develop this research, we have based our thinking on the theoretical framework of Bernstein (2000, 2003). A pedagogical practice is regulated by pedagogical discourse. Bernstein (2003) understands *pedagogical discourse* as a set of principles. These principles are analysed in terms of communication, that is, *what is* legitimate to speak and *how to speak* in a certain context, in this case, the school context.

According to Bernstein (2000), pedagogical discourse is a principle of appropriation of other discourses, a recontextualizing principle. Pedagogical discourse comprises a set of rules to embed and relate two discourses: the *instructional discourse* (the specialized discourse of the sciences that is expected to be transmitted at school) and the *regulative discourse* (a discourse associated with the values and standards in the pedagogical relationship). The instructional discourse is embedded in the regulative discourse.

In pedagogical practice, texts are individual productions of the agents that constitute the practice, in this chapter, the teacher and her students. Texts can be based on instructional discourse and regulative discourse. Text is understood in terms of communication, and for this reason we should not restrict writing to the conception of the term, but extend it to any communicative act, such as a gesture, a way of expression, a glance. Similarly, we can speak of different types of text, such as verbal, written or gestural (Luna et al. 2011). For Bernstein (2000), legitimate text is a text, produced by the agents, that presents relevant meaning to the context of a given pedagogical practice.

We will take two principles that control the pedagogical practices, which are named by Bernstein (2000) as *classification* and *framing*. As for the hierarchy, *classification* can be stronger or weaker. Stronger classification means separation with well-defined borders between agents (teachers and students); weaker classification indicates that there is some interaction between the agents or disciplines. *Classification* governs *power relations* involving the spaces of each agent hierarchically. The second principle is *framing*, which refers to the way and place of communication among the agents, between the teachers and students. The *framing* involves *control relations* governing communication practices within social relations. Therefore, *control relations* establish legitimate ways of communication for each group. *Framing* can be stronger or weaker. When the relations of control on the teacher and students are weakened, the latter have the opportunity to produce their texts, which can be legitimised by the first; however, when *control relations* are strengthened, the text is produced predominantly by the teacher.

45.3 Method, Participants and Research Context

Qualitative research can be defined as the analysis of material forms of life, beliefs, discursive production, individuals' practice, etcetera, because the data analysed in qualitative research are texts, not numbers (Denzin and Lincoln 2005). Thus, the investigation conducted is qualitative because it aims at determining what types of texts are produced in mathematical modelling activities. We captured this discursive production by using an observation method for the collection of data. We used observation as a method of data collection because we intended to identify the types of texts produced in developing modelling activities at the time they were created. Observation is the procedure for data collection, and its main characteristic is to allow the understanding of the study theme in the temporal moment of data constitution (Angrosino 2005). The participants whose production of texts was analysed were 22 students (aged 9–10 years old) who attended the fourth grade of a private primary school and a teacher who developed the modelling activity on the topic *Virtual Water*. Virtual water represents all the water used for the production of certain foods, from cultivation to manufacture.

45.4 Data Analysis

The modelling activity on virtual water was performed on three consecutive days. The lessons were videotaped, totalling 6 h of recording, and transcribed in their entirety. The written material produced by the students along with their texts and the teacher's texts were selected and categorized. The process of categorization revealed different types of mathematical texts of instructional discourse; however, in this chapter, we will present only three of these, those we consider representative of the development of modelling activities in the early years. The following texts were selected: *contextualised mathematical text*, *critical mathematical text* and *interdisciplinary mathematical text*. Furthermore, although regulative discourse is inserted in instructional discourse, this chapter will focus on the analysis of text production of instructional discourse.

45.5 The Modelling Environment

The teacher started the development of the modelling activity by asking students if they knew what virtual water would be. Based on a selection of the answers given, the teacher showed a video[1] on the topic to the students. The modelling activity, in this chapter, is understood as pedagogical practice regulated by principles.

[1] The video was extracted by the teacher from http://www.youtube.com/ watch?v = HCXkGETpmwm.

After reviewing and discussing the concept of virtual water, the teacher gave the students a text that reported the amount of water used in the production of various food products, such as soya, tomato, coffee and rice. The following texts refer to a specific discussion about the total amount of water used in the annual soybean production.

Teacher: According to the text we have just read, Brazil exports soya to other countries and spends 45 million cubic metres of water yearly.

Student 1: Teacher, what does cubic mean?

Teacher: It is a measure; the depth of a pool, for example.

Tcacher: On the internet here, I see that a swimming pool, like those in which Gustavo Borges[2] swims, I mean those Olympic ones. It says here it has the capacity to hold 2,700 m^3 of water.

Teacher: How about calculating it? I would like to know how many times 2,700 m^3 fits in 45,000,000 m^3.

Student 2: What type of operation is it? Multiplication or division?

Student 3: [Using a calculator, the student divides 45,000,000 by 2,700 and finds 16,373 for a result.]

Teacher: Kids, look what we found. How many pools of water are spent in the annual soya beans production in Brazil?

Student 1: 16,373.

Teacher: Is this little or a lot?

Students: It's a lot.

In this excerpt, we identify that the control relations are weakened, because both the teacher and the students produce texts, not only the teacher. This enables the production of texts of instructional discourse by both the teacher and students. We note that sometimes the teacher assumes the role of producing legitimate texts, which occurs when the same student responds to a question about cubic metres. However, the production of legitimate texts is also performed by the student, when he/she chooses and elaborates the mathematical procedure to be adopted. The weakening of control relations in modelling activities enables and recognises the students' own ideas as valid in the resolution and conducting of appropriate calculations for understanding the problem. The students seek clarification with the teacher, but they also propose mathematical procedures, find answers, and interpret the results.

The fact that the students can also produce legitimate texts indicates that the power relations are weakened in the excerpt analysed. The role of producing legitimate mathematical texts is also exercised by the students. The weakening of control relations in modelling activities, namely the ability of students to interact with each other, propose their responses to the questions studied, etcetera, is documented in some studies, such as those conducted by English (2013).

[2] Gustavo Borges is a popular Brazilian Olympic swimmer.

However, in this excerpt, we note that the student is given a possibility to propose ways to understand the question and legitimate mathematical procedures to solve it. Moreover, the teacher recognises, that is, agrees with and accepts the students' ideas as valid. This fact questions the understanding of this being a role exclusive to the teacher, and that, therefore, the texts of instructional discourse are produced exclusively by the teacher. The production of texts, by the teacher and students, previously mentioned, indicates that modelling activities can enable students to indicate and produce legitimate texts too, not only the teacher; therefore, the teacher's role is not restricted to assessing the discursive production of students. In a strengthened power relation, for instance, students can produce their texts, but only the teacher can assess or give legitimacy to such production.

With control relations also weakened, one of the students questions the teacher about the meaning of cubic metre. The teacher compares the idea of cubic metre to the capacity of a pool. To make the students understand this quantity, the teacher related the capacity of an Olympic swimming pool with the total amount of water required for the country's annual soya bean production.

We classify these texts produced by the teacher and students as *contextualized mathematical texts* of instructional discourse. In this case, the teacher and students produce texts about the capacity of a pool when they discussed the concept of cubic metre, a concept of instructional discourse of the school, that is, the instructional discourse of school mathematics. The teacher used an example, a context, to teach the students about the concept of cubic metre.

After the discussion on *Virtual Water*, the teacher asked the students to perform an individual virtual water research, specifically on the virtual water that may be consumed in their daily meals. The following is the production of texts related to this task.

Teacher: What did you like best in your research? What did you find most interesting?

Student 5: We thought we spent little water, but we spend a lot, and when we have a look at the water bill, it does not show exactly how much water we spend to make the products.

Teacher: That's it.

Student 5: The water bill only shows the quantity we spend at home.

Teacher: You have said something interesting. Take that package of rice over there to check if the label shows how much virtual water is spent. Will there be this information?

Student 5: I don't think so.

In this excerpt, the students question the fact that the consumption and billing of expenditures on water does not include the actual expenditure of a family on water, because it does not show the amount of water spent to produce the food consumed. This is indicative of a second type of text of instructional discourse – the *critical mathematical text*. This text is characterised by criticism about the presence of mathematics in a given text used in society. In the analysis performed here, the text analysed by the students was the one of the water bill of their homes.

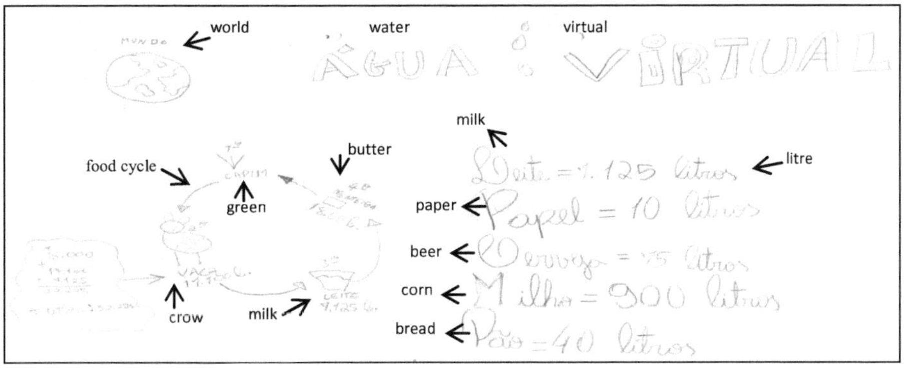

Fig. 45.1 Student texts on the *Virtual Water* consumption for butter production

In this case, the control relations are still weakened, but the power relation becomes stronger because the teacher indicates which production she considers legitimate – “You have said something interesting”, and encourages the students to reflect about the importance of having information of the total water consumption on food packaging. Thus we can observe that the production of critical mathematical texts was initially prompted by a child and later legitimised and emphasised by the teacher.

Next, we introduce a scheme produced by one of the students in response to the teacher’s request. This scheme refers to the third and final type of production of texts analysed in this chapter.

In the texts shown in Fig. 45.1, we can identify that the students sought to present the total amount of water required for the production of butter, calculating the amount of water consumed to maintain the pasture, raise the cows, and produce the milk. The students have drawn a cycle of the food chain and accounted for the water consumption involved in the entire production of butter, thus relating to the instructional science text studied at school.

The texts produced by the students have led us to identify a third type of text of instructional discourse – the *interdisciplinary mathematical text*. The production of this text indicates that modelling activities can prompt students to make their own relations between the instructional discourse of school mathematics and other instructional discourses addressed in the school context, in the temporal moment of realisation of the modelling activity. In this case, the instructional discourse of sciences – the food cycle.

45.6 Final Considerations

The theoretical framework of Basil Bernstein (2000, 2003) has allowed us to understand the development of modelling activities in the school context as regulated by principles. Thus, in this chapter, we analysed the development of a modelling activity from the standpoint of the social relations that are formed in that environment, that is, we did not analyse just a modelling activity and the responses the students elaborated about it. Thus, this framework extended our analysis to the principles that permeate social relations in modelling activities. In this chapter, they were analysed with respect to the rules of the school environment, the curriculum, the role of agents in practice, and teaching strategies, amongst others. Therefore, we believe that modelling activities developed in the school context are governed by *relations of power and control*. There are instructional discourses that are widely legitimate and accepted by the agents that constitute the pedagogical practice and there are also ways of producing these discourses. In other words, there is legitimacy in "what to speak" and "how to speak", and this legitimacy should be considered when we analyse the development of a modelling activity in the school context.

The texts produced by the teacher and students enabled us to identify that this legitimacy is regulated by the relations of power and control that govern the pedagogical practice. Relations of power and control are not fixed; they are social, that is, are constituted by their agents, in this case, the teacher and her students. In the modelling activity analysed in this chapter, these relations were weaker and/or stronger depending on the circumstance. The variations of power and control relations observed in the excerpts analysed in this chapter allowed us to identify legitimate texts produced by the students. The students produced critical mathematical texts of instructional discourse and interdisciplinary mathematical texts.

This production points to the matter of the nature of students' participation in modelling activities. Do the students participate only when they propose their ideas, argue about their answers, and look for ways to solve the problems?

The texts in this chapter produced by the students indicate that they can also have the opportunity of preparing legitimate texts of the instructional discourse, that is, they can propose mathematical procedures, select mathematical concepts, develop their answers to the problems posed in modelling activities and so on, because of the weakening of relations of power and control.

The production of contextualised mathematical text of instructional discourse produced by the teacher indicates the pedagogical strategies that the teacher uses, or can use, to prompt the production of instructional discourse of school mathematics by the students. The teacher chose to produce contextualised texts, texts not directly related to instructional discourse, but used in comparison with this discourse. Is this pedagogical strategy peculiar to the production of instructional discourse with young students? What other strategies are peculiar to this level of education?

It is possible that the specific production of these texts is a peculiar feature of pedagogical practice instituted in the use of modelling in primary school. There is a need for further studies that identify these and other features of implementation of mathematical modelling in primary school.

References

Angrosino, M. V. (2005). Recontextualizing observation: Ethnography, pedagogy, and the prospects for the progressive political agenda. In N. K. Denzin & Y. S. Lincoln (Eds.), *Handbook of qualitative research* (3rd ed., pp. 729–745). Thousand Oaks: Sage.

Barbosa, J. C. (2006). Mathematical modelling in classroom: a critical and discursive perspective. *ZDM – The International Journal on Mathematics Education, 38*(3), 293–301.

Barbosa, J. C. (2010). The students' discussions in the modelling environment. In R. Lesh, P. L. Galbraith, C. Haines, & A. Hurford (Eds.), *Modeling students' mathematical modeling competencies* (1st ed., pp. 365–372). New York: Springer.

Bernstein, B. (2000). *Pedagogy, symbolic control and identity: Theory research critique* (Rev. ed.). London: Taylor & Francis.

Bernstein, B. (2003). *Class, codes and control, Vol. IV: The structuring of pedagogic discourse.* Londres: Routledge.

Biembengut, M. S. (2007). Modelling and applications in primary education. In W. Blum, P. L. Galbraith, H.-W. Henn, & M. Niss (Eds.), *Modelling and applications in mathematics education* (pp. 451–456). New York: Springer.

Denzin, N. K., & Lincoln, Y. S. (Eds.). (2005). *Handbook of qualitative research* (3rd ed., pp. 729–745). Thousand Oaks: Sage.

English, L. D. (2009). Modeling with complex data in the primary school. In R. Lesh, P. L. Galbraith, C. R. Haines, & A. Hurford (Eds.), *Modeling students' mathematical modeling competencies* (pp. 287–300). New York: Springer.

English, L. D. (2010). Young students' early modelling with data. *Mathematics Education Research Journal, 22*(2), 24–47.

English, L. D. (2013). Complex modelling in the primary and middle school years: An interdisciplinary approach. In G. A. Stillman, G. Kaiser, W. Blum, & J. P. Brown (Eds.), *Teaching mathematical modelling: Connecting to research and practice* (pp. 491–505). Dordrecht: Springer.

English, L., & Sriraman, B. (2010). Problem solving for the 21st century. In B. Sriraman & L. English (Eds.), *Theories of mathematics education: Seeking new frontiers* (pp. 263–290). Berlin: Springer.

Kaiser, G., & Sriraman, B. (2006). A global survey of international perspectives on modelling in mathematics education. *ZDM_The International Journal on Mathematics Education, 38*(3), 302–310.

Lesh, R., & Fennewald, T. (2010). Introduction to part I modeling: What is it? Why do it? In R. Lesh, P. L. Galbraith, C. R. Haines, & A. Hurford (Eds.), *Modeling students' mathematical modelling competencies* (pp. 5–10). New York: Springer.

Lesh, R., Galbraith, P. L., Haines, C. R., & Hurford, A. (Eds.). (2010). *Modeling students' mathematical modelling competencies.* New York: Springer.

Luna, A. V. A., Barbosa, J. C., & Morgan, C. (2011, July). Mathematical modelling and pedagogical recontextualisation of in-service teachers. *A selection of papers: ICTMA 15* [CDROM] (8 pp.). Melbourne: Australian Catholic University.

Chapter 46
A Differential Equations Course for Engineers Through Modelling and Technology

Ruth Rodríguez Gallegos

Abstract This chapter presents an experience of an educational practice in a private university in México involving a different way to teach a Differential Equations course for future engineers based on a didactical proposal developed through mathematical modelling. This proposal emphasizes that mathematics is a human activity that answers problems of a different nature, and throughout this problem solving activity it is likely that the emergence of mathematical concepts, notions and procedures occurs. Modelling is an important foundation to the design of this proposal. Three modelling situations are presented that are performed in a class of differential equations designed to ensure that students transitioning between different stages of the modelling cycle put into play different skills that are generally poorly treated in a traditional setting. Some preliminary results are outlined.

46.1 Introduction

The purpose of this chapter is to share the experience of an educational practice in a private university in México. This experience involves a different way to teach Calculus and Differential Equations courses for future engineers based on a proposal developed by the Mathematics Faculty over 14 years. This proposal originated from the idea of re-designing the scholarly mathematical discourse present in the Integral and Differential Calculus courses for engineers. It emphasizes that mathematics is, above all, a human activity that answers several problems of a different nature, and throughout this problem solving activity it is likely that the emergence of mathematical concepts, notions and procedures occurs. The teaching

R. Rodríguez Gallegos (✉)
Department of Mathematics, Tecnológico de Monterrey, Av. Eugenio Garza Sada, 2501, Col. Tecnológico C.P. 64849, Monterrey, Nuevo León, Mexico
e-mail: ruthrdz@itesm.mx

G.A. Stillman et al. (eds.), *Mathematical Modelling in Education Research and Practice*, International Perspectives on the Teaching and Learning of Mathematical Modelling, DOI 10.1007/978-3-319-18272-8_46

of mathematics is an important goal to prepare critical citizens. These citizens should develop the correct skills to identify and solve problems in whatever context they encounter them as well as be able to express, test, revise or reject even their own ways of thinking (Alsina 2007). To achieve this goal requires the development of activities that allow students to recognize the importance of mathematics in everyday life situations.

In our institution, the DE course is the last formal course in basic mathematics. It is intended that the students are able to use this knowledge in later subjects of their specialty but that does not happen automatically or successfully. This course is currently taught in 25 different engineering programs. Since in the current context it is important to prepare future engineers who are able to solve problems in their disciplines, important background to this study is the redesign of the mathematics curriculum that the Engineering school started in 1999 (Salinas and Alanís 2009). The questions posed then were not only how to teach mathematics – in terms of methodologies and didactic techniques – but also what and why to teach such content. Students first learn the instrumental aspect of mathematical notions. It is in this spirit that our proposal takes over the beaten path and enriches it with a proposal for the Differential Equations (DE) course: a mathematical modelling approach (see Sect. 46.3); thus, emphasizing the fact that mathematical objects (the DE in our case) are seen primarily as tools to model various phenomena in different contexts. To do this, the work of Rodríguez (2009, 2010), which justifies the choosing of DE as an ideal tool to model phenomena of different natures, is extended. Students should find meaning within the mathematical object itself. The concept of the mathematical modelling process is explained in the following section. The work is still in progress; however, there has been great advance in the curriculum design for Differential Equations.

46.2 Designing a New Proposal: The Case of a Differential Equations Course

We have worked on the design and implementation of an innovative course of differential equations (DE). We want to demonstrate how DE can be taught to future engineers from the perspective of mathematical modelling and with the use of specific technologies.

46.2.1 Problems in the Teaching and Learning of DE

In universities worldwide and specifically in Mexican universities, the teaching of differential equations predominantly focuses on analytical methods (Artigue 1996) rather than on qualitative and numerical methods. This has been reported despite the wealth of both approaches in the teaching of DE (Arslan et al. 2004). This, and other developments, have been evidenced in the community of mathematics education for over 20 years, yet little changes have been reported in daily classroom activities. While successful innovative proposals have been documented on teaching DE (especially internationally) over the past few years (e.g., Blanchard 1994; Kallaher 1999; Rasmussen and Whitehead 2003) and some other research on the subject has been published, only few changes can be observed in classrooms and academic programs at various universities nationwide in México, particularly in the area of engineering. This proposal aims to acknowledge the importance of the changes in registers (algebraic, numeric and graphic), the modelling approach, and the effective use of technology in the teaching/learning process of DEs.

46.2.2 Re-designing a DE Course, a Collegiate Experience

In brief, the background of this departmental educational project started in August 2008 in our Institution, Tecnológico de Monterrey (TEC, Mexico). The main purpose was to re-design the DE course, implemented in our institution. It was a collegiate work of the faculty members of the Academy of Differential Equations. The early findings demonstrated that there were several educational proposals claiming the urgent need to reformulate the way DE was taught. As a result and with the active participation of the faculty members of this group, a new syllabus for the DE course at the undergraduate level was approved systemwide (for the 32 campuses of the TEC system).

46.2.3 A Mathematical Modelling Perspective in the DE Course

The study of mathematical modelling began at least four decades ago (Niss et al. 2007). It aimed to bridge school mathematics and the mathematics used in other settings. From six perspectives identified for mathematical modelling by Kaiser and Sriraman (2006), this chapter focuses on the realistic or applied. This perspective stresses the importance of teaching through mathematical modelling as a more pragmatic approach to solve real problems and to develop certain competencies in students, particularly those of modelling.

After a detailed study of several authors (e.g., Blum and Niss 1991; Niss et al. 2007), we proposed to visualize the mathematical modelling into a scholar setting, and we chose to further this study adopting the description of this process in terms of stages (e.g., Real Situation RS) and transitions (e.g., *a*, *b*) between stages as shown in Fig. 46.1. This proposal explicitly incorporates two important elements: the inclusion of a physical domain which is modelled and the importance given to

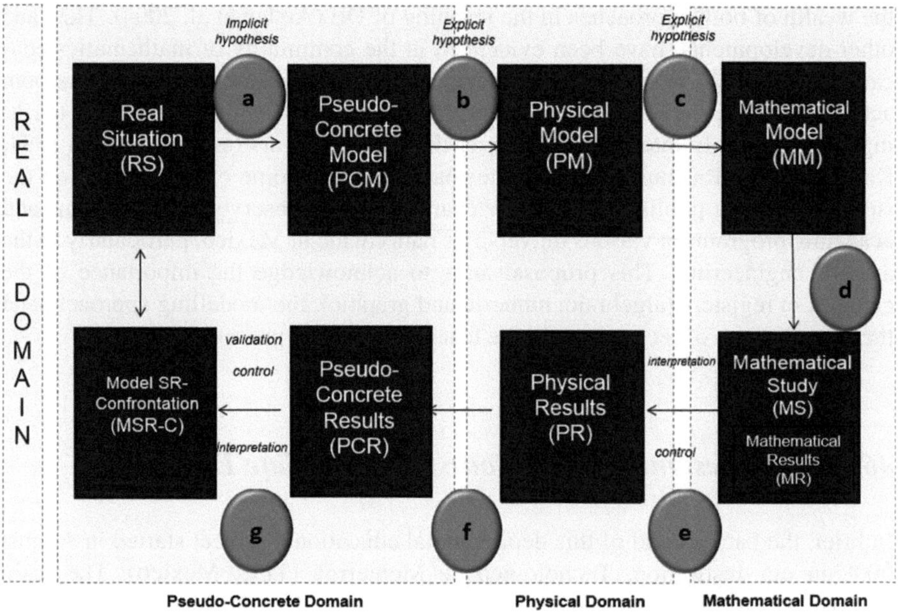

Fig. 46.1 Modelling schema (Rodríguez 2010, p. 194)

the pseudo-concrete domain as the difficult transition for students in the modelling process. It is important to note that the mathematical model is understood as the different graphical representations of the DE: solution graph, the DE itself as an analytical model, and a table of data that can eventually be modelled by a DE and/or its solution. The literature analysis has made it clear that mathematical modelling has allowed multiple benefits when used with students at different educational levels (Blomhφj and Carreira 2009). The benefits have translated into the achievement of connections between school mathematics and everyday life mathematics, reduction of anxiety towards the subject, promotion of communication and collaborative work, and the development of mathematical skills (Lombardo and Jacobini 2009). In more recent studies (Rodríguez 2010; Rodríguez and Quiroz 2015; Zavala et al. 2013), a precise theoretical approach has been formulated to implement mathematical modelling in the classroom, which includes, among other elements, the role of technology in learning mathematics, the importance of collaborative learning, and the development of modelling competencies according to the specific levels in the development of such competencies.

46.2.3.1 Development of Modelling Skills in a Differential Equations Course

Together with the above-mentioned, it is important to highlight that interest in the development of generic skills has been a matter of concern for different national and international institutions such as the Tuning Latin American Project (Beneitone et al. 2007), the Accreditation Board for Engineering and Technology (2011) and the 2015 Mission and Vision statements of Tecnológico de Monterrey, all of which base their views on the development of students proficient in solving problems in their settings and acquiring technological and collaborative skills. Likewise, the Organization for Economic Cooperation and Development [OECD] in the PISA Study (OECD 2003) presents the development of mathematical competencies, and more specifically those in modelling, as important. In an effort to define these competencies the PISA Study shows in the definition of mathematical literacy that mathematical modelling skills include the abilities to perform the modelling process adequately and to focus it properly. It is also concerned with the possibility of setting those skills into action. Based on the above approach, this chapter intends to present the design of activities to facilitate the transition between the various stages of mathematical modelling by having students face situations that are related to the contexts that students commonly experience in engineering.

46.3 Methodology in the Design and Implementation of Innovative Material in the DE Course

Innovative material (hands-on activities, laboratory practices, modelling and simulation practices, worksheets or spreadsheets) has been developed for the DE course. Its main focus is concerned with the modelling of biological, physical or chemical phenomena. Several researchers (Blanchard 1994; Kallaher 1999; Rassmusen and Whitehead 2003) have shown the need to change the way DE is taught, from the "traditional" way, which emphasizes analytical methods, to an integrative mode, which uses graphical and numerical methods. This integrative mode should enable students to identify and recognize a DE in its different representations; and thus, improve the learning of DEs as mathematical objects. The student should not only learn how to use techniques to solve DEs but also learn the application of the DE as a tool to model several problems. This is also strengthened through the use of specific technology and software such as CAS; simulations, and laboratory practices with sensors in the classroom to better model and understand the phenomenon of study: temperature, an RC circuit, or a spring-mass system. The student should be capable of integrating mathematical knowledge (DEs) with practical skills through modelling. Different active learning environments play an important role in promoting the implementation of the course with hands-on, modelling, and simulation activities, and the development of communication, problem solving and modelling skills.

46.3.1 Using Technology in a DE Course

Within the available software to teach mathematics, there are several educational technological products meant as tools with no explicit disciplinary content. The assorted technology available is considered as follows: specific mathematical CAS software such as Mathematica, or other resources freely available on the internet such as Wolfram Alpha; graphing calculators for Symbolic Computation; the use of sensors (temperature, voltage, motion) so that the student can perform some experimental practices in the classroom and see the illustrated form of a DE model in a real phenomenon, taking real data and adjusting, suggesting a model and verifying; Open Educational Resources (OER) such as *PhET* simulators from the University of Colorado and the *DE Tools*. The technology described above enables the design of activities in which the modelling skills may be eventually developed and foster meaningful learning of DE as a mathematical object as well as a tool.

46.3.2 Designing Modelling Activities

We have designed several activities for the DE course. Three are overviewed emphasizing two design aspects as follows: as models to represent various phenomena and as mathematical objects. In particular, some of the modelling stages defined in Fig. 46.1 (see transitions between stages) are explored with the idea that the student, through these tasks and the use of appropriate technology, develops modelling skills through the DE object. We want to share with the reader how the sequences of these activities have been chosen taking account of a specified theoretical perspective of mathematical modelling (Fig. 46.1) and how we recognize the importance to the student to have experiences in the transitions through several stages of the modelling process.

46.3.2.1 Modelling the Temperature of Hot Water

This activity is useful to show the students how experimentation could give meaning to the DE model and help them to build an empirical deduction of a model. In the *change in temperature of an object activity*, the situation is to measure how the temperature of liquid changes (boiling water, oil or alcohol) versus time, by using the temperature sensor EasyTemp. Students are asked to establish the way in which temperature changes with regards to time by using graphical and numerical representations provided by the TI-Nspire CX CAS. In other words, through a real situation students can understand the underlying DE associated with the phenomenon (see Fig. 46.2).

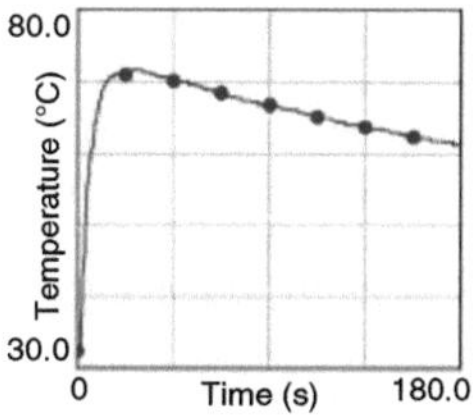

Graphical Representation (data provided by the sensor Easy Temp)
Modelling transition $a \rightarrow b \rightarrow c$

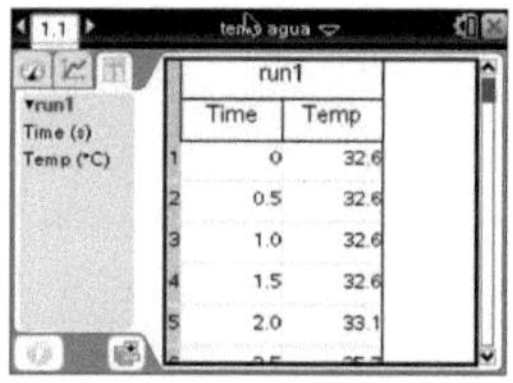

Numerical Representation (data provided by the sensor Easy Temp)

Modelling transition $a \rightarrow b \rightarrow c$

The DE is:

$$\frac{dT(t)}{dt} = k(T(t) - T_m)$$

(Newton law)
The general solution is:

$$T(t) = T_m + Ce^{kt}$$

With the "dsolve" TI command or Wolfram Alpha
Algebraic Representation
Modelling transition d

Fig. 46.2 Steps and transitions in the modelling cycle and different representations of the DE of temperature

In this activity, students have shown great interest in recognizing a theoretical model such as Newton's "law" on data they have. They have had to adjust the theoretical model to their data and are pleasantly surprised how effective their proposal may be. Students recognize in this activity the importance of working a proposed model with different data. Some issues related to why the sensor data does not behave exactly like the theoretical model data (DE) have appeared bringing up a topic for discussion in class on issues that are usually little or never addressed.

46.3.2.2 Modelling the Mixing of Water and Salt in a Tank

This activity was designed to show how simulation is useful to recreate the situation for students to set the DE associated with the phenomenon. In the study of *how salinity changes in a tank of water*, the situation is to measure the amount of salt in a tank with regards to time, using the *PhET* simulator (see Fig. 46.3). Students are asked to establish the "way in which the amount of salt in a tank of water changes in regards to time" by the initial use of an activity and the simulator. In other words, through a real situation, students can propose a DE to model the phenomenon.

To implement the modelling activity of the mixing problem with differential equations the following steps are considered: step 0, organization in teams of three students; step 1, creation of awareness to the phenomenon through a video showing the hypothesis to the problem; step 2, three-part activity begins. This step 2 is separated into Part I: *Identification of potential variables to study.* It aims to highlight the step in the modelling diagram from the "real" or pseudo-concrete (Text of the exercise) situation to a graph or qualitative representation of the evolution of the amount of interest. Part II: *Identification of the Mix Mathematical Model in its analytical representation.* Through the writing of the activity, the students are guided in the theoretical explanation of the tank model (chart to DE). What is mainly discussed is the way in which the concentration changes over time according to the law of conservation of matter. In this step, based on the modelling

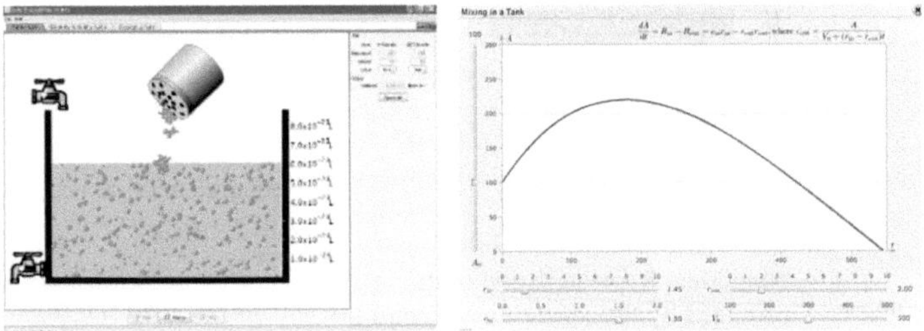

The DE is:

$$\frac{dx}{dt} = Ce * fe - \left(\frac{x}{V(t)}\right) * fs$$

With the "dsolve" command of the IT or Wolfram Alpha

Analytical Representation

Example of the Physical (Virtual) Model with PhET
Modelling transition a → b → c

Graphical Representation
Modelling transition a → b → c

Modelling transition d

Fig. 46.3 Steps and transitions in the modeling cycle and different representations of the DE of a mixing context

diagram, the student transitions from the Pseudo Concrete Model (statement) to the physical model (tank diagram) and / or the "virtual physical" model (simulator) to the mathematical model as a first order DE. Afterwards, students are asked to analyse the type of DE (ordinary) and its order (first order); to solve it by previously identifying the method to use (linear model); and to finally sketch the graph. Part III corresponds to the part in the modelling process to *establish a mathematical model in its analytical representation (DE) and / or graphical representation* (curve of the solution function). It is important to note that this part of the DE resolution is an essential aspect of the course; however, this teaching proposal is only one part of the modelling process. This part will eventually be facilitated through a technological resource (see Fig. 46.3). The final step 3 is the learning affirmation done in three ways: assigning for homework to solve some problematic mixing situations, in similar contexts and/or slightly different from those treated in class; or giving a monthly examination or the final examination containing a mixing problem.

This activity is important in the DE course because we explain to the student the situation and they need to build a mathematical model (DE). It is a challenging situation for them but due to the design of the activity into three parts they must think what variables are most interesting to watch (Part I, little generally requested) and propose a model taking into account a mass balance (Part II). It is an intuitive context for them that allows them to at least try to build a mathematical model in the classroom. The third part of the class was successfully achieved but the rest presented major difficulties in a central aspect of the model: the concentration of salt $x(t)/V(t)$.

46.3.2.3 Modelling the Change of Charge in an RC Circuit

In this activity, we propose an experimentation and simulation to recreate a situation for students to become familiar with the electrical phenomenon and to verify the DE associated with the phenomenon. We are interested in measuring how

Fig. 46.4 Physical model (**a**), lab practice (**b**) and simulation in a DE course (**c**)

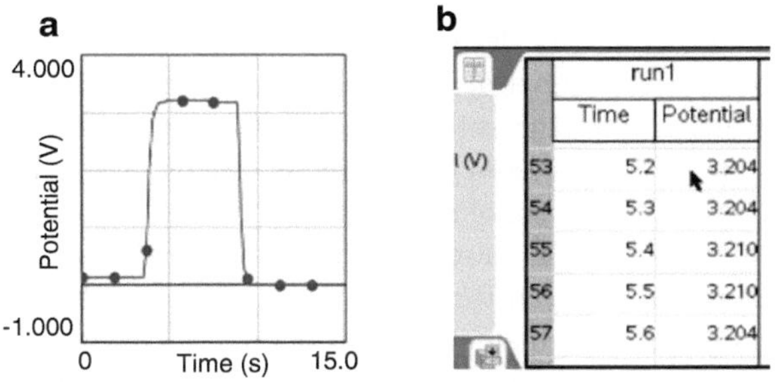

Graphical representation

Numerical Representation

c

The DE is:

$$R\frac{dq(t)}{dt}+\frac{1}{C}q(t)=E_0$$

The general solution is:

$$q(t)=\frac{E_0}{R}-\frac{E_0}{R}e^{\frac{-t}{RC}}$$

Analytic Representation

Fig. 46.5 Different representations of the DE in the setting of an RC circuit

the charge *q(t)* changes in a capacitor *C* over time, using the voltage sensor. Students are asked to establish “the way in which the charge in Capacitor *C* changes over time *t*” by using graphical and numerical representations using a graphing calculator (see Fig. 46.4). Then students are asked to model an RL circuit, but due to the lack of material in the laboratory (in particular, coil *L*), the *PhET* simulation is used to accomplish this assignment (see Fig. 46.5). It is a good moment to introduce a variable, a voltage source with an AC (sinusoidal). Therefore, the way to solve it is analyzed in terms of steady and transient state.

In summary, the implementation of this class activity was designed as follows: (1) Discussion in groups of previous knowledge regarding the DE analytical model that models the change of the load of a capacitor in an RC circuit. (2) Assembly of an electrical RC circuit and measurement of the load capacitor through a voltage sensor. The material used for the activity consisted of a circuit (a bulb, a capacitor, a set of four batteries and connectors) for each computer, a TI sensor of voltage, a TI browser and the worksheet for the practice. (3) Analysis of the graph generated by the sensor and recognition of its shape; analysis of the behaviour with regards to the modelled real phenomenon. (4) Solving analytically the DE of a RC circuit with constant voltage input with the linear DE method previously studied in class by three-member teams. (5) Solving a DE of an RC circuit of variable voltage input analytically and individually.

This activity is designed to be useful in bringing students to an unfamiliar context that is intuitive for a few of them: the electrical phenomenon. Some results observed were that students benefited by understanding more what is mathematically modelled and interpreted over the sensor results and the model in terms of the response of the RC circuit to a specified source of voltage (DC/AC). The use of simulators was very useful to simulate RL and RC circuits in the classroom.

46.4 Conclusion

This chapter aimed at presenting the activity design in a DE course, particularly showing the theoretical foundations that support the design and eventual implementation of activities in the classroom. The role of Student-Centred Learning Environments (SCLE) such as the ACE classroom (ACE are the initials in Spanish for SCLE, see Zavala et al. 2013) and a technological classroom as well as the proper use of appropriate technology used in these activities (calculator, sensors, software, simulators) are discussed in-depth in Rodríguez and Quiroz (2015, this volume). It is important to highlight the role that technology plays in a Differential Equations course geared at future engineers. The emphasis given to DE as models to represent different phenomena in nature, and further emphasis on showing the students' modelling stages (where the object DE is one of many ways to represent a phenomenon) shows a large potential to use technology in the classroom not only to simplify the algorithmic part but also to solve through *dsolve*, view graphic solutions, inspect behaviour in data tables, analyse the meaning of parameters in the solution graphs and above all show the experiment and data collection prior to the establishment of a DE. Given the portability of current technological devices (e.g., calculator and sensors) partners can take them to a classroom with the infrastructure and minimum equipment necessary to implement this in class. The idea is not to teach another discipline in mathematics classes, but to encourage future engineers to give context to first- or second-order DEs and to re-signify the parameters, variables or behaviours seen in experiments in the proposed model.

Finally, technology use aims to highlight the richness of the new course to show students the range of possible multiple representations of the same mathematical object (DE). It is expected that at the end of this course, the future engineers will have a better idea of the meaning of DE in both models to represent various phenomena and solutions to be the same over a catalogue of analytical methods in solving DE. It is important to note the enormous access to simulators with which students relate real phenomena and key concepts of DE. Finally, this chapter concludes by emphasizing the importance of the teaching of mathematics in general and DE through modelling, especially the work of designing activities based on the stages of the process and the contribution of each to leading students to move between key stages and modelling.

References

Accreditation Board for Engineering and Technology [ABET]. (2011). *Criteria for accrediting engineering technology programs.* http://abet.ams.abet.org/uploadedFiles/Accreditation/Accreditation_Process/Accreditation_Documents/Current/abet-tac-criteria-2011-2012.pdf

Alsina, C. (2007). Teaching applications and modelling at tertiary level. In W. Blum, P. L. Galbraith, H.-W. Henn, & M. Niss (Eds.), *Modelling and applications in mathematics education: The 14th ICMI study* (pp. 469–474). New York: Springer.

Arslan, S., Chaachoua, H., & Laborde, C. (2004). Reflections on the teaching of differential equations. What effects of the teaching of algebraic dominance? *En Memorias del X Congreso Internacional de Matemática Educativa (ICME XI).* Copenhagen: ICME.

Artigue, M. (1996). Teaching and learning elementary analysis. *Memorias del VIII Congreso Internacional de Matemática Educativa (ICME VII), Selected Lectures,* (pp. 15–29). Sevilla: ICME.

Beneitone, P., Esquetini, C., González, J., Maletá, M., Siufi, G., & Wagenaar, R. (2007). *Reflexiones y perspectivas de la Educación Superior en América Latina, Informe Final Proyecto Tuning América Latina 2004–2007.* Madrid: Universidad de Deusto.

Blanchard, P. (1994). Teaching differential equations with a dynamical systems viewpoint. *The College Mathematics Journal, 25,* 385–393.

Blomhϕj, M., & Carreira, S. (2009). *Mathematical applications and modelling in the teaching and learning of mathematics* (pp. 47–60). Roskilde: Roskilde University.

Blum, W., & Niss, M. (1991). Applied mathematical problem solving, modeling, applications, and links to other subjects – State, trends and issues in mathematics instruction. *Educational Studies in Mathematics, 22*(1), 37–68.

Kaiser, G., & Sriraman, B. (2006). A global survey of international perspectives on modelling in mathematics education. *ZDM, 38*(3), 302–310. doi:10.1007/BF02652813.

Kallaher, J. M. (1999). Revolutions in Differential Equations, exploring ODES with modern technology. *Mathematical Association of America Notes, No. 50.* Washington, DC: MAA.

Lombardo, D. H., & Jacobini, O. R. (2009). Mathematical modelling: From classroom to the real world. In M. Blomhϕj & S. Carreira (Eds.), *Mathematical applications and modelling in the teaching and learning of mathematics* (pp. 35–46). Roskilde: Roskilde University.

Niss, M., Blum, W., & Galbraith, P. (2007). Introduction. In W. Blum, P. L. Galbraith, W.-H. Henn, & M. Niss (Eds.), *Applications and modelling in mathematics education: ICMI study 14* (pp. 3–32). New York: Springer.

Organization for Economic Cooperation and Development [OECD]. (2003). *Informe PISA 2003.* (OECD, Ed.). *Learning.* Paris: OECD Publishing.

Rasmussen, C., & Whitehead, K. (2003). *Learning and teaching ordinary differential equations.* Washington, DC: The Mathematical Association of America.

Rodríguez, R. (2009). Differential equations as a tool for mathematical modelling in physics and mathematics courses. A study of textbooks and modelling processes of high-school senior students. In M. Blomhøj & S. Carreira (Eds.), *Mathematical applications and modelling in the teaching and learning of mathematics* (pp. 19–34). Roskilde: Roskilde University.

Rodríguez, R. (2010). Aprendizaje y Enseñanza de la Modelación: el caso de las ecuaciones diferenciales. *Revista Latinoamericana de Matemática Educativa, 13*(4-I), 191–210.

Rodríguez, R. & Quiroz, S. (2015). Developing modeling competencies through the use of technology. In G. Stillman, W. Blum, & M.S. Biembengut (Eds.), *Mathematical modelling in education research and practice: Cultural, social and cognitive influences* (pp. 443–452). Cham: Springer.

Salinas, P., & Alanís, J. A. (2009). Hacia un paradigma en la enseñanza del cálculo dentro de una institución educativa. *Revista Latinoamericana de Matemática Educativa, 12*(3), 355–382.

Zavala, G., Domínguez, A., & Rodríguez, R. (2013). ACE: Innovative educational model for teaching physics and mathematics to engineering students. *Proceedings of the 120th annual conference of the American society of engineering education (ASEE).* Atlanta: ASEE. http://www.asee.org/public/conferences/20/papers/7988/view

Chapter 47
Contributions of Mathematical Modelling in Education of Youth and Adults

Jonson Ney Dias da Silva, Taise Sousa Santana, and Carlos Henrique Carneiro

Abstract This chapter discusses a case study in Education of Youth and Adults that attempts to capture critical pedagogical practices in a multi-disciplinary approach to improving student learning. The aim was to show possible contributions of a mathematical modelling environment for using mathematics to understand phenomena in other knowledge areas, including students' prior knowledge and everyday experiences. A class of youth and adults who were part of a project of Youth and Adult Education in a state college in Brazil participated. Student involvement in the activity as well as their difficulties in interpreting the information mathematically in order to solve the questions posed were observed. The teacher acted as a mediator between the real situation and the mathematical objects used in mathematising to facilitate students refining their mathematisations.

47.1 Introduction

The Education of Youth and Adults (EYA) in Brazil began in the colonial period when the Jesuits Society of Jesus came to Brazil in order to teach the indigenous people to read and write, because only then would the Christian faith spread among them. According to Gadotti and Romão (2005), the EYA was instituted only after the completion of the First International Conference on Adult Education in 1949 in Denmark, when it was possible to give another direction to this type of education,

J.N.D. da Silva (✉) • T.S. Santana
Department of Exact and Technological Sciences State, State University of Southwest Bahia, Road of Bem Querer, km 4, PO Box 95, Vitória da Conquista, Bahia 45083-900, Brazil
e-mail: jonsonney@yahoo.com.br; taisess@yahoo.com.br

C.H. Carneiro
Department of Exact Sciences, State University of Feira de Santana, Avenue Transnordestina, S/N, Novo Horizonte, Feira de Santana, Bahia 44016-900, Brazil
e-mail: chcwillis@hotmail.com

G.A. Stillman et al. (eds.), *Mathematical Modelling in Education Research and Practice*, International Perspectives on the Teaching and Learning of Mathematical Modelling, DOI 10.1007/978-3-319-18272-8_47

making it a kind of moral education. Since then, every decade, policy makers and education professionals have sought public policies and better methodologies for teaching contextualisation of the reality of life for these students in order to enable more meaningful ways of teaching, serving not only the purpose of raising the level of education of youth and adults, but also human development and citizenship for many of them who for a variety of social and educational reasons had no access to education when they were at school age (Brasil Ministério Da Educação, Secretaria De Educação Fundamental 2002).

For Arroyo (2006), the richness of the EYA student lies in the fact that they have different backgrounds, ages, professional experiences, educational and learning pace. These students are typically people who work, have families that depend on their work and whose ethical and moral values come from their experience, from their environment and cultural reality in which they are entered and none of these should be accounted for in the educational process.

Thus, in an attempt to develop this EYA proposal, the Government of the State of Bahia in Brazil, inserted this kind of program in schools, in order to make learning something as close as possible to the contextual realities of these students by using everyday situations to address and discuss the content of the school subjects, not valuing only the content but also the skills necessary for their formation, in a qualitative process of skills and competencies evaluation by these students. To meet these specific needs it is proposed to overcome the multidisciplinary paradigm for developing students' learning not by discipline (i.e., school subject) but by knowledge area (languages, the study of society and nature, human sciences and natural sciences and mathematics) to contribute as an instrument capable of accounting for or explaining the issues of social practice. Thus, the use of non-mathematical contexts to teach mathematics can enable everyday situations to be problematised in the classroom.

In the literature different ways are presented to organize the mathematics classes in which knowledge from other areas such as chemistry, biology, physics, or situations of everyday life, are moved into the contexts of mathematics classes (Barbosa et al. 2007; Blum et al. 2007). In particular, the use of modelling in the context of EYA is little discussed in Brazil and internationally appears as a gap in the literature, including the works of ICTMA.

This approach is called, in mathematics education, mathematical modelling (Barbosa et al. 2007; Bassanezi 2002; Blum et al. 2007; Carrejo and Marshall 2007). These authors argue that the use of modelling[1] in mathematics classes enables students to understand the problems of daily life and from different areas of knowledge. It also encourages students to investigate how mathematics is used in society and science and how mathematics is used in making decisions (Skovsmose 2007). According to Bassanezi (2002), modelling was conceived as drawing out mathematical models, being initially a practice performed only in professional

[1] To avoid repeating mathematical modelling, whenever we use the term modelling we are referring to mathematical modelling.

areas and/or applied mathematics, for example, the mathematical models of population growth, economic growth and forecasting that can provide the number of those infected during the outbreak of an epidemic.

In the educational context, according to Barbosa (2009), the modelling has to be performed with the purpose of learning mathematics. This concern was to facilitate mathematical learning from students' everyday themes. Furthermore, the modelling environment allows the teacher to develop activities that can support students in understanding how mathematics is used in social practices (Barbosa 2003). From this perspective – called socio-critical by Barbosa (2003) – modelling has an emphasis on analysis of the role of mathematics in social practices and how to create opportunities for student performance in social debates in society.

In this perspective, the present work aims to contribute to the debate on the inclusion of the modelling environment in the context of EYA both in Brazil and elsewhere (see, e.g., Mouwitz 2013; Noss and Hoyles 2010), analysing the possible contribution of this environment to use of mathematics to study and understand phenomena in other disciplines, whilst making room for the inclusion of extracurricular knowledge of students. For this purpose, we developed a modelling activity with a class composed of young adults who were part of the Time Formative – Axis VII of the State College Agostinho Froes da Mota, in the city of Feira de Santana – Bahia – Brazil. This type of teaching for EYA appreciates mathematics through projects addressing cross-cutting issues[2] proposed by the National Curriculum as well as everyday situations of the student. It is thus a multi-disciplinary approach to student learning. In the following sections the context and methodology used to develop the study will be presented.

47.2 Methodology and Participants

The present study falls within what might be called a qualitative case study because, according to Bogdan and Biklen (1992), qualitative research is interested in investigating such problems as they manifest themselves in activities, procedures and daily interactions, using the naturalistic environment as the direct source of data. The class chosen for the development of the activity consisted of 40 students, aged 18–30 years old, coming from the rural and urban zone of the city of Feira de Santana and two students with special needs, who had an accompanying specialist teacher of Braille.

[2] According to Ministério da Educação (MEC – Education Department), cross-cutting issues "are issues that are facing the understanding and the construction of social reality and the rights and responsibilities related to personal and collective life and the affirmation of the principle of political participation". Based on this idea, the ME set some cross-cutting themes that address values related to citizenship: Ethics, Health, Environment, Sexual Orientation, Work and Consumption and Cultural Plurality.

This activity was undertaken in four lessons lasting 50 min each, which were distributed equally over 2 days. The first day was dedicated to the invitation of the teacher (i.e., scene setting) and the organization and development of activities by the groups. The following day, the activity ended with sharing and discussion of the results developed by students.

A group of students was selected, by drawing lots, for observation which was carried out by the authors in the naturalistic classroom environment where the students' utterances were produced. This group consisted of four students who had no previous contact with the particular living or discussion on the subject. In order to preserve identity students will be referred to as Student 1, 2, 3 and 4. The observations took place during all classes and were recorded through filming. Copies of all written materials developed by students during the activity were collected.

In this case study we chose to use non-structured observation, a method in which, according to Adler and Adler (1994), the behaviours to be observed are not predetermined. They are observed and reported on how they occur, in order to describe and understand what is occurring in a given situation. Given the objectives of the research, we are interested in the verbal interactions occurring among students and between them and the teacher during the development of a modelling activity. The following section describes the moments of activity development, as well as the analysis of the observed data.

47.3 Activity: Study of Recycling of Aluminium Cans

On the first day, the class was arranged in a semicircle and the classes were divided into two periods, the *invitation* and *group activity*. The invitation began with the presentation of a short video that featured the theme of recycling in Brazil, as the country occupies the top position in the world ranking for recycling. Subsequently, students were asked with the aid of slides: What is recycling? Have you ever recycled some material? What influence does this process have in your life? What products can be recycled? These questions prompted an initial discussion starting on the topic, which allowed students to participate by presenting their lived experiences about the issue at hand. The importance of working with non-mathematical topics, which allows students to express how they see things and understand certain concepts was noted, as some students presented themselves as professional scavengers.

At that time, the written handout for the activity which contained information on the topic was distributed. The teacher asked students to read aloud to the class monitor. Students discussed the information contained in the text mediated by the teacher who encouraged students to debate on the topic.

D'Ambrosio (1986) states that modelling can stimulate the ability the student has to analyse a situation of global reality in which it appears, and, from the knowledge that they have at their disposal, extract the necessary tools to understand and act on this situation. Modelling also enables the students to learn about the possible role of mathematics in society, when they become familiar with the analysis of real situations mathematically sustained, and to realize that mathematical arguments are used, for example, to give support to political decisions (Almeida and Dias 2004; Barbosa 2003; Jacobini and Wodewotzki 2006).

Continuing on, the teacher asked students to form groups in order to solve the proposed problem. Then, there was a reading of each question slowly discussing questions that arose. The first question asked of the students was to try to find the amount of cans produced using the maximum possible recycling process from 1 kg of cans. Later, they were asked: What is the cost benefit of this whole process for the recycling can collector? Students, a priori, armed with the information contained in the text and with the support of the discussions developed earlier, began talking about what would be considered for the development of this issue.

Thus, during the discussion it was realised that the groups were developing a strategy that did not represent the given situation. The question asked of students was to find out what would be the value of the loss of material during the recycling process. The text of the activity presented information that 1 kg cans represent 75 cans and that after the first recycling process 62 new cans are produced. With these data the students launched into a subtraction finding the value 13, which represents the amount of cans lost in the process, as can be seen in the discussions of the students below:

Student 1: The recycling process is per pound.
Student 2: So this [...]
Student 3: One kilogram.
Student 2: That's not what he wants? That [is] after recycling, how can [we do that]?
Student 1: Oh! 62.
Student 2: That's it.
Student 4: And 75? Are 75?
Student 1: Look 75 every 75 units you produce 62 cans. Got it? Student 4?
Student 2: So, ask him to come here to explain to us. I think he means that [...]
Student 1: How many times [...] of 62 [...] Then we will reduce how many? From 75 [...]
Student 4: In this case, it's 13, out of 75 to 62 [...]
Student 1: 13, from 62 to 75 will decrease of 13, by 13, won't it? [Doing the calculations on the notebook.]
Student 1: We made the following calculation here, as it is shown here in the text that 75 units of cans with 75 recycled cans, 62 cans are done. We made the following calculation: gave 75 minus 62 is 13. So this 13 is what we were decreasing from each [process] of can recycling ... here was

> a process according to the second, third, fourth, and fifth processes of recycling, just 10 cans left so these 10 cans can still be reused so in [this] case, we have six processes of recycling.

In this excerpt, first the students try to understand how the amount of cans decays after the recycling process. They concluded that this decay occurs in a linear manner, that is, every recycling 13 cans diminish. Considering this strategy, they began to sequentially subtract the value of 13 each result found, disregarding that the loss would not be represented by an integer, for a variable value represented by a portion of the whole, which would change according to the number of cans to be recycled. Thus, this loss could be better represented by percentage, that is, a percentage of loss or recovery of material.

Following Silva and Barbosa (2011), one can say that the strategy used by the students in the calculation of the loss of material of the cans in the recycling process, shows that somehow the "real" problem situations brought to the students seem to be approached by being structured in terms of objects of school mathematics. In other words, phenomena external to school mathematics seem to be converted into objects of school mathematics. For the authors, this situation characterises the formatting power of mathematics, in other words, students use mathematical objects of the school context to create representations of some reality (Skovsmose 1994). These representations are structured by mathematical objects used. In the activity analysed, for example, the calculation of the amount of cans lost in the recycling process was directly related to subtraction, considering that the students were working with the idea of loss. According to Skovsmose (1994), the formatting power of mathematics is also revealed in the process of representing a certain phenomenon.

To help students understand this situation, the teacher decided to use a sheet of paper to explain that the loss was given by the recycling ratio, which was determined from the initial value that was being recycled. At this point in time the teacher tore off a part of the sheet that would represent a loss in the recycling process. With the part remaining, he asked the students, if the piece was recycled again, would the loss of the whole be the same. Students said "No", because the whole had diminished in size, then the loss would be smaller than the previous one. After this explanation, groups realized that the loss of recycling material was proportional to the amount of recycled material and that this quantity could be represented by a percentage. Continuing this activity, groups were able to find the loss value in the process, thus completing the partial resolution of the problem posed.

To bring students to this realisation, the teacher adopted a style of open communication (Barbosa 2011), since he formulated an approach that challenged and encouraged the students to analyse the strategy adopted by them for resolving the situation. For Oliveira and Campos (2007), conducting modelling activity in practice demands certain teacher strategies and actions to be adopted for facilitating this environment. This moment shows how the teacher can play the role of mediator between the mathematical objects and their application in mathematising, which strengthened the confidence of the students.

After finding the result of the maximum amount of process 1 kg of aluminium cans could suffer, the students worked on solving the other question posed to them which aimed to find the cost value benefit to the collector of the cans. With calculations they found a value which was too low and not expected regarding this cost benefit, which generated a great discussion between all members of the groups, who considered it the value received by the collector. Also at the time, the experiences and knowledge of students about the situation with reference to reality was evident in their exchanges as can be seen in the following:

Student 2: And one thing is that the tin is varied [in] value, [you] have the times that the cost is higher and have times cost is lower.

Student 3: And is it?

Student 2: Yeah, and in the can of iron, of iron a kilo is five cents.

Student 3: And why does it cost [...] this costs R$3.50?

Student 2: There is a party that can increase R$0.10. Now that one is the recycling process that lasts longer, of aluminium, it costs [...] R$2.50. so in the parties [time] it increases.

Student 2: Hey teacher, but the price isn't fixed because the can, there is a kind of can that at parties time costs R$3.50. When it's not a period of parties it costs from R$2.00 to R$2.50.

In this excerpt, student 2 reported how the trade of purchasing and selling cans fluctuates as there are price variations and weight changes with metals, such knowledge comes from her everyday life experiences. This was a time of many discussions, where students wanted to express their opinions on the topic. Thus, it was necessary to ask these students to continue the activity and discussion of these results at the presentation time.

It is noted at this point that the social system offered students a symbolic representation of a certain reality, which allowed for the interpretation and organization of data gathered from the real world experience (Lerman 2001). This reinforces the point of Araújo and Barbosa (2005) that upon entering the school the students consider the parameters of school practice, but school fails to consider their other experiences.

At the end of the first meeting students were asked to organize group results for presenting at the next meeting. At this point, we wanted to know how those presentations would be planned and how we could organise and enable the participation of all groups involved.

For the classes on the second day, the seating in the room was organised in a semicircle in an attempt to create opportunities for the participation of all. Initially, the teacher took up the theme of the work with the presentation of the first group. The solution for this group of the first question generated some discussion, since the group of students chose to solve it by finding the percentage of utilization, whilst others used the percentage representing the loss. This provided an opportunity for the groups to discuss their findings on the first question. At the end of the presentation the teacher asked which of these solutions could be considered a recovery or a loss. The teacher then showed the students both solutions gave the same end result.

In the second presentation, the group discussed the calculations realised to find the value received by the collector. This generated much discussion in the room as the students judged the result to be a very low value. The students mentioned reports of people who worked as scavengers, and the strategies they used to change the weight of the cans, which favoured an increase in the amount received. A student reported that he puts small stones inside the dented can to change the amount of weight during weighing.

47.4 Discussion and Final Considerations

During all stages of the development of the activity, full student involvement was evident, which may have been due to the topic being part of students' interests and part of their everyday life as found by others (e.g., Yoshimura and Yanagimoto 2013). On the other hand, the difficulty of students interpreting the information contained in the activity was striking. In the development of the activity it was realised that students could not understand how to interpret the situation in mathematical terms in order to solve the questions proposed in the activity. This finding has been reflected in many contexts with beginning modellers and is an exemplification of the paradox raised by Niss (2010, p. 57), "How can students learn to anticipate putting mathematical knowledge to work in modeling before they have learnt modeling?"

This difficulty in translating the problem situation into mathematical terms is, according to Silva and Barbosa (2011) and others (e.g., Stillman et al. 2010), because it is a far more complex modelling process and because of the difficulties presented by the students in the interpretation and explanation of the data. Silva and Barbosa, in keeping with the Niss (2010) paradox, believe that this is due to lack of experience of the students, who often cannot distinguish which data should be used and how to use them for the development of the resolution of the situation. In this case study the teacher adopted the pedagogical practice of taking on the role of mediator between the real world and mathematical objects used by students in representations of elements of it to encourage students in refining their mathematising.

In terms of satisfying the objective of conducting this research, namely to investigate the possible contribution of a modelling environment to use of mathematics by youth and adults to study and understand phenomena in other disciplines, as well as making room for the inclusion of extra school knowledge for students, we note that the modelling environment allowed the teacher to develop activities that allowed students to develop an understanding of how mathematics is used in social practices (Barbosa 2003). This modelling perspective—called *socio-critical* by Barbosa (2003)—places an emphasis on the analysis of the role of mathematics in social practices and how to nurture the performance of the student as a member of society in social debates. Thus, we have demonstrated through this case study that modelling activity in the context of an educational proposal for EYA enables the students to work with their life experiences that can both contribute to the process of acquiring knowledge and to providing learning. Additionally, the activity has allowed us to learn how to reorient the teacher's pedagogical practice in EYA aiming at literacy and mathematical literacy as an enabler to reduce the distance between interdisciplinary knowledge and appreciation of students' previous experiences of the particular teaching context and how they perceive and relate mathematics in society experiences with the school knowledge environment.

References

Adler, P. A., & Adler, P. (1994). Observational techniques. In N. K. Denzin & Y. S. Lincoln (Eds.), *Handbook of qualitative research* (pp. 377–392). Thousand Oaks: Sage.

Almeida, L. M. W., & Dias, M. R. (2004). Um studo sobre o uso de modelagem matemática como estratégia de ensino e aprendizagem. *Bolema: Boletim De Educação Matemática, 22*, 19–35.

Araújo, J. L., & Barbosa, J. C. (2005). Face a face com a modelagem matemática: Como os alunos interpretam essa atividade? *Bolema: Boletim De Educação Matemática, 23*, 79–95.

Arroyo, M. (2006). Educação de Jovens e Adultos: Um campo de direitos e de responsabilidade pública. In M. A. Giovanetti, N. L. Gomes, & L. Soares (Eds.), *Diálogos no educação de jovens e adultos* (pp. 19–50). Belo Horizonte: Autêntica.

Barbosa, J. C. (2003). Modelagem matemática na sala de aula. *Perspectiva, Erechim, 27*(98), 65–74.

Barbosa, J. C. (2009). *Integrando Modelagem Matemática. Nas Práticas Pedagógicas. Educação Matemática Em Revista* (pp. 17–2). Brasília, No. 26.

Barbosa, J. C., Caldeira, A. D., & Araujo, J. D. L. (Eds.). (2007). *Modelagem matemática na educação matemática Brasileira: Pesquisas e práticas educacionais*. Recife: Sbem.

Bassanezi, R. C. (2002). *Ensino-aprendizagem com modelagem matemática*. São Paulo: Contexto.

Blum, W., Galbraith, P. L., Henn, H.-W., & Niss, M. (Eds.). (2007). *Modelling and applications in mathematics education: The 14th ICMI study*. New York: Springer.

Bogdan, R. C., & Biklen, S. K. (1992). *Qualitative research in education: An introduction to theory and methods*. Boston: Allyn & Bacon.

Brasil Ministério Da Educação, Secretaria De Educação Fundamental. (2002). *Proposta curricular para a educação de jovens e adultos: Segundo segmento do ensino fundamental – 5ª A 8ª Sé ries*. Brasília: MEC/SEF.

Carrejo, D., & Marshall, J. (2007). What is mathematical modelling? Exploring prospective teachers' use of experiments to connect mathematics to the study of motion. *Mathematics Education Research Journal, 19*(1), 45–76.

D'Ambrosio, U. (1986). Matemática e Desenvolvimento. In *Da Realidade A Ação: Reflexões Sobre Educação E Matemática* (5th ed., pp. 13–25). São Paulo: Summus.

Gadotti, M., & Romão, J. E. (Eds.). (2005). *Educação De Jovens E Adultos: Teoria, Prática. E Proposta* (7th ed.). São Paulo: Cortez: Instituto Paulo Freire.

Jacobini, O. R., & Wodewotzki, M. L. (2006). Uma reflexão sobre a modelagem matemática no contexto da educação matemática crítica. *Bolema: Boletim De Educação Matemática, 25*, 71–88.

Lerman, S. (2001). Cultural, discursive psychology: A sociocultural approach to studying the teaching and learning of mathematics. *Educational Studies in Mathematics, 46*(1–3), 87–113.

Mouwitz, L. (2013). Rationality of practice and mathematical modelling – On connections, conflicts, and codifications. In G. A. Stillman, G. Kaiser, W. Blum, & J. P. Brown (Eds.), *Teaching mathematical modelling: Connecting to research and practice* (pp. 563–572). Dordrecht: Springer.

Niss, M. (2010). Modeling a crucial aspect of students' mathematical modelling. In R. Lesh, P. L. Galbraith, C. R. Haines, & A. Hurford (Eds.), *Modelling students' modelling competencies* (pp. 43–59). New York: Springer.

Noss, R., & Hoyles, C. (2010). Modeling to address techno-mathematical literacies in work. In R. Lesh, P. L. Galbraith, C. R. Haines, & A. Hurford (Eds.), *Modeling students' mathematical modeling competencies* (pp. 75–86). New York: Springer.

Oliveira, A. M. P., & Campos, I. S. (2007). As estratégias do professor a partir do "Convite Inicial" nas atividades de modelagem matemática. *Fifth Conferência Nacional Sobre Modelagem Na Educação Matemática*, Ouro Preto (Vol. 1, pp. 239–252) [CD ROM]. Ouro Preto: Ufop/Ufmg.

Silva, J. N. D., & Barbosa, J. C. (2011). Modelagem matemática: As discussões técnicas e as experiências prévias de um grupo de alunos. *Bolema: Boletim De Educação Matemática, 24*, 197–218.

Skovsmose, O. (1994). *Towards a philosophy of critical mathematics education*. Dordrecht: Kluwer.

Skovsmose, O. (2007). *Educação crítica: Incerteza, matemática, responsabilidade. Tradução de Maria Aparecida Viggiani Bicudo*. São Paulo: Cortez.

Stillman, G., Brown, J., & Galbraith, P. (2010). Identifying challenges within transition phases of mathematical modeling activities at year 9. In G. Stillman, J. Brown, & P. Galbraith (Eds.), *Modeling students' mathematical modeling competencies* (pp. 385–398). New York: Springer.

Yoshimura, N., & Yanagimoto, A. (2013). Mathematical modelling of a social problem: Pension tax issues. In G. A. Stillman, G. Kaiser, W. Blum, & J. P. Brown (Eds.), *Teaching mathematical modelling: Connecting to research and practice* (pp. 563–572). Dordrecht: Springer.

Chapter 48
Pre-service Mathematics Teachers' Experiences in Modelling Projects from a Socio-critical Modelling Perspective

Mónica E. Villarreal, Cristina B. Esteley, and Silvina Smith

Abstract Partial findings from a research project aimed at characterising experiences in mathematical modelling (MM) projects carried out by pre-service mathematics teachers are reported. The non-mathematical themes selected and the mathematical content used in designing and developing free MM projects are reported. Aspects of the socio-critical modelling perspective present in such projects are also analysed. This analysis revealed that: (a) the selected themes could be categorized as: socio-economic, ecological, personally relevant, didactical and mathematically focused; (b) the mathematical content involved in the projects were associated with statistics, probability, and analysis. The study of a single project about trash and recyclable collection reveals characteristics of socio critical modelling perspective, difficulties of the MM process, and educational reflections.

48.1 Mathematical Modelling and Teacher Education

The incorporation of applications of mathematical models into real world problems (in the sense of Blum et al. 2003) or the development of modelling activities in mathematics classes at different educational levels is a trend that has spread in recent decades in the international context (Kaiser et al. 2013; Stillman et al. 2013b).

M.E. Villarreal (✉)
National Council of Scientific and Technical Research (CONICET) – Faculty of Mathematics, Astronomy and Physics – University of Córdoba, Medina Allende s/n, Ciudad Universitaria, Córdoba 5000, Argentina

Faculty of Mathematics, Astronomy and Physics – University of Córdoba, Medina Allende s/n, Ciudad Universitaria, Córdoba 5000, Argentina
e-mail: mvilla@famaf.unc.edu.ar

C.B. Esteley • S. Smith
Faculty of Mathematics, Astronomy and Physics – University of Córdoba, Medina Allende s/n, Ciudad Universitaria, Córdoba 5000, Argentina
e-mail: esteley@famaf.unc.edu.ar; smith@famaf.unc.edu.ar

G.A. Stillman et al. (eds.), *Mathematical Modelling in Education Research and Practice*, International Perspectives on the Teaching and Learning of Mathematical Modelling, DOI 10.1007/978-3-319-18272-8_48

More recently, extracurricular modelling events for students have gained popularity in some places (Stillman et al. 2013a). Although there is agreement that modelling should play an important role in mathematics education, resistance and obstacles related to MM are still present in school and university (Silveira and Caldeira 2012).

In our local context,[1] various (national or state) curriculum documents give some recommendations for working with MM at secondary school level and during mathematics teacher education. In the Curriculum Design for Secondary Education of the state of Córdoba (Ministerio de Educación del Gobierno de la Provincia de Córdoba 2011), it is suggested to consider modelling for solving problems, external and internal to mathematics and to encourage the study of limits of the mathematical model to explain a problem or phenomenon. The curriculum documents also emphasize the relationships between the real world and mathematics through modelling processes; however, most of the activities that are found in textbooks and classrooms are illustrative applications (Muller and Burkhardt 2007) to solve semi-real problems (Skovsmose 2001). Although there are some local experiences at secondary schools working with open MM projects (Villarreal et al. 2010; Villarreal and Esteley 2013), the engagement of teachers, pre-service teachers or students in active modelling is still scarce.

Many mathematics educators believe pre-service teachers should experience MM activities if we intend that modelling become an extended trend at school level (Doerr 2007; Lingefjärd 2007; Widjaja 2013). According to Doerr, "Pre-service teachers need to encounter modelling experiences that provide for a range of contexts and tools and that engage them in meta-level analyses of their modelling activity" (p. 77). Lingefjärd states that "in order for modelling to become a part of a teacher's functioning and practice, experiences provided for them in the course of their own mathematics learning should assist them in constructing an image of the teaching and learning that is enhanced by modelling" (p. 477). In agreement with these ideas, Widjaja accomplished an exploratory case study with Indonesian pre-service teachers to build awareness of MM, having in mind to provide teachers with knowledge and skills to make possible student-centred lessons.

If we turn our attention to teacher education in Argentina, we can find statements coming from recent official documents, espousing the importance and necessity of introducing MM activities during initial education of mathematics teachers. Meanwhile, the teaching of mathematics at our university for pre-service teachers, usually gives little room for active modelling. It seems that in order to turn around this situation and give account of the curriculum demands, it is necessary to review and act on teacher education.

[1] In the Argentinean school system, even though there are national curriculum orientations, the design and implementation of the curriculum are the responsibility of each state.

In agreement with the ideas of Doerr (2007) and Lingefjärd (2007), and considering our local context we decided to propose a MM scenario in which pre-service mathematics teachers have the opportunity of experiencing a complete MM process. A MM scenario is characterised by the presence of a set of spaces, situations, conditions, materials, actions and interactions that give sense to the MM process transforming it into an experience that aims to introduce into educational contexts MM as a pedagogical approach and as a mathematical activity (Esteley 2010).

In our proposal we promoted the creation of MM scenarios characterised by: (a) the open nature of the activities, due to the free choice of a real world theme to study and posing of questions, and the absence of pre-determined mathematical content to be taught; (b) the interdisciplinary nature of the work; (c) the promotion of reflections about mathematics itself, the models created, and the social role of mathematics and MM, and (d) the domain of the whole modelling process. The creation of such scenarios usually encourages and promotes the treatment of critical social aspects. Modelling activities with these characteristics, can be recognized as belonging to the *socio-critical modelling perspective* (Kaiser and Sriraman 2006), to which our work belongs. For these authors, such a perspective "emphasises the role of mathematics in society and claims the necessity to support critical thinking about the role of mathematics in society, about the role of and nature of mathematical models and the function of mathematical modelling in society" (p. 306). A main teaching objective for this perspective is the promotion of critical thinking and students' debates during MM. This particular perspective has roots in Brazil where it is closely related to the ethnomathematics movement. In addition, the socio-critical perspective is strongly related to thematic project work as developed by Skovsmose (1994, 2001) within the framework of Critical Mathematics Education. Literature within this perspective is increasing in the international mathematics education community. Some examples are Araújo (2012), Greer et al. (2007), Julie and Mudaly (2007) and Barbosa (2006) that bring discussions about learners' or teachers' engagement with the MM of social issues and how modelling could be used as a tool for critical analysis of various unjust or discriminatory situations or social problematic issues in their contexts. Our work adds to the socio-critical perspective considering the study of MM scenarios for pre-service mathematics teachers, differing from the provision of MM experiences by other teacher educators such as Widjaja (2013) by focusing on development of this perspective in the pre-service teachers.

In summary we decided to create a MM scenario for pre-service teachers and consider such experiences from an investigative point of view with the aims of (a) analyzing mathematical content that they used to create a mathematical model, the thematic issues that they selected in designing and developing MM projects, and the relations with social concerns, and (b) studying characteristics and difficulties of the socio-critical modelling perspective present in the modelling projects. Regarding this last aim, in this chapter we will restrict the analysis to one particular modelling project due to space limitations.

48.2 Methodological Approach

In order to give account of our aims, we developed a qualitative research, based on data from a larger study on the professional development of pre-service teachers in modelling scenarios. In this chapter, we focus on the analysis of 11 MM projects developed by pre-service teachers from the Faculty of Mathematics, Astronomy and Physics at the University of Córdoba (Argentina). In this institution, the graduate course lasts 4 years with 66 % of the syllabus being mathematics courses taught by mathematicians. The remainder of the course deals with educational issues, including mathematics education. During the mathematics courses, the pre-service teachers do not work with authentic MM processes. Mainly, these courses focus on pure mathematics with few applications of ready-made models to extra-mathematical contexts.

The context for our research was within a regular annual course on mathematics education, conducted by two teacher educators, always one of them being a researcher of our team. During this course, the teacher educators and pre-service teachers discussed several trends in mathematics education: problem solving, critical mathematics education, uses of technology in mathematics education, ethnomathematics and mathematical modelling. With the aim of allowing the pre-service teachers to experience a complete process of MM, we asked them to develop, in small groups, MM projects starting from a free choice of a real world theme of interest. At the end, they had to elaborate a written report and they had to present their work to the whole class. The decision for carrying out MM activities in a course of mathematics education was motivated not only by the richness and opportunities that such an environment opens to the future teachers, but also by consideration of the future curriculum demands that they would face as mathematics teachers. This experience was repeated with three cohorts of pre-service teachers, 2010–2012.

Our main data sources were the written reports produced by 11 groups of pre-service mathematics teachers belonging to cohorts 2010–2012, the videotapes of oral presentations produced in 2011 and our field notes.

In our analysis, we first focus on the 11 modelling projects considering the following dimensions: the mathematical content that the pre-service teachers used, the thematic issues, and their relations with social problems. Finally, we study in depth a modelling project developed by one group according to the socio-critical modelling perspective. We also refer to some emergent difficulties and challenges which are relevant for the future professional practice of the pre-service teachers.

48.3 Results and Discussion

In order to present our results, we first offer a general analysis of the 11 modelling projects. Later we focus on a particular project.

48.3.1 The Mathematical Modelling Projects

From a general analysis of the data, specially the written reports, there was diversity in the themes proposed by the pre-service teachers as well as on the issues that seem to motivate the selection of these themes. In order to systematize the data, the projects were first analyzed and then classified according to mathematical content used, thematic issues and their relations with social concerns. Table 48.1 displays the four main mathematical content areas that were evident in the 11 projects considered. As the table shows, the pre-service teachers addressed their problems appealing mostly to statistics, probability or analysis. Within these areas, in general, only a reduced set of tools was used. The projects, in which statistics or probability was used, were based on surveys and frequency tables. Those which appealed to analysis, limited its use almost exclusively to functions from $\mathfrak{R}$ to $\mathfrak{R}$, mostly linear or affine, although some other functions appeared: two variable affine functions in a project about water harvesting in dry areas, exponential and logarithmic functions in a project regarding investment recovery for a given business, and inverse proportionality functions when trash and recyclable collection was studied.

A second analysis allowed us to classify the projects taking into consideration the thematic issues selected by the pre-service teachers. We were able to distribute the projects into five groups, as shown in Table 48.2. These categories were socio-economic issues, ecological issues, personally relevant issues, didactical proposals and mathematically focused.

When we related the themes and problems posed by the pre-service teachers with social issues, we noticed that the projects included in the first two categories had authentic social concerns. We infer this from the nature of the issues posed and the ways the pre-service teachers analyzed these themes in their written or oral presentations. The pre-service teachers who developed the project focused on lack of water, considered such a question as a societal problem and they particularized the study in a rural area of the state of Córdoba which is characterized by serious

Table 48.1 Mathematical content in the projects

Mathematical content	Pre-service teachers' MM projects
Statistics	Household electric energy consumption
	Household water consumption
Probability	Lottery games
	Genetic transmission and human characteristics
Linear functions	Waiting time at the university dining hall
	Bottled gas supply in a countryside area
	Soy consumption
	Trash and recyclable collection
Non-linear functions	Investment recovery for a given business (exp, log)
	Trash and recyclable collection ($f(x) = c/x$)
	Water harvesting in dry areas (function of two variables)
Linear programming	Travel costs for end-of-year school trip

Table 48.2 Classification of projects according to thematic issues

Thematic issues	Pre-service teachers' MM projects
Socio-economic	The lack of water as a social issue
Ecological conscience	Household water consumption
	Household electric energy consumption
	Trash and recyclable collection
A problematic that affects them personally or as a group	Waiting time at the university dining hall
	Bottled gas supply in a countryside area
	Human genes– Interactions
	Soy consumption
Didactical: a teaching proposal	Travel costs for end-of-year school trip
Mathematics	Lottery games
	Investment recovery for a given business

problems of access to water. The constructed model was then critiqued on the basis of the cultural, economical or natural conditions of the selected region. The pre-service teachers who focused their study on issues related to ecological conscience proposed a topic that can be considered open enough as it affects all societies; however, they treated the issues from a local point of view. Whilst two projects particularized the study on water or electrical consumptions in their homes, the third one extended the issue, considering trash and recyclable collection in their city (Córdoba). This distinction conferred a greater relevance to this last project due to the variety of data as well as to the implications of the results in terms of proposals and critiques. We analyze this project in the next sub-section.

We observe that the themes selected highlight different perspectives and relationships between pre-service teachers and the contexts that surround them. Some projects are tightly related to their everyday contexts whilst others broaden their local perspectives and contexts. Some projects focus on micro aspects of society, whilst others on macro aspects. These issues not only become evident in the type of selected environments for the projects but also in the imposed conditions, the data collection and the scope of their conclusions and proposals. These differences derived from the commitment that pre-service teachers are able to establish with their own socio-environmental contexts and with the ways they understand the ideas related to the MM socio-critical perspective as treated during the mathematics education course.

48.3.2 The Project About Trash and Recyclable Collection

In this section, we concentrate on one particular case, a project related to trash and recyclable collection. Such a project illustrates the MM process followed by a group of two pre-service teachers, Irene and Rose. We analyze a piece of their project work. All figures and quotations used belong to the oral presentation that they prepared to show their work to the whole class. This presentation was videotaped.

Irene and Rose referred to the difficulties they had in selecting a theme and posing problems to solve. They mentioned all the themes that came to mind as a kind of "brain storm": petroleum reserves and types of uses, car accidents, afforestation and deforestation, transportation, music and mathematics and trash and recyclable collection, the last being the selected theme. The reasons for their final choice were twofold: they recognized that recyclable collection has a social function and that it was a local problematic of our city, Córdoba. After selecting the theme, they posed several questions: "How much recyclable trash is collected per day/week? Is it classified? What are the classes? How many people have this service? Does everyone classify the trash?" They then looked for information to answer their questions. Separate trash collection and the environmental and social aspects related to it, including the benefits of recycling were investigated. They interviewed an employee of the city government in charge of the city recycling program and visited the website for this program. Finally, the results of the 2010 census of population and housing in Córdoba were obtained.

Considering the questions they had raised, Irene and Rose were concerned with the mathematical complexity that would be involved in their models. During their oral presentation, they said that at the beginning of their project they wanted to obtain a sophisticated formula, or to apply more complicated mathematics, not just linear models. Then, after a very thorough search of the internet and discussion of many issues concerning trash, they decided that the main aim of their project would be "Modelling to raise awareness" and not modelling "to obtain a super formula". This decision is evidence of the pre-service teachers' concerns with social issues related to the trash. In this case, mathematics is subordinated to a social aim, it is a tool to understand the phenomenon and think about it.

Considering all the data and information they collected, the pair raised the following hypotheses:

- The inhabitants of Córdoba aren't aware of the amount of recyclable trash they produce.
- The trash that one person produces, each year, is more than his own weight.
- A large percentage of trash that could be recycled is thrown away with regular trash.

After establishing their first hypotheses, they considered the following data: the city of Córdoba had 1,329,604 inhabitants (2010 census); the inhabitants of Córdoba produce 1,500,000 kg of trash per day; 31 % of the population sort the trash; 28 t of trash are collected per day. Based on these data, Irene and Rose raised an additional hypothesis: *Suppose that every person produces a similar amount of recyclable trash.* With this assumption and using all the available information, Irene and Rose started to produce their linear models in the form of numerical indicators, such as that shown in Fig. 48.1a, considering the quotient between the

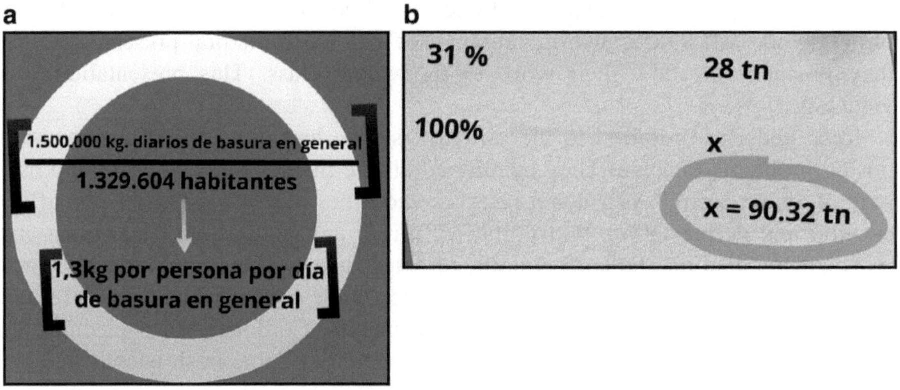

Fig. 48.1 Numerical indicators shown by Irene and Rose in their oral presentation

Fig. 48.2 The sequence of the video

total amount of trash collected per day and the total population of the city of Córdoba. In this way, they obtained as a first numerical indicator: the inhabitants of Córdoba produce 1.3 kg of trash per day per person.

Figure 48.1b shows another indicator: if 31 % of population sorts the trash and that means that we have 28 t of recyclable trash per day, then, supposing that every person produces the same amount of recyclable trash per day, the inhabitants of Córdoba produce 90.32 t of recyclable trash per day. After this calculation, Irene and Rose raised the following question: "Do you have any notion of how much is 90.32 t?" In order to answer this question, they made a video. The video presents a visual representation intended to give a notion of what percentage of the recyclable trash is being recycled in Córdoba (see Fig. 48.2).

The video starts by proposing the viewer thinks of a 90 kg man. Under this assumption, it would take 1,003 men to make 90.32 t. Then, a sequence of images of "trash men" (men wearing trash) begins to appear up to 1,003 units. After that, the following texts appear: "Only 31 % of those 90.32 t goes to recycle collection. The rest goes to regular refuse collection". Finally, the image of 1,003 trash men is divided representing the percentage of recycled men with the final text stating: "That means that only 311 'trash men' are recovered and recycled. The remaining 692 men are thrown away as waste". This video is a way of giving account of Irene and Rose's aim: modelling to raise awareness.

Irene and Rose concluded their oral presentation by saying that the theme they selected had many possibilities to be treated at the school level. They also previewed ways to continue studying the problematic of trash having in mind a new aim: "Modelling to encourage waste recycling".

During the analysis of Irene and Rose's work, we focused on two main aspects: (1) difficulties that they faced during the modelling projects and, (2) educational reflections that they made when they were immersed in a modelling project. Considering the difficulties, we could identify two. Firstly: the difficulty in selection of theme was evident in the way they presented this issue as a "Brain storm". The criteria for selection were related to social concerns. In their talk, they asserted: "We will concentrate on a local problematic with a social function". A second difficulty is that the pre-service teachers' strong mathematical background and the institutional context imposed certain restrictions regarding what mathematics to use. Irene said: "We thought that a model should be a complicated formula involving many variables. It took us some time to understand that the phenomenon could be described linearly..." From these words, we can infer that they felt that the kind of mathematical content they were using was not at the level expected for them by their mathematics teachers at the university. Regarding the second aspect, we will use their own words to note educational reflections that they made while they were immersed in a modelling project.

Rose: We were constantly moving from the role of students to the role of teacher. As students, we were doing a project but, at the same time, we wondered what may happen if we propose such a kind of project for a student (oral presentation).

Irene: I also thought that this could be perfectly done with children in a much more entertaining way than doing... those typical exercises of the paradigm of exercise that don't lead anywhere... I was interested in the project because you can see the mathematics of reality, you can interpret the mathematics in everyday reality... this is a critical reading of everyday mathematics... of what we do everyday... and there, you have the percentages (oral presentation).

The words of Irene and Rose illustrate a meta-level analysis of their modelling activity focusing on their future activity as teachers. Their experience with modelling seems to provide a view on "mathematics of reality" and its relation with a particular mathematical content (percentages). The reflection about their own mathematics let them imagine possible and future learning contexts related to everyday mathematics.

48.4 Concluding Remarks

Considering the 11 modelling projects produced by the pre-service teachers, the results have provided evidence and examples of the peculiarities of working in MM scenarios in which the pre-service teacher has the opportunity to freely select themes and pose problems. The different thematic issues considered for the MM projects offer to future teachers conditions to think about the use of mathematics in different real contexts, imagine MM scenarios for their future classes, and discuss the role of mathematics to treat social concerns.

Considering the work done by Irene and Rose, we would like to remark that, as other pre-service teachers in our study, they were able to build up a MM project developed from a socio-critical modelling perspective. Such a project promoted reflections about mathematics itself, the models created, and the social role of mathematics. In order to build up the project, Irene and Rose had to overcome two main obstacles: one of them was related to the selection of significant themes and problems and the other one was associated with their experience as students in the institutional context of the university. These obstacles were overcome when they realized that they were having a new learning experience. The new experience had to do with: (1) a new way of doing mathematics and understanding its implications for the real world; and (2) the possibility of imagining themselves as mathematics teachers.

Our research suggests that when pre-service teachers develop free MM projects in the context of a regular mathematics education course, they are able to discuss educational issues related to MM. Our MM scenario differs from other proposed for university students, such as the one studied by Caron and Bélair (2007). In their case, open-ended MM projects were carried out by the students as an assignment for a MM course in which they worked with real world themes selected by the instructor, and the focus was on the development of modelling competences among the students.

In line with the ideas of Doerr (2007) and Lingefjärd (2007), we consider that our findings bring evidence of the importance of MM experiences for future teachers, such as the work of Widjaja (2013) also brings (although in that case the pre-service teachers worked with a single assigned modelling task). The experiences we report also contribute to the discussion of social issues related to mathematics, which we consider is an important, but sometimes forgotten, dimension of mathematical education.

In summary, we can assert that experiences with MM during pre-service teacher education contribute to the final educational aim of promoting and extending MM as an important pedagogical trend and mathematical activity in school. In this sense, although it is not the focus of this chapter, we would like to emphasize that the MM experiences lived by our pre-service teachers encouraged some of them to create didactical proposals with a different kind of open modelling tasks to carry out during their supervised period of teaching practice at schools.

References

Araújo, J. (2012). Ser crítico em projetos de modelagem em uma perspectiva crítica de educação matemática. *Boletim de Educação Matemática-BOLEMA, 26*, 67–87.

Barbosa, J. (2006). Mathematical modelling in classroom: A socio-critical and discursive perspective. *ZDM – The International Journal of Mathematics Education, 38*(2), 294–301.

Blum, W., et al. (2003). ICMI Study 14: Applications and modelling in mathematics education—discussion document. *Educational Studies in Mathematics, 51*(1–2), 149–171.

Caron, F., & Bélair, J. (2007). Exploring university students' competencies in modelling. In C. Haines, P. Galbraith, W. Blum, & S. Khan (Eds.), *Mathematical modelling (ICTMA12): Education, engineering and economics* (pp. 120–129). Chichester: Horwood.

Doerr, H. M. (2007). What knowledge do teachers need for teaching mathematics through applications and modelling? In W. Blum, P. Galbraith, H. Henn, & M. Niss (Eds.), *Modelling and applications in mathematics education: The 14th ICMI study* (pp. 69–78). New York: Springer.

Esteley, C. (2010). *Desarrollo profesional en escenarios de modelización matemática: voces y sentidos*. Doctoral thesis. Córdoba: Universidad Nacional de Córdoba.

Greer, B., Verschaffel, L., & Mukhopadhyay, S. (2007). Modelling for life: Mathematics and children's experience. In W. Blum, P. Galbraith, H. Henn, & M. Niss (Eds.), *Modelling and applications in mathematics education: The 14th ICMI study* (pp. 89–98). New York: Springer.

Julie, C., & Mudaly, V. (2007). Mathematical modelling of social issues in school mathematics in South Africa. In W. Blum, P. Galbraith, H.-W. Henn, & M. Niss (Eds.), *Modelling and applications in mathematics education: The 14th ICMI study* (pp. 503–510). New York: Springer.

Kaiser, G., & Sriraman, B. (2006). A global survey of international perspectives on modelling in mathematics education. *ZDM – The International Journal on Mathematics Education, 38*(3), 302–310.

Kaiser, G., van der Kooij, H., & Wake, G. (2013). Educational interfaces between mathematics and industry at school level. In A. Damlanian, J. F. Rodrigues, & R. Sträßer (Eds.), *Educational interfaces between mathematics and industry* (pp. 263–270). Cham: Springer.

Lingefjärd, T. (2007). Modelling in teacher education. In W. Blum, P. L. Galbraith, H. Henn, & M. Niss (Eds.), *Modelling and applications in mathematics education* (pp. 475–482). New York: Springer.

Ministerio de Educación del Gobierno de la Provincia de Córdoba. (2011). *Diseño Curricular. Ciclo Básico de la Educación Secundaria 2011–2015*. Tomo 2. Retrieved October 8, 2014, from http://www.igualdadycalidadcba.gov.ar/SIPEC-CBA/publicaciones/EducacionSecundaria/LISTO%20PDF/TOMO%202%20Ciclo%20Basico%20de%20la%20Educacion%20Secundaria%20web%208-2-11.pdf

Muller, E., & Burkhardt, H. (2007). Applications and modelling for mathematics. In W. Blum, P. Galbraith, H. Henn, & M. Niss (Eds.), *Modelling and applications in mathematics education: The 14th ICMI study* (pp. 267–274). New York: Springer.

Silveira, E., & Caldeira, A. (2012). Modelagem na sala de aula: resistências e obstáculos. *Boletim de Educação Matemática-BOLEMA, 26*, 249–275.

Skovsmose, O. (1994). *Towards a philosophy of critical mathematics education*. Dordrecht: Kluwer.

Skovsmose, O. (2001). Landscapes of investigations. *ZDM – The International Journal on Mathematics Education, 33*(4), 123–132.

Stillman, G. A., Brown, J. P., & Galbraith, P. (2013a). Challenges in modelling challenges: Intents and purposes. In G. A. Stillman, G. Kaiser, W. Blum, & J. P. Brown (Eds.), *Teaching mathematical modelling: Connecting to research and practice* (pp. 217–228). Dordrecht: Springer.

Stillman, G. A., Kaiser, G., Blum, W., & Brown, J. P. (Eds.). (2013b). *Teaching mathematical modelling: Connecting to research and practice*. New York: Springer.

Villarreal, M., & Esteley, C. (2013). Escenarios de modelización y medios: acciones, actividades y diálogos. In M. Borba & A. Chiari (Eds.), *Vinte anos de GPIMEM: um mosaico de pesquisas em movimento* (pp. 273–308). São Paulo: Editora Livraria da Física.

Villarreal, M., Esteley, C., & Mina, M. (2010). Modeling empowered by information and communication technologies. *ZDM – The International Journal on Mathematics Education, 42*(3–4), 405–419.

Widjaja, W. (2013). Building awareness of mathematical modelling in teacher education: A case study in Indonesia. In G. A. Stillman, G. Kaiser, W. Blum, & J. P. Brown (Eds.), *Teaching mathematical modelling: Connecting to research and practice* (pp. 583–593). Dordrecht: Springer.

Chapter 49
A Mathematical Modelling Challenge Program for J.H.S. Students in Japan

Akira Yanagimoto, Tetsushi Kawasaki, and Noboru Yoshimura

Abstract We are in the process of carrying out modelling challenge programs for junior high school and high school students in Kyoto for 3 years. The first year of the program for junior high school students was carried out in February 2013. Two courses which consist of scientific and social content were prepared in the program. One course dealt with a problem concerning the phenomenon of a cart with a sail rolling down a slope, and the other course dealt with a problem concerning the income and expenditure of an electric power company. As a result, students showed strong interest in mathematics and became aware of the possibility of applying mathematics more freely to real-world problems.

49.1 Introduction

There are several challenge programs for mathematical modelling and application being carried out around the world, for example, the mathematical modelling challenge program in Hamburg University (Kaiser and Stender 2013), HiMCM in U.S.A., A-lympiad in the Netherlands (Andersen 1998), A B Patterson Gold Coast Modelling Challenge (Stillman et al. 2013), and so forth. Galbraith et al. (2010) took an analytical approach to the design and selection of modelling and applications tasks that involve topics found to interest adolescents in secondary school. Maaß (2010) also developed a comprehensive classification for modelling tasks.

A. Yanagimoto (✉)
Kyoto University of Education, 612-8522 1-Fukakusa Fujinomori, Fushimi, Kyoto, Japan
e-mail: ayanagi@kyokyo-u.ac.jp

T. Kawasaki
Gifu University, 501-1193 1-1 Yanagido, Gifu, Japan
e-mail: tetsushi@gifu-u.ac.jp

N. Yoshimura (✉)
Tennoji J.H.S. attached to Osaka Kyoiku University, 543-0054 4-88 Minamikawahori-cho, Tennoji-ku, Osaka, Japan
e-mail: noboru@cc.osaka-kyoiku.ac.jp

G.A. Stillman et al. (eds.), *Mathematical Modelling in Education Research and Practice*, International Perspectives on the Teaching and Learning of Mathematical Modelling, DOI 10.1007/978-3-319-18272-8_49

These give us many suggestions to consider when designing modelling tasks in challenge programs for secondary students.

Yanagimoto (2003), Yanagimoto and Yoshimura (2001, 2013), Yoshimura and Yanagimoto (2013) and Yoshimura (2015) have successfully introduced modelling tasks into the classroom for Japanese secondary students which had social or ecological content. However, the content of mathematical modelling is generally not taught in school mathematics in Japan. Most Japanese students do not know the term "mathematical modelling" or the concept. Of course there were no mathematical modelling challenge programs in Japan previously; but, the new Japanese course of study attaches great importance to making use of practical mathematics and mathematical activities. We think that mathematical modelling should be taught in regular mathematics classes in Japanese junior high schools and high schools, and that some challenge programs on mathematical modelling should be prepared. Therefore, we planned to carry out modelling challenge programs for secondary school students in Kyoto over a 3-year period. The first program for junior high school students was carried out in February 2013. Two courses, consisting of scientific and social content, were prepared in the program. The detailed content of the program and the responses of the students participating in the program will be discussed in Sects. 49.4 and 49.5.

49.2 Aims and Design of Teaching Units

Our aims in carrying out modelling challenge programs for Japanese students are to clarify what mathematical modelling problems they could concentrate and work on, and what kind of mathematical modelling problems they are interested in, that is, which do they prefer to solve, scientific problems or social problems. Furthermore, we intend to investigate the perception of the students about mathematics, the learning of mathematics and their interests and motivations through this modelling challenge. This challenge program could allow Japanese teachers to recognize mathematical modelling more widely. We intend to examine the relation of the mathematical modelling in the challenge programs and the classroom in the future.

We considered the following when we designed the teaching units:

- It is possible for Japanese ninth grade students to work on the problems using learnt mathematics.
- Students can perform multiple condition setting. The success of this technique with Japanese students was demonstrated by Yanagimoto and Yoshimura (2013).
- Students can carry out any actual experiments using graphic calculators in the scientific course. Yanagimoto and Yoshimura (2013) have shown how graphic calculators were useful enablers of students at lower levels of schooling being able to treat more advanced mathematical models.

- The problem is a contemporary serious problem in Japan in the social course. This was the motivation for the choice of task in previous work (e.g., Yoshimura and Yanagimoto 2013) which has been shown to motivate some students. It is also one of the reasons students choose particular topics during modelling challenges where the students choose the topic (Stillman et al. 2013).

For the first year of the program, the problems chosen related to a rolling cart and the finances of an electric power company. The example of a rolling cart with a sail is a classical physics problem, but it is easy for us to experiment under several conditions in the scientific course. Furthermore, it can be regarded partly as a quadratic function or a linear function. On the other hand the income and expenditure of electric power company is now a very serious problem, and many people are interested in it especially in Japan.

Galbraith et al. (2010) outlined six principles as suitability criteria for modelling problems, that is, relevance and motivation, accessibility, feasibility of approach, feasibility of outcome, validity, and didactical flexibility. The problems we chose for our two courses adopt most to these principles. Furthermore, Blum (2011) stated, "By modelling, mathematics becomes more meaningful for learners" and pointed out that "mathematical modelling is meant to help students' to better understand the world and to contribute to an adequate picture of mathematics, and so on". These teaching units are chosen under the consideration of such an aim, though it is only a short 1 day program and not sufficient to ensure the sustainability of this outcome (see Bracke and Geiger 2011).

49.3 Framework of the Challenge Program in 2013

The challenge program was carried out for the junior high school ninth grade students on Saturday February 16th 2013 at Kyoto University of Education. Eight students from two schools in Kyoto participated in the challenge program. Two teams, one consisting of five students, tackled the scientific problem, and one team, consisting of three students, tackled the social problem.

This was a 1-day challenge program and took 5 h. Students worked on the basic challenge activities for 2 h in the morning and the advanced challenge activities for 2 h in the afternoon. Two teachers supported their activities during these two program courses and some university student staff members helped with the challenge program.

The students in the scientific course carried out a physics experiment with PAScar Dynamics Systems which combines a high quality aluminium track with polycarbonate plastic carts, Graphic Calculator fx-9860II, Data Analyzer EA2000 and Motion Sensor EA-2. When a cart rolled down a 1.81° slope, the time and distance were measured and then students tried to analyze the functional relation.

On the other hand, in the social course the students used only simple calculators. In this course, students took on the challenge of mathematical modelling regarding the problem of income and expenditure of an electric power company. Students needed to use calculators to deal with the complex calculation.

49.4 Description of the Modelling Examples

49.4.1 The Scientific Problem Course

Experiment A Firstly we treated a quadratic function approximation in regards to the phenomenon of a PAScar rolling down the slope naturally. The incline of the slope is about 1.81°. Students were asked to examine the relation between time (x second) and distance (y metre) when a PAScar rolled down the slope.

In experiment I, they measured the change in time and distance when the cart rolled down for a distance of 2 m. The students drew the graph from the data and mathematized the graph. The students were required to answer the following problem:

PAScar Task A
When the cart rolls down the slope in the same condition, how do you predict the following items (i) the distance the cart rolls in 5 s (ii) the time it takes for the cart to roll 6 m?

In experiment II, after the students made a mathematical prediction, they performed a measurement experiment in which the cart rolls 6 m. They confirmed how big the error was and how accurate the prediction was. Figure 49.1 shows the graphs of experiment I and II.

Experiment B Secondly, we took up an experiment that examined the relations of time (x second) and distance (y metre) when the PAScar with a 25 cm by 25 cm square sail rolled down a slope. In experiment III, students measured the time and distance as the cart rolled down for a distance of 4 m. The students drew the graph from the data again and mathematized the graph. The students were required to answer the following problem:

PAScar Task B
When the cart rolls down under the same conditions, how do you predict the following items (i) the distance the cart rolls in 7 s (ii) the time it takes the cart to roll 8 m?

In experiment IV, after the students made a mathematical prediction, they performed a measurement experiment in which the cart rolled 8 m. In addition,

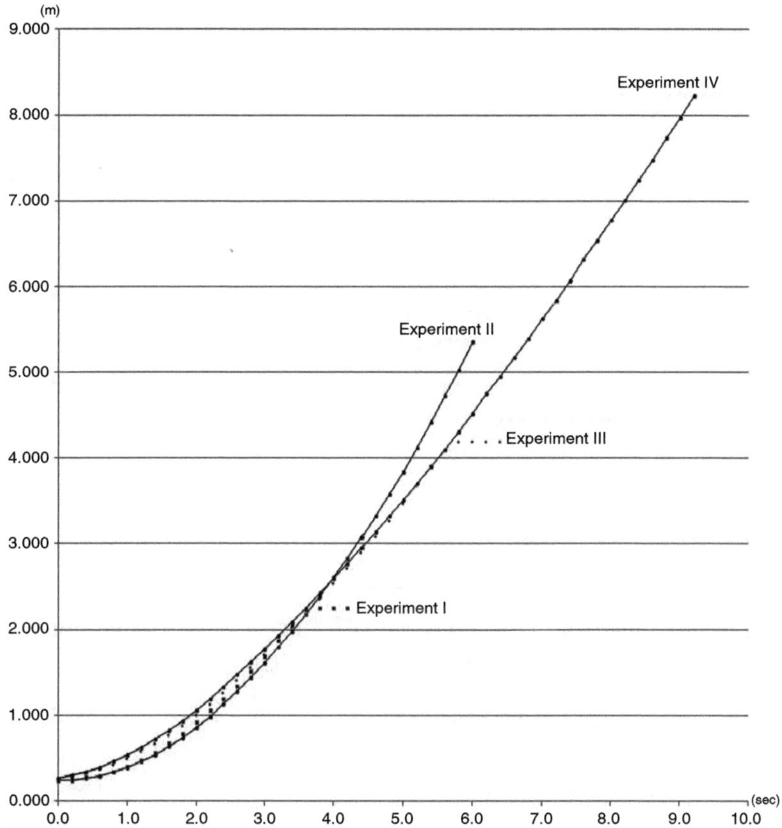

Fig. 49.1 The graph of experiments I, II, III and IV

they confirmed how big the error was and how accurate the prediction was. In this case, the teaching staff helped the students with a stopwatch and measured the time manually, because the motion sensor could not measure such a long distance. Figure 49.1 shows the graphs of experiments III and IV. Figure 49.2 shows a diagram of the experiments.

For *PAScar Task A,* one group came up with the formula $y = 0.15x^2 + 0.3$, and predicted 7.56 m for (i) and 7.16 s for (ii). Another group came up with the formula $y = 0.164x^2 + 0.25$. For *PAScar Task B,* at first, the students expected that the curve of the graph would become gentler than in experiment A, and supposed that the relation formula becomes the function $y = ax^2 + b$. However, they concluded that the graph might not show a parabola because the value of a became smaller due to air resistance as the value of x became bigger. The students did not think that the graph was divided into two parts and that the first half looked like a parabola and that the latter half was similar to a straight line, though we teachers assumed so.

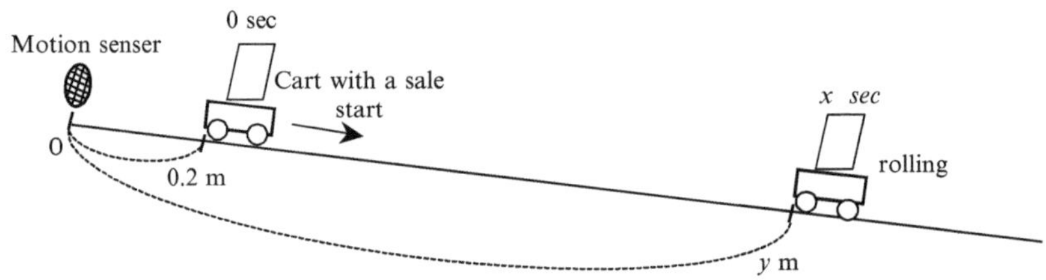

Fig. 49.2 The diagram of experiments III and IV

49.4.2 The Social Problem Course

In the second course, firstly students studied how to solve the following *Home Electricity Generation Task* concerning the home generation of electricity from sunlight for about 2 h in the morning.

Home Electricity Generation Task
In Giraud's house, the monthly average electricity consumption is 500 kWh, and the monthly electricity bill is 12,000 yen. The electricity consumption during the daytime is between 20 % and 30 % of the total. His family talked about energy saving and made a plan to attach 24 pieces of 190 W sun panels with an output of 4.56 kW this month for a cost of two million yen for equipment. In the case of the Giraud family, the subsidies from the state, the prefecture, and the city amounts to 400,000 yen. The electric power company buys the excess electricity generated during the daytime for 42 yen per 1 kWh for the time being. There is no battery charger. As for the total sum (including the cost of equipment) of the electricity bill, how do you think it will change?

1. In the following two cases, draw graphs for the electricity bill y yen after x month, and compare the two graphs:

   (a) when they do not attach the sun panels
   (b) when they attach the sun panels.

2. How many months after the installation of the solar panels will the total sum of the electricity bill become lower than before the installation?

Secondly, in the afternoon the students worked in groups on the following *Problems A, B, C* of income and expenditure of the electric power company sequentially.

Electric Power Company Finances
Real Situation: Electric power company K generates 150 billion kWh of electricity a year. 30 % of this is generated by photovoltaic power generation

(continued)

and bought by electric power company K for 42 yen per 1 kWh. 70 % is generated through thermal power generation with a fuel cost of 7.4 yen per 1 kWh. In addition, it costs 1,200 million yen a year for other expenditures (personnel expenses, cost of equipment, etc.). On the other hand, they sell 40 % of the annual generation for 20 yen/kWh to general families, and sell the remaining 60 % for 12 yen/kWh to businesses.

Problem A: When there is a change in income:

1. How much is the annual income and expenditure settlement of accounts?
2. If we change only the general home sale rate to x yen/kWh and the settlement of accounts is y 100 million yen, what will the function be?
3. If we change only the sale rate for businesses to x yen/kWh and the settlement of accounts is y 100 million yen, what will the function be?
4. Company K raises the price to x yen/kWh for general families and the sale rate to y yen/kWh for companies, and expects the annual accounts to be zero. What is the equation? What kind of price rise would you suggest?

Problem B: When there is a change in expenditure:

1. When the other expenditures are reduced by half, what percentage of photovoltaic power generation will be needed for getting rid of the deficit?
2. When the other expenditures are reduced by 25 %, what should the purchase price of photovoltaic power generation be to cancel the deficit?
3. Let the other expenditures be reduced x % and the percentage of photovoltaic power generation be y %. What is the equation for cancelling the deficit?

Problem C: When we change both the conditions of the income and the expenditure, suggest your plan for breaking even.

As for *Problems A* and *B* students had an animated discussion and were able to find the solutions enthusiastically in a group. Due to the complexity of the problems, they sometimes seemed to fall into a panic in the process. However, they patiently wrestled with the problems and were able to come up with the answers. For *Problem C*, one group of students produced the following analysis:

The current deficit is 1,587 billion yen. Items suppressing expenditures are the photovoltaic power generation purchase price of 42 yen/kWh and other expenditures of 1,200 million yen. The students supposed a 30 % cut to be the limit of the other expenditures. As for increasing the income, the only method is to raise the sales prices. Fixed are the total power consumption of 150 billion kWh, the selling ratio 2:3 between general families and businesses, and the fuel cost of 7.4 yen/kWh.

Let the selling price for general families be a yen, the selling price to businesses be b yen, the photovoltaic power generation purchase price be x yen/kWh, and the ratio of the photovoltaic power generation be y. Then we could formulate the following equation:

$$600a + 900b - 1500(1 - y) \times 7.4 - 1500xy - 12000 \times 0.7 = 0$$
$$\text{that is} \quad 2a + 3b + 37y - 5xy = 65$$

Here we suppose that $y = 35$, $a = 28$, and $b = 18$. Then $x = 33$. (omit it after this)

49.5 Response of the Students

Eight students participated in this modelling challenge at Kyoto University of Education. The perception of the participating students was examined by carrying out a pre-test before this challenge program and a post-test afterwards. The tests showed the following results.

Reasons for Participating and Satisfaction Afterwards Four students participated because of recommendations by teachers, two students responded that they were interested in this program, and the remaining two students participated because of recommendations by friends. However, all students stated that it was very good to participate in this challenge program. Some student responses were:

"Because modelling is usually not taught in school and we learned a lot from today's experience."
"Because new discoveries were made when we made numerical values through real experiments and not through figures in text books."
"I was able to sense the applied power of mathematics firsthand."

Purpose of Learning Mathematics Six students liked mathematics, so most of the participants were interested in this challenge program. We examined the students' attitudes regarding the benefits of learning mathematics before the challenge. Seven students considered mathematics to be 'training of the thought', five students chose 'enjoyment' and 'necessity for entrance examination', four students chose 'beneficial for future career', and three students chose that it was 'cultural'. In our past examination of ninth grade students in a national school, which consists of top half students, the percentage of students who liked mathematics was approximately 30 %, the percentage of students who dislike mathematics was approximately 30 %, and the percentage of student who said they neither like nor dislike mathematics was approximately 40 %. So it appeared that all eight students who participated in this challenge program had a favourable attitude toward mathematics unlike what might occur in a typical classroom.

Usefulness of Mathematics The number of students who answered that mathematics was very helpful when applied to a real-life problem increased to six after this challenge, as opposed to four students before this challenge. Before the challenge some students said: "We can't make use of what we learned in mathematics in our everyday life", or "I think that there is a limit to what we are learning". But after this challenge they said, "By expressing things in formulae, mathematics is helpful for us to think about things logically", "I realized that I could use mathematics in today's example and even in other cases", "I think that I can understand problems better with a knowledge of mathematics", and so forth.

Interest in this Challenge Program All the students who participated in this mathematical modelling challenge stated that it had been a very good experience. Typical responses were:

"I was able to think by myself and was able to present my thoughts with my friends."
"I was able to come up with an answer after having thought it through thoroughly."
"I was able to feel my ability to apply mathematics."
"We usually do not readily have such an opportunity and we were able to learn a lot through today's experience".

In addition, almost all the students were very interested in the challenge of solving problems.

Image of Mathematics The images that many students generally have of junior high school mathematics in Japan were as follows: "Because we work on problems with only one concrete answer, we can develop our ability to think, but we don't acquire much social application ability", "Mathematics in junior high school is the foundation of high school and university", "It is something which just cannot be applied to a real problem", "Its practical use is limited", and so forth. However, after this modelling challenge, the students' impression of mathematics was transformed:

"I realized that mathematics is something you adapt and apply to problems and that I didn't have the ability to do so. It was interesting and I was able to develop my ability to apply mathematics and enjoy it at the same time."
"At first, I felt it would be impossible, but I was able to think from various viewpoints. Coming up with new ideas was really enjoyable."
"Though I was working on 'a function', I really didn't feel as if I were. I felt that this was the mathematical view and way of thinking. I am happy that I participated in this program."

49.6 Conclusion

In Japan, it is difficult to press ninth graders to participate in a mathematical modelling challenge program, because they are busy studying for entrance examinations, do not understand mathematical modelling and do not have any background at all in mathematical modelling. However, it was confirmed that such a program was effective in improving the perception of mathematics by junior high students. In other words, they realized that mathematics becomes more useful when dealing with real-world problems and that they can use mathematics more freely.

It was not possible to clarify what mathematical modelling problems students could concentrate and work on, or whether they prefer to solve scientific problems or social problems, because of too few participants in the program this time. However every student concentrated and worked on problems given in both courses. So it is thought that the four points we considered when we designed the teaching units were effective. Furthermore, it was clarified that it is possible to carry out a day long program of the mathematical modelling challenge in Japan on a holiday. That will be generally better for eight or tenth grade students, because there is an entrance examination for ninth grade students in Japan.

It became clear through the scientific course that Japanese students are not used to dividing a phenomenon into its every domain and considering each by an approximate function. So we suggest teachers should handle such learning activities more in regular mathematics lessons. This is an important modelling skill, but there is insufficient teaching of this skill in Japan. We generally have to handle algebraic expressions including many variables as in this social course while we work on modelling tasks. Therefore it is clarified that more complex algebraic expressions should be handled in mathematics classrooms.

We intend to open the program to tenth grade students and carry out this mathematical modelling challenge program again, and investigate problems that occur when conducting this type of program in Japan.

References

Andersen, S. A. (1998). Danish students compete in the A-lympiads. In P. Galbraith, W. Blum, G. Booker, & I. D. Huntley (Eds.), *Mathematical modelling: Teaching and assessment in a technology-rich world* (pp. 301–306). Chichester: Horwood.

Blum, W. (2011). Can modelling be taught and learnt? Some answers from empirical research. In G. Kaiser, W. Blum, R. B. Ferri, & G. Stillman (Eds.), *Trends in teaching and learning of mathematical modelling* (pp. 15–30). New York: Springer.

Bracke, M., & Geiger, A. (2011). Real-world modeling in regular lessons: A long-term experiment. In G. Kaiser, W. Blum, R. B. Ferri, & G. Stillman (Eds.), *Trends in teaching and learning of mathematical modelling* (pp. 529–549). New York: Springer.

Galbraith, P., Stillman, G., & Brown, J. (2010). Turning ideas into modelling problems. In R. Lesh, P. Galbraith, C. R. Haines, & A. Hurford (Eds.), *Modelling students' mathematical competencies* (pp. 133–144). New York: Springer.

Kaiser, G., & Stender, P. (2013). Complex modelling problems in co-operative, self-directed learning environments. In G. A. Stillman, G. Kaiser, W. Blum, & J. P. Brown (Eds.), *Teaching mathematical modelling: Connecting to research and practice* (pp. 277–293). Dordrecht: Springer.

Maaß, K. (2010). Classification scheme for modelling tasks. *Journal für Mathematik-Didaktik (Journal for Didactics of Mathematics), 31*(2), 285–311.

Stillman, G. A., Brown, J. P., & Galbraith, P. (2013). Challenges in modelling challenges: Intents and purposes. In G. A. Stillman, G. Kaiser, W. Blum, & J. P. Brown (Eds.), *Teaching mathematical modelling: Connecting to research and practice* (pp. 241–251). Dordrecht: Springer.

Yanagimoto, A. (2003). Environmental problems and mathematical modelling. In S. J. Lamon, W. A. Parker, & S. K. Houston (Eds.), *Mathematical modelling: A way of life ICTMA11* (pp. 53–60). Chichester: Ellis Horwood.

Yanagimoto, A., & Yoshimura, N. (2001). Mathematics of traffic safety – Composite real mathematics approach. In J. F. Matos, W. Blum, S. K. Houston, & S. P. Carreira (Eds.), *Modelling and mathematics education: Applications in science and technology* (pp. 401–410). Chichester: Ellis Horwood.

Yanagimoto, A., & Yoshimura, N. (2013). Mathematical modelling of a real-world problem: The decreasing number of bluefin tuna. In G. A. Stillman, G. Kaiser, W. Blum, & J. P. Brown (Eds.), *Teaching mathematical modelling: Connecting to research and practice* (pp. 229–239). Dordrecht: Springer.

Yoshimura, N. (2015). Mathematical modelling of a social problem in Japan: The income and expenditure of an electric power company. In G. A. Stillman, W. Blum, & M. S. Biembengut (Eds.), *Mathematical modelling in education research and practice: Cultural, social, and cognitive influences* (pp. 251–261). Cham: Springer.

Yoshimura, N., & Yanagimoto, A. (2013). Mathematical modelling of a real-world problem: Pension tax issues. In G. A. Stillman, G. Kaiser, W. Blum, & J. P. Brown (Eds.), *Teaching mathematical modelling: Connecting to research and practice* (pp. 241–251). Dordrecht: Springer.

Part VII
Modelling and Applications in the Lived Environment

Chapter 50
Modelling the Wall: The Mathematics of the Curves on the Wall of Colégio Arquidiocesano in Ouro Preto

Daniel Clark Orey and Milton Rosa

Abstract In this chapter, the authors share results initiating from a conversation during a walk along a street in Ouro Preto, in the state of Minas Gerais, Brazil. The simple observation of a common everyday phenomenon and its mathematical potential initiated a debate that turned the observation into an exploration of the mathematics inherent in a common architectural detail. There are many similar situations in any city that can easily be used to encourage conversation, develop models, and explore the relationships between mathematics, procedures, and practices developed by members of a community; in so doing, the authors of this chapter studied mathematical models of the curves along a brick wall and sought to verify if these curves were related to exponential, parabolic or catenary curves.

50.1 Introduction

As a visiting professor and now a professor at the Universidade Federal de Ouro Preto, professor Orey began a project which has come to be known as the *Trilha de Matemática de Ouro Preto* (Ouro Preto Mathematics Trail). The project began in 2005–2006 during which time the Prefeitura (*City Hall*) of Ouro Preto developed a program called *O Museu Aberto* (*The Open Museum*) that was designed to encourage people to take pride in their city. In this program, the city developed historical routes that encouraged people to walk in different neighbourhoods of the city and to read the informative plaques placed on important homes, structures, buildings, and historical points of interest throughout the city. Figure 50.1 shows the view of Ouro Preto, in the state of Minas Gerais, Brazil.

D.C. Orey (✉) • M. Rosa
Centro de Educação Aberta e a Distância (CEAD), Universidade Federal de Ouro Preto (UFOP), Campus Morro do Cruzeiro-Bauxita, Ouro Preto, MG 35.400-000, Brazil
e-mail: oreydc@gmail.com; milton@cead.ufop.br

G.A. Stillman et al. (eds.), *Mathematical Modelling in Education Research and Practice*, International Perspectives on the Teaching and Learning of Mathematical Modelling, DOI 10.1007/978-3-319-18272-8_50

Fig. 50.1 View of Ouro Preto (Source: Daniel C. Orey)

In 2005, a course on mathematical modelling was taught at the *Universidade Federal de Ouro Preto* by Prof. Orey who had been documenting numerous opportunities for ethnomathematics and modelling research. In collaborating with university students and with the Ouro Preto Municipal Schools, an interesting pattern was noticed on the wall of the Colégio Arquidiocesano along Rua Alvarenga. The observation of the curves on this wall created potential for an interesting debate that quickly turned this observation into an exploration of the mathematics inherent in a common architectural detail.

There are many similar situations that can be used to encourage the development of models to explore the relationships between mathematical ideas, procedures, and practices developed by the members of a community to academic mathematics (D'Ambrosio 1990). This context allowed us to develop models related to the curves along a wall of a school and to verify if the shapes of these curves were related to exponential, parabolic, or catenary curves. The curves on the wall provide an interesting insight into the mathematics of an architectural feature. Figure 50.2 shows the curves on the wall of the Colégio Arquidiocesano.

In this context, mathematical modelling is the study of problems or situations taken from reality for understanding, simplification, and resolution with insights for a possible forecast or modification of the studied objects (Bassanezi 2002). Here, the learning of mathematics is not unlike that of learning a second language, and the development of mental concepts that incorporate individual reality and context (Rosa and Orey 2007). Thus, any study of mathematical modelling represents a powerful means for validating contextualized mathematical situations. This perspective forms the basis for significant contributions of a Freirean-based mathematical perspective in re-conceiving the discipline of mathematics in a pedagogical practice.

Fig. 50.2 Curves on the wall of the Colégio Arquidiocesano (Source: Daniel C. Orey)

50.2 Mathematical Modelling and Its Relation to the Community Context

An effective mathematics learning environment encourages deep understanding, which allows users to debate and interpret their own ideas and findings. Previous experiences that students possess take-on new forms of knowledge by evolving new and increasingly complex contexts, ideas and procedures as they learn to solve problems that exist in their own communities (Rosa and Orey 2010).

This often stands as a creation incorporating the historical-cultural nature of mathematical concepts, by assisting educators and students to reflect in context on the mathematical processes that they use (D'Ambrosio 1990). As they gain more confidence in developing new ideas for themselves, they show the presence of mathematics in their daily lives through dialogue and respect. The construction of mathematical knowledge must incorporate situations, problems, and phenomena that occur in the cultural context of the school community (Orey 2000).

50.2.1 Ethnomodelling: The Cultural Aspects of Mathematical Modelling

Studying the mathematical action on a system allows it to become an object of critical analysis (Bassanezi 2002). The process by which we consider, analyze, and make these critical reflections on a system is named ethnomodelling (Bassanezi 2002). This allows for the development of *fluency* in mathematics by acquiring understanding, and representations of problems. The relations between systems and their representations help in validating findings and occur through triangulation, dialogue and ongoing analyses of models. In this regard, ethnomodelling is much more than standard transference of knowledge.

The use of ethnomodelling as pedagogical action for the teaching and learning of mathematics values students' prior knowledge by developing their capacity to assess the process of elaborating a mathematical model in its different applications starting with their social context, reality, and interests (Rosa and Orey 2010). In this regard, ethnomodelling is much more than the standard transference of knowledge. Here, teaching becomes an activity that introduces the creation of knowledge. This transformative approach in mathematics education is the antithesis of turning students into containers to be filled with information (Freire 1970).

To pursue this direction, it is necessary to start with problems and situations found in the day to day lives in the community. This approach allows educators to develop skills as facilitators and to create conditions for increased involvement making mathematics accessible for all (MEC 1996; National Council of the Teachers of Mathematics 1989). It is important to emphasize that this process enables learners to learn how to engage in a process of continual modification, to alter their reflections, to develop mature arguments, to create ongoing analysis and to develop solid conclusions regarding their models and allows them to experience the scientific method within the context of learning mathematics.

50.3 Modelling the Wall: Searching for Mathematical Models

What happened in Ouro Preto was something altogether surprising, and in the end, the final results, important as they are, were eclipsed by the actual opportunity we had to discuss and debate aspects related to exponential curves, parabolas, and catenaries as we determined ways to relate functions to patterns found on the wall. Not unlike archeologists or ongoing restoration efforts in Ouro Preto, as our work progressed and the group had discussions as data were gathered, we worked on the models and discussed mathematical concepts and patterns that many had never taken much notice of along Rua Alvarenga.

By observing the architectonic design on the wall of the Colégio Arquidiocesano, we were trying to relate these features to several curves. Initially, we tried to check the similarity that seems to exist between these shapes and the exponential curve when we considered the shapes on the wall as a whole. Further, by analyzing these shapes, individually, we were trying to relate them to exponential curves, parabolas, and catenaries. Thus, in getting to the object of our study, it was interesting to discuss if the curves had an exponential, parabolic or catenary shape. In order to have the necessary arguments to answer this conjecture, some mathematical models were elaborated and analyzed, and discussed.

50.3.1 Data Collection and Mathematical Models

Curves were randomly selected on the wall of the school and an *X*-*Y* coordinate system was constructed using string to determine the *X* and *Y* axes. After a brief discussion, it was decided that the origin of the *X*-*Y* coordinate system would be placed on the lowest point of one of the curves. At that point, there was no certainty if the lowest point was the *vertex* or the *apex* of the curve. Figure 50.3 shows the curve on part of the wall of Colégio Arquidiocesano with the strings that represent the *X* and *Y* axes. The points placed on the selected curves on the wall were organized in tables for further analyses. Table 50.1 shows the coordinates of the points that were placed on the selected curve on the wall.

The first mathematical model elaborated was related to the adjustment of these points to an exponential curve; the second model was related to its adjustment to a parabola, and the third one to a catenary. On the other hand, we also obtained a quadratic function (i.e., the graph with highest *y*-intercept) whose graph is also similar to the graph of the catenary function. When we visualized the shapes in only one part of the curve on the wall, we could observe the existing similarities between the graphs of all the functions in the restricted domain. In this regard, after examining the data collected when we measured various curves on the wall of the Colégio Arquidiocesano and tried to fit them to exponential and quadratic functions through mathematical models we came to the conclusion that the curves on the wall best fit that of a *catenary curve* (Fig. 50.4).

Fig. 50.3 Curve on the wall of Colégio Arquidiocesano with the string *X* and *Y* axes (Source: Daniel C. Orey)

Table 50.1 Points placed on the selected curve (Source: Daniel C. Orey)

x (in metres)	*y* (in metres)	Points
−0.925	0.23	A (−0.925, 0.23)
−0.745	0.15	B (−0.745, 0.15)
0	0	C (0, 0)
0.7	0.125	D (0.7, 0.125)
0.91	0.2	E (0.91, 0.2)

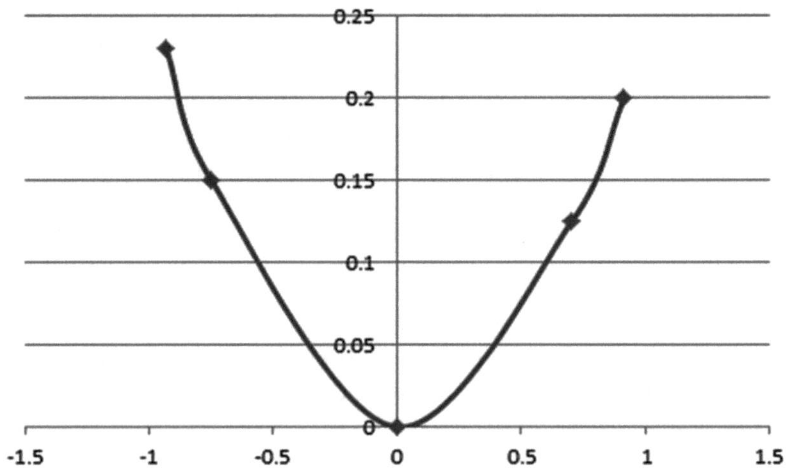

Fig. 50.4 Graph of the curve containing points in Table 50.1

50.4 Similarities and Differences Between Parabolas and Catenaries

After the students' reflections on the elaborated models, we guided a discussion about similarities and differences between parabolas and catenaries because these curves are frequently used in architecture. For example, catenary shapes can be applied in complex architectural buildings because they are very efficient in carrying heavy loads while parabolic shapes are frequently applied in suspension bridges because in general the roadway is very nearly a uniformly distributed load, and each cable hangs down in a curve closely approaching that of a parabola. On the other hand, we decided to not have a discussion about exponential curves because they are mainly applied to exponential growth and decay such as population, interest, and depreciation.

In this context, we discussed that parabolas and catenaries are two similar and yet different curves. Both are symmetrical and have a cup shape; which goes up infinitely on either side of a minimum value. In Figure 50.5, the left graph represents the parabola and the higher right graph represents the catenary function, graphed together here for readers to see their similarities and differences.

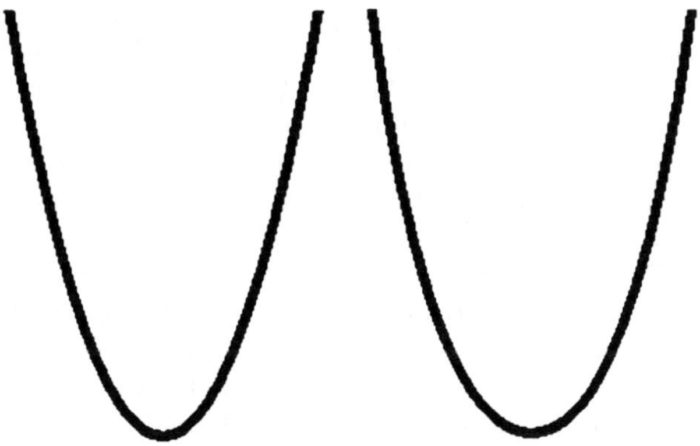

Fig. 50.5 The graph of a parabola and the graph of a catenary (Source: Daniel C. Orey)

The mathematical equation $y = \frac{(\cosh(x)-1)}{(\cosh(1)-1)}$ is the hyperbolic cosine function of the catenary. When we move the vertex from (0, 1) down to the origin, it agrees with the parabola when $x=1$. The catenary is just slightly more flat at the bottom and rises faster than the parabola for large values of x. In fact, when we graph parabolas and catenaries for larger domains, the catenary is different from the parabola.

While parabolas possess the shape of the curve $y=x^2$, catenaries take on the shape of the graphs of the hyperbolic cosine function $y=cosh(x)$ graphs. Parabolas and catenaries are mathematically distinct types of curves, with similar properties. When both curves open upward, they have a single low point. In this case, the lowest point of the parabola is called its *vertex*, and the lowest point of the catenary is called its *apex*. They also have a vertical line of symmetry and appear to be continuous and differentiable throughout the shape of the curve. Both curves have slopes that are steeper as they move away from the lowest point where the curves never become vertical lines and they may be generated by hanging a flexible cable, wire, or chain between two fixed points.

The difference between the two curves is related to the ways that the weight is distributed along the length of each cable, chain, string, or wire. For example, if the weight is distributed evenly along the length of the chain, then the curve is a catenary and if the weight is distributed evenly along a horizontal line, then the curve is a parabola. The reason the St. Louis Gateway Arch, a hanging cable, or the curves on the wall take the shape of a catenary, whilst the cables on a suspension bridge form a parabola is a result of the physics of each situation. By applying calculus, we are able to verify that the catenary is the solution to a differential equation that describes a shape that directs the force of its own weight along its own curve, so that, if hanging, it is pulled into that shape, and, if standing upright, it can support itself.

It is important to emphasize that, in mathematics; the catenary is the shape of a curve described by an ideal string that is suspended by its two endpoints. In this case, the ideal means that the string is perfectly flexible, that is, the curve is inextensible, with no thickness, with uniform density, and is subjected to only the influence of gravity. This shape imparts great strength to structures when built in the shape of a catenary. On the other hand, the parabola does not have the same property, but is the solution of other important equations that describe other situations. For example, in nature, approximations of parabolas are found in any number of diverse situations. In physics there is the trajectory of a particle or body in motion under the influence of a uniform gravitational field without air resistance such as the parabolic trajectory of a baseball and projectiles.

Parabolic shapes are also found in several physical situations such as parabolic reflectors commonly observed in microwave or satellite dish antennas. The mathematical properties of parabolas make them excellent models for physical objects in which a focusing component is essential. It can be shown that parallel lines drawn on the inside of any parabola are reflected from the curve of the parabola to its focus. Thus, many telescopes and satellite television receivers are designed using parabolic reflection properties. Parabolas also model the motion of a body in free fall towards the surface of the Earth and are used in the design of bridges and other structures involving arches. After looking at the designs on the wall, we decided to compare them with what we knew about parabolic or catenary functions and how they related to suspension bridges.

50.4.1 The Specific Case of Suspension Bridges

In light of the facts discussed previously, the following question was formulated and debated with the students: *Do cables of suspension bridges have a catenary shape?* Figure 50.6 shows the construction proceeding on the Brooklyn Bridge in New York, in 1881.

After discussing differences between parabolas and catenaries we decided that the answer to this question is *no*. In this specific case, it is interesting to note that when suspension bridges are constructed, before the suspension cables are tied to the deck below them, they initially have a hyperbolic cosine function shape, that is, the shape of a catenary. This happens when the structure of the bridge is being built and when the main cables are attached to the towers and then the cables are attached to the deck with hangers. In so doing, the cable of a suspension bridge is under tension from holding up the bridge. The cable is also under the influence of a uniform load, that is, the deck of the bridge, which deforms the cable into a parabola. This deformation happens because the weight of the deck is equally distributed along the curve. In this regard, it is possible to conclude that the catenary curves under its own weight while the parabola curves under both its own weight and by holding up the weight of the deck.

Fig. 50.6 Construction of the Brooklyn Bridge (Source: Museum of New York)

50.5 Some Concluding Remarks About the Ethnomodelling Process

In order to gather more information about this investigation, we had a meeting with the principal of the Colégio Arquidiocesano, which was founded in 1927 and started its activities in Ouro Preto in 1933. In this meeting, we learned that there was no architectural drawing of the wall. According to this context, it is important to emphasize that the main issue being investigated was to understand how the builders had managed to capture the shape of a hanging chain as it was built into a wall.

In this regard, from the outset we presumed that the procedure may have been done by the builders hanging a chain between two support posts as well as using it as a template to build into the wall. In so doing, if one hangs a chain between two support posts along a wall, while making sure the chain can glide and find its own equilibrium, one may discover that the shape of the chain depends nearly uniquely on its own length. In so doing, the catenary curve searches for its own point of equilibrium, that is, the point where the sum of the weight of the two hanging ends equals the weight of the chain hanging in between the support posts. This means that in order to produce the most stable wall, it was necessary that its curves had catenary shapes. This approach ensured that the majority of the gravitational forces acting on the wall were directed downwards rather than sideways. This means that the tension passing through the curves on the wall chain is equivalent to the compression in a standing wall. In this regard, the weight of the wall is distributed through its structure.

After discussions related to this issue, we agreed that it would be unlikely that any other practical method would have been used by the builders so quadratic or exponential shapes are unlikely. Consequently, the purpose of applying ethnomodelling was to understand how the builders may have managed to obtain such a feature in the wall.

50.6 Final Considerations

Any study of mathematical modelling represents a powerful means for validating contextualized mathematical situations. The use of ethnomodelling as pedagogical action for the teaching and learning of mathematics values the previous knowledge of the mathematical practices of the members of the community by developing students' capacity to assess the process of elaborating a mathematical model in its distinct applications and contexts by having started with the social context, reality and interests of the students.

In this chapter, we have shared the results from an activity from our walks with students and colleagues along streets in Ouro Preto. The context here in which this particular model arose was the curves found along the wall of the Colégio Arquidiocesano. By observing the design found along the wall of the school, we attempted to relate them to several curve functions. By analysis of these shapes, we related them to exponential curves, parabolas, and catenaries.

From the results of this investigation, we found that the mathematics that defined the required shapes of the curves comes from real physical laws of a parabola and a catenary, while the exponential curve is much more of an idealized mathematical abstraction, and was not really consistent with what we observed along the wall. Discussions related to the application of mathematical modelling in a cultural and common context allowed us to see that there are good physical and mathematical reasons why curves used in architecture would be more like catenaries than parabolas.

References

Bassanezi, R. C. (2002). *Ensino-aprendizagem com modelagem matemática* [Teaching and learning with mathematical modelling]. São Paulo: Editora Contexto.

D'Ambrosio, U. (1990). *Etnomatemática* [Ethnomathematics]. São Paulo: Editora Ática.

Freire, P. (1970). *Pedagogia do Oprimido* [Pedagogy of the oppressed]. Rio de Janeiro: Paz e Terra.

MEC. (1996). *Lei de diretrizes e bases da educação nacional* [Law of guidelines and basis for the national education]. Brasília: Ministério da Educação e Cultura – MEC.

National Council of the Teachers of Mathematics. (1989). *Curriculum and evaluation standards for school mathematics*. Reston: NCTM.

Orey, D. C. (2000). The ethnomathematics of the Sioux tipi and cone. In H. Selin (Ed.), *Mathematics across culture: The history of non-western mathematics* (pp. 239–252). Dordrecht: Kluwer.

Rosa, M., & Orey, D. C. (2007). Cultural assertions and challenges towards pedagogical action of an ethnomathematics program. *For the Learning of Mathematics, 27*(1), 10–16.

Rosa, M., & Orey, D. (2010). Ethnomodelling as a pedagogical tool for the ethnomathematics program. *Revista Latinoamericana de Etnomatemática, 3*(2), 14–23.

Refereeing Process

To ensure the scholastic quality of the chapters in this book, all proposed chapters submitted to the editors for consideration in a timely manner have been subjected to a strict impartial and peer review independent of the editors. Only those chapters that have been accepted after this process of review and revision as being of appropriate quality (given differences of geographical location) and consistent with the intended aims of the book have been included. The editors wish to thank the following international reviewers:

Jonas Bergman Ärlebäck Linkoping University, Sweden
Jussara de Loiola Araújo Universidade Federal de Minas Gerais, Brazil
Michael Besser Leuphana University of Lüneburg, Germany
Morten Blomhøj Roskilde University, Denmark
Rita Borromeo Ferri University of Kassel, Germany
Jill P Brown Australian Catholic University (Melbourne), Australia
Susana Carreira Universidade De Lisboa, Portugal
Dirk De Bock Katholieke Universiteit Leuven & Hogeschool-Universiteit Brussels, Belgium
Helen Doerr Syracuse University, USA
Ann Downton Australian Catholic University (Melbourne), Australia
Xiaojun Duan National University of Defence Technology, P.R. China
Lyn English Queensland University of Technology, Australia
Cristina Esteley Universidad Nacional de Córdoba, Argentina
Peter Frejd Linkoping University, Sweden
Peter Galbraith University of Queensland, Australia
Solomon Garfunkel COMAP, The Consortium for Mathematics and Its Applications, USA
Vincent Geiger Australian Catholic University (Brisbane), Australia
Gilbert Greefrath University of Munster, Germany
Chris Haines City University (London), United Kingdom

G.A. Stillman et al. (eds.), *Mathematical Modelling in Education Research and Practice*, International Perspectives on the Teaching and Learning of Mathematical Modelling, DOI 10.1007/978-3-319-18272-8

Toshikazu Ikeda Yokohama National University, Japan
Cyril Julie University of the Western Cape, South Africa
Tinne Kjeldsen Roskilde University, Denmark
Janeen Lamb Australian Catholic University (Brisbane), Australia
Dominik Leiss Leuphana Universität Lüneburg, Germany
Richard Lesh Indiana University, USA
Issic K.C. Leung The Hong Kong Institute of Education, Hong Kong
Thomas Lingefjard Göteborgs Universitet, Sweden
Nick Mousoulides Cyprus University of Technology, Cyprus
Kit Ee Dawn Ng Nanyang Technological University, Singapore
Mogens Niss Roskilde University, Denmark
Daniel Orey Universidade Federal de Ouro Preto, Brazil
Irit Pelid University of Haifa, Israel
Jacob Perrenet Technische Universiteit Eindhoven, Netherlands
Xenia Reit Geothe-University Frankfurt, Germany
Ruth Rodriguez Tecnológico de Monterrey (Monterrey), Mexico
Milton Rosa Universidade Federal de Ouro Preto, Brazil
Stanislaw Schukajlow University of Paderborn, Germany
Björn Schwarz Universität Hamburg, Germany
Hans-Stefan Siller University of Salzburg, Austria
Maxwell Stephens University of Melbourne, Australia
Monica Villarreal University of Córdoba, Argentina
Pauline Vos University of Agder, Norway
Geoffrey Wake University of Nottingham, United Kingdom
Dag Wedelin Chalmers University of Technology, Sweden
Mark Winter University of the Witwatersrand (Johannesburg), South Africa
Jinxing Xie Tsinghua University (Beijing), P.R. China
Akira Yanagimoto Kyoto University of Education, Japan

Index

G.A. Stillman et al. (eds.), *Mathematical Modelling in Education Research and Practice*, International Perspectives on the Teaching and Learning of Mathematical Modelling, DOI 10.1007/978-3-319-18272-8

N

O

P

Q

R

S

MIX
Papier aus verantwortungsvollen Quellen
Paper from responsible sources
FSC® C105338

If you have any concerns about our products,
you can contact us on
ProductSafety@springernature.com

In case Publisher is established outside the EU,
the EU authorized representative is:
Springer Nature Customer Service Center GmbH
Europaplatz 3, 69115 Heidelberg, Germany

Printed by Libri Plureos GmbH
in Hamburg, Germany